T0317578

Time and Frequency Domain Solutions of EM Problems Using Integral Equations and a Hybrid Methodology

WILEY SERIES IN MICROWAVE AND OPTICAL ENGINEERING

KAI CHANG, Editor
Texas A&M University

A complete list of the titles in this series appears at the end of this volume.

Time and Frequency Domain Solutions of EM Problems Using Integral Equations and a Hybrid Methodology

B. H. Jung, T. K. Sarkar, S. W. Ting, Y. Zhang, Z. Mei, Z. Ji, M. Yuan, A. De, M. Salazar-Palma, and S. M. Rao

IEEE PRESS

A JOHN WILEY & SONS, INC., PUBLICATION

For general information on our other products and services or for technical support, please contact our Customer Care Department within the United States at (800) 762-2974, outside the United States at (317) 572-3993 or fax (317) 572-4002.

Wiley also publishes its books in a variety of electronic formats. Some content that appears in print may not be available in electronic formats. For more information about Wiley products, visit our web site at www.wiley.com.

Library of Congress Cataloging-in-Publication Data is available.

ISBN 978-0-470-48767-9

10 9 8 7 6 5 4 3 2 1

CONTENTS

Preface

Numerical solutions of electromagnetic field problems constitute an area of paramount interest in academia, industry, and government. Many numerical techniques exist based on the solutions of both the differential and integral forms of Maxwell's equations. It is not recognized very often that for electromagnetic analysis of both conducting and practical linear piecewise homogeneous isotropic bodies, integral equations are still one of the most versatile techniques that one can use in both efficiency and accuracy to solve challenging problems including analysis of electrically large structures.

In this book, our emphasis is to deal primarily with the solution of time domain integral equations. Time domain integral equations are generally not as popular as frequency domain integral equations and there are very good reasons for it. Frequency domain integral equations arise from the solution of the unconditionally stable elliptic partial differential equations of boundary-value problems. The time domain integral equations originate from the solution of the hyperbolic partial differential equations, which are initial value problems. Hence, they are conditionally stable. Even though many ways to fix this problem have been proposed over the years, an unconditionally stable scheme still has been a dream. However, the use of the Laguerre polynomials in the solution of time domain integral equations makes it possible to numerically solve time domain problems without the time variable, as it can be eliminated from the computations associated with the final equations analytically. Of course, another way to eliminate the time variable is to take the Fourier transform of the temporal equations, which gives rise to the classical frequency domain methodology. The computational problem with a frequency domain methodology is that at each frequency, the computational complexity scales as $\mathcal{O}(N^3)$ [$\mathcal{O}(\cdot)$ denotes of "the order of", where N is the number of spatial discretizations]. Time domain methods are often preferred for solving broadband large complex problems over frequency domain methods as they do not involve the repeated solution of a large complex matrix equation. The frequency domain solution requires the solution of a large matrix equation at every frequency step, whereas in the solution of a time domain integral equation using an implicit method, one only needs to solve a real large dense matrix once and then use its inverse in the subsequent computations for each timestep. So that, a time domain method requires only a vector-matrix product of large dimension at each time step. Thus, in this case, at each timestep, the computational complexity is $\mathcal{O}(N^2)$. The use of the orthogonal associated Laguerre functions for the approximation of the transient responses not only has the advantage of an implicit transient solution technique as the computation time is quite small, but also separates the space and the time variables, making it possible to eliminate the time variable from the final computations analytically. Hence, the numerical dispersion due to time discretizations does not exist for this methodology. This methodology has been presented in this book. The conventional marching-on-in-time method is also presented along with its various modifications.

The frequency domain solution methodology is also included, so that one can verify the time domain solutions by taking an inverse Fourier transform of the frequency domain data. This will provide verification for the accuracy and stability of the new Laguerre-based time domain methodology, which is called the marching-on-in-degree/order technique.

Finally, a hybrid method based on the time and frequency domain techniques is presented to illustrate that it is possible to go beyond the limitations of the computing resources through intelligent processing. The goal of this hybrid approach is to generate early-time information using a time domain code and low-frequency information using a frequency domain code, which is not computationally demanding. Then an orthogonal series is used to fit the data both in the early-time and low-frequency domains, and then it is shown that one can use the same coefficients to extrapolate the response simultaneously both for the late-time and high-frequency regions. In this approach, one is not creating any new information but is using the existing information to extrapolate the responses simultaneously in the time and frequency domains. Three different orthogonal polynomials have been presented, and it is illustrated how they can be utilized to achieve the stated goal of combining low-frequency and early-time data to generate a broadband solution. This forms the core of the parametric approach. The fact that Fourier transforms of the presented polynomials are analytic functions allows us to work simultaneously with time and frequency domain data. A priori error bounds are also provided and a processing-based methodology is outlined to know when the data record is sufficient for extrapolation. A nonparametric approach is also possible using either a Neumann series approximation or through the use of the Hilbert transform implemented by the fast Fourier transform. This hybrid methodology may make it possible to combine measured vector network analyzer data with a time domain reflectometer measurements in a seamless and accurate fashion without using any of the approximate inaccurate methodologies to process the measured data that currently are implemented in most commercial time domain reflectometer (TDR) and network analyzer measurement systems.

This essentially provides the unique salient features of this book. To make this book useful to practicing engineers and graduate students trying to validate their own research, sample computer programs are presented at the end of this book. For more sophisticated versions of the programs, an interested reader may contact the authors.

Next we provide a detailed description of the contents of the book.

The introduction of Chapter 1 will provide a summary of the different types of spaces including the concept of mapping and projections leading to the formulation of operator equations. The definition of an ill-posed and well-posed problem is discussed, leading to the solution of an operator equation. Next the various methods for the solution of an operator equation are presented, including the method of moments, variational methods, Galerkin's method, and the method of least squares followed by iterative methods, particularly the method of conjugate directions, leading to the conjugate gradient method. The mathematical subtleties of the various methods are also presented including the appropriate scientific choice for the basis and testing functions.

Chapters 2–4 discuss the solution of frequency domain integral equations using the popular triangular-shaped discretizations for the surface of the objects. Chapter 2 describes the frequency domain integral equations for the analysis of conducting objects, and Chapter 3 presents the corresponding methodology for the dielectric objects, including the combined field integral equation along with its various derivatives. Chapter 4 presents the analysis of composite structures, consisting of both conducting and dielectric bodies in the frequency domain.

Chapters 5–9 describe time domain integral equations and illustrate how to solve them using the classical marching-on-in-time and the new marching-on-in-degree methodologies. This latter method has been very successful in solving general dielectric body problems even though some specific-case-dependent fixes are available for specific class of problems for the former scheme. Chapter 5 describes the transient scattering from wire objects. Chapter 6 presents the time domain integral equations for conducting objects, and Chapter 7 presents a similar methodology for the analysis of dielectric objects.

In Chapters 8 and 9, a new improved version of the marching-on-in-degree (MOD) methodology is presented that can not only guarantee stability of the solution but also can improve significantly the computational speed at least by an order of magnitude. This is because in this MOD technique, in the final computations, not only do the space and the time variables in the retardation time expressions separate, but also the temporal variable and its associated derivatives can be eliminated analytically from the final computations. This is achieved through the use of the orthogonal associated Laguerre functions approximating the functional temporal variations. Therefore, in this methodology, there is no need to have a sampling criterion like the Courant–Friedrich–Levy condition linking the spatial to the temporal sampling. Chapter 8 presents the theory of the new, improved methodology substantiating the claims by numerical examples in Chapter 9.

In Chapter 10, it is illustrated how to go beyond the limitations of the hardware resources of a computer to solve challenging electrically large electromagnetic field problems. This is accomplished by employing a hybrid methodology by combining both the solutions generated for early-time using a time domain code and low-frequency information using a frequency domain code. The basic philosophy is that the early time provides the missing high-frequency response, whereas the low-frequency provides the missing late-time response, and hence, through this hybrid methodology, one can stitch the two diverse domain data. In this methodology, one is not extrapolating the data but simply extracting the given information through intelligent processing. Because modest computational resources are required to generate the early-time and the low-frequency response, through this methodology, it is possible to address solutions of electrically large electromagnetic field problems. Thus, it also provides a technique to combine experimental data obtained with a vector network analyzer along with a time domain reflectometer to generate data that can go beyond the measurement capabilities of either.

Finally, user-oriented electromagnetic analysis computer codes in both time and frequency domains for the analysis of conducting structures are presented in the Appendix for researchers and graduate students to get some experience with frequency and time domain codes.

Every attempt has been made to guarantee the accuracy of the materials in the book. We, however would appreciate readers bringing to our attention any errors that may have appeared in the final version. Errors and/or any comments may be emailed to (On behalf of all the authors: tksarkar@syr.edu).

Acknowledgments

We gratefully acknowledge Carlos Hartmann (Syracuse University, Syracuse, NY), Michael C. Wicks, Darren M. Haddad, and Gerard J. Genello (Air Force Research Laboratory, Rome, NY), John S. Asvestas, Saad Tabet and Oliver E. Allen (NAVAIR, Patuxent River, MD), Eric Mokole (Naval Research Laboratory, Washington, DC), Steven Best (Mitre Corporation, Bedford, MA) and Dev Palmer (Army Research Office, Raleigh, NC) for their continued support in this endeavor. Thanks also go to Yeoungseek Chung and Wonwoo Lee for their help with this book. We gratefully acknowledge Raviraj Adve, Peter Vanden Berg, and Jinhwan Koh, for their help and suggestions in preparing some of the materials of the book, specifically those related to Chapter 10.

Thanks are also due to Ms. Robbin Mocarski, Ms. Brenda Flowers, and Ms. Maureen Marano (Syracuse University), for their expert typing of the manuscript. We would also like to express sincere thanks to Xiaopeng Yang, Hongsik Moon, LaToya Brown, Woojin Lee, Mary Taylor, Daniel García Doñoro, Weixin Zhao, and Walid Diab for their help with the book.

B. H. Jung, T. K. Sarkar, S. W. Ting, Y. Zhang, Z. Mei, Z. Ji, M. Yuan, A. De,
M. Salazar-Palma, and S. M. Rao
March 2010

List of Symbols

A	Magnetic vector potential
E	Electric field
$\mathbf{E}^i$	Incident electric field
$\mathbf{E}^s$	Scattered electric field
F	Electric vector potential
H	Magnetic field
$\mathbf{H}^i$	Incident magnetic field
$\mathbf{H}^s$	Scattered magnetic field
J	Electric current density
M	Magnetic current density
f , **g**	Triangular patch vector functions
l	Position vector on wire
n	Unit normal
r	Observation point
$\mathbf{r}'$	Source point
u , **v**	Hertz vectors
ρ	Position vector on triangular patch
G	Green's function
I	Integral term
J_n	Expansion coefficient for electric current
$J_n(t)$	Transient electric current coefficient
$L(t)$	Laguerre polynomials
L_n	The nth wire segment
M	Integer
M_n	Expansion coefficient for magnetic current
$M_n(t)$	Transient magnetic current coefficient
N	Integer
Q	Electric charge density
Q_e	Electric charge density
Q_m	Magnetic charge density
$Q_n(t)$	Transient coefficient for the charge density
R	Distance between field points; defined as $R = \left| \mathbf{r} - \mathbf{r}' \right|$
S	Surface

S_0	Surface without singularity
T_n	The nth triangular patch
a_n	Area of the nth triangular patch
c	Light Speed in free space
c_v	Wave velocity in a media v
f	Frequency
i, j	Iteration variable, $i = 1, 2, \ldots \infty$, $j = 1, 2, \ldots \infty$
k	Wave number; defined as $k = \omega \sqrt{\mu \varepsilon}$
l_n	Length, e.g. length of an edge or a wire segment
m, n	Integer, e.g. the mth and nth edges
p, q	Positive or negative components ($+$ or $-$)
r	Magnitude of vector $\mathbf{r}$
s	Scaling factor for temporal basis function
t, τ	Time
$u_n(t)$, $v_n(t)$	Transient coefficient for Hertz vector $\mathbf{u}(\mathbf{r}, t)$, $\mathbf{v}(\mathbf{r}, t)$
$u_{n,j}$, $v_{n,j}$	Expansion coefficient for transient coefficient $u_n(t)$, $v_n(t)$
Ψ	Magnetic scalar potential
Φ	Electric scalar potential
$\phi(t)$	Associated Laguerre function
θ, ϕ	Directions for the Spherical Coordinates
μ	Permeability
ε	Permittivity
ω	Angular frequency
η	Wave impedance
υ	Interior or exterior region (1 or 2)
σ	Radar Cross Section
π	(pi) = 3.1415.........
κ, α, β	Linear combination coefficients, $[0,1]$
$\mathcal{O}(M^n)$	Order of M^n

Acronyms

3D	Three dimensions
CEM	Computational electromagnetics
CFIE	Combined field integral equation
CG	Conjugate gradient
CPU	Central processing unit
EM	Electromagnetics
EFIE	Electrical field integral equation
FD	Frequency domain
FFT	Fast Fourier transform
HOBBIES	**H**igher **O**rder **B**asis function **B**ased **I**ntegral **E**quation Solver
IDFT	Inverse Discrete Fourier Transform
MOD	Marching-on-in-degree
MFIE	Magnetic field integral equation
MoM	Method of moments
MOT	Marching-on-in-time (method)
MSE	Mean squared error
PEC	Perfect electric conductor
PMCHW	Poggio–Miller–Chang–Harrington–Wu
RCS	Radar cross section
RWG	Rao–Wilton–Glisson
SVD	Singular value decomposition (method)
TDIE	Time domain integral equation
TD	Time domain

1

MATHEMATICAL BASIS OF A NUMERICAL METHOD

1.0   SUMMARY

In this chapter, we discuss the mathematical basis of a numerical method. We first start by defining the concepts of elements and spaces followed by the definitions of a set, a field, and a space. Then different types of spaces are discussed, namely metric and vector (linear) spaces. Associated with a vector space is a normed space, an inner product space and various combinations of the two. A hierarchy of the various vector spaces is also presented. The concept of mapping and projections are introduced, leading to the definition of operator equations. The definitions of an ill-posed and a well-posed problem are discussed, leading to the solution of an operator equation. Next the various methods for the solution of an operator equation are presented, including the method of moments, variational methods, Galerkin's method, and the method of least squares followed by iterative methods, particularly the method of conjugate directions leading to the conjugate gradient. The mathematical subtleties of the various methods are also presented. It is shown that if the inner products in the respective numerical method are approximated by a quadrature rule, then the resulting numerical method degenerates into a weighted point-matched solution irrespective of the technique that one started with. Finally, it is demonstrated that the application of an iterative method to the solution of a discretized operator equation is different from applying an iterative method directly to solve an operator equation symbolically and then discretizing the various expressions for numerically computing the expressions at each step to reach the final solution. This difference is not visible in the solution of an integral equation-based boundary value problem in the frequency domain but is noticeable in the time domain formulation of a similar boundary value problem.

Because many new quantities are introduced in this chapter, a list of all the various terminologies and symbols are also listed as follows:

- *Scalars*:   Represented by lower case italicized quantities; a, b, x, y, α, β, ...
- *Vectors*:   Represented by bold roman letters: $\mathbf{a}$, $\mathbf{b}$, $\mathbf{x}$, $\mathbf{y}$, $\mathbf{A}$, $\mathbf{B}$, ...
- *Matrices*:   Represented by italicized quantities placed inside brackets; $[a]$, $[b]$, $[A]$, $[B]$, ...
- *Functions*:   Represented by italicized roman letters or bold corresponding to scalar or vector, respectively; $f(x)$, $v(y)$, $\mathbf{f}(x)$, $\mathbf{g}(z)$, $\mathbf{v}(y)$, ...
- *Set or Fields*:   Represented by letters in the capital Euclid Fraktur font; $\mathfrak{A}$, $\mathfrak{B}$, $\mathfrak{F}$, $\mathfrak{R}$, $\mathfrak{C}$, ...
- *Spaces*:   Represented by letters in the capital Euclid Math Two font, $\mathbb{S}$, $\mathbb{D}$, $\mathbb{R}$, $\mathbb{M}$, $\mathbb{W}$, ...
- *Functionals*:   Represented by letters in the Lucida Handwriting font; $\mathcal{F}$, $\mathcal{R}$, $\mathcal{P}$, $\mathcal{p}$, ...
- *Operators*:   Represented by letters in the capital Lucida Calligraphy font; $\mathcal{A}$, $\mathcal{B}$, $\mathcal{C}$, $\mathcal{L}$, ...

1.1   ELEMENT AND SPACES

1.1.1   Elements

An *element* [1,2] is a definable quantity. Some examples of elements are:

(a) *Scalar*—a number that can be either real or complex. We normally denote scalars by italicized lowercase Greek or Roman letters, such as α, β, γ, ... a, b, c.

(b) *n-tuple vectors*—a group of scalars considered as a linear array. We denote vectors either by boldface lower case letter, $\mathbf{x}$, $\mathbf{y}$, $\mathbf{z}$ ..., or by uppercase letter, such as $\mathbf{X}$, $\mathbf{Y}$, $\mathbf{Z}$. They can also be represented by matrices, which are characterized by bracketed letters, such as $[x]$, $[y]$, $[z]$, ... or as $[X]$, $[Y]$, $[Z]$, ... Vectors can be represented either as columns

$$\mathbf{x} = [x] = \begin{bmatrix} x_1 \\ x_2 \\ \vdots \\ x_n \end{bmatrix} \tag{1.1a}$$

or as rows
$$\mathbf{x}^T = [x]^T = [x_1, x_2, \cdots, x_n] \tag{1.1b}$$

where the superscript T denotes transpose of a matrix, which is obtained by interchanging rows and columns. The numbers x_1, x_2, ..., x_n are called the components of a vector.

(c) *Matrix*—a group of scalars considered as a rectangular array. We denote matrices by bracketed letters $[A]$, $[B]$, $[C]$, ... The scalar components of a matrix $[A]$ are labeled a_{ij} and are represented by

$$[A] = \begin{bmatrix} a_{11} & a_{12} & \cdots & a_{1n} \\ a_{21} & a_{22} & \cdots & a_{2n} \\ \vdots & \vdots & \vdots & \vdots \\ a_{m1} & a_{m2} & \vdots & a_{mn} \end{bmatrix} \tag{1.2}$$

Column and row matrices generally will be considered as vectors. Other matrices in this chapter will be represented by capital letters.

(d) *Function*—a rule whereby to each scalar x, there corresponds another scalar $f(x)$. A function can also be a vector with an argument, which may be a scalar or a vector. In either case, a vector function will be characterized as $\mathbf{f}(x)$ or $\mathbf{f}(\mathbf{x})$.

This hierarchy of elements can be extended indefinitely. For example, there are matrices of vectors, matrices of functions, functions of vectors, and so on.

1.1.2 Set

A *set* [2] is a collection of elements considered as a whole. By a set $\mathfrak{S}$ of elements a, c, d, e, we mean a collection of well-defined entities. By $a \in \mathfrak{S}$, we mean that a is an element of the set $\mathfrak{S}$, and $b \notin \mathfrak{S}$ implies that b does not belong to $\mathfrak{S}$. We usually denote sets by placing $\{\ \}$ around the elements, (e.g., $\mathfrak{S} = \{a,c,d,e,...\}$ or $\{a_i\}$).

1.1.3 Field

A collection of elements of a set of $\mathfrak{F}$ (scalars) are said to form a *field* $\mathfrak{F}$ [1] if they possess two binary operations, referred to as the sum and the product, such that the results of these operations on every two elements of $\mathfrak{F}$ lead to elements also in $\mathfrak{F}$ and the following axioms are satisfied. If $\alpha \in \mathfrak{F}$ and $\beta \in \mathfrak{F}$, then their sum will be denoted by $\alpha + \beta$ and their product by $\alpha\beta$. The properties $(\alpha + \beta) \in \mathfrak{F}$ and $\alpha\beta \in \mathfrak{F}$ are referred to as *closure* under addition and multiplication, respectively. So, the axioms are as follows:

1. A binary operation, referred to as addition, is defined for every two elements of $\mathfrak{F}$ and satisfies
   (a) $\alpha + \beta = \beta + \alpha$ (commutativity of addition)
   (b) $\alpha + (\beta + \gamma) = (\alpha + \beta) + \gamma$ (associativity of addition)
   (c) An element θ exists in $\mathfrak{F}$ called *additive identity* (also called zero) if $\alpha + \theta = \alpha$.

   (d) For each $\alpha \in \mathfrak{F}$, there is an element $-\alpha \in \mathfrak{F}$ such that $\alpha + (-\alpha) = \theta$. The element $-\alpha$ is called the *additive inverse* of α or simply the negative of α.

2. A binary operation, referred to as the product, is defined for every two elements of $\mathfrak{F}$ and satisfies the following:

   (a) $\alpha\beta = \beta\alpha$ (commutativity of multiplication)
   (b) $\alpha(\beta\gamma) = (\alpha\beta)\gamma$ (associativity of multiplication)
   (c) An element e exists called *multiplicative identity* and is denoted by e such that as $\alpha e = \alpha$, $e \neq \theta$. When we deal with real or complex numbers, e will be defined as 1.
   (d) For each element $\alpha \neq \theta$ in $\mathfrak{F}$, an element $(\alpha^{-1}) \in \mathfrak{F}$ exists called the *multiplicative inverse* of α such that $\alpha(\alpha^{-1}) = e$.

3. The binary operation referred to as the product is distributive with respect to the addition operation so that $\alpha(\beta + \gamma) = \alpha\beta + \alpha\gamma$.

Thus, a collection of objects forms a *field* when the concepts of addition, multiplication, identity, and inverse elements are defined subject to the previous axioms. It can be shown that $\mathfrak{F}$ contains a unique zero element called the additive identity and a unique unity element called the multiplicative identity.

The two simplest examples of fields are the field of real numbers, which is called the *real field* and is defined by $\mathfrak{R}$ and the field of complex numbers that is called the *complex field* and is denoted by $\mathfrak{C}$. In these cases, the binary operations are ordinary addition and multiplication, and the zero element and the identity element are, respectively, the numbers 0 and 1. Generally, we will be dealing with the field of real numbers $\mathfrak{R}$ and complex numbers $\mathfrak{C}$. However, it is important to remember that the previous definition is valid for any general field of scalars. For example, the set of all rational numbers, real numbers, or complex numbers of the form, $c + \sqrt{-1}\, d$, where c and d are real numbers, also forms a *field*. However, the set of integers does not form a field because the property 2(d) stated earlier is not satisfied. In addition, any 2×2 matrices of the form:

$$\begin{bmatrix} a & -b \\ b & a \end{bmatrix}$$

also form a field under usual matrix addition and multiplication when the numbers a and b are real, but the existence of singular matrices reveals that not all sets of 2×2 matrices form a field.

1.1.4   Space

A *space* [1,2] is a set for which some mathematical structure is defined. We denote spaces by the Euclid Math Two font as $\mathbb{A}$, $\mathbb{B}$, $\mathbb{C}$, $\mathbb{U}$, $\mathbb{V}$,.... Some examples of specific spaces are:

(a) *Space of the scalar field*, denoted as $\mathbb{F} = \{\alpha, \beta, \gamma, ...\}$ where the elements α, β, γ, ... are scalar. This space must also contain the sum, difference, product, and quotient of any two scalars in the field (division by zero excluded). In the space of scalar field, the only two fields that we need are the real field $\mathfrak{R}$ and the complex field $\mathfrak{C}$. When the field is $\mathfrak{R}$ (real numbers), we call $\mathbb{F}$ a linear space over the real field or a *real linear space*. When the field is $\mathfrak{C}$ (complex scalars), we call $\mathbb{F}$ a linear space over the complex field, or a *complex linear space*.

(b) *Space of real vectors of n components*, denoted $\mathbb{R}^n = \{\mathbf{x}, \mathbf{y}, \mathbf{z}, ...\}$ where each vector is an array of the form of Eq. (1.1*a*) or (1.1*b*).

(c) *Space of complex vectors of n components*, $\mathbb{C}^n = \{\mathbf{x}, \mathbf{y}, \mathbf{z}, ...\}$ where each vector is an array, Eq. (1.1*a*), or (1.1*b*) with the x_i as complex scalars.

(d) *Space of functions* continuous on the interval $a \le x \le b$, denoted by $[a, b] = \{f, g, h, ...\}$ where the elements are functions. We use the convention that closed intervals $a \le x \le b$ are denoted by $[a, b]$, open intervals $a < x < b$ by (a, b), and so on. If the functions are real, then we call the space real $\mathbb{R}[a, b]$, and if they are complex, then we call it complex $\mathbb{C}[a, b]$.

Let us now start with the general concept of a space.

1.2   METRIC SPACE

A nonempty set of elements (or points) $\mathbb{X}$ containing u, v, w is said to be a *metric space* [1–3], if to each pair of elements u, v a real number $\mathscr{D}(u, v)$ is associated, of two arguments called *distance* between u and v satisfying

$$\mathscr{D}(u, v) > 0 \text{ for } u, v \in \mathbb{X} \text{ and are distinct}$$
$$\mathscr{D}(u, v) = 0 \text{ if and only if } u = v$$
$$\mathscr{D}(u, v) = \mathscr{D}(v, u)$$
$$\mathscr{D}(u, w) \le \mathscr{D}(u, v) + \mathscr{D}(v, w) \qquad \text{(triangle inequality)}$$

This function $\mathscr{D}$ is called the *metric* or *distance function*. Once a distance is available, one can define the notion of a neighborhood for an element, and that leads up to the concept of convergence. A sequence of points $\{u_k\}$ converges to u if and only if the sequence of real numbers $\{\mathscr{D}(u, u_k)\}$ converges to 0. We write the following:

$$\lim_{k \to \infty} u_k = u$$

or $u_k \to u$ and say that $\{u_k\}$ converges to u or that $\{u_k\}$ has the limit u, if for each $\varepsilon > 0$, there exists an index N exists such that $\mathscr{D}(u, u_k) \le \varepsilon$ whenever $k > N$. A sequence $\{u_k\}$ is called a *Cauchy sequence* if for each $\varepsilon > 0$ there exists an N such that $\mathscr{D}(u_m, u_p) \le \varepsilon$ whenever $m, p > N$. Therefore, a metric space $\mathbb{X}$ is complete if every Cauchy sequence of points from $\mathbb{X}$ converges to a limit in $\mathbb{X}$. As an example, consider the sequence of rational numbers, u_i,

$$u_1 = 1; \quad u_2 = 1 + \frac{1}{1!}; \quad \cdots; \quad u_n = 1 + \frac{1}{1!} + \frac{1}{2!} + \ldots + \frac{1}{(n-1)!}$$

with the metric defined by $\mathscr{D}(u_i, u_j) = |u_i - u_j|$, it can be shown that [2] the sequence u_i converges to a real number q, which is not a rational number. Therefore, the space of rational numbers is not complete. However, it can be made complete by enlarging this space to include irrational numbers.

For example, the real line becomes a metric space of the distance between two real numbers u and v defined as $|u - v|$, and this space is complete [1].

The set of all complex numbers $z_i = x_i + jy_i$ where $j = \sqrt{-1}$, is a metric space under the definition [1]

$$\mathscr{D}(z_1, z_2) = |z_1, z_2| = \sqrt{(x_1 - x_2)^2 + (y_1 - y_2)^2} \tag{1.3}$$

and the space is complete.

Consider the set of all real-valued continuous functions $u(x)$ and $v(x)$ defined on the interval $a \le x \le b$ with the distance function

$$\mathscr{D}_\infty(u, v) = \max_{a \le x \le b} |u(x) - v(x)| \tag{1.4}$$

So if the Cauchy sequence $\{u_n(x)\}$ converges to a function $u(x)$ that is necessarily continuous, then the space $\mathbb{R}[a, b]$ is complete, and this is known as completeness under the *uniform metric*.

However, if we consider the set of all real-valued continuous functions $u(x)$ and $v(x)$ on the bounded interval $a \le x \le b$ but now with the metric:

$$\mathscr{D}_2(u, v) = \sqrt{\int_a^b |u(x) - v(x)|^2 \, dx} \tag{1.5}$$

then it has been shown [1] that this formula generates a metric space. However, this space is not complete, as functions with jump discontinuities can be approximated in the mean-square sense (that is in metric $\mathscr{D}_2$) by continuous functions, like in a Fourier series expansion, which approximates a discontinuous function in the mean.

Just as in the case of rational numbers, we saw that the metric space was not complete, but it can be completed by including the irrational numbers. Here we also employ the same methodology; the space of complex-valued functions on $a \le x \le b$ can be completed in the metric $\mathscr{D}_2$ by generating the space $\mathbb{L}_2^{(r)}[a, b]$ of real-valued functions square integrable in the Lebesque sense [3, pp. 36–40].

1.3   VECTOR (LINEAR) SPACE

Any *linear space* can be called a *vector space* [1] but we usually use the term "vector space" to imply that the elements are *n*-tuple vectors of the form as in Eq. (1.1*a*) or (1.1*b*). We use the term *function space* to imply elements that are *functions*. The elements of a *linear space* also are called *points* of the space.

To define a vector space $\mathbb{V}$, it is necessary to have a set of objects $\{\mathbf{u}\}$ (usually referred to as elements, vectors, or points), a field of scalars $\mathfrak{F}$, and two binary operations defined on these scalars and vectors. These operations are called vector additions and scalar multiplications (i.e., the multiplication of a vector by a scalar).

In conformity with the mathematical literature, [1], with the terms "vector" or "point", we mean an element of a vector space. This is not to be confused with the concept of a vector as a directed line segment, commonly used in engineering terminology.

Let $\mathbf{u}$ and $\mathbf{v}$ be vectors of $\mathbb{V}$, and let α belong to a field $\mathfrak{F}$; then the sum of $\mathbf{u}$ and $\mathbf{v}$ denoted by $\mathbf{u} \oplus \mathbf{v}$ and the product of $\alpha \odot \mathbf{u}$ also must be elements of $\mathbb{V}$. This is called the *closure* property [1]. A collection of objects forms a vector space $\mathbb{V}$ over a field $\mathfrak{F}$ if the operations of vector addition and multiplication of a vector by a scalar taken from $\mathfrak{F}$ satisfy the following conditions:

1. For any arbitrary vectors $\mathbf{u}$, $\mathbf{v}$, $\mathbf{w}$ elements of the space $\mathbb{V}$ and any arbitrary scalars α, β belonging to a field $\mathfrak{F}$, the following properties need to be satisfied.
    (a) $\mathbf{u} \oplus \mathbf{v} = \mathbf{v} \oplus \mathbf{u}$ (commutativity of addition)
    (b) $(\mathbf{u} \oplus \mathbf{v}) \oplus \mathbf{w} = \mathbf{u} \oplus (\mathbf{v} \oplus \mathbf{w})$ (associativity of addition)
    (c) There is a null vector $\mathbf{0}$ in this space such that for any $\mathbf{u} \in \mathbb{V}$, we have $\mathbf{u} \oplus \mathbf{0} = \mathbf{u}$
    (d) To every vector $\mathbf{u} \in \mathbb{V}$ there corresponds an element $(-\mathbf{u}) \in \mathbb{V}$ such that $\mathbf{u} \oplus (-\mathbf{u}) = \mathbf{0}$
2. Let 1 denote the identity element of $\mathfrak{F}$, then:
    (a) $1 \odot \mathbf{u} = \mathbf{u}$
    (b) $(\alpha\beta) \odot \mathbf{u} = \alpha \odot (\beta \odot \mathbf{u})$
    It is important to note that a field must contain an identity element, but a vector space does not need to have a unity element.
3. (a) $(\alpha + \beta) \odot \mathbf{u} = (\alpha \odot \mathbf{u}) \oplus (\beta \odot \mathbf{u})$ (distributive property)
    (b) $\alpha \odot (\mathbf{u} \oplus \mathbf{v}) = (\alpha \odot \mathbf{u}) \oplus (\alpha \odot \mathbf{v})$

Thus, any collection of objects whose elements satisfy these axioms with two well-defined operations (addition of vectors and multiplication by scalars) forms a vector space. For the sake of mathematical clarity, here we have used the notation $(+,\cdot)$ for the operation of a field $\mathfrak{F}$ and the notation $(\oplus,\odot)$ for the operations of the vector space. Such a distinction in these operations is essential for the proper understanding of the concepts. Moreover, the operation of the addition of vector space generally coincides with the addition operation of the field of scalars. Thus, from now on, we will replace the symbol $\oplus$ by $+$ and will drop $\odot$ in favor of juxtaposition to denote scalar multiplication.

As an example, the field $\mathfrak{C}$ of complex numbers may be considered as a vector space $\mathbb{H}$ over the field of real numbers $\mathfrak{R}$. In addition, points of a two-dimensional real space $\mathbb{R}^2$ form a vector space over the field of real numbers $\mathfrak{R}$. Also, the set of all functions that are at least twice differentiable and satisfy the form of $d^2x/dt^2 + dx/dt - ax = 0$ for a is a constant and forms a linear vector space $\mathbb{L}^2$ over $\mathfrak{C}$. All real-valued functions $f(t)$ of the real variable t that are continuous in the interval $[a,b]$ form a vector space under the ordinary definition of addition and multiplication by a scalar. The null element of this space is a function that is identically zero in the same interval. However, the set of all polynomials of degree n (where n is a positive integer) does not form a vector space under the familiar definition for addition and multiplication by numbers as the sum of two polynomials of degree n and is not necessarily a polynomial of degree n, where n is the exponent of the leading nonzero term. The set of all polynomials of degree $\leq n$ forms a vector space.

1.3.1 Dependence and Independence of Vectors

A set of vectors $\mathbf{x}, \mathbf{y}, \mathbf{z}, \ldots, \mathbf{w}$ belonging to a vector space $\mathbb{V}$ defined over the field $\mathfrak{F}$ are said to be *linearly independent* [1] if

$$\alpha\mathbf{x} + \beta\mathbf{y} + \gamma\mathbf{z} + \cdots + \mu\mathbf{w} = \mathbf{0}; \quad \text{with } \alpha, \beta, \gamma, \ldots, \mu \in \mathfrak{F} \tag{1.6}$$

implies $\alpha = \beta = \gamma = \cdots = \mu = 0$. If a nontrivial relation of the previous type exists, where the coefficients do not vanish altogether, then the elements are said to be dependent.

1.3.2 Dimension of a Linear Space

If a vector space contains n linearly independent vectors (n is a finite positive integer) and if all $n + 1$ vectors of that space are linearly dependent, then the vector space is said to be of dimension n [1].

Any set of n-linearly independent vectors of $\mathbb{V}$ is said to form a basis for that space, if every vector of $\mathbb{V}$ can be expressed as a linear combination of elements of that set. If a basis in a linear space contains n-elements, then the space is called an n-dimensional space or simply an n-space [1].

1.3.3 Subspaces and Linear Manifolds

A subset of elements of a vector space $\mathbb{V}$ is said to form a *subspace* $\mathbb{W}$ of $\mathbb{V}$ if $\mathbb{W}$ is a vector space under the same law of addition and scalar multiplication defined for $\mathbb{V}$, over the same field $\mathfrak{F}$ [1]. So the null element is the smallest subspace of $\mathbb{V}$, and in a manner of speaking, $\mathbb{V}$ is the largest subspace of itself. Because any subspace of a finite dimensional vector space is by itself a vector space, it has some dimension, and the dimension of the subspace cannot exceed that of the original space.

If $\mathbf{x}$ and $\mathbf{y}$ are specified elements of a linear vector space $\mathbb{V}$ and if α and β are arbitrary numbers of the field $\mathfrak{F}$, then the set of elements $\alpha\mathbf{x} + \beta\mathbf{y}$ for all α and β in $\mathfrak{F}$ forms a linear subspace. We say that this subspace is generated or spanned by $\mathbf{x}$ and $\mathbf{y}$. Similarly, if $\{\mathbf{x}, \mathbf{y}, \mathbf{z}, \ldots, \mathbf{w}\}$ is a set of vectors $\mathbb{V}$, the collection of elements of the type $\alpha\mathbf{x} + \beta\mathbf{y} + \gamma\mathbf{z} + \cdots + \mu\mathbf{w}$; with $\alpha, \beta, \gamma, \ldots, \mu \in \mathfrak{F}$ forms a linear subspace $\mathbb{W}$. This subspace is called the *linear manifold* spanned by $\{\mathbf{x}, \mathbf{y}, \mathbf{z}, \ldots, \mathbf{w}\}$. The linear manifold spanned by these vectors is the linear subspace of the smallest dimension that contains them.

Once we are dealing with vectors, we can define it by its magnitude and the rotation or angle between any two vectors. Associated with these properties, one can define linear spaces based on these two unique properties of a vector.

1.3.4　　Normed Linear Space

A *normed linear space* [2] (real or complex) in which a real valued function $\|\mathbf{u}\|$ (known as the norm of $\mathbf{u}$) is defined to have the following properties:

(a) $\|\mathbf{u}\| > 0$; if $\mathbf{u} \neq \mathbf{0}$　　　　　　　　　(positivity)

(b) $\|\mathbf{u}\| = 0$; if and only if $\mathbf{u} = \mathbf{0}$　　　　(definiteness)

(c) $\|\alpha\mathbf{u}\| = |\alpha| \times \|\mathbf{u}\|$　　　　　　　　　　(linear scaling)

(d) $\|\mathbf{u} + \mathbf{v}\| \leq \|\mathbf{u}\| + \|\mathbf{v}\|$　　　　　　　(triangular inequality)

Here $|\alpha|$ denotes the absolute value for α. A space with a norm associated with every element of $\mathbf{u}$ in $\mathbb{V}$ is called a *normed space*. It is important to note that a *metric space*, in general, is different from a *normed space*. The various hierarchies of spaces are shown in Figure 1.1.

However, if the metric $\mathscr{D}(\mathbf{u}, \mathbf{v})$ is identified with the norm so that

$$\mathscr{D}(\mathbf{u}, \mathbf{v}) = \|\mathbf{u} - \mathbf{v}\| \tag{1.7}$$

then one obtains a *normed linear metric space* as illustrated in Figure 1.1. So here, $\|\mathbf{u}\|$ plays the same role as the length of $\mathbf{u}$ in an ordinary three-dimensional space. A *normed linear space that is complete in its natural metric*, where every *Cauchy sequence* converges is called a *Banach space,* as illustrated in Figure 1.1. The space of real numbers form a Banach space, but the set of rational numbers does not constitute a Banach space but a metric space. Similarly, the space of continuous functions $\mathbb{C}[a, b]$ is a Banach space, but the space of square integrable continuous functions $\mathbb{C}[a, b]$ does not form a Banach space.

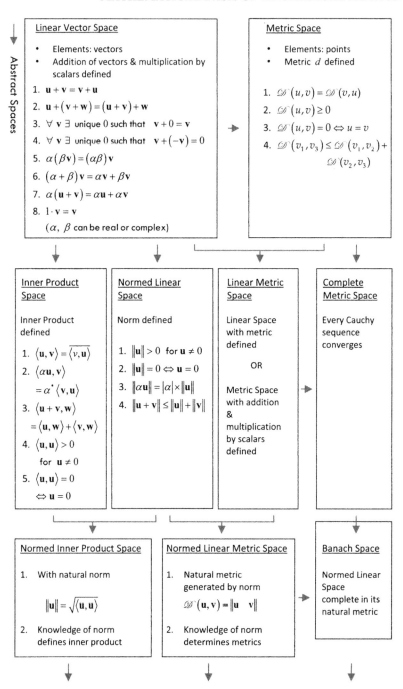

Figure 1.1. A summary of various spaces.

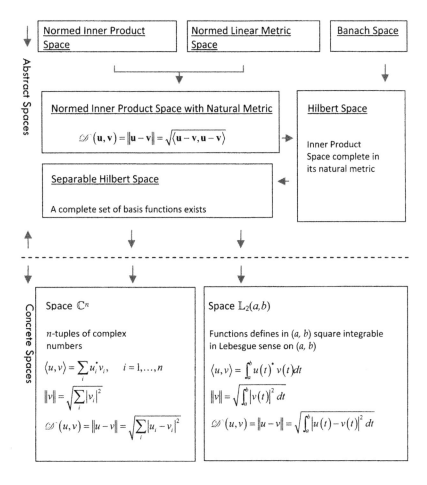

Note: Linear spaces may be finite or infinite dimensional.

1.3.5 Inner Product Space

In a normed linear space, a vector has length. We want to refine the structure further so that *the angle between two vectors* is also defined. In particular, we want a criterion for determining whether two vectors are orthogonal, and it is related to *the dot product between vectors*.

An *inner product* $\langle \mathbf{u}, \mathbf{v} \rangle$ [2] on a complex linear space $\mathbb{X}^{(c)}$ is a real/complex-valued function of an ordered pair of vectors $\mathbf{u}$, $\mathbf{v}$ with the properties

(a) $\quad \langle \mathbf{u}, \mathbf{v} \rangle = \overline{\langle \mathbf{v}, \mathbf{u} \rangle}$

where the overbar denotes the complex conjugate of an expression.

(b) $\langle \alpha \mathbf{u}, \mathbf{v} \rangle = \alpha \langle \mathbf{u}, \mathbf{v} \rangle$ and $\langle \mathbf{u}, \beta \mathbf{v} \rangle = \beta^* \langle \mathbf{u}, \mathbf{v} \rangle$

where * denotes a complex conjugate of a scalar.

(c) $\langle \mathbf{u} + \mathbf{v}, \mathbf{w} + \mathbf{y} \rangle = \langle \mathbf{u}, \mathbf{w} \rangle + \langle \mathbf{v}, \mathbf{y} \rangle + \langle \mathbf{v}, \mathbf{w} \rangle + \langle \mathbf{u}, \mathbf{y} \rangle$

(d) $\langle \mathbf{u}, \mathbf{u} \rangle > 0$ for $\mathbf{u} \neq \mathbf{0}$ and is a real value,

$\langle \mathbf{u}, \mathbf{u} \rangle = 0$ for $\mathbf{u} = \mathbf{0}$ (the inner product is positive definite)

(e) $|\langle \mathbf{u}, \mathbf{v} \rangle|^2 \leq \langle \mathbf{u}, \mathbf{u} \rangle \langle \mathbf{v}, \mathbf{v} \rangle$ (equality holds if and only if $\mathbf{u}$ and $\mathbf{v}$ are linearly dependent)

Two vectors are orthogonal if their inner product is zero. Orthogonality is an extension of the geometrical concept of perpendicularity in $\mathbb{R}^2$ or $\mathbb{R}^3$ space.

As an example, in a space with complex elements, the inner product is given by

$$\langle \mathbf{u}, \mathbf{v} \rangle = u_1 v_1^* + u_2 v_2^* + \cdots + u_n v_n^* \tag{1.8}$$

if $\langle \mathbf{u}, \mathbf{v} \rangle = 0$, then $\mathbf{u}$ is geometrically perpendicular to $\mathbf{v}$.

In the space of continuous functions, the inner product is defined in the complex function space as

$$\langle \mathbf{f}, \mathbf{g} \rangle = \int_a^b f(z) g^*(z) dz \tag{1.9}$$

So two complex functions are orthogonal if their average is zero; that is,

$$\langle \mathbf{f}, \mathbf{g} \rangle = \int_a^b f(z) g^*(z) dz = 0 \tag{1.10}$$

Sometimes it is convenient to define a weighted inner product as follows:

$$\langle \mathbf{f}, \mathbf{g} \rangle = \int_a^b w(z) f(z) g^*(z) dz \tag{1.11}$$

where $w(z) > 0$ on $[a, b]$ and is called the weight function. If $\mathbf{g}$ is real, then $g(z)$ will be used instead of $g^*(z)$.

A space with an inner product associated with any two elements of $\mathbf{u}$ and $\mathbf{v}$ in $\mathbb{V}$ is called *an inner product space* as shown in Figure 1.1.

One also could obtain a *normed inner product space* as shown in Figure 1.1 if we identify the norm with the inner product as;

$$\|\mathbf{u}\| = \sqrt{\langle \mathbf{u}, \mathbf{u} \rangle} \tag{1.12}$$

that is, the knowledge of the norm defines the inner product.

Finally, one could obtain a *normed inner product space* with its natural metric as illustrated in Figure 1.1 if we relate the norm, inner product, and the metric by

$$\mathscr{D}(\mathbf{u},\mathbf{v}) = \|\mathbf{u}-\mathbf{v}\| = \sqrt{\langle \mathbf{u}-\mathbf{v},\mathbf{u}-\mathbf{v}\rangle} \tag{1.13}$$

With the metric defined, we now can use the concept of convergence, which is repeated here for convenience: A sequence $\{u_k\}$ is said to converge to u if, for each $\varepsilon > 0$, there exists an $N > 0$ such that $\|\mathbf{u}_k - \mathbf{u}\| < \varepsilon$, with $k > N$.

An inner product space, which is complete in its natural metric, is called a *Hilbert space* as illustrated in Figure 1.1.

For complex-valued functions $u(x)$ defined in $a \le x \le b$, for which the Lebesque integral

$$\int_a^b |u(x)|^2\, dx$$

exists and is finite, we can define

$$\|\mathbf{u}\|_2 = \sqrt{\int_a^b |u(x)|^2\, dx} \tag{1.14a}$$

$$\mathscr{D}_2(\mathbf{u},\mathbf{v}) = \|\mathbf{u}-\mathbf{v}\|_2 = \sqrt{\int_a^b |u(x)-v(x)|^2\, dx} \tag{1.14b}$$

and the norm is called the $\mathscr{L}_2$ norm. The Cauchy–Schwartz inequality becomes

$$\left| \int_a^b u(x)v^*(x)dx \right| \le \sqrt{\int_a^b |u(x)|^2\, dx} \times \sqrt{\int_a^b |v(x)|^2\, dx} \tag{1.15}$$

The space $\mathbb{C}[a,b]$ of continuous functions (real- or complex-valued) on a bounded interval $a \le x \le b$ is a normed space under the definition

$$\|\mathbf{u}\|_\infty = \max_{a \le x \le b} |u(x)| \tag{1.16}$$

This norm generates the natural metric

$$\mathscr{D}_\infty(\mathbf{u},\mathbf{v}) = \|\mathbf{u}-\mathbf{v}\|_\infty = \max_{a \le x \le b} |u(x)-v(x)| \tag{1.17}$$

and $\mathbb{C}[a,b]$ can be shown to be a *Banach space* — a *complete normed space* [2], as this norm cannot be derived from an inner product. The reason that this space is not a Hilbert space but a Banach space is because the inner product cannot be generated for this metric.

Let Ω be a bounded domain in $\mathfrak{R}^n$ with a boundary Γ. Consider the set $\mathfrak{S}$ of real-valued functions that are continuous and have a continuous gradient on Ω. The following bilinear form

$$\langle \mathbf{p}, \mathbf{q} \rangle_{\mathfrak{S}} = \int_{\Omega} (\nabla u \cdot \nabla v) dx \tag{1.18}$$

is an admissible inner product on $\mathbb{M}$. The completion of $\mathbb{M}$ in the norm generated by this inner product is known as *the Sobolev Space* $\mathbb{H}^1(\Omega)$. This is a subset of a Hilbert space [2]. The term "completion of space $\mathbb{M}$" implies that the limit point of some sequence of functions is not a member of that space and that these elements need to be added so as to make the space complete. For example, if we take a sequence of square integrable functions that are continuous and forms a space $\mathbb{Q}$, then the limit point of the sequence of continuous functions may result in a discontinuous function. And these limit points, which are the discontinuous functions, do not belong to $\mathbb{Q}$. Hence, these extra elements that are the limit points need to be added to the space $\mathbb{Q}$ to make it *complete*.

1.4   PROBLEM OF BEST APPROXIMATION

If we wish to approximate an element $\mathbf{u}$ by an arbitrary set $\{\mathbf{u}_1, \mathbf{u}_2, ..., \mathbf{u}_k\}$, that is, we wish to find the element

$$\sum_{i=1}^{k} \alpha_i \mathbf{u}_i$$

which is closest to $\mathbf{u}$ in the sense of the metric of the space $\mathbb{H}$. Because the set of linear combinations of $\{\mathbf{u}_1, \mathbf{u}_2, ..., \mathbf{u}_k\}$ is a k-dimensional linear space $\mathbb{M}_k$, this approximation leads to finding the projection of $\mathbf{u}$ in that k-dimensional linear space $\mathbb{M}_k$. How to develop this procedure will be illustrated in the next section.

1.4.1   Projections

Consider a Euclidean space $\mathbb{E}$ of dimension n and a subspace $\mathbb{G}$ of dimension $m \leq n$. The projection theorem then states that an arbitrary element $\mathbf{f}$ in $\mathbb{E}$ can be expressed uniquely as

$$\mathbf{f} = \mathbf{g} + \mathbf{h} \tag{1.19}$$

where $\mathbf{g}$ is in $\mathbb{G}$ and $\mathbf{h}$ is orthogonal to $\mathbb{G}$ (orthogonal to every element of $\mathbb{G}$). Element $\mathbf{g}$ is called the *projection* (also called a *perpendicular projection*) of $\mathbf{f}$ on $\mathbb{G}$. The norm $\|\mathbf{h}\|$ is the minimum distance from $\mathbf{f}$ to any element of $\mathbb{G}$, called the distance from $\mathbf{f}$ to $\mathbb{G}$. So $\mathbf{g}$ is the element of $\mathbb{G}$ closest to $\mathbf{f}$ [1,2].

Because

$$\|\mathbf{f} - \mathbf{a}\|^2 = \|\mathbf{h}\|^2 + \|\mathbf{g} - \mathbf{a}\|^2 \tag{1.20}$$

it is clear that $\|\mathbf{f} - \mathbf{a}\|$ is a minimum when $\mathbf{g} = \mathbf{a}$, and this distance is $\mathbf{h}$. Alternatively, if $\{\mathbf{f}_1, \mathbf{f}_2, ..., \mathbf{f}_m\}$ forms a basis that generates $\mathbf{a}$ in $\mathbb{G}$, then $\mathbf{h}$ is orthogonal to $\mathbf{a}$. This implies,

$$\langle \mathbf{f}_i, \mathbf{h} \rangle = \langle \mathbf{f}_i, \mathbf{f} - \mathbf{g} \rangle = 0 \quad \text{or} \quad \langle \mathbf{f}_i, \mathbf{f} \rangle = \langle \mathbf{f}_i, \mathbf{g} \rangle, \text{ for each } \mathbf{f}_i \tag{1.21}$$

To calculate $\mathbf{g}$, let

$$\mathbf{g} = \sum_{k=1}^{m} \alpha_k \mathbf{f}_k$$

and substitute in Eq. (1.21) to obtain

$$\langle \mathbf{f}_i, \mathbf{f} \rangle = \sum_{k=1}^{m} \alpha_k \langle \mathbf{f}_i, \mathbf{f}_k \rangle \quad \text{for } i = 1, 2, ..., m \tag{1.22}$$

Therefore, $[\alpha]$, the matrix of the coefficients α_i is given by

$$[\alpha] = \left[\langle \mathbf{f}_i, \mathbf{f}_k \rangle \right]^{-1} \left[\langle \mathbf{f}_i, \mathbf{f} \rangle \right] \tag{1.23}$$

The matrix $\left[\langle \mathbf{f}_i, \mathbf{f}_k \rangle \right]$ is called the *gram matrix* of the basis $\{\mathbf{f}_i\}$. Because the determinant of this gram matrix is not zero, the coefficients α_i can be obtained from the solution of Eq. (1.23).

As an example, consider the function $f(x) = \sin(0.5\pi x)$ in $\mathfrak{R}[0,1]$ and the subspace $\mathbb{E}$, which is generated by $f_1(x) = 1$ and $f_2(x) = x$. We wish to calculate the projector of f on the subspace $\mathbb{G}$. So we have

$$\langle \mathbf{f}_1, \mathbf{f} \rangle = \int_0^1 \sin \frac{\pi x}{2} dx = 0.6366 \tag{1.24}$$

$$\langle \mathbf{f}_2, \mathbf{f} \rangle = \int_0^1 x \sin \frac{\pi x}{2} dx = 0.4053 \tag{1.25}$$

$$\langle \mathbf{f}_1, \mathbf{f}_1 \rangle = \int_0^1 dx = 1 \tag{1.26}$$

$$\langle \mathbf{f}_1, \mathbf{f}_2 \rangle = \langle \mathbf{f}_2, \mathbf{f}_1 \rangle = \int_0^1 x \, dx = \frac{1}{2} \tag{1.27}$$

$$\langle \mathbf{f}_2, \mathbf{f}_2 \rangle = \int_0^1 x^2 \, dx = \frac{1}{3} \tag{1.28}$$

Then from Eq. (1.22) we have

$$0.6366 = 1 \times \alpha_1 + 1/2 \times \alpha_2$$

$$0.4053 = 1/2 \times \alpha_1 + 1/3 \times \alpha_2 \tag{1.29}$$

which yields $\alpha_1 = 0.114$ and $\alpha_2 = 1.044$. Hence, the projection of $f(x)$ on $\mathbb{G}$ is $g(x) = 0.114 + 1.044x$. Here, $\mathbf{g}$ has been approximated under the $\mathscr{L}_2$ norm. The solution would be different if a different metric or inner product was used. Here, the projection of $\mathbf{f}$ on $\mathbf{a}$ is $g(x) = 0.114 + 1.044x$, and the function orthogonal to $\mathbf{a}$ is $h(x) = \sin(\pi x/2) - g(x)$.

In this case, the norm induced by the scalar product is

$$\|\mathbf{h}\|_2 = \|\mathbf{f} - \mathbf{g}\|_2 = \sqrt{\int_0^1 |f(x) - g(x)|^2 \, dx} \tag{1.30}$$

and hence $\|\mathbf{h}\|_2$ is the root-mean-squared error. The projection $\mathbf{g}$ is called the best root-mean-square approximation of $\mathbf{f}$.

One can illustrate the same principles using the principles of calculus to show that the projection of $\mathbf{f}$ on $\mathbf{a}$ is the element of $\mathbf{a}$ closest to $\mathbf{f}$. Let $\{\mathbf{f}_1, \mathbf{f}_2, \ldots, \mathbf{f}_m\}$ be a basis for $\mathbf{a}$; in which case, an arbitrary element $\mathbf{a}$ of $\mathbb{G}$ can be expressed as:

$$\mathbf{a} = \sum_{i=1}^m \alpha_i \mathbf{f}_i \tag{1.31}$$

The square of the distance from $\mathbf{a}$ to $\mathbf{f}$ is

$$\begin{aligned}
\mathbf{d}^2 = \|\mathbf{f} - \mathbf{a}\|^2 &= \langle \mathbf{f} - \mathbf{a}, \mathbf{f} - \mathbf{a} \rangle = \left\langle \mathbf{f} - \sum_i \alpha_i \mathbf{f}_i, \mathbf{f} - \sum_i \alpha_i \mathbf{f}_i \right\rangle \\
&= \langle \mathbf{f}, \mathbf{f} \rangle - \sum_i \alpha_i \langle \mathbf{f}_i, \mathbf{f} \rangle - \sum_i \alpha_k \langle \mathbf{f}, \mathbf{f}_k \rangle + \sum_i \sum_k \alpha_i \alpha_k \langle \mathbf{f}_i, \mathbf{f}_k \rangle
\end{aligned} \tag{1.32}$$

Necessary conditions for a minimum are

$$\frac{\partial \mathbf{d}^2}{\partial \alpha_i} = 0 \quad \text{for } i = 1, 2, \cdots, m \tag{1.33}$$

Performing the required differentiation and rearranging the terms, we have

$$-\langle \mathbf{f}_i, \mathbf{f} \rangle + \sum_i \alpha_k \langle \mathbf{f}_i, \mathbf{f}_k \rangle = 0 \quad \text{for } i = 1, 2, \ldots, m \tag{1.34}$$

This is identical to Eq. (1.22), which we obtained before, and the problem has a unique solution $\{\alpha_k\}$. That this solution provides a minimum value of the error and is not a maximum is evident from the expression for $\mathbf{d}^2$, which has no maximum, and its second derivative with respect to α_i is $\langle \mathbf{f}_i, \mathbf{f}_i \rangle$, which is positive.

1.5 MAPPING/TRANSFORMATION

A *mapping* or *transformation* [2] from a linear space $\mathbb{F}$ to a linear space $\mathbb{G}$ is a rule whereby to each element $\mathbf{f}$ in $\mathbb{F}$, there corresponds an element $\mathbf{g}$ in $\mathbb{G}$. This is symbolized by

$$\mathcal{M} : \mathbb{F} \to \mathbb{G} \quad \text{or equivalently by} \quad \mathcal{M}(\mathbf{f}) = \mathbf{g} \tag{1.35}$$

Some mappings of interest are

(a) *Function*, denoted by $f(x) = y$ maps from the scalar space $\mathbb{X}$ with element x to the scalar space $\mathbb{Y}$ with elements y

(b) *Functional* denoted by $\mathcal{L}(f) = y$, maps from the vector or function space $\mathbb{P}$ to the scalar space $\mathbb{Y}$ with elements y.

(c) *Operator* denoted by $\mathcal{M}(\mathbf{f}) = \mathbf{g}$ maps from a vector or function space $\mathbb{W}$ with elements $\mathbf{f}$ into itself, and $\mathbf{g}$ are also elements of $\mathbb{W}$.

The nomenclature is not uniform in all texts, some of which use the terms mapping, function, and operator as synonymous.

The element $\mathbf{g}_i$ corresponding to a particular $\mathbf{f}_i$ in Eq. (1.35) is called the *image* of $\mathbf{f}_i$. The space $\mathbb{W}$ to which the mapping applies is called the *domain* of $\mathcal{M}$, denoted by $\mathbb{D}(\mathcal{M})$. The set of $\mathbf{g}$ resulting from the mapping is called the *range* or *image* of $\mathcal{M}$ and is defined by $\mathbb{R}(\mathcal{M})$. If the range of $\mathcal{M}$ is the whole $\mathbb{W}$, then the mapping is said to be *onto* $\mathbb{W}$. If the range of $\mathcal{M}$ is a proper subset of $\mathbb{W}$, then the mapping is said to be *into* $\mathbb{W}$. If $\mathcal{M}(\mathbf{f}) = \mathbf{g}$ maps distinct elements $\mathbf{f}_i$ into distinct elements $\mathbf{g}_i$, then the mapping is said to be *one-to-one*. If n elements $\mathbf{f}_i$ map into the element $\mathbf{g}_i$, then the mapping is said to be *n-to-one*, and if each element $\mathbf{f}_i$ maps into n elements $\mathbf{g}_i$, then it is said to be *one-to-n*. These concepts are illustrated in Figure 1.2.

If a mapping is *one-to-one* and onto $\mathbb{W}$, then the inverse mapping

$$\mathcal{M}^{-1}(\mathbf{g}) = \mathbf{f} \tag{1.36}$$

exists, and the mapping is said to be invertible. The domain of $\mathcal{M}^{-1}$ is the image of $\mathcal{M}$, and the image of $\mathcal{M}^{-1}$ is the domain of $\mathcal{M}$.

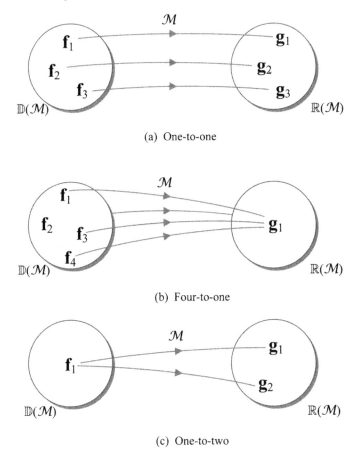

(a) One-to-one

(b) Four-to-one

(c) One-to-two

Figure 1.2. Pictorial representation of mapping.

If the range of a mapping is the domain of another mapping $\mathcal{P}$, then the product of these mappings is defined as the successive mappings of first $\mathcal{M}$ and then of $\mathcal{P}$ symbolized by

$$\mathcal{P}\mathcal{M}(\mathbf{f}) = \mathcal{P}(\mathcal{M}(\mathbf{f})) \tag{1.37}$$

The identity mapping is defined by $\mathcal{I}_{\mathcal{M}}(\mathbf{f}) = \mathbf{f}$ and the *null mapping* $\mathcal{N}_{\mathcal{M}}(\mathbf{f}) = \mathbf{0}$, (i.e., $\mathcal{N}_{\mathcal{M}}$ maps every $\mathbf{f}$ into the null element). Examples of some mappings are

$$\text{function: } f(x) = \sin x = y \tag{1.38}$$

$$\text{functional: } \phi(\mathbf{f}) = \int_0^1 f(x)\,dx = \int_0^1 \sin(x)\,dx = y \tag{1.39}$$

$$\mathbb{R}^2 \rightarrow \mathbb{R}; \quad \mathcal{M}(\mathbf{x}) = x_1 + 2x_2 = y \tag{1.40}$$

$$\mathbb{R}^3 \rightarrow \mathbb{R}; \quad \mathcal{M}(\mathbf{x}) = \begin{bmatrix} x_1 - x_2 \\ x_2 + 2x_3 \\ x_3 - x_1 \end{bmatrix} = y^2 \tag{1.41}$$

A transformation $\mathcal{A}$ is *bounded* on its domain, if for all $\mathbf{u}$ in $\mathbb{D}(\mathcal{A})$ there exists a constant $\mathcal{C}$ such that [3]

$$\|\mathcal{A}\mathbf{u}\| \le \mathcal{C}\|\mathbf{u}\| \tag{1.42}$$

Thus, the ratio of the "output" norm to the "input" norm is bounded. The smallest number $\mathcal{C}$ that satisfies the inequality for all $\mathbf{u}$ in $\mathbb{D}(\mathcal{A})$ is the norm of $\mathcal{A}$, written as $\|\mathcal{A}\|$, and is defined as follows:

$$\|\mathcal{A}\| = \sup_{\|\mathbf{u}\| \ne 0} \frac{\|\mathcal{A}\mathbf{u}\|}{\|\mathbf{u}\|} = \sup_{\|\mathbf{u}\|=1} \|\mathcal{A}\mathbf{u}\| \tag{1.43}$$

The prefix "sup" implies *supremum* (i.e., the maximum value attained for any $\mathbf{u}$). Differential operators are not bounded, whereas the integral operators are bounded [4, pp. 282–283].

A mapping $\mathcal{L}$ is linear if

$$\begin{aligned} \mathcal{L}(\alpha\mathbf{f}) &= \alpha\mathcal{L}(\mathbf{f}) \quad \text{and} \\ \mathcal{L}(\mathbf{f}_1 + \mathbf{f}_2) &= \mathcal{L}\mathbf{f}_1 + \mathcal{L}\mathbf{f}_2 \end{aligned} \tag{1.44}$$

for all $\mathbf{f}$ and α. The *null space*, $\mathbb{N}_\mathcal{L}$ of a mapping is the set $\{\mathbf{f}_i\}$ such that $\mathcal{L}(\mathbf{f}_i) = 0$. The null space of any linear mapping must contain the null element $\mathbf{f} = 0$. A linear mapping $\mathcal{L}$ on a finite-dimensional linear space has an inverse if and only if $\mathbf{f} = 0$ is the only solution to $\mathcal{L}(\mathbf{f}) = 0$.

An important property for differential operators is that they are usually not invertible unless boundary conditions are specified. As an example, consider the operator $\mathcal{D} = d/dx$ and the space $\mathbb{P}^n$ of polynomials of degree $n-1$ or less. If $\mathbf{f}$ is a constant α, then $\mathcal{D}\mathbf{f} = d\alpha/dx = 0$ and $\mathbf{f}$ is an element of the null space. Therefore, $\mathcal{D}$ is not of polynomials $f(x)$ for which $f(0) = 0$, and then $\mathcal{D}$ is invertible. The procedure can be described in two equivalent ways:

(a) The boundary condition $f(0) = 0$ is a part of the definition of the operator.
(b) The boundary condition $f(0) = 0$ restricts the domain of the operator.

Even if $\mathcal{A}$ is bounded, the supremum may not be attained for any element $\mathbf{u}$. If $\|\mathcal{A}\| = 0$, then $\mathcal{A}$ is the zero operator. If $\mathcal{A}$ is continuous at the origin, then it is continuous on all of $\mathbb{D}(\mathcal{A})$, and $\mathcal{A}$ is continuous if it is bounded. A linear transformation $\mathcal{A}$ *into* itself is characterized completely by its values through a fixed basis $\mathbf{h}_1, \mathbf{h}_2, \cdots, \mathbf{h}_n$. Indeed if we can write

$$\mathbf{u} = \sum_{j=1}^{n} \alpha_j \mathbf{h}_j \quad \text{and} \quad \mathbf{v} = \mathcal{A}\mathbf{u} = \sum_{j=1}^{n} \alpha_j \mathcal{A}\mathbf{h}_j$$

then the knowledge of n vectors $\mathcal{A}\mathbf{h}_1, \mathcal{A}\mathbf{h}_2, \ldots, \mathcal{A}\mathbf{h}_n$ enables us to calculate $\mathcal{A}\mathbf{u}$ or $\mathbf{v}$.

1.5.1 Representation by Matrices

Consider a linear mapping $\mathcal{L}$ from a space $\mathbb{W}$ with elements $\mathbf{f}$ to a space $\mathbb{G}$ resulting in elements $\mathbf{g}$, that is, $\mathcal{L}(\mathbf{f}) = \mathbf{g}$. Let $\{\mathbf{f}_1, \mathbf{f}_2, \ldots, \mathbf{f}_n\}$ be a basis for $\mathbb{W}$ and express an arbitrary element $\mathbf{f}$ in $\mathbb{W}$ as

$$\mathbf{f} = \sum_{i=1}^{n} \alpha_i \mathbf{f}_i \tag{1.45}$$

Substitute this equation into the linear mapping $\mathbf{g} = \mathcal{L}(\mathbf{f})$ and obtain

$$\mathbf{g} = \mathcal{L}[f] = \mathcal{L}\left[\sum_{i=1}^{n} \alpha_i \mathbf{f}_i\right] = \sum_{i=1}^{n} \alpha_i \mathcal{L}\mathbf{f}_i \tag{1.46}$$

Because every element in $\mathbb{G}$ can be written in this form, $\{\mathcal{L}\mathbf{f}_i\}$ must span $\mathbb{G}$. However, the $\{\mathcal{L}\mathbf{f}_i\}$ may be linearly dependent and, hence, may not form a basis for $\mathbb{G}$. We can obtain a basis $\{\mathbf{g}_1, \mathbf{g}_2, \ldots, \mathbf{g}_n\}$ for $\mathbb{G}$ from the $\{\mathcal{L}\mathbf{f}_i\}$ and express the $\mathbf{g}$ in this basis as

$$\mathbf{g} = \beta_1 \mathbf{g}_1 + \beta_2 \mathbf{g}_2 + \cdots + \beta_m \mathbf{g}_m \tag{1.47}$$

and express each $\mathcal{L}\mathbf{f}_i$ in this basis as

$$\begin{aligned}
\mathcal{L}\mathbf{f}_1 &= a_{11}\mathbf{g}_1 + a_{21}\mathbf{g}_2 + \cdots + a_{m1}\mathbf{g}_m \\
\mathcal{L}\mathbf{f}_2 &= a_{12}\mathbf{g}_1 + a_{22}\mathbf{g}_2 + \cdots + a_{m2}\mathbf{g}_m \\
&\cdots \\
\mathcal{L}\mathbf{f}_n &= a_{1n}\mathbf{g}_1 + a_{2n}\mathbf{g}_2 + \cdots + a_{mn}\mathbf{g}_m
\end{aligned} \tag{1.48}$$

In terms of these matrices, one can write that $[\mathcal{L}f]$ is the column matrix of $\mathcal{L}\mathbf{f}_i$, $[\mathbf{g}]$ is the column matrix of $\mathbf{g}_i$, and

$$[A] = \begin{bmatrix} a_{11} & a_{12} & \cdots & a_{1n} \\ a_{21} & a_{22} & \cdots & a_{2n} \\ \vdots & \vdots & \vdots & \vdots \\ a_{m1} & a_{m2} & \cdots & a_{mn} \end{bmatrix} \qquad (1.49)$$

Hence, one can write

$$[g]^T [\beta] = \left[[A]^T [g] \right]^T [\alpha] = [g]^T [A][\alpha] \qquad (1.50)$$

This is valid for all $[g]^T$, and therefore, the coefficient vectors $[\alpha]$ and $[\beta]$ are related by

$$[\beta] = [A][\alpha] \qquad (1.51)$$

Hence, for every linear mapping $\mathcal{L}(\mathbf{f}) = \mathbf{g}$, we can find a matrix representation $[A][\alpha] = [\beta]$. As there are infinitely many basis sets, hence, any linear mapping $\mathcal{L}(\mathbf{f}) = \mathbf{g}$ can be represented in infinite ways by the matrix equation $[A][\alpha] = [\beta]$.

1.5.2   Linear Forms

A linear mapping from a space $\mathbb{V}$ to the scalar field $\mathfrak{R}$ or $\mathfrak{C}$ is called a *linear form* or *linear functional*. If $\mathbb{V}$ is the space of n-tuples $\mathfrak{R}^n$ or $\mathfrak{C}^n$, then a linear form $p(\mathbf{x})$ can be written as

$$p(\mathbf{x}) = c_1 x_1 + c_2 x_2 + \cdots + c_n x_n \qquad (1.52)$$

where $c_1, c_2, \ldots, c_n$ are constants. In vector form, this can be written as follows:

$$p(\mathbf{x}) = [C]^T \mathbf{x} \qquad (1.53)$$

$[C]^T$ is the row vector of components c_i.

If $\mathbb{V}$ is a space of function $f(x)$ defined on the interval $a \leq x \leq b$, then a commonly encountered linear form is

$$p(f) = \int_a^b c(x) f(x) dx \qquad (1.54)$$

where $c(x)$ is a fixed function.

More generally, a linear form can be written as

$$p(f) = \int_a^b c(x) \mathcal{L}(f(x)) \, dx \qquad (1.55)$$

where $\mathcal{L}$ is a linear operator.

Let $\mathbb{M}$ be a linear manifold in the separable complex Hilbert space $\mathbb{H}$ (a space that contains an orthonormal spanning set, or in other words, a basis exists for that space). If to each $\mathbf{u}$ in $\mathbb{M}$ there corresponds a complex number, denoted by $\mathcal{L}(\mathbf{u})$, satisfying the condition.

$$\mathcal{L}(\alpha \mathbf{u} + \beta \mathbf{v}) = \alpha \mathcal{L}(\mathbf{u}) + \beta \mathcal{L}(\mathbf{v}) \quad \text{for all } \mathbf{u}, \mathbf{v} \text{ in } \mathbb{M} \qquad (1.56a)$$

and for all complex numbers α, β. We say that $\mathcal{L}$ is a *linear functional* in $\mathbb{M}$. A linear functional satisfies $\mathcal{L}(0) = 0$ and

$$\mathcal{L}\left[\sum_{j=1}^k \alpha_i \mathbf{u}_j\right] = \sum_{j=1}^k \alpha_j \mathcal{L}(\mathbf{u}_j) \qquad (1.56b)$$

A linear functional is bounded on its domain $\mathbb{M}$ if there exists a constant $\mathcal{T}$ such that for all $\mathbf{u}$ in $\mathbb{M}$, $|\mathcal{L}(\mathbf{u})| \leq \mathcal{T} \|\mathbf{u}\|$. The smallest constant $\mathcal{T}$ for which this inequality holds for all $\mathbf{u}$ in $\mathbb{M}$ is known as the norm of $\mathcal{L}$ and is denoted by $\|\mathcal{L}\|$.

A linear functional continuous at the origin is also continuous across its entire domain of definition $\mathbb{M}$. Boundedness and continuity are equivalent for linear functionals. A functional whose value at $\mathbf{u}$ is $\|\mathbf{u}\|$ is not linear. It is also of interest to note that to each continuous linear functional $\mathcal{L}$ defined on the space $\mathbb{H}$ corresponds an unambiguously defined vector $\mathbf{f}$ such that

$$\mathcal{L}(\mathbf{u}) = \langle \mathbf{u}, \mathbf{f} \rangle \quad \text{for every } \mathbf{u} \text{ in } \mathbb{H} \qquad (1.57)$$

1.5.3   Bilinear Forms

A mapping involving two elements, $\mathbf{x}$ in $\mathbb{V}_1$ and $\mathbf{y}$ in $\mathbb{V}_2$ ($\mathbb{V}_1$ and $\mathbb{V}_2$ may be the same space), to the scalar field $\mathfrak{R}$ or $\mathfrak{C}$ is called a bilinear form or a bilinear functional if it is linear in each element. In other words, $\mathcal{A}(\mathbf{x}, \mathbf{y})$ is a bilinear form if

$$\mathcal{A}(\alpha_1 \mathbf{x}_1 + \alpha_2 \mathbf{x}_2, \mathbf{y}) = \alpha_1 \mathcal{A}(\mathbf{x}_1, \mathbf{y}) + \alpha_2 \mathcal{A}(\mathbf{x}_2, \mathbf{y}) \quad \text{and} \qquad (1.58)$$

$$\mathcal{A}(\mathbf{x}, \alpha_1 \mathbf{y}_1 + \alpha_2 \mathbf{y}_2) = \alpha_1 \mathcal{A}(\mathbf{x}, \mathbf{y}_1) + \alpha_2 \mathcal{A}(\mathbf{x}, \mathbf{y}_2) \qquad (1.59)$$

for all $\mathbf{x}$ and $\mathbf{y}$. If $\mathbb{V}_1$ and $\mathbb{V}_2$ are complex spaces of n-tuples, then a bilinear form can be written as

$$\mathcal{A}(\mathbf{x},\mathbf{y}) = [x]^{H} [A][y] \tag{1.60}$$

where $[A]$ is a matrix and the superscript H denotes the conjugate transpose. Note that if $\mathbf{x}$ is a constant, then $\mathcal{A}(\mathbf{x},\mathbf{y})$ is a linear form in $\mathbf{y}$, and if $\mathbf{y}$ is a constant, then $\mathcal{A}(\mathbf{x},\mathbf{y})$ is a linear form in $\mathbf{x}$.

If $\mathbb{V}_1$ and $\mathbb{V}_2$ are function spaces, then a possible bilinear form in terms of a single variable is

$$\mathcal{A}(\mathbf{f},\mathbf{g}) = \int_{a}^{b} f(x) g(x) dx \tag{1.61}$$

More generally, f and g may be operated on by linear mappings. For example,

$$\mathcal{A}(\mathbf{f},\mathbf{g}) = \int_{a}^{b} dx \ f(x) \int_{a}^{b} g(x-y) dy \tag{1.62}$$

is a bilinear form. Again, note that if one function is kept constant, then the bilinear form reduces to a linear form in the other function.

1.5.4   Quadratic Forms

If both arguments of a bilinear form $\mathcal{A}(\mathbf{x},\mathbf{y})$ are the same, then we obtain a quadratic form or quadratic functional $q(\mathbf{x}) = \mathcal{A}(\mathbf{x},\mathbf{x})$. If $\mathbf{x}$ is an n-tuple, then the general quadratic form is

$$q(\mathbf{x}) = \mathcal{A}(\mathbf{x},\mathbf{x}) = [x]^{H} [A][x] \tag{1.63}$$

obtained by setting $\mathbf{x} = \mathbf{y}$. Note that $\mathcal{A}$ must now be a $n \times n$ matrix $[A]$. There is no loss of generality if we also assume $[A]$ to be symmetric.

If the argument of a quadratic form is a function, then we can also construct a symmetric bilinear form associated with it. In general, we have

$$q(\mathbf{f}+\mathbf{g}) = \mathcal{A}(\mathbf{f}+\mathbf{g},\mathbf{f}+\mathbf{g}) = \mathcal{A}(\mathbf{f},\mathbf{f}) + \mathcal{A}(\mathbf{f},\mathbf{g}) + \mathcal{A}(\mathbf{g},\mathbf{f}) + \mathcal{A}(\mathbf{g},\mathbf{g}) \tag{1.64}$$

If it is specified that $\mathcal{A}(\mathbf{f},\mathbf{g}) = \mathcal{A}(\mathbf{g},\mathbf{f})$, then

$$\mathcal{A}(\mathbf{f},\mathbf{g}) = \frac{1}{2}(q(\mathbf{f}+\mathbf{g}) - q(\mathbf{f}) - q(\mathbf{g})) \tag{1.65}$$

1.6   THE ADJOINT OPERATOR

The *adjoint operator* $\mathcal{A}^{H}$ is defined by [3,4]

$$\langle \mathcal{A} \mathbf{J}, \mathbf{W} \rangle = \langle \mathbf{W}, \mathcal{A}^H \mathbf{J} \rangle \tag{1.66}$$

for all $\mathbf{J} \in \mathbb{D}(\mathcal{A})$ and $\mathbf{W} \in \mathbb{D}(\mathcal{A}^H)$. The operator $\mathcal{A}$ is self-adjoint not only if $\mathcal{A} = \mathcal{A}^H$ but also requires $\mathbb{D}(\mathcal{A}) = \mathbb{D}(\mathcal{A}^H)$. If the two domains are not equal, then the operator $\mathcal{A}$ is called *symmetric*. The operator $\mathcal{A}$ tells us how a system will behave for a given external source. The adjoint operator, however, tells us how the system responds to sources in general. Thus, the adjoint operator provides physical insight into the system. An operator is said to be *self-adjoint* if $\langle \mathcal{A} \mathbf{x}, \mathbf{z} \rangle = \langle \mathbf{x}, \mathcal{A} \mathbf{z} \rangle$ for all functions in the domain of $\mathcal{A}$. A self-adjoint operator is said to be positive definite if $\langle \mathcal{A} \mathbf{z}, \mathbf{z} \rangle > 0$ for $\mathbf{z} \neq 0$.

For a differential operator, the adjoint boundary conditions are obtained from the given operator and its given boundary conditions. For example, consider the following second-order differential operator in the region between a and b [3, 4]:

$$\mathcal{A} = \frac{d^2 u(x)}{dx^2} \quad \text{with } \mathcal{B}_1\left(u(x=a)\right) = \alpha \text{ and } \mathcal{B}_2\left(u(x=b)\right) = \beta \tag{1.67}$$

then we have from the definition of the adjoint operator

$$\int_a^b \left(\mathbf{W} \mathcal{A} \mathbf{J} - \mathbf{J} \mathcal{A}^H \mathbf{W} \right) dz = \left[\mathcal{B}(\mathbf{J}, \mathbf{W}) \right]_a^b \tag{1.68}$$

By equating the term (called the bilinear concomitant) $[\mathcal{B}(\mathbf{J}, \mathbf{W})]_a^b = 0$, one obtains the adjoint boundary conditions for the given differential equation [3,4], and the expression for the adjoint operator is given in terms of the expression associated with $\mathcal{A}^H$.

For an integral operator, however, no adjoint boundary conditions are required. For example, for the given integral operator

$$
\begin{aligned}
\langle \mathcal{A} \mathbf{u}, \mathbf{v} \rangle &= \int_a^b v(z)\, dz \int_a^b K(z,x) u(x)\, dx \\
&= \int_a^b u(x)\, dx \int_a^b v(z) K(z,x)\, dx = \langle \mathbf{u}, \mathcal{A}^H \mathbf{v} \rangle
\end{aligned} \tag{1.69}
$$

Hence, if the *integral operator* is a *convolution operator* (i.e., $K(z,x) = z - x$), then the *adjoint operator* is the *correlation operator*. This subtle distinction is significant as we will observe later on in Section 1.12 when one wishes to find the convergence properties of a solution procedure.

Next we illustrate the interesting relationship between the domain and the range of the operator and its adjoint, in particular, from the orthogonal decomposition theorem [4],

$$\mathbb{D}(\mathcal{A}) = \mathbb{N}(\mathcal{A}) \overset{\perp}{\oplus} \overline{\mathbb{R}(\mathcal{A}^H)}; \text{ and } \mathbb{D}(\mathcal{A}^H) = \mathbb{N}(\mathcal{A}^H) \overset{\perp}{\oplus} \overline{\mathbb{R}(\mathcal{A})} \quad (1.70)$$

The symbol $\overset{\perp}{\oplus}$ denotes that these direct sums are orthogonal, and the only common element between the two spaces is the null element. The overbar over the two subspaces $\mathbb{R}(\mathcal{A})$ and $\mathbb{R}(\mathcal{A}^H)$ indicates the completion of these two linear manifolds. The two properties in Eq. (1.70) are illustrated in an abstract fashion by Figure 1.3. Here, for example, $\mathbb{N}(\mathcal{A}^H)$ denotes the null space of $\mathcal{A}^H$, (i.e., if $\mathcal{A}^H \mathbf{W} = 0$, then $\mathbf{W} \in \mathbb{N}(\mathcal{A}^H)$). It is thus clear from (1.70) and Figure 1.3 that $\mathbb{R}(\mathcal{A}) \subset \mathbb{D}(\mathcal{A}^H)$ (i.e., $\mathbb{R}(\mathcal{A})$ is contained in $\mathbb{D}(\mathcal{A}^H)$). In practice, usually it is easier to find $\mathbb{D}(\mathcal{A}^H)$ rather than $\mathbb{R}(\mathcal{A})$. $\mathbb{D}(\mathcal{A}^H)$ has a direct impact on the existence of a solution to an operator equation $\mathcal{A}\mathbf{f} = \mathbf{g}$, (i.e., the solution to this equation does not exist if $\mathbf{g} \notin \mathbb{D}(\mathcal{A}^H)$). The implication of this also will be illustrated by the conditions that the weighting functions need to satisfy in the solution of an operator equation by the method of moments. Next, we will look at the properties of the operator equation.

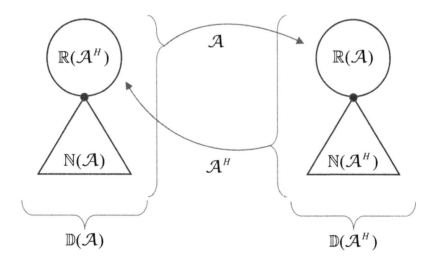

Figure 1.3 Orthogonal direct sum decomposition generated by the operator and its adjoint [4]. The only common element between the two orthogonal spaces is the null element.

1.7   CONCEPT OF ILL-POSED AND WELL-POSED PROBLEMS

Most problems of mathematical physics can be formulated in terms of an operator equation as follows:

$$\mathcal{A}\mathbf{x} = \mathbf{y} \tag{1.71}$$

where $\mathcal{A}$, in general, is a given linear integrodifferential operator and $\mathbf{x}$ is the unknown to be solved for a particular given excitation $\mathbf{y}$. Depending on the problem, either $\mathcal{A}$ or $\mathbf{x}$ may be defined as the system operator. For example, in a linear system identification problem, the goal is to determine the system operator given the response to a specific input. The system operator (i.e., the impulse response) would then be denoted by $\mathbf{x}$ and the input by $\mathcal{A}$. However, in a radiation problem, the objective might be to solve for the excitation currents on a structure given the radiated field $\mathbf{y}$. In this case, the excitation current would be denoted by $\mathbf{x}$ and the system operator (i.e., the Green's function operator) would be denoted by $\mathcal{A}$.

Example 1
Consider a linear system with impulse response $x(t)$ that is excited by an input $A(t)$ applied at $t = 0$. The system response is given by the integral equation

$$y(t) \triangleq \mathcal{A}\mathbf{x} = \int_a^b A(t - \tau) x(\tau) d\tau \tag{1.72}$$

Given the input $A(t)$ and the output $y(t)$, assume that the objective is to find the impulse response $x(t)$. Despite the fact that a solution for $x(t)$ may exist and be unique, the integral equation (1.72) is difficult to solve because of the smoothing action of the convolution operator. For example, denote the exact solution by

$$x_1(t) = x(t) \tag{1.73}$$

and a perturbed solution by

$$x_2(t) = x(t) + \mathcal{C} \sin(\omega t) \tag{1.74}$$

Even for very large values of $\mathcal{C}$, the frequency ω can be chosen high enough such that the difference between $\mathbf{y}_1 = \mathcal{A}\mathbf{x}_1$ and $\mathbf{y}_2 = \mathcal{A}\mathbf{x}_2$ is made to be arbitrarily small. This is demonstrated as follows.
    The difference between $\mathbf{y}_2$ and $\mathbf{y}_1$ is

$$\mathbf{y}_2 - \mathbf{y}_1 = \mathcal{A}\mathbf{x}_2 - \mathcal{A}\mathbf{x}_1 = \mathcal{C} \int_0^t A(t - \tau) \sin(\omega \tau) d\tau \tag{1.75}$$

If the input $\mathcal{A}$ is bounded, that is,

$$|\mathcal{A}(t)| \leq \mathcal{M} \quad \text{(a constant)} \tag{1.76}$$

it follows that

$$\mathbf{y}_2 - \mathbf{y}_1 \leq \mathcal{CM} \int_0^t \sin(\omega\tau) d\tau = \frac{\mathcal{CM} \left[1 - \cos(\omega t)\right]}{\omega} \tag{1.77}$$

We conclude that

$$\|\mathbf{y}_2 - \mathbf{y}_1\| < \frac{2\mathcal{CM}}{\omega} \tag{1.78}$$

Obviously, by selecting ω to be sufficiently large, the difference $\mathbf{y}_2 - \mathbf{y}_1$ can be made arbitrarily small. The ill-posedness of this example is evidenced by the fact that small differences in $\mathbf{y}$ can map into large differences in $\mathbf{x}$. This issue poses a serious problem because, in practice, measurement of $\mathbf{y}$ will be accompanied by a nonzero measurement error (or representation error in a finite precision digital system) δ. Use of the "noisy" data can yield a solution significantly different from the desired solution.

When a problem is ill-posed, an attempt should be made to regularize the problem. The solution to the regularized problem will be well-behaved and will offer a reasonable approximation to the solution of the ill-posed problem.

The concepts of well-posed problems and the regularization of ill-posed problems have been discussed in depth in the mathematical literature [5–12]. In this section, we illustrate some of the significant concepts involved by means of simple examples that arise in engineering applications.

1.7.1 Definition of a Well-Posed Problem

Hadamard [7,8] introduced the notion of a well-posed (correctly or properly posed) problem in the early 1900s when he studied the Cauchy problem in connection with the solution of Laplace's equation. He observed that the solution $x(t)$ did not depend continuously on the excitation $y(t)$. Hadamard concluded that something had to be wrong with the problem formulation because physical solutions did not exhibit this type of discontinuous behavior. Other mathematicians, such as Petrovsky, reached the same conclusion. As a result of their investigations, a problem characterized by the equation $\mathcal{A}\mathbf{x} = \mathbf{y}$ is defined to be well-posed provided the following conditions are satisfied:

(1) The solution $\mathbf{x}$ exists for each element $\mathbf{y}$ in the range space $\mathbb{Y}$; (this implies that either $\mathbb{N}(\mathcal{A}^H)$ is empty or else any nontrivial solution of $\mathcal{A}^H \mathbf{v} = 0$, with $\mathbf{v} \in \mathbb{N}(\mathcal{A}^H)$ should ensure $\mathbf{v}$ be orthogonal to $\mathbf{y}$.

(2) The solution $\mathbf{x}$ is unique; (This implies that $\mathbb{N}(\mathcal{A})$ is empty).

(3) Small perturbations in $\mathbf{y}$ result in small perturbations in the solution $\mathbf{x}$ without the need to impose additional constraints.

If any of these conditions are violated, then the problem is said to be ill-posed. Examples of ill-posed problems are now discussed. We first consider an example in which condition (1) is violated.

Example 2

Suppose it is known that

$$y = x_1 t_1 + x_2 t_2 \tag{1.79}$$

In addition, assume that the observation $\mathbf{y}$ is contaminated by additive noise. The observations and corresponding time instants are tabulated in Table 1.1.

TABLE 1.1. Data of Example 2

t_1	2	1	1
t_2	1	2	3
Observation of $y(t)$	3	3	5

Solution for x_1 and x_2 leads to the solution of the following matrix equation:

$$\mathcal{A}\mathbf{x} = \begin{bmatrix} 2 & 1 \\ 1 & 2 \\ 1 & 3 \end{bmatrix} \begin{bmatrix} x_1 \\ x_2 \end{bmatrix} = \begin{bmatrix} 3 \\ 3 \\ 5 \end{bmatrix} \tag{1.80}$$

Because Eq. (1.80) consists of three independent equations in two unknowns, a solution does not exist. Condition (1) is violated, and the problem is ill-posed.

Example 3

Violation of condition (2) is illustrated by the following example. Observe whether the operator $\mathcal{A}$ in Eq. (1.71) is singular (i.e., the solutions of the equation $\mathcal{A}\mathbf{x} = 0$ are nontrivial, and $\mathbf{x} \neq 0$), then Eq. (1.71) has multiple solutions. Such a case occurs when one is interested in the analysis of electro-magnetic scattering from a dielectric or conducting closed body at a particular frequency that corresponds to one of the internal resonances of the body, using either only the electric field or the magnetic field formulation.

The final example in this section illustrates a problem for which condition (3) is violated.

Example 4

Consider the solution of the matrix equation $[A][x] = [y]$ where $[A]$ is the 4×4 symmetric matrix of [13] as follows:

$$[A] = \begin{bmatrix} 36.86243 & 51.23934 & 53.50338 & 50.49425 \\ 51.23934 & 71.22350 & 74.37005 & 70.18714 \\ 53.50338 & 74.37005 & 77.66275 & 73.29752 \\ 50.49425 & 70.18714 & 73.29752 & 69.17882 \end{bmatrix} \qquad (1.81)$$

and

$$[y]^T = \begin{bmatrix} 192.09940 & 267.02003 & 278.83370 & 263.15773 \end{bmatrix}$$
$$[x]^T = \begin{bmatrix} 1 & 1 & 1 & 1 \end{bmatrix} \qquad (1.82)$$

However, it can be shown that $[A]$ is an ill-conditioned matrix in the sense that the ratio of its maximum to minimum eigenvalues is on the order of 10^{18}.

Let us study the effect on the solution x_i when only one component of y_i is changed in the fifth decimal place. Specifically, we obtain the results presented in Table 1.2. Clearly, extremely small perturbations in $[y]$ result in large variations in $[x]$. Condition (3) is violated, and the problem is ill-posed.

TABLE 1.2. Data of Example 4

$[y]^T \rightarrow [$	192.09939	267.02003	278.83370	263.15773 $]$
$[x]^T \rightarrow [$	$-6,401,472,429$	$3,866,312,299$	$1,607,634,613$	$-953,521,374$ $]$
$[y]^T \rightarrow [$	192.09940	267.02002	278.83370	263.15773 $]$
$[x]^T \rightarrow [$	$3,866,312,299$	$-2,335,145,694$	$-970,966,842$	$575,900,539$ $]$
$[y]^T \rightarrow [$	192.09940	267.02003	278.83369	263.15773 $]$
$[x]^T \rightarrow [$	$1,607,634,615$	$-970,966,842$	$-403,733,529$	$239,462,717$ $]$
$[y]^T \rightarrow [$	192.09940	267.02003	278.83370	263.15772 $]$
$[x]^T \rightarrow [$	$-953,521,374$	$574,900,539$	$239,462,717$	$-142,030,294$ $]$

1.7.2   Regularization of an Ill-Posed Problem

Most inverse problems of mathematical physics are ill-posed under the three conditions of Hadamard. In a humorous vein, Stakgold [3, p. 308] pointed out that there would likely be a sharp drop in the employment of mathematicians if this were not the case.

Given an ill-posed problem various schemes are available for defining an associated problem that is well-posed. This approach is referred to as regularization of the ill-posed problem. In particular, an ill-posed problem may be regularized by

(a) Changing the definition of what is meant by an acceptable solution
(b) Changing the space to which the acceptable solution belongs
(c) Revising the problem statement
(d) Introducing regularizing operators
(e) Introducing probabilistic concepts so as to obtain a stochastic extension of the original deterministic problem

These techniques now are illustrated by a series of examples. Technique (a) is demonstrated in Example 5.

Example 5
Once again, consider the problem introduced in Example 2. This resulted in Eq. (1.80) for which a solution did not exist. Nevertheless, an approximate solution is possible. One of several possible approximate solutions for any over-determined linear system is the Moore Penrose generalized inverse [14], which is a least-squares solution to Eq. (1.80) and is given by

$$[x] = \left[[A]^H [A] \right]^{-1} [A]^H [y] = \begin{bmatrix} 0.8 \\ 1.3142 \end{bmatrix} \tag{1.83}$$

where the superscript H denotes the transpose conjugate of a matrix. Note that $[x]$ does not exactly satisfy any of the three equations in Eq. (1.80). Yet, $[x]$ is a reasonable approximate solution. We see that the original problem, which was ill-posed because a classical solution did not exist, has been regularized by redefining what is meant by an acceptable solution. Technique (b) is illustrated in Example 6.

Example 6
Relative to Lewis–Bojarski inverse scattering [15–17], the relationship at all frequencies and aspects between the backscattered fields $y(k)$ to the size and shape of the target $x(r)$ is given by

$$y(k) = \int_{-\infty}^{\infty} x(r) e^{-jkr} dr \tag{1.84}$$

This relation is valid under the assumption that the target currents are obtained using a physical optics approximation. In Eq. (1.84), $\mathcal{A}$ is the Fourier transform operator. The solution to Eq. (1.84) is given by the inverse Fourier transform

$$x(r) = \frac{\mathcal{N}}{\sqrt{2\pi}} \int_{-\infty}^{\infty} y(k) e^{+jkr} dk \tag{1.85}$$

where, for convenience, the constant $\mathcal{N} = 1/\sqrt{2\pi}$ has been introduced.

It is now demonstrated that this problem is ill-posed when solutions are allowed in the metric space for which the following norm is defined as

$$\left\| \mathbf{x}_1 - \mathbf{x}_0 \right\|_\infty \triangleq \max_r \left| x_1(r) - x_0(r) \right| \tag{1.86}$$

On the other hand, the problem is well-posed in the space for which the norm is defined as:

$$\left\| \mathbf{x}_1 - \mathbf{x}_0 \right\|_2 \triangleq \sqrt{\int_{-\infty}^{\infty} \left(x_1(r) - x_0(r) \right)^2 \, dr} \tag{1.87}$$

The norm in Eq. (1.86) is referred to as the "sup" norm, whereas the norm in Eq. (1.87) is known as the well-known $\mathcal{L}_2$ norm.

Given a scattered field function $y_0(k)$, consider the perturbed scattered field function $y_1(k)$ such that

$$y_1(k) = y_0(k) + \varepsilon e^{-\alpha k^2} \tag{1.88}$$

where ε and α are arbitrary positive constants. Then

$$
\begin{aligned}
x_1(r) &= \frac{\mathcal{N}}{\sqrt{2\pi}} \int_{-\infty}^{\infty} e^{jkr} \left(y_0(k) + \varepsilon e^{-\alpha k^2} \right) dk \\
&= x_0(r) + \frac{\mathcal{N}\varepsilon}{\sqrt{2\pi}} \int_{-\infty}^{\infty} e^{jkr - \alpha k^2} \, dk = x_0(r) + \frac{\mathcal{N}\varepsilon}{\sqrt{2\alpha}} e^{-\frac{r^2}{4\alpha}}
\end{aligned}
\tag{1.89}
$$

Using the sup norm, the perturbations in the expressions for $y_1(k)$ in Eq. (1.88) and $x_1(r)$ in Eq. (1.89) are bounded by

$$
\begin{aligned}
\max_k \left\| \mathbf{y}_1 - \mathbf{y}_0 \right\| &= \max_k \left| y_1(k) - y_0(k) \right| \le \varepsilon \\
\max_r \left\| \mathbf{x}_1 - \mathbf{x}_0 \right\| &= \left| x_1(r) - x_0(r) \right| \le \frac{\mathcal{N}\varepsilon}{\sqrt{2\alpha}}
\end{aligned}
\tag{1.90}
$$

Obviously, for a fixed small perturbation in $\mathbf{y}_0$, the perturbation in $\mathbf{x}_0$ can be made arbitrarily large depending on the choice of α. Hence, the third condition of Hadamard is violated, and the inverse Fourier transform is ill-posed in the metric space using the sup norm. Physically, this implies that the inverse Fourier transform is ill-posed under a pointwise convergence criterion.

On the other hand, using the $\mathcal{L}_2$ norm,

$$\|\mathbf{y}_1 - \mathbf{y}_0\| \triangleq \sqrt{\int_{-\infty}^{\infty} (y_1(k) - y_0(k))^2 \, dk} = \sqrt{\int_{-\infty}^{\infty} \varepsilon^2 e^{-2\alpha k^2} \, dk} = \varepsilon \left(\frac{\pi}{2\alpha}\right)^{1/4} \quad (1.91)$$

and

$$\|\mathbf{x}_1 - \mathbf{x}_0\| \triangleq \sqrt{\int_{-\infty}^{\infty} (x_1(r) - x_0(r))^2 \, dr} = \sqrt{\frac{\mathcal{N}\varepsilon}{2\alpha} \int_{-\infty}^{\infty} e^{\frac{-r^2}{2\alpha}} \, dr} = \mathcal{N}\varepsilon \left(\frac{\pi}{2\alpha}\right)^{1/4} \quad (1.92)$$

Because the $\mathcal{L}_2$ norm of the perturbation for $x_0(r)$ and $y_0(k)$ differ only by the constant $\mathcal{N}$, small perturbations in $y_0(k)$ result in small perturbations $x_0(r)$. Hence, the inverse Fourier transform is well-posed under the $\mathcal{L}_2$ norm, which implies that convergence in the mean square for the shape of the structure can be obtained.

      In this example, we have illustrated how a problem can be made well-posed just by changing the space of the solution. Physically, this implies that the problem is well-posed when we want an "average" description of the target, and the problem is ill-posed when we demand an accurate precise definition of the target.

Example 7
In this example, a problem is regularized by means of technique (c) (i.e., revising the problem statement). Consider a linear system identification problem where $A(t)$, $x(t)$, and $y(t)$ denote the input, impulse response, and output, respectively. Using the convolution integral the output is

$$y(t) = \int_{-\infty}^{\infty} x(t - \tau) A(\tau) \, d\tau \quad (1.93)$$

Application of the Fourier transform theory results in the solution for the impulse response as

$$x(t) = \frac{1}{2\pi} \int_{-\infty}^{\infty} \frac{\tilde{y}(\omega)}{\tilde{A}(\omega)} e^{j\omega t} \, d\omega \quad (1.94)$$

where $\tilde{y}(\omega)$ and $\tilde{A}(\omega)$ are the Fourier transforms of $y(t)$ and $A(t)$, respectively. In practice, the ideal output is contaminated by additive noise. Denote the measured response by $d(t)$ where

$$d(t) = y(t) + n(t) \quad (1.95)$$

and $n(t)$ represents a stationary zero mean additive noise process. The autocorrelation function of $n(t)$ is denoted by $\phi(\tau)$, and its Fourier transform is denoted by $\phi(\omega)$. When the measured noisy response $d(t)$ is used in place of the ideal output $y(t)$, the solution $x(t)$ becomes a random process. The variance of $x(t)$ is given by [18]

$$\sigma_x^2 = \frac{1}{2\pi} \int_{-\infty}^{\infty} \frac{\tilde{\phi}(\omega)}{|\tilde{A}(\omega)|^2} d\omega \tag{1.96}$$

This problem is considered to be well-posed only if σ_x^2 is suitably small.

Typically, the noise is assumed to contain a background component of white noise. With this assumption, the power spectral density $\tilde{\phi}(\omega)$ approaches a nonzero constant $\mathcal{K}$, whereas $|\omega|$ approaches infinity. When σ_x^2 is infinite, the problem becomes obviously ill-posed.

The problem can be made well-posed by revising the problem statement. From a mathematical point of view, the input could be restricted to signals whose Fourier transform $\tilde{A}(\omega)$ is a rational function with a numerator polynomial of a higher degree than the denominator polynomial. Then, even with a white noise background, σ_x^2 remains finite, and the problem becomes well-posed. However, this regularization implies the presence in $A(t)$ of singularity functions such as impulses, doublets, and so on. Therefore, the regularization is achieved at the expense of requiring $A(t)$ to be an unrealizable signal having infinite energy.

A preferable approach is to remove the commonly used assumption of a white noise background. Provided the power spectral density of the noise $\tilde{\phi}(\omega)$ falls off in frequency at a rate that is at least as fast as that of

$$\left|\tilde{A}(\omega)\right|^2 / \omega^{1+\varepsilon} \quad (\varepsilon > 0, \text{ arbitrary})$$

σ_x^2 will be finite. This demonstrates that the noise must be modeled carefully if the problem is to be well-posed. This principle has direct implications in the solution of time domain electromagnetic field problems.

Example 8

Technique (d), the application of regularizing operators, is discussed next. In example 2, the three simultaneous equations were shown to be inconsistent, and therefore, an exact solution did not exist. In example 5, an approximate solution was obtained using the least-squares approach. The solution was given by Eq. (1.83).

As the dimensionality of the problem is increased, there is a tendency for the solution to oscillate and increase in magnitude. Therefore, the problem becomes ill-posed as the dimensionality increases. This difficulty may be overcome by introducing regularizing operators that impose additional constraints on the solution. In particular, instead of solving the problem $\mathcal{A}\mathbf{x} = \mathbf{y}$ directly, attention is focused on the problem of minimizing $\|\mathcal{A}\mathbf{x} - \mathbf{y}\|^2$ under the constraint

$$\|\hat{L}\mathbf{x}\|^2 = \mathcal{C}$$

where $\hat{L}$ is a suitably chosen linear operator. Equivalently, one finds

$$\min_{\mathbf{x}} \left\{ \|\mathcal{A}\mathbf{x} - \mathbf{y}\|^2 + \mu^2 \|\hat{L}\mathbf{x}\|^2 \right\}$$

where μ plays the role of a Lagrange multiplier. If $\hat{L}$ is the identity operator, then this approach will result in that solution $\mathbf{x}$, which carries out the minimization of $\|\mathcal{A}\mathbf{x} - \mathbf{y}\|^2$ having a specified value of $\|\mathbf{x}\|^2$. If $\hat{L}$ is the derivative operator, then this approach will result in that solution $\mathbf{x}$, which carries out the minimization of $\|\mathcal{A}\mathbf{x} - \mathbf{y}\|^2$ with a priori degree of smoothness for the solution (i.e., a specified value of $\|d\mathbf{x} / dt\|^2$).

The parameter μ is determined by the constraint (e.g., the specified value of $\|\mathbf{x}\|^2$, or the specified degree of smoothness or a combination of both) [8, chs. II, III]. Many choices of the operator $\hat{L}$ are possible, and one may simultaneously apply several constraints with different corresponding Lagrange multipliers [8, chs. II, III].

The solution of the problem posed in Eq. (1.71) is equivalent to the solution of the matrix equation

$$\left[[A]^H [A] + \mu^2 [\hat{L}]^H [\hat{L}] \right][x] = [A]^H [y] \tag{1.97}$$

and is given by

$$[x] = \left[[A]^H [A] + \mu^2 [\hat{L}]^H [\hat{L}] \right]^{-1} [A]^H [y] \tag{1.98}$$

This approach is known as the Tykhonov regularizing scheme [8,18,19]. Note, if $\hat{L}$ is the null operator (i.e., if no constraint is imposed), then the solution in (1.97) reduces to the classical least-squares solution of (1.83). The Tykhonov regularizing scheme has been used to regularize the ill-posed antenna pattern synthesis problem and the image-processing problem [5].

The last approach to be discussed for regularizing an ill-posed problem is through the application of technique (e). In this approach, a stochastic extension of an otherwise deterministic problem is obtained by introducing probabilistic concepts. Once again, consider the solution of $\mathcal{A}\mathbf{x} = \mathbf{y}$, where $\mathcal{A}$ and $\mathbf{y}$ are deterministic quantities. As mentioned previously, in some instances, perfect measurement or observation of **y** is impossible. For those cases, the use of the noisy measurements for **y** results in an ill-posed problem. With technique (e), the problem is made well-posed by recognizing that uncertainty in the observation of **y** causes uncertainty in the resulting solution for **x**. Consequently, the solution is viewed as a random process. An error criterion is specified, and a stochastically optimum solution is obtained. The solution is stochastically optimum in the sense that repetition in measurements of **y** produces solutions for **x**, which on the average, are optimum according to the specified error criterion. The details of this approach are available in references [5,20,21].

1.8   NUMERICAL SOLUTION OF OPERATOR EQUATIONS

If it is assumed that the solution to (1.71) exists for the given excitation **y**, then symbolically, the solution to (1.71) can be written as

$$\mathbf{x} = \mathcal{A}^{-1}\mathbf{y} \tag{1.99}$$

Observe whether the operator $\mathcal{A}$ is singular (i.e., the solutions of the equation $\mathcal{A}\mathbf{x} = 0$ are nontrivial, $\mathbf{x} \neq 0$) because then (1.99) has multiple solutions. We will assume, at the present moment, that such a situation does not develop. If one is really interested in the analysis for the case when $\mathcal{A}\mathbf{x} = 0$ has a nontrivial solution, then the present techniques can be modified to treat those special cases. Finally, we assume $\mathcal{A}^{-1}$ is a bounded operator. This implies that small perturbations of the excitation only produce a small perturbation of the solution as:

$$\|\mathbf{x}\| \leq \|\mathcal{A}^{-1}\|\|\mathbf{y}\| \tag{1.100}$$

where $\|\mathbf{x}\|$ defines either the $\mathscr{L}_2$ norm, $\mathscr{L}_1$ norm, or the $\mathscr{L}_\infty$ norm defined by;

$$\mathscr{L}_2\!: \ \|\mathbf{x}\|_2 = \int_{e_1}^{e_2} |x(z)|^2 \, dz$$

$$\mathscr{L}_1\!: \ \|\mathbf{x}\|_1 = \int_{e_1}^{e_2} |x(z)| \, dz \tag{1.101}$$

$$\mathscr{L}_\infty\!: \ \|\mathbf{x}\|_\infty = \max_{e_1 \leq z \leq e_2} |x(z)|$$

In the text, we will use only the mean square or the $\mathscr{L}_2$ norm because it is easy to prove convergence of numerical techniques, even for discontinuous solutions.

In this section, we wish to solve the operator equation $\mathcal{A}\mathbf{x} - \mathbf{y} = 0$. The solution **x** is called a *classical* solution if (1.71) is satisfied everywhere in the space $\mathbb{S} = \mathbb{D}(\mathcal{A})$. A *generalized* solution for **x** is the one that satisfies $\langle \mathcal{A}\mathbf{x} - \mathbf{y}, \varphi \rangle = 0$ for all possible functions φ in a given class.

Next we discuss methods on how to obtain the solution for **x**.

1.8.1   Classical Eigenfunction Approach

Historically, the first technique developed is the eigenvector expansion. In this case, we choose a set of right-hand and left-hand eigenvectors, ϕ_i and ξ_i, respectively for the operator $\mathcal{A}$. A Hermitian operator $\mathcal{A}$ is defined as $\mathcal{A} = \mathcal{A}^H$. This implies $\phi_i = \xi_i$. Therefore, by definition

$$\mathcal{A}\,\phi_i = \lambda_i\,\phi_i \quad \text{and} \quad \xi_i^H\,\mathcal{A} = \lambda_i\,\xi_i^H \tag{1.102}$$

where λ_i is the eigenvalue corresponding to the eigenvectors ϕ_i and ξ_i. We represent the solution **x** as follows:

$$\mathbf{x} = \sum_{i=1}^{\infty} a_i\phi_i \tag{1.103}$$

where a_i are the unknowns to be solved for. We substitute Eqs. (1.102) and (1.103) in the operator equation (1.71) and obtain the following:

$$\sum_{i=1}^{\infty} a_i\mathcal{A}\phi_i = \sum_{i=1}^{\infty} a_i\lambda_i\phi_i = \mathbf{y} \tag{1.104}$$

Because right-hand and left-hand eigenvectors corresponding to different eigenvalues form a biorthogonal set (i.e., $\langle\phi_k,\xi_m\rangle = 0$ for $k \neq m$), we can multiply both sides of Eq. (1.104) by the left-hand eigenvector ξ_k and integrate the product over the domain of the operator $\mathcal{A}$ and represent it in the form of inner product as in Eq. (1.9), that is, with

$$a_k\lambda_k\langle\phi_k,\xi_k\rangle = \langle\mathbf{y},\xi_k\rangle; \quad \text{or} \quad a_k = \frac{1}{\lambda_k}\frac{\langle\mathbf{y},\xi_k\rangle}{\langle\phi_k,\xi_k\rangle} \tag{1.105}$$

Once the unknown coefficients are given by Eq. (1.105), the solution is obtained as follows:

$$\mathbf{x} = \sum_{i=1}^{\infty} \frac{\langle\mathbf{y},\xi_i\rangle}{\langle\phi_i,\xi_i\rangle}\frac{\phi_i}{\lambda_i} \tag{1.106}$$

It can be observed that the solution process is straightforward provided that the series in Eq. (1.106) converges. The series will converge if λ_i does not approach zero. If λ_i approaches zero, then this is equivalent to the statement that $\left\|\mathcal{A}^{-1}\right\|$ is not bounded and the problem is ill-posed. This particular situation occurs when the operator $\mathcal{A}$ is an integral operator with a kernel K that is square integrable, that is,

$$\int_{e_1}^{e_2} K(p,q)x(q)dq = y(p) ; \quad \text{with} \tag{1.107}$$

$$\int_{e_1}^{e_2} dp \int_{e_1}^{e_2} |K(p,q)|^2 \, dq < \infty \tag{1.108}$$

and p and q are variables of the kernel K. The major problem with Eq. (1.106) is the determination of the eigenvectors. For simple geometries and other geometries that conform to one of the separable coordinate systems, the eigenfunctions are relatively easy to find. However, for arbitrary geometries, the determination of the eigenvectors themselves is a formidable task. For this reason, we take recourse to the computer for evaluating certain integrals and so on. Once we set the problem up on a computer, we can no longer talk about an exact solution, as the computer has only finite precision. The next generation of numerical techniques under the generic name of *method of moments* (MoM) revolutionized the field of computational electromagnetics. Problems that were difficult to solve by eigenvector expansion and problems with arbitrary geometries now can be solved with ease.

1.8.2 Method of Moments (MoM)

In the *method of moments* [22], we choose a set of expansion functions ψ_i, which need not be the eigenfunctions of $\mathcal{A}$. We now expand the unknown solution x in terms of these basis functions ψ_i weighted by some unknown constants a_i to be solved for. So

$$\mathbf{x} \approx \mathbf{x}_N = \sum_{i=1}^{N} a_i \psi_i \tag{1.109}$$

The functions ψ_i need not be orthogonal. The only requirement on ψ_i is that we know them precisely. By using the expression in Eq. (1.109), the solution to an unknown functional equation of Eq. (1.71) is reduced to the solution of some unknown constants in which the functional variations are known. This significantly simplifies the problem from the solution of an unknown function to the solution of a matrix equation with some unknown constants. The latter is a much easier computational problem to handle in practice.

We now substitute Eq. (1.109) in Eq. (1.71) and obtain the following:

$$\mathcal{A}\mathbf{x} \approx \mathcal{A}\mathbf{x}_N = \sum_{i=1}^{N} a_i \mathcal{A}\psi_i \approx \mathbf{y} \tag{1.110}$$

This results in an equation error

$$\mathbf{E}_N = \mathcal{A}\mathbf{x}_N - \mathbf{y} = \sum_{i=1}^{N} a_i \mathcal{A}\psi_i - \mathbf{y} \tag{1.111}$$

We now weigh this error $\mathbf{E}_N$ to zero with respect to certain weighting functions, $W_k(z)$, that is,

$$\langle \mathbf{E}_N, \mathbf{W}_k \rangle = \int_{e_1}^{e_2} E_N(z) W_k(z)\, dz = 0 \tag{1.112}$$

This results in

$$\sum_{i=1}^{N} a_i \langle \mathcal{A}\psi_i, \mathbf{W}_k \rangle = \langle \mathbf{y}, \mathbf{W}_k \rangle \text{ for } k = 1, 2, \ldots, N \tag{1.113}$$

Equation (1.113) is the key equation of the method of moments. The method gets its name from Eq. (1.112) in which we are taking the functional moment of the error $\mathbf{E}_N$ with respect to the weighting functions and equating them to zero.

Notice that the upper limit in Eq. (1.109) is N and not ∞ because it is not practically feasible to do an infinite sum on the computer. Whenever we solve any problem on a computer, we always talk in terms of an approximate solution, (i.e., we are solving the problem in a finite N-dimensional space instead of the original problem, which has the setting in an infinite-dimensional space). As an example, consider the partial sum S_n of the following series: $S_n = 1 + 1/2 + 1/3 + \cdots + 1/n$. Suppose we set the problem up on the computer and instruct the computer to stop summing when the relative absolute difference of two neighboring partial sums is less than a very small value (i.e., $|S_{n+1} - S_n| \leq 10^{-20}$). The computer in all cases will give a convergent result. However, in the final sum of this series, $S_n \to \infty$ as $n \to \infty$. It is clear from this example, that the approximate solution yielded by the computer may be nowhere near the exact solution. The natural question that now arises, like the example of the summation of the previous series, is what guarantee do we have for the approximate solution $\mathbf{x}_N$ to converge to the exact solution $\mathbf{x}_{\text{exact}}$ as $N \to \infty$. This is still an open question for the mathematicians. But suffice it to say that the convergence of $\mathbf{x}_N \to \mathbf{x}_{\text{exact}}$ depends on the boundedness of the inverse operator $\mathcal{A}^{-1}$ and on the choice of the expansion functions ψ_i. In our text, we will avoid answering the question as to what happens when N is ∞.

Observe that there is essentially a doubly infinite choice of expansion and weighting functions. However, the choices are not arbitrary, and they must satisfy certain specific criteria which are discussed next.

1.8.2.1   On the Choice of Expansion Functions. The expansion functions ψ_i chosen for a particular problem have to satisfy the following criteria:

(1) The expansion functions should be in the domain of the operator $\mathbb{D}(\mathcal{A})$ in some sense, (i.e., they should satisfy the differentiability criterion and must satisfy the boundary conditions for an integrodifferential operator $\mathcal{A}$ [23–28]). It is not at all necessary for each expansion function to exactly satisfy the boundary conditions. What is required is that the total solution must satisfy the boundary conditions at least in some distributional sense. The same holds for the differentiability conditions. When the boundary and the differentiability conditions are satisfied exactly, we have a classical solution, and when the previous conditions are satisfied in a distributional sense, we have a distributional solution.

(2) The expansion functions must be such that $\mathcal{A}\psi_i$, form a complete set for $i = 1, 2, 3,...$ for the range space $\mathbb{R}(\mathcal{A})$ of the operator. It really does not matter whether the expansion functions are complete in the domain of the operator; what is important is that $\mathcal{A}\psi_i$ must be chosen in such a way that $\mathcal{A}\psi_i$ is complete, as is going to be shown later on. It is interesting to note that when $\mathcal{A}$ is a differential operator, ψ_i have to be linearly dependent for $\mathcal{A}\psi_i$ to form a complete set. This is an absolute necessity, as illustrated by the following example [23–28].

Example 9
Consider the solution to the following differential equation [23]:

$$-\frac{d^2x(z)}{dz^2} = 2 + \sin(z) \quad \text{for } 0 \leq z \leq 2\pi \qquad (1.114)$$

with the boundary conditions: $x(z = 0) = 0 = x(z = 2\pi)$. A normal choice for the expansion functions would be to take $\sin(iz)$ for all i, that is,

$$x_i(z) = \sum_{i=1}^{\infty} a_i \sin(iz) \qquad (1.115)$$

These functions satisfy both the differentiability conditions and the boundary conditions. Moreover, they are orthogonal. The previous choice of expansion functions leads to the solution

$$\mathbf{x} = x(z) = \sin(z) \qquad (1.116)$$

It is clear that Eq. (1.116) does not satisfy Eq. (1.114), and, hence, is not the solution. Where is the problem? Perhaps the problem may be that the functions $\sin(iz)$ do not form a complete set in Eq. (1.114) even though they are orthogonal in the interval $[0, 2\pi]$. Therefore, in addition to the *sin* terms we add the constant and the *cos* terms in Eq. (1.115). This results in the following:

$$\mathbf{x} = x(z) = a_0 + \sum_{i=1}^{\infty} (a_i \cos(iz) + b_i \sin(iz)) \qquad (1.117)$$

where a_i and b_i, are the unknown constants to be solved for. Now the total solution given by Eq. (1.117) has to satisfy the boundary conditions of Eq. (1.114). Observe that Eq. (1.117) is the classical Fourier series solution, and hence, it is complete in the interval $[0, 2\pi]$. Now if we solve the problem again and choose the weighting functions to be same as the basis functions, then we find the solution still to be given by Eq. (1.115), and we know that it is not the correct solution. What exactly is still incorrect? The problem is that even though a_0, $a_i \cos iz$, and $b_i \sin iz$ form a complete set for $\mathbf{x}$, $\mathcal{A}\psi_i$ (the operator $\mathcal{A}$ operating on the expansion function ψ_i) do not form a complete set. This is because $\mathcal{A}\psi_i$, for $i = 1, 2, 3,...$ are merely $c_i \cos iz$ and $d_i \sin iz$, where c_i and d_i are certain constants. Note that from the set $\mathcal{A}\psi_i$, the constant term is missing. Therefore, the representation of $\mathbf{x}$ given by Eq. (1.117) is not complete and, hence, does not provide the correct solution.

From the previous discussion, it becomes clear that we ought to have the constant term in $\mathcal{A}\psi_i$, which implies that the representation of $\mathbf{x}$ must be of the following form:

$$\mathbf{x} = x(z) = \sum_{i=1}^{\infty} [a_i \cos(iz) + b_i \sin(iz)] + a_0 + cz + dz^2 \qquad (1.118)$$

and the boundary conditions in Eq. (1.114) have to be enforced on the total solution. Observe that the expansion functions $\{1, z, z, \sin(iz), \text{ and } \cos(iz)\}$ in the interval $[0, 2\pi]$ and in the limit $N \to \infty$ form a linearly dependent set. This is because the set $1, \sin(iz)$, and $\cos(iz)$ can represent any function such as z and z^2 in the interval $[0, 2\pi]$. The final solution when using the expansion of (1.118) is obtained as:

$$\mathbf{x} = x(z) = z(2\pi - z) + \sin(z) \qquad (1.119)$$

which turns out to be the exact solution. This simple example illustrates the mathematical subtleties that exist with the choice of expansion functions in the MoM and hence the satisfaction of condition (2) by the basis functions described earlier is essential.

1.8.2.2   On the Choice of Weighting Functions.
It is important to point out that the weighting functions $W_j(z)$ in the MoM have to satisfy certain conditions also

[23–28]. Because the weighting function weights the residual $\mathbf{E}_N$ to zero, we have the following:

$$\langle \mathbf{E}_N, \mathbf{W}_j \rangle = \left\langle \sum_{i=1}^{N} a_i \mathcal{A}\psi_i - \mathbf{y}, \mathbf{W}_j \right\rangle = \langle \mathbf{y}_N - \mathbf{y}, \mathbf{W}_j \rangle = 0 \qquad (1.120)$$

As the difference $\mathbf{y}_N - \mathbf{y}$ is made orthogonal to the weighting functions $W_j(z)$, it is clear that the weighting functions should be able to reproduce $\mathbf{y}_N$ and to some degree $\mathbf{y}$.

From approximation theory [4,24–28], it can be shown using the development of projection theory of section 1.4 that, for a unique minimum norm representation, $\mathbf{y}_N$ must be orthogonal to the error $\mathbf{y} - \mathbf{y}_N$ as shown abstractly in Figure 1.4. By Eq. (1.120), the weighting functions $\mathbf{W}_j$ are also enforced to be orthogonal to $\mathbf{y} - \mathbf{y}_N$, and because they are not orthogonal to $\mathbf{y}_N$, it follows that the weighting functions should be able to reproduce $\mathbf{y}_N$ and to some degree $\mathbf{y}$.

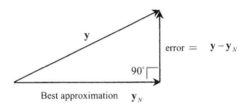

Figure 1.4. Best approximation of $\mathbf{y}$ by $\mathbf{y}_N$.

To recapitulate, in Eq. (1.71) the operator $\mathcal{A}$ maps the elements $\mathbf{x}$ from the domain of $\mathcal{A}$ denoted by $\mathbb{D}(\mathcal{A})$ to the elements $\mathbf{y}$ in the range of $\mathcal{A}$, denoted by $\mathbb{R}(\mathcal{A})$. This been depicted in Figure 1.3. Because the objective is to solve for $\mathbf{x}$, which is approximated by a linear combination of the expansion functions $\{\psi_i\}$ for $i = 1, 2, ..., N$, it is clear that the expansion functions need to be in $\mathbb{D}(\mathcal{A})$. Only then can the expansion functions represent any element $\mathbf{x}_N$ in $\mathbb{D}(\mathcal{A})$. Then, by definition

$$\mathbf{y}_N = \sum_{i=1}^{N} a_i \mathcal{A}\psi_i \qquad (1.121)$$

is in $\mathbb{R}(\mathcal{A})$ because it is obtained by applying $\mathcal{A}$ to $\mathbf{x}_N$, and the set $\mathcal{A}\psi_i$ must span $\mathbb{R}(\mathcal{A})$, as any $\mathbf{y}_N$ in $\mathbb{R}(\mathcal{A})$ can be written as a linear combination of $\{\mathcal{A}\psi_i\}$. It is the duty of the weighting functions $\mathbf{W}_k$ to make the difference $\mathbf{y} - \mathbf{y}_N$ small.

Also, as the weighting functions approximate $\mathbf{y}_N$, they must be in the range of the operator. In other words, the weighting functions must form a basis in the range of the operator, and in the limit $N \to \infty$ must be able to represent any excitation $\mathbf{y}$. When the weighting functions $\mathbf{W}_k$ cannot approximate $\mathbf{y}$ to a high degree of accuracy, the solution $\mathbf{x}_N$ obtained by the technique of minimizing the residuals may not produce a solution that resembles the true solution. Any method that enforces the inner product to zero in Eq. (1.120) must satisfy the previous criterion. In short, it is seen by relating $\langle \mathbf{y}_N - \mathbf{y}, \mathbf{W}_k \rangle = 0$ in Eq. (1.120) to the theory of functional approximations that

(a) The weighting function $\mathbf{W}_k$ must be in the range of the operator or, more generally, in the domain of the adjoint operator.
(b) Because the weighting functions are orthogonal to the error of the approximation, the set $\{\mathbf{W}_k\}$ should span $\mathbf{y}_N$.
(c) As $N \to \infty$, $\mathbf{y}_N \to \mathbf{y}$. Therefore, the weighting function should be able to represent the excitation $\mathbf{y}$ in the limit.

If the weighting functions do not satisfy all of these criterions, then a meaningful solution may not be obtained by the method of moments. Hence, if the weighting functions are not in $\mathbb{D}(\mathcal{A}^H)$, then they cannot be in $\mathbb{R}(\mathcal{A})$. Therefore, to have a meaningful result, the weighting functions must necessarily be in $\mathbb{D}(\mathcal{A}^H)$. It is also clear from Figure 1.3 that if $\mathbf{y}$ of Eq. (1.71) is not an element $\mathbb{D}(\mathcal{A}^H)$, then Eq. (1.71) does not possess any classical mathematical solution whatsoever. This also precludes the existence of even a least-squares solution because, in this case, the equivalent problem of $\mathcal{A}^H \mathcal{A} \mathbf{x} = \mathcal{A}^H \mathbf{y}$ is solved, and if $\mathbf{y}$ is not in $\mathbb{D}(\mathcal{A}^H)$, then $\mathcal{A}^H \mathbf{y}$ is not mathematically defined. However, one could still find a solution numerically, in which the solution will diverge as shown from a mathematical point of view in [4,24–28]. In numerical electromagnetics, this situation occurs when one solves a radiation problem with a delta gap excitation. In this case, the solution seems to converge initially, but as more and more expansion (basis) functions are chosen, ultimately, the solution seems to diverge. This type of convergence is known as asymptotic convergence [29].

There are three different variations of the method of moments depending on the choice of the testing functions. When $W_k(z) = \delta_k(z)$, a delta function, we obtain the point matching method. When $W_k(z) = \psi_k(z)$, we obtain the Galerkin's method. And finally, when $W_k(z) = \mathcal{A}\psi_k(z)$, we obtain the method of least squares. They are special cases of the more general method of minimizing residuals [22–28].

1.9   DEVELOPMENTS OF THE VARIATIONAL METHOD (RAYLEIGH-RITZ), GALERKIN'S METHOD, AND THE METHOD OF LEAST SQUARES

So, in the present discussions, we assume that the operator equation in Eq. (1.71) is well posed. Next, we investigate the nature of the solution yielded by the three techniques—*variational method* (Rayleigh-Ritz), *Galerkin's method,* and the *method of least squares*. We also present the conditions under which the three techniques yield meaningful results and discuss their rates of convergence.

These three techniques proceed by assuming the nature of the solution a priori in terms of known expansion functions. They start with the idea that the approximate solution $\mathbf{x}_N$ of $\mathbf{x}$ can be expressed in the following form:

$$\mathbf{x}_N = \sum_{n=1}^{N} a_n \mathbf{u}_n \tag{1.122}$$

where a_n are certain unknown constants to be solved for. The expansion functions $\mathbf{u}_n$ are a complete set of functions in an N-dimensional space. In other words, the sequence $\{\mathbf{u}_n\}$ for $n = 1, 2, ..., N$ is a linearly independent set and thereby forms a basis for the N-dimensional space in which the approximate solution $\mathbf{y}_N$ is sought. It is not necessary that $\{\mathbf{u}_n\}$ for $n = 1, 2, ..., N$ form an orthogonal set of functions. The exact solution

$$\mathbf{x} = \sum_{n=1}^{\infty} a_n \mathbf{u}_n \tag{1.123}$$

lies in an infinite dimensional space. In summary, by the term convergence, then, we imply convergence in the mean, $\|\mathbf{x}_N - \mathbf{x}\| \to 0$ as $N \to \infty$ (i.e., in the limit N approaches infinity), the mean-squared difference between $\mathbf{x}_N$ and $\mathbf{x}$ approaches zero. In the remaining sections, we discuss the restrictions that each of the three methods impose on the expansion functions $\mathbf{u}_n$ and how an approximate solution $\mathbf{x}_N$ is obtained. We also address the question of convergence of $(\mathbf{x}_N - \mathbf{x}) \to 0$ as $N \to \infty$ for the three methods. It is important to note that there are four different types of convergence. They are [3]:

(1) *Uniform convergence* (*pointwise convergence* implies that, given a fixed $\varepsilon > 0$, we can find a number N such that the absolute value

$$\|x(z) - x_N(z)\| < \varepsilon \quad \text{for all } z \in [a,b] \tag{1.124}$$

(2) *Strong convergence* (convergence in the norm or *convergence in the mean*) implies

$$\|\mathbf{x} - \mathbf{x}_N\| < \varepsilon \; ; \text{ that is, } \sqrt{\int_a^b \left(x(z) - x_N(z)\right)^2 dz} < \varepsilon \tag{1.125}$$

(3) *Convergence in energy* requires that

$$\left\langle \mathcal{A}(\mathbf{x} - \mathbf{x}_N), (\mathbf{x} - \mathbf{x}_N) \right\rangle < \varepsilon \tag{1.126}$$

$\langle \mathcal{A}\mathbf{x}, \mathbf{x} \rangle$ in some physical problems represents energy, hence the name energy norm, whereas convergence in the mean implies that the converging sequence $\mathbf{x}_N$ may not approach the limit function $\mathbf{x}$ at every point in the domain of $\mathcal{A}$, but the region in which the converging sequence $\mathbf{x}_N$ differs from $\mathbf{x}$ becomes arbitrarily small, as $N \to \infty$. Convergence in energy implies similar things happen to the derivative of the sequence as well. This convergence only can occur for a positive definite operator $\mathcal{A}$.

(4) *Weak convergence* (*convergence in a distributional sense*) implies that $\mathbf{y}_N$ converges weakly to $\mathbf{x}$ if

$$\lim_{N \to \infty} \langle \mathbf{x}_N, \boldsymbol{\varphi} \rangle = \langle \mathbf{x}, \boldsymbol{\varphi} \rangle \qquad (1.127)$$

or all $\boldsymbol{\varphi}$ in the space. The method of moments often yields solutions that converge weakly to a generalized solution.

Finally, a set of functions $\mathbf{u}_i$ is said to be *complete* in a space if any function $\mathbf{x}$ in that space can be expanded in terms of the set $\{\mathbf{u}_i\}$, so that

$$\left\| \mathbf{x} - \sum_{i=1}^{\infty} a_i \mathbf{u}_i \right\| < \varepsilon$$

A set of functions is said to be *complete in energy* if

$$\left\langle \mathcal{A}\left(\mathbf{x} - \sum_{i=1}^{\infty} a_i \mathbf{u}_i \right), \left(\mathbf{x} - \sum_{i=1}^{\infty} a_i \mathbf{u}_i \right) \right\rangle < \varepsilon$$

for a positive definite operator $\mathcal{A}$.

1.9.1   Variational Method (Rayleigh-Ritz)

In the variational method, the solution to Eq. (1.71) is attempted by minimizing the functional $\mathcal{F}(\mathbf{z})$ defined by [23–25, 30–33]

$$\mathcal{F}(\mathbf{z}) = \langle \mathcal{A}\mathbf{z}, \mathbf{z} \rangle - \langle \mathbf{y}, \mathbf{z} \rangle - \langle \mathbf{z}, \mathbf{y} \rangle \qquad (1.128)$$

In Eq. (1.128), it is assumed that the operator $\mathcal{A}$ is a self-adjoint operator {i.e., $\mathcal{A} = \mathcal{A}^H$ (adjoint operator) and $\mathbb{D}(\mathcal{A}) = \mathbb{D}(\mathcal{A}^H)$ }. So, by definition, a self-adjoint operator satisfies the following equality: $\langle \mathcal{A}\mathbf{x}, \mathbf{z} \rangle = \langle \mathbf{x}, \mathcal{A}\mathbf{z} \rangle$. For non-self-adjoint operators, $\mathcal{A}$ and $\mathbf{y}$ in Eq. (1.128) may be replaced by $\mathcal{A}^H \mathcal{A}$ and $\mathcal{A}^H \mathbf{y}$, respectively, and the following discussion still holds. The technique then becomes the method of least squares. Assume that $\mathbf{z}$ is the estimated solution and $\mathbf{x}$ is the exact solution of $\mathcal{A}\mathbf{x} = \mathbf{y}$. Now observe that Eq. (1.128) can be rewritten as

$$F(z) = \langle \mathcal{A}(z-x),(z-x) \rangle - \langle y,x \rangle \tag{1.129}$$

Because $\langle y,x \rangle$ is a constant, minimization of $F(z)$ directly implies minimization of $\|z-x\|$ if and only if the operator $\mathcal{A}$ is positive definite. It is thus clear that, for an operator to be positive definite, it has to be self-adjoint first. This is because

$$\|z-x\|^2 \leq \langle \mathcal{A}(z-x),(z-x) \rangle \tag{1.130}$$

This is the philosophy behind solving Eq. (1.71) in a roundabout fashion by minimizing the functional $F(z)$ of Eq. (1.128).

We now replace z in Eq. (1.128) with x_N (as defined in Eq. (1.122)) to obtain

$$F(x_N) = \left\langle \sum_{n=1}^{N} a_n \mathcal{A} u_n, \sum_{i=1}^{N} a_i u_i \right\rangle - \left\langle y, \sum_{n=1}^{N} a_n u_n \right\rangle - \left\langle \sum_{i=1}^{N} a_i u_i, y \right\rangle \tag{1.131}$$

We solve for the unknowns a_n by minimizing $F(x_N)$ of Eq. (1.131). This is achieved by taking the partial derivatives for $\partial F(x_N)/\partial a_i$ for $i = 1, 2, \dots, N$ and equating each of the partial derivatives to zero. This results in a set of equations

$$\sum_{n=1}^{N} a_n \langle \mathcal{A} u_n, u_k \rangle - \langle y, u_k \rangle \quad \text{for } k = 1, 2, \dots, N \tag{1.132}$$

The unknowns a_n are solved from Eq. (1.132). In the following subsections, we develop the finer structures and limitations of this technique.

1.9.1.1 Does a Functional $F(z)$ Always Exist?

According to [3,23,25], the solution x of a boundary value problem can also be characterized as the element for which a related functional $F(z)$ is stationary. Such a stationary characterization is a variational principle whose Euler–Lagrange equation is the differential equation of the original boundary value problem. The variational problem may also provide guidance to the existence of a solution for the boundary value problem.

If $\mathcal{A}x = y$ has a solution x_0, then this solution minimizes the functional $F(z)$ (as defined in equation (1.128)). Conversely, if an element $x_0 \in \mathbb{D}(\mathcal{A})$ exists that minimizes the functional $F(z)$, then x_0 satisfies $\mathcal{A}x = y$. This method of solving a boundary value problem that consists of replacing $\mathcal{A}x = y$ with the problem of minimizing a functional $F(z)$ is also called *the energy method* (as it minimizes the energy of the system). This implies that a variational method finds a solution with finite energy, or mathematically $\langle y,x_0 \rangle$ or $\langle \mathcal{A}x_0,x_0 \rangle$ is bounded. As a first example, consider the self-adjoint operator $\mathcal{A}$ as

$$\mathcal{A}\mathbf{x} = -\frac{d^2 x(z)}{d z^2};$$

with $x(z = 0) = 0$ and $x(z = 1) = 0$. It has been shown in reference [24] that this operator is positive definite and, hence, has a bounded inverse. Moreover, the homogeneous equation has the trivial solution. Now if we select $\mathbf{y} = y(z) = -0.25z^{-3/2}$, then the exact solution is given by $\mathbf{x} = x(z) = z - \sqrt{z}$. We observe $\mathcal{F}(\mathbf{z}) \rightarrow \infty$ as

$$\langle \mathbf{y}, \mathbf{x} \rangle = \int_0^1 \left(-\frac{z^{-1.5}}{4} \right) \left(z - \sqrt{z} \right) dz \rightarrow \infty .$$

Hence, the variational method is defined only for a class of problems in which $\langle \mathbf{y}, \mathbf{x} \rangle$ and $\langle \mathbf{y}, \mathbf{x}_N \rangle$ are bounded.

The discussion so far does not address the question of the existence of the functional $\mathcal{F}(\mathbf{z})$ nor does it show how a solution $\mathbf{x}_0$ can be obtained. However, such information may be obtained if the operator $\mathcal{A}$ is positive/negative definite or for a semi-definite operator if $\langle \mathbf{x}_0, \mathbf{y} \rangle$ is bounded. In both cases, it is assumed that $\mathbf{y}$ is square integrable. Thus, if $\langle \mathbf{x}_0, \mathbf{y} \rangle$ is unbounded, then a variational method cannot be set up for the solution of $\mathcal{A}\mathbf{x} = \mathbf{y}$, as the functional $\mathcal{F}(\mathbf{z})$ become infinite. This is true for any self-adjoint operator.

As an analogy, consider the vector electric field $\mathbf{E}$. Can this electric field $\mathbf{E}$ always be derived from a scalar potential function V? If the curl of the vector is zero (i.e., $\nabla \times \mathbf{E} = 0$), then only the vector can be represented in terms of a potential (i.e., $\mathbf{E} = \nabla V$). When the curl is zero, the integral of $\mathbf{E}$ along a path in space $\mathbb{Z}$ is independent of the actual path taken and depends only on the end points z_1 and z_2, that is,

$$\int_{z_1}^{z_2} \mathbf{E} \cdot d\mathbf{z} = V_2 - V_1$$

This provides an equivalent method of determining whether a vector field can be derived from a scalar potential. So, from these elementary concepts of vector calculus, it follows that if we regard the Euler equation in a variational method as the gradient of a functional $\mathcal{F}(\mathbf{z})$ analogous to a potential, then we should not expect every operator equation to be derivable from a potential independent of the path integral.

Consider again the second-order differential operator given by the last example. We know that the domain of the operator $\mathbb{D}(\mathcal{A})$ consists of the space of twice differentiable and square integrable functions that satisfy the boundary conditions. Now if we consider the solution to $\mathcal{A}\mathbf{x} = \mathbf{y}$, in which $\mathcal{A}$ is the second order differential operator and $\mathbf{y}$ is a square integrable function, then our starting point for solution is $\mathcal{F}(\mathbf{z})$, as given by Eq. (1.123). In this case, $\mathcal{F}(\mathbf{z})$ is bounded below on $\mathbb{D}(\mathcal{A})$ so that $\mathcal{F}(\mathbf{z})$ has a minimum, say $\wp$, on $\mathbb{D}(\mathcal{A})$. If there is an element $\mathbf{x}_0 \in \mathbb{D}(\mathcal{A})$, for which $\mathcal{F}(x_0) = \wp$, then the infinimum is attained on

$\mathbb{D}(\mathcal{A})$ so that the problem for minimizing $\mathcal{F}(\mathbf{z})$ on $\mathbb{D}(\mathcal{A})$ has the solution $\mathbf{x}_0$. There are two reasons for considering the minimum of $\mathcal{F}(\mathbf{z})$ on a larger domain $\mathbb{M}$ (as yet unspecified). First, $\mathcal{F}(\mathbf{z})$ may not have a minimum on $\mathbb{D}(\mathcal{A})$. If $\mathbf{y}$ is only piecewise continuous and if $\mathcal{A}$ is defined by the second-order differential operator of the last example, then the solution $\mathbf{x}_0$ to $\mathcal{A}\mathbf{x} = \mathbf{y}$ may not be in the space of square integrable, twice differentiable functions so that $\mathcal{A}\mathbf{x} = \mathbf{y}$ has no solution. Hence, $\mathcal{F}(\mathbf{z})$ does not have a minimum on $\mathbb{D}(\mathcal{A})$. The trouble here is simply that $\mathbb{D}(\mathcal{A})$ is not a suitable domain on which to analyze the boundary value problem when $\mathbf{y}$ is not continuous. The second reason is more subtle. Even if $\mathcal{F}(\mathbf{z})$ has a minimum on $\mathbb{D}(\mathcal{A})$, it may be hard to prove the fact directly. It turns out to be easier to prove the existence of a minimum on the larger domain $\mathbb{M}$ and then to show that the minimizing element lies in $\mathbb{D}(\mathcal{A})$. Thus, we shall try to enlarge $\mathbb{D}(\mathcal{A})$ so that the minimum for $\mathcal{F}(\mathbf{z})$ can be guaranteed to be in the new domain. It is reasonable to hope for such an existence because the expression for $\mathcal{F}(\mathbf{z})$ involves only the first derivative of $x(z)$, whereas an element of $\mathbb{D}(\mathcal{A})$ has a continuous derivative of order 2.

We now complete $\mathbb{D}(\mathcal{A})$ to form the larger domain $\mathbb{M}$. This is achieved by completing $\mathbb{D}(\mathcal{A})$ with respect to a new inner product $\langle \mathbf{x}, \mathbf{x} \rangle_{\mathcal{A}} \triangleq \langle \mathcal{A}\mathbf{x}, \mathbf{x} \rangle$ with $\mathbf{x} \in \mathbb{M}$. The completion of $\mathbb{D}(\mathcal{A})$ with respect to this inner product results in a new domain $\mathbb{N}$. Thus, all elements of $\mathbb{M}$ are with finite energy as $\langle \mathcal{A}\mathbf{x}, \mathbf{x} \rangle$ is bounded. If the operator $\mathcal{A}$ is positive definite, then all elements of $\mathbb{M}$ satisfy $\langle \mathcal{A}\mathbf{x}, \mathbf{x} \rangle \geq \mathcal{C}^2 \langle \mathbf{x}, \mathbf{x} \rangle = \mathcal{C}^2 \|\mathbf{x}\|^2$ where $\mathcal{C}$ is a predetermined positive constant. Therefore, for a positive-definite operator, the following inequality holds:

$$\|\mathbf{x}\|^2 = \langle \mathbf{x}, \mathbf{x} \rangle \leq \frac{\langle \mathcal{A}\mathbf{x}, \mathbf{x} \rangle}{\mathcal{C}^2} = \frac{\langle \mathbf{x}, \mathbf{x} \rangle_{\mathcal{A}}}{\mathcal{C}^2}$$

and all elements of $\mathbb{M}$ are in $\mathbb{D}(\mathcal{A})$. But in general, $\mathbb{M}$ is larger than $\mathbb{D}(\mathcal{A})$. Now we investigate the conditions for the existence of the functional $\mathcal{F}(\mathbf{z})$ on $\mathbb{M}$. For a positive-definite operator,

$$\langle \mathbf{x}, \mathbf{y} \rangle^2 \leq \|\mathbf{y}\|^2 \times \|\mathbf{x}\|^2 \leq \frac{\|\mathbf{y}\|^2}{\mathcal{C}^2} \langle \mathcal{A}\mathbf{x}, \mathbf{x} \rangle \quad \text{for } \mathbf{x} \in \mathbb{M} \qquad (1.133)$$

holds for any square integrable function $\mathbf{y}$, and $\langle \mathbf{x}, \mathbf{y} \rangle$ is always bounded on $\mathbb{M}$. Thus, the variational method exists for a positive-definite operator, and the operator equation has a solution for any square integrable function.

However, when $\mathcal{A}$ is semi-definite, inequalities like Eq. (1.133) do not exist. Hence, for the variational method to exist, it is required that $\langle \mathbf{x}, \mathbf{y} \rangle$ be bounded for a functional $\mathcal{F}(\mathbf{z})$ to exist.

Because $\langle \mathbf{x}, \mathbf{y} \rangle$ is now bounded, according to reference [3] a unique element $\mathbf{u}_0 \in \mathbb{M}$ exists such that $\langle \mathbf{x}, \mathbf{y} \rangle = \langle \mathbf{x}, \mathcal{A}\mathbf{x}_0 \rangle$ for $\mathbf{x} \in \mathbb{M}$. This element $\mathbf{x}_0$ also minimizes the functional $\mathcal{F}(\mathbf{z})$, and the minimum value is given by $\mathcal{F}(\mathbf{x}_0) = -\langle \mathbf{x}_0, \mathbf{x}_0 \rangle_{\mathcal{A}} = -\langle \mathcal{A}\mathbf{x}_0, \mathbf{x}_0 \rangle$. For a positive-definite operator, it is also possible to obtain bounds for $F(\mathbf{z})$ on both sides. Let $\mathbf{x}_0$ be the exact solution and $\mathbf{x}_N$ be the Nth approximation; then it can be shown that

$$F(x_N) - \frac{\|\mathcal{A}\mathbf{x}_N - \mathbf{y}\|^2}{L_A} \leq F(x_0) \leq F(x_N) - \frac{\|\mathcal{A}\mathbf{x}_N - \mathbf{y}\|^2}{U_A} \qquad (1.134)$$

where L_A and U_A are the lower and upper bounds for $\|\mathcal{A}\|$; that is,

$$U_A \|\mathbf{x}\|^2 \geq \langle \mathcal{A}\mathbf{x}, \mathbf{x} \rangle \geq L_A \|\mathbf{x}\|^2 \qquad (1.135)$$

When the solution $\mathbf{x}_0$ constructed by the variational method lies in the subset $\mathbb{D}(\mathcal{A})$ of $\mathbb{M}$ and satisfies $\mathcal{A}\mathbf{x} = \mathbf{y}$, we call $\mathbf{x}_0$ the classical solution. It may happen that $\mathbf{x}_0$, which minimizes the functional $F(\mathbf{z})$ on the domain $\mathbb{M}$, may not be in $\mathbb{D}(\mathcal{A})$. In that case, $\mathbf{x}_0$ is referred to as the generalized solution.

These mathematical concepts can be explained by the following example. Let us view the minimization of a functional from a point of generating a soap film on a deformed wire for producing soap bubbles. The wire is bent in any odd shape, dipped into a soap solution, and slowly removed. The equation for the surface of the soap film generated usually gives the solution to the problem of minimizing a functional. The functional, in general, represents the energy associated with the soap film. There are three possible alternatives, which are as follows:

(1) The case in which the surface of the soap film minimizes the energy and also the functional is representable in terms of known functions. The solution to this problem is straightforward and yields the classical solution.

(2) In this case, the surface minimizes the energy of the film, but the solution to the surface is not representable by elementary known functions in the domain of the operator. That is, we know the functional $F(\mathbf{x}_N)$ exists, but we do not know how to obtain $\mathbf{x}_N$. In this case, we obtain the generalized solution of the problem.

(3) For this case, we consider a surface of the soap film, where it is difficult to see that the film indeed minimizes the energy; this occurs, for example, when three surfaces meet at a point. For this class of problems, the variational method is not clearly defined.

In this section, we have shown that case 3 occurs when $\langle \mathbf{y}, \mathbf{x} \rangle$ is unbounded. When the element $\mathbf{x}_N$, which minimizes $F(\mathbf{x}_N)$, does not lie in the domain of the operator, case 2 occurs. When $\mathbf{x}_N$ is in the domain of $\mathcal{A}$, we get case 1.

For a non-self-adjoint operator, if we do not intend to replace $\mathcal{A}$ by $\mathcal{A}^H \mathcal{A}$ and $\mathbf{y}$ by $\mathcal{A}^H \mathbf{y}$ in Eq. (1.129) and proceed with the classical approach, then serious problems of convergence may result. This is illustrated by the nature of the convergence.

1.9.1.2   Does Minimization of $F(\mathbf{x}_N)$ Lead to a Monotonic Convergence of the Approximate Solution $\mathbf{x}_N$ to the Exact One? The answer is yes, only if $\mathcal{A}$ is a positive/negative-definite operator. Otherwise, there is no guarantee that the error in the solution $\|\mathbf{x}_N - \mathbf{x}\|$ becomes smaller as the order of the approximation is increased. We illustrate this by an example.

Example 10

Consider the solution of the following 3×3 matrix equation:

$$\begin{bmatrix} 3 & -4 & 2 \\ -4 & 3 & 6 \\ 2 & 6 & 2 \end{bmatrix} \begin{bmatrix} x_1 \\ x_2 \\ x_3 \end{bmatrix} = \begin{bmatrix} 1 \\ 5 \\ 10 \end{bmatrix} \tag{1.136}$$

In this problem, $\mathcal{A}$ is the 3×3 square matrix and $\mathbf{y}$ is the 3×1 column matrix. We would like to obtain an approximate solution $\mathbf{x}_N$ by the variational method. Of course, when $N = 3$, we should get the exact solution of Eq. (1.136). In this problem, we study the nature of the various orders of approximations to the solution. By the terms of the problem, we form the functional

$$F(x) = \begin{bmatrix} x_1 & x_2 & x_3 \end{bmatrix} [A] \begin{bmatrix} x_1 \\ x_2 \\ x_3 \end{bmatrix} - 2 \begin{bmatrix} x_1 & x_2 & x_3 \end{bmatrix} \begin{bmatrix} 1 \\ 5 \\ 10 \end{bmatrix}$$

$$= 3x_1^2 - 8x_1x_2 + 3x_2^2 + 4x_1x_3 + 12x_2x_3 + 2x_3^2 - 2x_1 - 10x_2 - 20x_3 \tag{1.137}$$

We now choose the expansion functions $\mathbf{u}_n$ (orthonormal) as

$$\mathbf{u}_1^T = \begin{bmatrix} 1 & 0 & 0 \end{bmatrix}^T \quad \mathbf{u}_1^T = \begin{bmatrix} 0 & 1 & 0 \end{bmatrix}^T \quad \mathbf{u}_1^T = \begin{bmatrix} 0 & 0 & 1 \end{bmatrix}^T \tag{1.138}$$

We now seek the first approximate solution to $\mathbf{x}$ as $\mathbf{x}_1^T = a_1\mathbf{u}_1^T = \begin{bmatrix} a_1 & 0 & 0 \end{bmatrix}^T$ and solve for $\mathbf{x}_1$ by the variational method by using Eq. (1.132) to obtain $\mathbf{x}_1^T = \begin{bmatrix} 1/3 & 0 & 0 \end{bmatrix}^T$. Because the exact solution to the problem is $x_i = 1$ for $i = 1, 2, 3$, the error in the solution is

$$\|\mathbf{x} - \mathbf{x}_1\| = \sqrt{(1 - 1/3)^2 + (1 - 0)^2 + (1 - 0)^2} = 1.56$$

We now solve for $\mathbf{x}_2$, which is the second-order approximation to $\mathbf{x}$. So

$$\mathbf{x}_2^T = a_1\mathbf{u}_1^T + a_2\mathbf{u}_2^T = \begin{bmatrix} a_1 & 0 & 0 \end{bmatrix}^T + \begin{bmatrix} 0 & a_2 & 0 \end{bmatrix}^T = \begin{bmatrix} a_1 & a_2 & 0 \end{bmatrix}^T \tag{1.139}$$

and we obtain the solution for $\mathbf{x}_2$ by using Eq. (1.132) as $\mathbf{x}_2^T = \begin{bmatrix} -23/7 & -19/7 & 0 \end{bmatrix}^T$. Observe that the error in the solution for the second-order approximation is

$$\|\mathbf{x} - \mathbf{x}_2\| = \sqrt{(1-23/7)^2 + (1-19/7)^2 + (1-0)^2} = 5.76$$

By comparing $\|\mathbf{x} - \mathbf{x}_2\|$ with $\|\mathbf{x} - \mathbf{x}_1\|$, it is clear that there is a 269% increase in the value of the error as the order of the approximation has been increased from $N = 1$ to $N = 2$. Also, note that the residuals $\|\mathcal{A}\mathbf{x}_1 - \mathbf{y}\| = 11.28$ and $\|\mathcal{A}\mathbf{x}_1 - \mathbf{y}\| = 32.86$ do not decrease either as N increases. However, when $N = 3$, we would obtain the exact solution.

   The error in the solution increases as we increased the order of the approximation because $\mathcal{A}$ is not a positive-definite matrix even though $\|\mathcal{A}^{-1}\|$ is bounded. So, if in a problem, the operator $\mathcal{A}$ is self-adjoint but not positive definite, then the variational method will not yield a more accurate answer, as the order of the approximation is increased. Only for the solution of a scalar problem of charge distribution on a wire or a plate caused by a constant potential, is the operator self-adjoint and semi-definite. In almost all other cases, $\mathcal{A}$ is a complex symmetric operator. Hence, unless we have an operator that is positive definite, increasing the order of the approximation in a variational method may not yield a more accurate solution.

   It is important to point out that the symmetric inner product is a mere mathematical artifact and does not give rise to a natural definition of the norm. Also, one cannot comment on the eigenvalues of a complex symmetric operator. The symmetric product is more akin to the definition of reciprocity, and the normal inner product is related to the definition of power, which is a physical quantity. Thus by using the symmetric product, it is very difficult to comment on the properties of the operator, as a symmetric product naturally does not give rise to a norm, and hence, one cannot talk about convergence.

1.9.1.3   What Are the Restrictions on the Expansion Functions u_n?   The expansion functions need not be in the domain of the operator. However, it is essential that the expansion functions form a complete set with respect to the following inner product $\langle \mathcal{A}\mathbf{u}_i, \mathbf{u}_j \rangle$. It is totally immaterial whether the expansion functions form a complete set with respect to the inner product $\langle \mathbf{u}_i, \mathbf{u}_j \rangle$. For a positive-definite-second-order differential operator $\mathcal{A}$, physically the previous requirements imply that the expansion functions must be capable of representing the solution $\mathbf{x}$, as well as its first derivative. Thus, the expansion functions need not be required to approximate the second derivative of the solution $\mathbf{x}$. This principle has been illustrated with an example in the Section 1.9.1.2 on the choice of expansion functions.

1.9.1.4   Does Minimization of the Functional $F(\mathbf{x}_N)$ Guarantee the Convergence of the Residuals? The question can be restated mathematically as follows: We know that when the operator $\mathcal{A}$ is positive definite and $F(\mathbf{x}_{N+1}) < F(\mathbf{x}_N)$, then we have $\|\mathbf{x} - \mathbf{x}_{N+1}\| < \|\mathbf{x} - \mathbf{x}_N\|$. Does this imply $\|\mathcal{A}\mathbf{x}_{N+1} - \mathbf{y}\| < \|\mathcal{A}\mathbf{x}_N - \mathbf{y}\|$?

The answer is that there is no guarantee whatsoever that the above inequality would be satisfied. In fact, it is not even certain that the residuals will decrease to zero as N increases. This is because in defining the expansion functions $\mathbf{u}_n$, we have not specified that they belong to the domain of the operator. If the expansion functions do not belong to the domain of the operator, then they do not satisfy all the boundary conditions and the requisite number of differentiability conditions for the problem. Hence, if $\mathbf{u}_n \in \mathbb{D}(\mathcal{A})$, then $\mathcal{A}\mathbf{u}_n$, and hence, $\mathcal{A}\mathbf{u}_N$ lacks any meaning. So, in a variational method, it is not at all necessary that

$$\lim_{N \to \infty} \|\mathcal{A}\mathbf{u}_N - \mathbf{y}\| \to 0$$

If $\mathcal{A}$ is a second-order differential operator, then what the variational method guarantees is a solution such that the solution $\mathbf{x}_N$ and its first derivative converge in the mean to the exact solution $\mathbf{x}$ and its derivative $\mathbf{x}'$, respectively, as $N \to \infty$. However, the second derivative $\mathbf{x}_N''$ may not converge to $\mathbf{x}''$ in any sense. However, if the sequences $\mathcal{A}\mathbf{u}_n$ are bounded, then it can be shown [24,25,30–32] that the residual $\mathcal{A}\mathbf{x}_N - \mathbf{y}$ converges weakly to zero, in which the definition of weak convergence is given by Eq. (1.127).

1.9.1.5   Epilogue Observe that, in the previous sections, we have talked of convergence with respect to the norm defined by Eq. (1.125) (i.e., convergence in the mean). Thus, the class of problems for which the convergence of a variational method is mathematically justified is the class of problems that has a square integrable solution $\mathbf{x}$. The problems then for which the convergence of a variational method cannot be physically or mathematically justified are the charge distribution on a rectangular plate, the current distribution on an open-ended cylinder irradiated by a plane wave, and so on. These are problems that do not have a square integrable solution, as one component of the current has a square root singularity.

However, a variational method may be extended to problems that do not have a square integrable solution $\mathbf{x}$ by redefining the inner product in Eq. (1.9) with Eq. (1.11) as

$$\langle \mathbf{u}_i, \mathbf{u}_k \rangle_w \triangleq \int_a^b w(z) u_i(z) u_k^*(z)\, dz$$

where $w(z)$ is generally a nonnegative weight function. If we find a solution $\hat{\mathbf{x}}_N(z)$ with the previously redefined inner product, then we know from the previous discussions that we cannot guarantee that either of the two conditions,

$$\lim_{N\to\infty} \int_a^b \left(\widehat{\mathbf{x}}_N(z) - \mathbf{x}(z)\right)^2 dz \to 0 \text{ and } \lim_{N\to\infty} \int_a^b \left(\mathcal{A}x_N(z) - y(z)\right)^2 dz \to 0$$

will be satisfied. So, the application of a variational method to a problem with non-square integrable solutions with the redefined inner product may yield a mathematical solution that may not have any correlation to the actual physical problem.

The other important point to remember is that, in a variational method, we minimize the functional $\mathcal{F}(\mathbf{z})$ in Eq. (1.128). Moreover, we have not chosen the expansion functions to be in the domain of the operator (i.e., satisfy both the differentiability conditions and the boundary conditions). So there may be a possibility that the element $\mathbf{u}_0$, which minimizes the functional $\mathcal{F}(\mathbf{z})$, may not even be in the domain of the operator. This often happens [23]. Under those circumstances, the solution obtained by a variational method is called the generalized solution to the problem. (A generalized solution is one that obeys $\langle \boldsymbol{\phi}, \mathcal{A}\mathbf{x}_N - \mathbf{y}\rangle = 0$ for all possible $\boldsymbol{\phi}$ in a given class.)

In summary, the crux of a variational method lies in setting up the functional $\mathcal{F}(\mathbf{z})$, as it may not always exist. The proper selection of a complete set of expansion functions $\mathbf{u}_n$ is very important. Because the expansion functions need not be in the domain of the operator, there is no guarantee that

$$\lim_{N\to\infty} \left\|\mathcal{A}\mathbf{x}_N - \mathbf{y}\right\| \to 0$$

If $\mathcal{F}(\mathbf{x}_N)$ exists, then

$$\lim_{N\to\infty} \left\|\mathbf{x}_N - \mathbf{x}\right\| \to 0$$

if and only if $\langle \mathbf{y}, \mathbf{x}\rangle$ and $\langle \mathbf{y}, \mathbf{x}_N\rangle$ are bounded for all N

1.9.2   Galerkin's Method

In Galerkin's method, we select a complete set of expansion functions $\{\mathbf{u}_n\}$ that are in the domain of the operator, to approximate $\mathbf{x}_N$ in Eq. (1.122). This implies that the set $\{\mathbf{u}_n\}$ satisfies not only all the boundary conditions but also the differentiability conditions of the problem. We now substitute Eq. (1.122) in Eq. (1.71) to form

$$\sum_{i=1}^N a_i \mathcal{A}\mathbf{u}_i = \mathbf{y} \tag{1.140}$$

Now $\mathcal{A}\mathbf{u}_i$ means $\mathbf{u}_i \in \mathbb{D}(\mathcal{A})$. Next the residual

$$\sum_{i=1}^{N} a_i \mathcal{A} \mathbf{u}_i - \mathbf{y}$$

is formed, and it is equated to zero with respect to the functions $\mathbf{u}_k$, $k = 1, 2, \cdots, N$. This yields a set of equations [23–31]

$$\left\langle \mathcal{A} \sum_{i=1}^{N} a_i \mathbf{u}_i, \mathbf{u}_k \right\rangle = \left\langle \mathbf{y}, \mathbf{u}_k \right\rangle, \ k = 1, 2, \cdots, N \qquad (1.141)$$

Observe that, in Eq. (1.141), $\mathcal{A}$ can even be nonlinear and non-self-adjoint. A variational formulation for a nonlinear operator is complex. When $\mathcal{A}$ is a linear operator, Eq. (1.141) simplifies to

$$\sum_{i=1}^{N} a_i \left\langle \mathcal{A} \mathbf{u}_i, \mathbf{u}_k \right\rangle = \left\langle \mathbf{y}, \mathbf{u}_k \right\rangle, \ k = 1, 2, \cdots, N$$

$$\text{or } [G][a] = [Y] \qquad (1.142)$$

from which the unknowns a_i are solved for. The only requirement in Eq. (1.142) is that $\left\langle \mathbf{y}, \mathbf{u}_k \right\rangle$ be finite. Because in Eq. (1.142), we weigh the residuals to zero, Galerkin's method is also known as the method of weighted residuals. Even though Eq. (1.132) of the variational method is identical in form to Eq. (1.142), there are significant differences in obtaining these expressions. For example, in Eq. (1.132), $\mathcal{A}$ needs to be a self-adjoint operator, whereas in Eq. (1.142), $\mathcal{A}$ can be non-self-adjoint. Equation (1.132) is not applicable to non-self-adjoint operators. For a non-self-adjoint operator, a variational principle has to be formulated not only for the original operator, but also for its adjoint [23–25,29–33]. A physical interpretation for the mathematical differences between the variational method and the Galerkin method is given by Singer in [34].

If the expansion functions $\mathbf{u}_n$ for Galerkin's method are chosen such that the boundary conditions are not satisfied, then additional constraints have to be imposed on the solution procedure. For example, consider for the operator equation $\mathcal{A}\mathbf{x} = \mathbf{y}$, we have the following boundary conditions, which are $x(z = g) = 0$ and $x(z = h) = 0$. Now if we choose the approximation for $\mathbf{x}$ as

$$\mathbf{x} \simeq \mathbf{x}_N = \sum_{i=1}^{N} a_i \mathbf{u}_i$$

and the expansion functions $\mathbf{u}_i$ do not satisfy the boundary conditions, then we have to solve the following set of equations [30–32]

$$\sum_{i=1}^{N} a_i \left\langle \mathcal{A} \mathbf{u}_i, \mathbf{u}_k \right\rangle = \left\langle f, \mathbf{u}_k \right\rangle \qquad (1.143)$$

$$\sum_{i=1}^{N} a_i \left\langle \mathbf{u}_i \big|_{z=g}, \mathbf{u}_k \right\rangle = 0 \qquad \text{for } k = 1, 2, \ldots, N.$$

$$\sum_{i=1}^{N} a_i \left\langle \mathbf{u}_i \big|_{z=h}, \mathbf{u}_k \right\rangle = 0$$

1.9.2.1   Is it Always Possible to Find a Set of Approximate Solutions that Guarantees the Weak Convergence of the Residuals? In general, Eq. (1.142) of Galerkin's method guarantees weak convergence of the residuals in a finite dimensional space (i.e., $\left\langle \mathcal{A}\mathbf{x}_N - \mathbf{y}, \mathbf{u}_k \right\rangle = 0$ for all $\mathbf{u}_k$; $k = 1, 2, \ldots, N$). The question then is raised if $N \rightarrow \infty$, whether this weak convergence of the residuals still is guaranteed. It has been shown in [24,25,30,32,35] that, if $\left\| \mathcal{A}\mathbf{x}_N \right\| \leq \mathcal{C}$ where $\mathcal{C}$ is a constant independent of N for a particular choice of expansion functions $\mathbf{u}_k$, then as $N \rightarrow \infty$, weak convergence of the residuals is guaranteed. These concepts will be illustrated later with examples.

1.9.2.2   Does the Weak Convergence of the Residuals Guarantee Its Strong Convergence? In other words, does weak convergence of the residuals imply strong convergence (i.e., convergence in the norm) of the residuals? Again, it is shown in [24,25,30,32,35] that, if $\mathcal{A}$ is a bounded operator, then strong convergence of the residuals is always guaranteed. However, if $\mathcal{A}$ is an unbounded operator, then for the strong convergence of the residuals, either of the following two conditions must be satisfied:

(1) The projection operator $\mathcal{P}_N$ has to be bounded. The projection operator is defined as

$$\mathcal{P}_N \left(\mathcal{A}\mathbf{x} - \mathbf{y} \right) = \mathcal{A}\mathbf{x}_N - \mathcal{P}_N\mathbf{y} = \sum_{i=1}^{N} b_i \mathbf{u}_i ,$$

It is the operator that projects the problem from an infinite dimensional space to a finite dimensional space. Thus, the properties of the projection operator depend on the choice of the expansion function $\mathbf{u}_i$. So the boundedness of the projection operator implies $\left\| \mathcal{P}_N \right\| \leq \mathcal{C}$, where $\mathcal{C}$ is a constant for all N. It is easy to show that a projection operator is always bounded if it is an orthogonal projection (i.e., $\left\| \mathcal{P}_N \right\| = 1$ for all N). However, if $\mathcal{P}_N$ is not an orthogonal projection as for a polynomial basis function (i.e., $\mathcal{P}_N \left(\mathcal{A}\mathbf{x} - \mathbf{y} \right) = \sum_{i=1}^{N} b_i \mathbf{x}^i$), then it can be shown that the norm of the projection operator can be given by [24,25,30,32] $\left\| \mathcal{P}_N \right\| \leq \mathcal{C} \ln(N)$. In this case, the projection operator is unbounded.

(2) If it is not possible to obtain a bounded projection operator, then select an operator $\mathcal{B}$ that is similar to the original operator $\mathcal{A}$ with $\mathbb{D}(\mathcal{B}) = \mathbb{D}(\mathcal{A})$. For example, if

$$\mathcal{A}x = a\,\frac{d^2x(z)}{dz^2} + b\,\frac{d\,x(z)}{dz} + c\,x(z) \text{ with } x(z=0)=0=x(z=1) \quad (1.144)$$

and if $\mathcal{B}x = \dfrac{d^2x(z)}{dz^2}$, with $x(z=0)=0=x(z=1)$, then operators $\mathcal{A}$ and

$\mathcal{B}$ are similar. We select the operator $\mathcal{B}$ in such a way that $\mathcal{B}$ has discrete eigenvalues. If the eigenfunctions of $\mathcal{B}$ are chosen as the expansion functions $\mathbf{u}_n$, then the residuals will converge to zero even though the operator $\mathcal{A}$ is unbounded. This implies that the expansion functions in this case are with respect to $\langle \mathcal{B}\mathbf{u}_i, \mathbf{u}_j \rangle$ (i.e., the expansion functions are capable of representing not only the exact solution but also its first derivative).

The example of Stephens [35] illustrates these principles [33].

1.9.2.3   *Does the Sequence of Solutions $\|\mathbf{x}_N - \mathbf{x}\|$ Converge to Zero as* $N \to \infty$? We observe that

$$\|\mathbf{x}_N - \mathbf{x}\| \le \|\mathcal{A}^{-1}\| \times \|\mathcal{A}\mathbf{x}_N - \mathcal{A}\mathbf{x}\| \le \|\mathcal{A}^{-1}\| \times \|\mathcal{A}\mathbf{x}_N - \mathbf{y}\|$$

So if for a particular problem $\|\mathcal{A}\mathbf{x}_N - \mathbf{y}\|$ goes to zero, and the inverse operator is bounded, then the sequence of approximate solutions converges to the exact solution. As is clear from this equation, if we have no guarantee about the convergence of the residuals and we do not know the exact solution, then it is extremely difficult, if not impossible, to comment on the accuracy of the approximate solutions obtained by Galerkin's method [36–38].

1.9.2.4   *Is Galerkin's Method Always Applicable?* If the same expansion functions do not form the basis of both the domain and the range of the operator, then meaningful results may not be obtained by Galerkin's method. However, for some self-adjoint operators, Galerkin's method reduces to the method of least squares.

Example 11
Consider the solution of the following problem:

$$\mathcal{A}x = -\frac{d^2x(z)}{dz^2} = \cos z = \mathbf{y} \text{ with } x(z=0)=0=x(z=\pi) \quad (1.145)$$

We observe that Galerkin's method for this self-adjoint operator yields the least-squares solution if we choose the expansion functions to be the eigenfunctions of the operator.

A complete set of expansion functions for this problem would be $\sin(nz)$. These expansion functions are the eigenfunctions for the operator and satisfy the boundary conditions for this problem. If

$$\mathbf{x}_N = \sum_{n=1}^{N} a_n \sin(nz) , \text{ then } \mathcal{A}\mathbf{x}_N = \sum_{n=1}^{N} n^2 a_n \sin(nz)$$

Observe that the functions $\sin(nz)$ in the expression for $\mathcal{A}\mathbf{x}_N$ cannot span $\cos(z)$. Indeed it does not, if we talk about pointwise (uniform) convergence. Hence, one may jump to the conclusion that Galerkin's method may not be applicable for this problem. However, the solution do converge in the mean. The convergence in the mean is guaranteed by the discussions of this subsection, and we will show that it does. In general, it is difficult, if not impossible, to know whether a certain set of expansion functions will span the range of the operator.

By solving for the unknowns a_n by Galerkin's method, we get

$$\mathbf{x}_N = \frac{2}{\pi} \sum_{k=1}^{N} \frac{\sin(nz)}{k\left(4k^2 - 1\right)}.$$ It can be shown that, in the limit as $N \to \infty$, both $\mathbf{x}_N$

and $\dfrac{d\mathbf{x}_N}{dz}$ converge to $\mathbf{x}$ and $\dfrac{d\mathbf{x}}{dz}$ uniformly in the region of $0 \le z \le \pi$. However,

$\left\| \mathcal{A}\mathbf{x}_N - \mathbf{y} \right\|^2$ does not converge uniformly but converges in mean to zero as $N \to \infty$. This is clear from the following equation:

$$\lim_{N \to \infty} \left\| \mathcal{A}\mathbf{x}_N - \mathbf{y} \right\|^2 = \lim_{N \to \infty} \left| \frac{\pi}{2} - \frac{32}{\pi} \sum_{k=1}^{N} \frac{k^2}{\left(4k^2 - 1\right)^2} \right|^2 \to 0 \qquad (1.146)$$

1.9.2.5 Epilogue. In summary, the convergence of the residuals is guaranteed only if the expansion functions are chosen in a particular fashion. And because the inverse operator is bounded, the convergence of the residuals also implies that the sequence of solutions converges to the exact solution. In Galerkin's method, the expansion functions must satisfy all boundary conditions and differentiability conditions of the problem. This may not be easy for the case of differential operators. So, variational methods are suitable for differential operators (as the expansion functions do not have to satisfy any boundary conditions), and Galerkin's method is suitable for integral operators, as the latter are often bounded, and hence, strong convergence of the residuals is always guaranteed.

Finally, we discuss the numerical stability of the Galerkin's method. Because, in this solution procedure of Eq. (1.142), we have to invert a matrix $[G]$, the computational stability of this procedure is then dictated by $Cond[G]$. The term $Cond$ defines the condition number of a matrix. The condition number is the ratio of the maximum eigenvalue to the minimum eigenvalue of a matrix, and it specifies the number of significant digits that are required in performing any functional evaluation of this matrix. It can be shown [29] that the condition number of the Galerkin matrix $[G]$ is bounded by:

$$Cond[G] \leq Cond[A] \times Cond[E]$$

where

$Cond[A]$ is the condition number of the of the operator $\mathcal{A}$ in the finite dimensional N space in which the problem is solved, and

$Cond[E]$ is the condition number of the Gram matrix whose elements are defined by the inner product between the various basis functions, so that its elements are given by $E_{ik} = \langle \mathbf{u}_i, \mathbf{u}_k \rangle$.

It is also important to note that the operator $\mathcal{A}$ may not have any eigenvalues in an infinite dimensional space, but it has at least an eigenvalue on a finite dimensional space [3, p. 332].

1.9.3   Method of Least Squares

In the variational method (Rayleigh-Ritz) and Galerkin's method, the convergence of the residuals $\lim_{N \to \infty} (\mathcal{A}\mathbf{x}_N - \mathbf{y}) \to 0$ is guaranteed only under special circumstances. The method of least squares, however, always yields a strong convergence of the residuals even when $\|\mathcal{A}\|$ is unbounded. However, the expansion functions $\mathbf{u}_n$ are required to be in the domain of $\mathcal{A}$, and $\mathcal{A}\mathbf{u}_k$ must span the range of $\mathcal{A}$. So, for a second-order operator, the expansion functions must be able to approximate second derivatives in the mean, whereas in the variational method, it was only necessary to approximate first derivatives. However, if $\mathcal{A}$ is bounded, then any complete set of expansion functions will be adequate.

In the method of least squares, the solution of $\mathcal{A}\mathbf{x} = \mathbf{y}$ is attempted by trying to minimize the functional $F^1(\mathbf{x}_N)$, which is given by

$$F^1(\mathbf{x}_N) = \|\mathcal{A}\mathbf{x}_N - \mathbf{y}\|^2 = \langle \mathcal{A}\mathbf{x}_N - \mathbf{y}, \mathcal{A}\mathbf{x}_N - \mathbf{y} \rangle \tag{1.147}$$

We now substitute Eq. (1.122) in Eq. (1.147) and solve for the unknowns a_n by equating the partial derivative of $F^1(\mathbf{x}_N)$ with respect to a_i equal to zero. This results in a set of equations

$$\sum_{n=1}^{N} a_n \langle \mathcal{A}\mathbf{u}_n, \mathcal{A}\mathbf{u}_k \rangle = \langle \mathbf{y}, \mathcal{A}\mathbf{u}_k \rangle \text{ for } k = 1, 2, \cdots, N \qquad (1.148)$$

By comparing Eq. (1.148) with Eqs. (1.132) and (1.142), it is clear that the method of least squares requires more work than either Galerkin's or the Rayleigh-Ritz method. However, the reward of doing more work lies in obtaining strong convergence of the residuals (i.e., as $N \to \infty$, the residuals converge to zero in a mean square sense).

It is interesting to note that when the expansion functions $\mathbf{u}_n$ are chosen to be the eigenfunctions of the operator $\mathcal{A}$, then Eq. (1.148) reduces to Eq. (1.142) of Galerkin's method. Under these restrictions, the two methods yield identical sets of equations.

In the following subsections, the finer points of the method of least squares are explained.

1.9.3.1 *What Are the Characteristics of the Expansion Functions?* The expansion functions $\mathbf{u}_n$ must be such that $\mathcal{A}\mathbf{u}_n$ must form a linearly independent set and, therefore, should span $\mathbf{y}$. In other words, $\mathbf{u}_n$ must be complete with respect to the $\langle \mathcal{A}\mathbf{u}_n, \mathcal{A}\mathbf{u}_k \rangle$ norm. It does not make any difference whether $\mathbf{u}_n$ forms a complete set with respect to the $\langle \mathbf{u}_n, \mathbf{u}_k \rangle$ norm. The reason behind this is that if the residuals must converge to zero, then a linear combination of $\mathcal{A}\mathbf{u}_n$ should be able to reproduce $\mathbf{y}$. This is the subtlety behind the method of least squares.

Example 12

Consider the operator equation $\dfrac{d\,x(z)}{d\,z} = 1$ with $x(z = 0) = 0$ and $0 \le z \le 1$. If we choose the expansion functions as $\sin(n\pi z/2)$, then we observe

$$\frac{d\,x_N(z)}{d\,z} = \frac{d}{d\,z}\left(\sum_{n=1}^{N} a_n \sin\frac{n\pi z}{2}\right) = \sum_{n=1}^{N} \frac{a_n\, n\pi}{2}\cos\frac{n\pi z}{2}$$

In this case, the method of least squares yields a solution that is the Fourier series solution. The solution of the previous operator equation is obtained from Eq. (1.148) as

$$\mathbf{x} = x(z) = \sum_{n=1}^{\infty} \frac{8}{n^2\pi^2}\sin\frac{n\pi z}{2}$$

Observe that the previous solution converges uniformly (pointwise) to the exact solution $\mathbf{x} = x(z) = z$. However,

$$\lim_{N\to\infty} \frac{d\mathbf{x}_N}{dz}$$

does not converge uniformly to 1 but converges in the mean; that is,

$$\lim_{N\to\infty} \frac{dx_N(z)}{dz} = \lim_{N\to\infty} \int_0^1 \left(1 - \frac{dx_N(z)}{dz}\right)^2 dz = 0 \text{ and } \lim_{N\to\infty} \frac{d\mathbf{x}_N}{dz} \neq 1$$

So, even though $\|\mathcal{A}\|$ is unbounded, we have obtained strong convergence of the residuals.

1.9.3.2   Do the Residuals Converge Monotonically as $N \to \infty$? In the method of least squares, it is always true that the residuals $\mathcal{A}\mathbf{x}_N = \mathbf{y}$ converge in the mean to zero monotonically even when $\mathcal{A}$ is unbounded. This is seen in Eq. (1.148). As the order of the approximation $\mathbf{x}_N$ is increased, $F^1(\mathbf{x}_N)$ in Eq. (1.147) will always decrease monotonically.

1.9.3.3   Does the Sequence of Approximate Solutions Monotonically Converge to the Exact Solution? Because $\|\mathbf{x}_N - \mathbf{x}\| \leq \|\mathcal{A}^{-1}\| \times \|\mathcal{A}\mathbf{x}_N - \mathbf{y}\|$ and because $\|\mathcal{A}^{-1}\|$ is bounded, monotonic convergence of the residuals will always guarantee monotonic convergence of the sequence of approximate solutions $\mathbf{x}_N$ to the exact solution. Therefore, it is always true that

$$\lim_{N\to\infty} \|\mathbf{x}_N - \mathbf{x}\| \to 0$$

1.9.3.4   Epilogue. In summary, the method of least squares always generates monotonic convergence of the residuals (in the mean) if the expansion functions are in the domain of $\mathcal{A}$ and if $\mathcal{A}\mathbf{u}_n$ span $\mathbf{y}$. The disadvantage of this method is that it requires more work than the other two methods. Finally, the sequence of approximate solutions $\mathbf{x}_N$ monotonically approaches $\mathbf{x}$ as $N \to \infty$.

It is also important to note that if $\mathbf{y} \notin \mathbb{R}(\mathcal{A})$, then the sequence of solution $\mathbf{x}_N$ generated by the various methods provides asymptotic convergence. In an asymptotic convergence, the solution tends to converge as more and more expansion functions are chosen, but then the solution starts diverging [29]. This situation occurs in the solution of electromagnetic problems in which the excitation source is a delta function generator. As the delta function is not in the range of the operator, the sequence of solutions provides asymptotic convergence. The physics of it lies in the fact that one is trying to approximate the current at a delta gap, which may have infinite capacitance.

1.9.4   Comparison of the Rates of Convergence of the Various Methods

To compare the rate of convergence [23,33] between the variational method (Rayleigh-Ritz) and the method of least squares, we assume that we have used the same expansion functions $\mathbf{u}_n$ in solving $\mathcal{A}\mathbf{x} = \mathbf{y}$. Let $\mathbf{x}_N^V$ and $\mathbf{x}_N^L$ be the approximate solutions obtained by the variational method and the method of least squares, respectively, so that

$$\mathbf{x}_N^V = \sum_{n=1}^{N} a_n^V \mathbf{u}_n \quad \text{and} \quad \mathbf{x}_N^L = \sum_{n=1}^{N} a_n^L \mathbf{u}_n \tag{1.149}$$

Observe that, for the same number of expansion functions, the approximate solutions are different. This is because the variational method minimizes the functional

$$F\left(\mathbf{x}_N^V\right) = \left\langle \mathcal{A}\mathbf{x}_N^V, \mathbf{x}_N^V \right\rangle - \left\langle \mathbf{y}, \mathbf{x}_N^V \right\rangle - \left\langle \mathbf{x}_N^V, \mathbf{y} \right\rangle \tag{1.150}$$

whereas the method of least squares minimizes

$$F^1\left(\mathbf{x}_N^L\right) = \left\| \mathcal{A}\mathbf{x}_N^L - \mathbf{y} \right\|^2 = \left\langle \mathcal{A}\mathbf{x}_N^L - \mathbf{y}, \mathcal{A}\mathbf{x}_N^L - \mathbf{y} \right\rangle \tag{1.151}$$

So, by the terms of the problem, we have $F\left(\mathbf{x}_N^V\right) \le F\left(\mathbf{x}_N^L\right)$. It is important to note that

$$\left\langle \mathcal{A}(\mathbf{x}_N^V - \mathbf{x}), (\mathbf{x}_N^V - \mathbf{x}) \right\rangle \le \left\langle \mathcal{A}(\mathbf{x}_N^L - \mathbf{x}), (\mathbf{x}_N^L - \mathbf{x}) \right\rangle$$

Hence, the rate of convergence of the approximate solutions for the variational method is at least as fast as the method of least squares, if not better. The equality in the rates of convergence is achieved when the expansion functions are the eigenfunctions of the operator $\mathcal{A}$.

So even though the method of least squares requires more work, it has a slower rate of convergence than the variational method. On the other hand, the method of least squares guarantees monotonic convergence of the residuals. Hence, when the exact solution is not known, it is much easier with the method of least squares to establish relative error estimates of the solution. It is important to note that when the expansion functions are the eigenfunctions of the operator, the method of least squares and Galerkin's method become identical.

Next we observe the effect of the numerical approximations used in generating the final equation of the various solution procedures.

1.10   A THEOREM ON THE MOMENT METHODS

The inner product involved in the moment methods is usually an integral, which is evaluated numerically by summing the integrand at certain discrete points. In connection with this inner product, a theorem is proved, which states that the overall number of points involved in the integration must not be smaller than the number of unknowns involved in the moment method. If these two numbers are equal, then a point-matching solution is obtained, irrespective of whether one has started with Galerkin's method or the least-squares method. If the number of points involved in the integration is larger than the number of the unknowns, then a weighted point-matching solution is obtained [39–41].

1.10.1 Formulation of the Theorem

The inner products in Eq. (1.113) expressed in the functional form $\langle \mathbf{f}, \mathbf{g} \rangle$ are usually in the form of integrals of the product of functions $f(z)$ and $g(z)$. In very few cases, the inner products in Eq. (1.113) can be evaluated analytically, and in most practical problems, it is evaluated numerically. This involves only samples of the integrand at certain points. In other words, the numerical integration formulas used to evaluate the inner product can be written in the general form as

$$\int_D p(Z) \, dD = \sum_{k=1}^{M} b_k \, p(z_k) \tag{1.152}$$

where D is the domain over which the integration is performed, Z is a point in that domain, b_k are the weighting coefficients, and z_k are the points at which the samples of the function $p(Z)$ are evaluated. The function $p(Z)$, in our case, equals $f(z)g*(z)$.

If the same integration formula is applied to both the inner products of Eq. (1.113), then we have

$$\sum_{i=1}^{N} a_i \sum_{k=1}^{M} w_j^*(z_k) b_k \mathcal{A} u_i(z_k) = \sum_{k=1}^{M} w_j^*(z_k) b_k y(z_k) \quad \text{for } j = 1,2,\dots,N \tag{1.153}$$

Let us introduce the following matrices:

$$[F] = \left[\mathcal{A} u_i(z_k) \right]_{N \times M} ; \quad [W] = \left[w_j^*(z_k) \right]_{N \times M} ; \quad [B] = \operatorname{diag}(b_1, \dots, b_M);$$

$$[G] = \left[y(z_k) \right]_{M \times 1} ; \quad [A] = [a_i]_{N \times 1} \tag{1.154}$$

The system (1.153) can now be written in a compact form as

$$[W][B][F]^T [A] = [W][B][G] \tag{1.155}$$

where the superscript T denotes the transpose. Let us also denote $[V] = [W][B]$. Now we have the following equation instead of Eq. (1.155):

$$[V][F]^T [A] = [V][G] \tag{1.156}$$

The matrix $[V]$ can be considered as a weighting matrix, which multiplies the system of linear equations $[F]^T[A] = [G]$

If $M > N$, then the system Eq. (1.156) is, generally, over determined. Note that Eq. (1.156) are, essentially, point-matching equations, which are obtained by postulating that the approximate Eq. (1.153) is satisfied at points $z = z_k$, $k = 1, 2, ..., M$. The purpose of multiplying the system Eq. (1.156) by $[V]$ is to obtain a system of N equations in N unknowns. The solution to Eq. (1.113) can be regarded as a weighted point-matching solution. Note that it is not necessary that the same integration formula be used for each of equations (1.113) [39–41].

Let us consider Eq. (1.156). For this equation to have a unique solution for $[A]$, the matrix $[V][F]^T$ has to be regular, and the matrix $[V][F]^T$ will be regular if $\text{rank}[V] = \text{rank}[F] = N$. If this condition is violated, then a unique solution does not exist (although a minimum-norm solution can be found, which might be useful in certain cases [40]).

Thus, we have proved that if the integrals representing the inner product in a moment method solution to a linear operator equation are evaluated numerically, then the overall number of points involved in the integration must not be smaller than the number of the unknown coefficients.

1.10.2   Discussion

As the first consequence of the previous theorem, let us consider the special case when $M = N$. If the inverse matrix $[V]^{-1}$ exists, then both sides of Eq. (1.156) can be multiplied by $[V]^{-1}$, and an $N \times N$ system of linear equations is obtained. This matrix system is, essentially, a system of point-matching equations, and hence, the solution to Eq. (1.156) is identical to the point-matched solution. An example, in which such a situation can occur, is illustrated by the following construct. Let us adopt the expansion in Eq. (1.113) to be a piecewise-constant approximation (usually referred to as a pulse approximation). In that case, $\mathbf{u}_i = u_i(z)$ and is zero everywhere except over a small domain of z where it is a constant (usually equal to unity). If the Galerkin method is used, then the weighting functions are $\mathbf{W}_i = \mathbf{u}_i$. The weighting functions being nonzero only over a small domain, the inner products in Eq. (1.113) are sometimes evaluated by using the midpoint rule (i.e., by using only one integration point per inner product). Obviously, the final result is identical to a point-matched solution, for which the matching points coincide with the integration points in the previous Galerkin procedure. A similar result can be obtained if the least-squares technique is used. In this case, we have $\mathbf{W}_i = \mathcal{A}\mathbf{u}_i$. If the pulse approximation is adopted, then we have to evaluate the integrals representing the inner products over the whole domain of z where the original Eq. (1.71) is to hold, unlike the Galerkin procedure, in which the integration has to be performed only over the domain where $\mathbf{u}_i$ is nonzero. However, the overall number of the integration points must be larger than N, unless we wish to obtain a point-matched solution (for $M = N$). Of course, this result is valid whatever weighting functions are used (triangular, piecewise-sinusoidal, etc.), as long as the inner products are evaluated by using a numerical quadrature formula to evaluate the inner products.

The need for taking only a few integration points can result in not only increasing the speed of the computations, but also because of certain problems associated with the kind of the approximation adopted for the solution, which are not always clearly recognized. For example, in solving a wire-antenna problem, a piecewise-constant approximation of the current distribution can be used. If the exact kernel is taken, then $\mathcal{A}\mathbf{u}_i$ has a nonintegrable singularity at the edge of the domain where $\mathbf{u}_i$ is nonzero. This precludes the use of both the Galerkin and the least-squares technique because the resulting inner products diverge! Yet, if the integration in evaluation of an inner product is confined to the interior of the domain where $\mathbf{u}_i$ is nonzero, then acceptable results might be obtained, although the accuracy of the results can be much worse than with a point-matching solution (in addition to requiring a much longer CPU time), and the final result, obviously is not a Galerkin or a least-squares solution because such a solution does not exist. A singularity, though square integrable, also occurs if a piecewise-linear approximation (i.e., triangular approximation) is adopted, which can have an adverse effect especially to a least-squares solution.

In summary, an analogous theorem can be formulated in connection with the moment method solution of any integral equation of the general form [39–41]. Namely, if the unknown function $\mathbf{x}$ is approximated according to Eq. (1.113), then the overall number of points involved in the numerical integration must not be smaller than N.

1.11   PHILOSOPHY OF ITERATIVE METHODS

For the method of moments, the expansion functions are selected a priori, and the numerical procedure has to do with the solution for the unknown coefficients that multiply the expansion functions. For the iterative methods, the form of the solution is not selected a priori but evolves as the iteration progresses. This is very important from a philosophical point of view. In the MoM, it is essential to choose the expansion and the weighting functions in a particular fashion. If this is not done carefully, then the method may not work. This was pointed out by Harrington in his classic paper [42].

For the iterative methods, we do not have to worry about the proper selection of the expansion and weighting functions. Because the solution evolves as the iteration progresses, one need not worry about the proper choice of expansion functions. Also, for the iterative methods, the weighting functions are dictated by the expansion functions, and hence, if the iterative method is handled properly, then convergence is guaranteed [43–51]. In addition to this philosophical difference, there are two additional practical differences:

(a) In MoM, the maximum number of unknowns is limited by the largest size of the matrix that one can handle on the computer. Therefore, we would like to propose a solution procedure in which we arrive at the solution without ever having to explicitly form the square matrix in Eq. (1.142). One could also achieve this goal by computing each element of the square matrix in Eq. (1.142) as needed, but this procedure may be inefficient.

(b) Once we have solved for the unknown in Eq. (1.142) by the conventional techniques, we do not have any idea about the quality of the solution (i.e., we do not know the degree of accuracy with which the computed solution satisfies Eq. (1.71) or how close it is to the exact solution). Moreover, with the conventional techniques [33] when we increase the number of unknowns in a non-self-adjoint problem from N to $2N$, we are not guaranteed to have a *better* solution.

It is primarily for these two reasons that we propose to use a class of iterative methods in which some *error criterion* is minimized at each iteration, and so we are guaranteed to have an improved result at each iteration. Moreover, with the proposed methods, one is assured to have the exact solution in a finite number of steps. In summary, we prefer to use the new class of finite step iterative methods over the conventional matrix methods because the new techniques address the three very fundamental questions of any numerical method, namely,

(a) the establishment of convergence of the solution
(b) the investigation of the rapidity of convergence
(c) an effective estimate of the error

to our satisfaction.

An attempt to answer these three questions was made in [43–51], in which they tried to answer the questions in terms of the solution of a matrix equation. They used Eq. (1.142), the discretized form of the original operator Eq. (1.71) as the matrix equation.

In this section, we essentially follow the approaches of [43–51] and present a class of iterative methods, which always theoretically converge to the solution in the absence of round-off and truncation errors in a finite number of steps, irrespective of what the initial guess is. These classes of iterative methods are known as the method of conjugate directions. Both the conjugate gradient method and the well-known Gaussian elimination techniques belong to this class. In this section, we will confine our attention only to the class of conjugate gradient methods.

1.11.1 Development of Conjugate Direction Methods

The method of conjugate directions is a class of iterative methods that always converges to the solution at the rate of a geometric progression in a finite number of steps, irrespective of the initial starting guess [43,44] barring round-off and truncation errors in the computations. The fundamental principle of the method of conjugate directions is to select a set of vectors $\mathbf{p}_i$ such that they are $\mathcal{A}$-conjugate or $\mathcal{A}$-orthogonal, that is,

$$\langle \mathcal{A}\mathbf{p}_i, \mathbf{p}_k \rangle = 0 \quad \text{for } i \neq k \tag{1.157}$$

The practical significance of Eq. (1.157) is explained as follows: Suppose we want to approximate the solution $\mathbf{x}$ by $\mathbf{x}_N$ so:

$$\mathbf{x} = \mathbf{x}_N = \sum_{i=1}^{N} \alpha_i \mathbf{p}_i \tag{1.158}$$

then

$$\mathcal{A}\mathbf{x} = \mathcal{A}\mathbf{x}_N = \sum_{i=1}^{N} \alpha_i \mathcal{A}\mathbf{p}_i = \mathbf{y} \tag{1.159}$$

The vectors $\mathbf{p}_i$ span the finite dimensional space in which the approximate solution is sought. By taking the inner product of both sides of Eq. (1.158) with respect to $\mathbf{p}_i$, we obtain the following:

$$\alpha_i \langle \mathcal{A}\mathbf{p}_i, \mathbf{p}_i \rangle = \langle \mathbf{y}, \mathbf{p}_i \rangle; \quad \text{or } \alpha_i = \frac{\langle \mathbf{y}, \mathbf{p}_i \rangle}{\langle \mathcal{A}\mathbf{p}_i, \mathbf{p}_i \rangle} \tag{1.160}$$

as $\langle \mathcal{A}\mathbf{p}_i, \mathbf{p}_k \rangle = 0$ for $i \neq k$ and, therefore

$$\mathbf{x} = \sum_{i=1}^{N} \frac{\langle \mathbf{y}, \mathbf{p}_i \rangle}{\langle \mathcal{A}\mathbf{p}_i, \mathbf{p}_i \rangle} \mathbf{p}_i \tag{1.161}$$

It is clear from Eq. (1.161) that if we could select a set of $\mathcal{A}$-orthogonal vectors $\mathbf{p}_i$, then the construction of the numerical solution would be straightforward. The computational complexity of this algorithm then reduces to computing some inner products and a scalar ratio.

Equation (1.161) actually represents the fundamental basis of the conjugate direction methods. Now, so far, we have not said how to choose the functions $\mathbf{p}_i$. When the functions $\mathbf{p}_i$ are selected a priori from the unit coordinate vectors, we get the direct methods like Gaussian elimination, and when the vectors $\mathbf{p}_i$, are determined iteratively, we get an iterative method. Before we present the iterative version of the conjugate directions, we discuss a particular direct form of the conjugate direction method, namely Gaussian elimination. In the next section, we present the generalized version (applicable to arbitrary operator equations) of Gaussian elimination and demonstrate where efficiency/accuracy in the solution procedure can be achieved. We also point out the weak point of Gaussian elimination and the reason why we would like to pursue an iterative method.

It is important to point out that when $\mathcal{A}$ is a matrix, Eq. (1.161) actually is the computational form of Gaussian elimination. For this case, the vectors $\mathbf{p}_i$ are selected as follows. We start with a set of coordinate column vectors $\mathbf{q}_i$, such that $\mathbf{q}_i$ has a 1 in location i and a 0 elsewhere. From the $\mathbf{q}_i$ vectors, we get the $\mathcal{A}$-orthogonal $\mathbf{p}_i$ vectors by using the Gram-Schmidt decomposition procedure. So we let:

$$[p_1]^T = [q_1]^T = [1 \quad 0 \quad 0 \quad \cdots \quad 0]^T \tag{1.162}$$

and obtain

$$[p_2] = [q_2] - \frac{\langle \mathcal{A}\mathbf{q}_2, \mathbf{p}_1 \rangle}{\langle \mathcal{A}\mathbf{p}_1, \mathbf{p}_1 \rangle} [p_1] \tag{1.163}$$

and so on. In general,

$$[p_{i+1}] = [q_{i+1}] - \sum_{k=1}^{i} \frac{\langle \mathcal{A}\mathbf{q}_{k+1}, \mathbf{p}_k \rangle}{\langle \mathcal{A}\mathbf{p}_k, \mathbf{p}_k \rangle} [p_k] \tag{1.164}$$

The vectors $\mathbf{p}_i$, thus generated are substituted in Eq. (1.161), and we obtain the solution yielded by Gaussian elimination for matrices. This development can be easily extended to the general operator equation by treating $\mathbf{q}_i$'s as functions rather than as vectors. Even though this computation process is straightforward, it has the following advantages/disadvantages:

(1) One can obtain any meaningful solution only after one has gone through the computation of the N terms of the series prescribed by Eq. (1.161). This can be very time consuming, particularly if N is large. However, this method would be advantageous if N is small (e.g., $N \leq 500$).

(2) Second, even after all the terms have been computed to yield the solution, we are not sure about the accuracy of the solution. When N is large, the round-off error in the computation of $\mathbf{p}_i$ in Eq. (1.164) can build up to cause erroneous results. There is no way to rectify these round-off errors, and they propagate as N gets large. However, one can partially rectify the round-off errors if the vectors $\mathbf{q}_i$, are computed in an iterative fashion.

In short, for small N, the direct method is efficient (e.g., for $N \leq 500$ and for well-conditioned matrix $[\mathcal{A}]$, it is well known that Gaussian elimination is a very efficient way to solve equation (1.142)). However, if N is large, we would prefer a method in which we know the error in the computed solution. That is why we lean toward an iterative method, in which a suitable error criterion is minimized at each iteration and so, if any truncation or round-off error is generated in the computation of $\mathbf{p}_i$, then it is corrected.

An almost infinite number of ways exist to select $\mathbf{p}_i$. However, in our approaches, we select $\mathbf{p}_i$ from the residuals computed after each iteration. Because the residual

$$\mathbf{r} = \mathbf{y} - \mathcal{A}\mathbf{x} \tag{1.165}$$

is proportional to the gradient of the quadratic functional $F(\mathbf{x})$

$$F(\mathbf{x}) = \langle \mathbf{r}, \mathbf{r} \rangle = \langle \mathbf{y} - \mathcal{A}\mathbf{x}, \mathbf{y} - \mathcal{A}\mathbf{x} \rangle = \| \mathbf{y} - \mathcal{A}\mathbf{x} \|^2 \tag{1.166}$$

to be minimized, this particular choice of $\mathbf{p}_i$ from the residuals $\mathbf{r}$ is called the method of conjugate gradient. In the next section, we present the basic philosophy of the generalized conjugate gradient method and develop two specialized cases, which are computationally straightforward.

1.11.2   A Class of Conjugate Gradient Methods

For the class of conjugate gradient methods, which will be discussed next, we propose to deal with the following functional:

$$F(\mathbf{x}) = \langle \mathbf{r}, S\mathbf{r} \rangle \tag{1.167}$$

where S is a Hermitian positive-definite operator (assumed known but has not been defined yet). So at the $i+1$ iteration, we minimize the functional

$$F(\mathbf{x}_i) = \langle \mathbf{r}_i, S\mathbf{r}_i \rangle \tag{1.168}$$

where

$$\mathbf{r}_i = \mathbf{y} - \mathcal{A}\mathbf{x}_i \tag{1.169}$$

We propose to minimize $F(\mathbf{x}_{i+1})$ by incrementing $\mathbf{x}_i$ in the following way:

$$\mathbf{x}_{i+1} = \mathbf{x}_i + a_i \mathbf{p}_i \tag{1.170}$$

where a_i is chosen to minimize $F(\mathbf{x}_{i+1})$. This is achieved by equating the following:

$$\frac{\partial F(\mathbf{x}_{i+1})}{\partial a_i} = 0 = \frac{\partial}{\partial a_i} \langle (\mathbf{y} - \mathcal{A}\mathbf{x}_i - a_i \mathcal{A}\mathbf{p}_i), S(\mathbf{y} - \mathcal{A}\mathbf{x}_i - a_i \mathcal{A}\mathbf{p}_i) \rangle \tag{1.171}$$

This results in

$$\frac{\partial}{\partial a_i} \left(\langle \mathbf{r}_i, S\mathbf{r}_i \rangle - a_i \langle \mathcal{A}\mathbf{p}_i, S\mathbf{r}_i \rangle - a_i \langle \mathbf{r}_i, S\mathcal{A}\mathbf{p}_i \rangle + |a_i|^2 \langle \mathcal{A}\mathbf{p}_i, S\mathcal{A}\mathbf{p}_i \rangle \right) = 0 \tag{1.172}$$

The functional $F(\mathbf{x}_{i+1})$ is minimized if the following is true:

$$a_i = \frac{\langle \mathbf{r}_i, S\mathcal{A}\mathbf{p}_i \rangle}{\langle \mathcal{A}\mathbf{p}_i, S\mathcal{A}\mathbf{p}_i \rangle} \qquad (1.173)$$

We also observe that the residuals can be generated recursively by

$$\mathbf{r}_{i+1} = \mathbf{y} - \mathcal{A}\mathbf{x}_{i+1} = \mathbf{r}_i - a_i\mathcal{A}\mathbf{p}_i \qquad (1.174)$$

By comparing Eqs. (1.173) and (1.174), it is clear that

$$\langle \mathbf{r}_k, S\mathcal{A}\mathbf{p}_i \rangle = 0 \text{ for all } i \neq k \text{ and } k > i; \text{ or } \langle S\mathbf{r}_k, \mathcal{A}\mathbf{p}_i \rangle = 0 \qquad (1.175)$$

Equation (1.175) implies that certain orthogonality conditions must be satisfied if the residuals are to be computed recursively by using Eqs. (1.170) and (1.173).

So far we have not specified how to generate the vectors $\mathbf{p}_i$. We obtain the search direction vectors $\mathbf{p}_i$, recursively, as:

$$\mathbf{p}_{i+1} = \mathcal{K}\mathbf{g}_{i+1} + b_i\mathbf{p}_i \qquad (1.176)$$

where $\mathbf{g}_{i+1}$ are certain functions yet undetermined. $\mathcal{K}$ is an arbitrary Hermitian positive-definite operator yet is undefined. We now select the parameter b_i such that the vectors $\mathbf{p}_i$ are $\mathcal{A}$-orthogonal, with respect to the following inner product:

$$\langle \mathcal{A}\mathbf{p}_i, S\mathcal{A}\mathbf{p}_k \rangle = 0 \text{ for } i \neq k. \qquad (1.177)$$

Equation (1.177) guarantees that the method will converge in a finite number of steps. Hence, the parameter b_i is derived from the concept of $\mathcal{A}$-orthogonalization. By enforcing Eq. (1.177) in Eq. (1.176), we get

$$b_i = -\frac{\langle \mathcal{A}^H S\mathcal{A}\mathbf{p}_i, \mathcal{K}\mathbf{g}_{i+1} \rangle}{\langle \mathcal{A}\mathbf{p}_i, S\mathcal{A}\mathbf{p}_i \rangle} \qquad (1.178)$$

where $\mathcal{A}^H$ is the adjoint operator for $\mathcal{A}$ and is defined by $\langle \mathcal{A}\mathbf{u}, \mathbf{v} \rangle = \langle \mathbf{u}, \mathcal{A}^H\mathbf{v} \rangle$.

Equations (1.169), (1.170), (1.173), (1.174), (1.176), and (1.178) describe the generalized conjugate direction method.

We have yet to specify the operators S and $\mathcal{K}$ and the vectors $\mathbf{g}_i$. When we select the vectors $\mathbf{g}_i$ as the coordinate vectors (i.e., $\mathbf{g}_i$ has a 1 in the ith position and zero elsewhere), we get the generalized Gaussian elimination method. When we choose the following:

$$\mathbf{g}_i = \mathcal{A}^H S\mathbf{r}_i \qquad (1.179)$$

we get the class of conjugate gradient methods.

The conjugate gradient methods described in Eqs. (1.169), (1.170), (1.173), (1.174), (1.176), and (1.179) all converge to the solution in a finite number of steps starting with any initial guess [43–50].

We see that, for the class of conjugate gradient methods, the functional $F(\mathbf{x})$ is minimized at each iteration. Using the definition of the functional $F(\mathbf{x})$ of Eq. (1.168) we observe that it gets reduced at each iteration as

$$F(\mathbf{x}_i) - F(\mathbf{x}_{i+1}) = \frac{\langle \mathbf{r}_i, S\mathcal{A}\mathbf{p}_i \rangle^2}{\langle \mathcal{A}\mathbf{p}_i, S\mathcal{A}\mathbf{p}_i \rangle} > 0 \qquad (1.180)$$

In Eq. (1.180), the right-hand side of the equation is always positive, as S is a positive definite operator, so that the error is minimized at each iteration and the sequence $\mathbf{x}_i$ converges.

Of course, it is well and good to know that the sequence $\mathbf{x}_i$ converges, but it is much better to know how it converges. If we introduce an operator T such that

$$T = \mathcal{K}\mathcal{A}^H S\mathcal{A} \qquad (1.181)$$

and let

$q =$ infinimum [spectrum (T)] = smallest eigenvalue of operator T, when T is a positive-definite matrix

$Q =$ supremum [spectrum (T)] = largest eigenvalue of operator T, when T is a positive-definite matrix

then it can be shown that the functional $F(\mathbf{x})$ is minimized at each iteration and that the rate of convergence at ith iteration is given by [43–45]

$$F(\mathbf{x}_i) \leq \left[\frac{2(1-\alpha)^i}{(1+\sqrt{\alpha})^{2i} + (1-\sqrt{\alpha})^{2i}} \right]^2 F(\mathbf{x}_0) \qquad (1.182)$$

and that the sequence of solutions $\mathbf{x}_i$ converges to the exact solution $\mathbf{x}$ so that $\|\mathbf{x}_i - \mathbf{x}\|$ tends to zero faster than $\left(\dfrac{1-\sqrt{\alpha}}{1+\sqrt{\alpha}} \right)^i$ where $\alpha = q/Q$. In the next section, we present two versions (special cases) of the conjugate gradient method. Each special case has certain advantages. The reason we chose these two special cases is because these two versions in particular lead to the simplest algorithm. We also derive the rate of convergence of each of these techniques. Other special cases are also possible, and they are presented elsewhere [51].

1.11.2.1 Minimization of the Residuals (Conjugate Gradient Method A). For this special case, we set

$$S = \mathcal{K} = I = \text{Identity operator,}$$

and so, from Eq. (1.181)

$$\mathcal{T} = \mathcal{A}^H \mathcal{A} \tag{1.183}$$

Also, one obtains

$$\mathcal{F}(\mathbf{x}_i) = \|r_i\|^2 \tag{1.184}$$

The conjugate gradient method for this case starts with an initial guess $\mathbf{x}_0$ and defines the following:

$$\mathbf{r}_0 = \mathbf{y} - \mathcal{A}\mathbf{x}_0 \tag{1.185}$$

$$\mathbf{p}_0 = \mathcal{A}^H \mathbf{r}_0 \tag{1.186}$$

For $i = 1, 2, \ldots,$ let:

$$a_i = \frac{\left\|\mathcal{A}^H \mathbf{r}_i\right\|^2}{\left\|\mathcal{A}\mathbf{p}_i\right\|^2} \tag{1.187}$$

$$\mathbf{x}_{i+1} = \mathbf{x}_i + a_i \mathbf{p}_i \tag{1.188}$$

$$\mathbf{r}_{i+1} = \mathbf{r}_i - a_i \mathcal{A}\mathbf{p}_i \tag{1.189}$$

$$\mathbf{p}_{i+1} = \mathcal{A}^H \mathbf{r}_{i+1} + b_i \mathbf{p}_i \tag{1.190}$$

$$b_i = \frac{\left\|\mathcal{A}^H \mathbf{r}_{i+1}\right\|^2}{\left\|\mathcal{A}^H \mathbf{r}_i\right\|^2} \tag{1.191}$$

By letting q and Q be the lower and upper spectral bounds for $\mathcal{T} = \mathcal{A}^H \mathcal{A}$ and by setting $\alpha = q/Q$, the characteristics of this method can be summarized as follows:

(1) The iterations described by Eqs. (1.185)–(1.191) is such that $\mathbf{x}_i$ converges to $\mathbf{x}$ for all initial guesses.

(2) $\|\mathbf{r}_i\|^2$ is the least possible in i steps by using any iterative scheme of the form Eq. (1.188).

(3) $\|\mathbf{x} - \mathbf{x}_i\|$ decreases at each iteration even though we do not know the exact solution $\mathbf{x}$.

(4) The following bound on the functional holds:

$$F(\mathbf{x}_i) \leq \left[\frac{2(1-\alpha)^i}{\left(1+\sqrt{\alpha}\right)^{2i} + \left(1-\sqrt{\alpha}\right)^{2i}} \right]^2 F(\mathbf{x}_0) \qquad (1.192)$$

Because $\|\mathbf{r}_i\|^2$ is minimized, this particular method essentially gives a sequence of least-squares solutions to $\mathcal{A}\mathbf{x} = \mathbf{y}$, such that at each iteration, the estimate $\mathbf{x}_i$ gets closer to $\mathbf{x}$. This is the most widely used technique.

1.11.2.2   Minimization of solution Error (Conjugate Gradient Method B). For this case, we let

$$S = \left(\mathcal{A}\mathcal{A}^H \right)^{-1} \qquad (1.193a)$$

with

$$\mathcal{K} = \mathcal{A}^H \mathcal{A} \quad \text{and} \quad \mathcal{T} = \mathcal{A}^H \mathcal{A} \qquad (1.193b)$$

so that

$$F(\mathbf{x}_i) = \|\mathbf{x} - \mathbf{x}_i\|^2 \qquad (1.194)$$

Observe in Eq. (1.194) that we are minimizing the error between the exact and the approximate solution even though we do not know the exact solution. Substitution of Eq. (1.193) into the generalized conjugate gradient algorithm yields this special version. Given $\mathbf{x}_0$, let

$$\mathbf{r}_0 = \mathbf{y} - \mathcal{A}\mathbf{x}_0 \qquad (1.195)$$

$$\mathbf{p}_0 = \mathcal{A}^H \mathbf{r}_0 \qquad (1.196)$$

For $i = 1, 2, \ldots$, let

$$a_i = \frac{\|\mathbf{r}_i\|^2}{\|\mathbf{p}_i\|^2} \qquad (1.197)$$

$$\mathbf{x}_{i+1} = \mathbf{x}_i + a_i \mathbf{p}_i \qquad (1.198)$$

$$\mathbf{r}_{i+1} = \mathbf{r}_i - a_i \mathcal{A}\mathbf{p}_i \tag{1.199}$$

$$\mathbf{p}_{i+1} = \mathcal{A}^H \mathbf{r}_{i+1} + b_i \mathbf{p}_i \tag{1.200}$$

$$b_i = \frac{\|\mathbf{r}_{i+1}\|^2}{\|\mathbf{r}_i\|^2} \tag{1.201}$$

By defining q and Q to be the lower and upper spectral bounds of $\mathcal{T} = \mathcal{A}^H \mathcal{A}$ and $\alpha = q/Q$, we get the following results:

(1) The iteration is such that $\mathbf{x}_i$ converges to $\mathbf{x}$ for all i.
(2) $\|\mathbf{x} - \mathbf{x}_i\|$ is the least possible in i steps for all possible iterative methods of the form Eq. (1.198) even though we do not know the exact solution.
(3) $\|\mathbf{r}_i\|^2$ may not decrease; in fact, at each iteration, it may even increase. Often, it oscillates and does not get reduced until the last few iterations.
(4) The functional at each iteration is minimized, and the bound given by Eq. (1.192) still holds.

Other special cases are available elsewhere [51].

Examples of the two iterative schemes and the type of results they generate are discussed in the next section.

1.11.3   Numerical Examples

As an example, we consider the analysis of electromagnetic scattering from a 0.5λ long antenna of radius 0.001λ, irradiated by a plane wave of amplitude 1 V/m. The antenna is subdivided into 30 segments. Methods A and B are now applied directly to the solution of the Pocklington's equation for the thin wire antenna [46]. So, in this formulation, we bypass the matrix formulation stage. Both methods A and B have been applied to solve the problem. The numerical computations obtained from the two techniques are presented in Table 1.3.

   Observe that, for method A, the magnitude of the residuals decrease monotonically. For method B, even though the absolute error between the true solution and the solution at the end of each iteration decrease, the residuals actually increase and do not go down until the very end. Even though, from a theoretical point of view, method B is attractive but from a numerical standpoint, it seems both methods perform equally well. However, from a psychological point of view, method A is preferred, as at each iteration, a user can observe the monotonic convergence of the residuals as the iterative method progresses. The CPU time taken by both techniques is almost the same. Observe both the methods even converged into a *good* solution in 15 iterations, starting from a null initial guess. The reason, these methods converged in 15 iterations is because the current distribution on the structure is symmetrical about the midpoint of the antenna.

Hence, the operator would have two sets of eigenvalues that are equal. Because the conjugate gradient method yields the solution in at most M steps, where M is the number of independent eigenvalues of the operator (15 in this case), the method converged in 15 iterations.

Table 1.3. Errors of the two versions of the conjugate gradient method (A and B)

Number of iterations	Normalized error $\dfrac{\|\mathcal{A}\mathbf{x}_n - \mathbf{y}\|^2}{\|\mathbf{y}\|^2}$	
N	Method A	Method B
1	1.000	1.000
2	0.984	61.558
3	0.972	77.731
4	0.961	88.222
5	0.951	94.835
6	0.942	98.433
7	0.933	99.625
8	0.925	98.909
9	0.916	96.716
10	0.907	93.419
11	0.898	89.361
12	0.889	84.002
13	0.881	100.192
14	0.849	24.200
15	0.590	1.9308
16	1.045×10^{-17}	1.013×10^{-17}

However, one disadvantage of the conjugate gradient method for a non-self-adjoint operator is that, even though the method minimizes the quantity $\left\|\mathcal{A}^H \mathbf{r}_i\right\|^2$ at each iteration, it does not guarantee that $\|\mathbf{r}_i\|^2$, the residual is minimized (i.e., how well the original equation $\mathcal{A}\mathbf{x} = \mathbf{y}$ is satisfied). In fact, even though $\left\|\mathcal{A}^H \mathbf{r}_i\right\|^2$ may be small at a particular iteration, $\|\mathbf{r}_i\|^2$ could be very large. Hence, the number of iterations taken by the iterative conjugate gradient method for some problems can be large to minimize $\|\mathbf{r}_i\|^2$ to an acceptable level and, therefore, can be less efficient than Gaussian elimination with pivoting. However, when solving extremely large problems [52] in which one needs to have good confidence on the final solution, it may be useful to follow the result obtained by the Gaussian elimination with pivoting with a few steps of the conjugate gradient method.

1.12 THE DIFFERENCE BETWEEN APPLICATION OF THE CONJUGATE GRADIENT METHOD (CGM) TO SOLVE A MATRIX EQUATION AS OPPOSED TO DIRECT APPLICATION OF CGM TO SOLUTION OF AN OPERATOR EQUATION

From the previous sections, it is clear that there are fundamental philosophical differences between applying the conjugate gradient method to solve the moment matrix equation of Eq. (1.113), as opposed to applying it to minimize Eq. (1.71) recursively [50]. The difference is that the latter guarantees a strong convergence of the residuals and the solution (depending on case A or case B [46]), whereas the former guarantees weak convergence. The question now arises: Does this difference still exist when we look at the problem from a computational point of view? The answer is YES, and it is demonstrated by a simple example.

To illustrate this point, we start with an operator equation,

$$\mathcal{A}\mathbf{J} = \mathbf{y} \qquad (1.202)$$

and through discretization for computational purposes, we convert it to a matrix equation using the method of moment principles as

$$[A_M][J_M] = [Y_M] \qquad (1.203)$$

where $[A_M]$, $[J_M]$, and $[Y_M]$ are the matrix version of the continuous operator equation. If we now solve the matrix version in Eq. (1.203) by the conjugate gradient method, then we are essentially solving the following equations, known as the normal equations:

$$[A_M]^H [A_M][J_M] = [A_M]^H [Y_M] \qquad (1.204)$$

where $[A_M]^H$ is the adjoint matrix of $[A_M]$ (which is simply the conjugate transpose). Now, if we apply the conjugate gradient method directly to the solution of the operator equation, then we solve

$$\mathcal{A}^H \mathcal{A}\mathbf{J} = \mathcal{A}^H \mathbf{Y} \qquad (1.205)$$

Here $\mathcal{A}^H$ is the adjoint operator. For numerical computation, Eq. (1.205) has to be discretized to a matrix equation and we obtain

$$[A_D^H][A_D][J_D] = [A_D^H][Y_D] \qquad (1.206)$$

It is seen that $[A_D] = [A_M]$, $[J_D] = [J_M]$, and $[Y_D] = [Y_M]$. Therefore, during numerical implementation, the basic difference between Eqs. (1.204) and (1.206) will be in how the continuous adjoint operator $\mathcal{A}^H$ has been discretized to $[A_D^H]$

and whether the "matricized" adjoint operator $[A_D^H]$ is identical to the conjugate transpose of the discretized original operator, $[A_M]^H$. If the two matrices (namely the discretized adjoint operator $[A_D^H]$ and the conjugate transpose of the "matricized" operator $[A_M]^H$) are not identical, then there will be a difference between the applications of the conjugate gradient method to the solution of a matrix equation, as opposed to the solution of the operator equation. For many problems, Eqs. (1.204) and (1.206) are not identical.

For example, consider the convolution equation [3,50]

$$\int_{-\infty}^{\infty} x(t-u)v(u)\,du = y(t) \text{ for } 0 \le t < \infty \tag{1.207}$$

where $x(t)$ and $y(t)$ are assumed to be known. We redefine Eq. (1.207) as $\mathcal{A}v = y$. This operator equation can now be written in the matrix form, using the method of moments concepts of pulse basis functions (piecewise-constant functions) for the unknown v, and with impulse weighting, resulting in $[A_M][V_M] = [Y_M]$, or in a matrix form as

$$\begin{bmatrix} x_1 & 0 & \cdots & 0 \\ x_2 & x_1 & \cdots & 0 \\ \vdots & \vdots & \vdots & \vdots \\ x_N & x_{N-1} & \cdots & x_{N-M} \end{bmatrix}_{N \times M} \begin{bmatrix} v_1 \\ v_2 \\ \vdots \\ v_M \end{bmatrix}_{M \times 1} = \begin{bmatrix} y_1 \\ y_2 \\ \vdots \\ y_2 \end{bmatrix}_{N \times 1} \tag{1.208}$$

If we now apply the conjugate gradient method to solve this matrix equation, we will be solving $[A_M]^H[A_M][V_M] = [A_M]^H[Y_M]$. The adjoint matrix $[A_M]^H$ in this case is

$$\begin{bmatrix} x_1 & x_2 & \cdots & x_N \\ 0 & x_1 & \cdots & x_{N-1} \\ \vdots & \vdots & \vdots & \vdots \\ 0 & 0 & \cdots & x_{N-M} \end{bmatrix} \tag{1.209}$$

We get a solution for $[V]$ (in this case) by solving Eq. (1.204) with the conjugate gradient method. Now, let us apply the conjugate gradient method directly to the operator equation as Eq. (1.207). We consider the adjoint operator for the integral equation as Eq. (1.207). The adjoint operator, $\mathcal{A}^H$, of $\mathcal{A}$, is defined by $\langle \mathcal{A}V, Z \rangle = \langle V, \mathcal{A}^H Z \rangle$, or equivalently, by

$$\int z(t)\,dt \int x(t-u)v(u)\,du = \int v(u)\,du \int z(t)x(t-u)\,dt \tag{1.210}$$

which is not a *convolution* but a *correlation* operation. We have observed in section 1.6 that the adjoint of the convolution operator is the correlation operator, which is similar to an advance convolution operator. By comparing Eqs. (1.209) and (1.210), it is apparent that $[A_M]^H$ is a restricted version of the actual adjoint operator in Eq. (1.210). This is because the adjoint operator is given by

$$\mathcal{A}^H \mathbf{z} = \int z(t) x(t-u) dt \qquad (1.211)$$

This continuous operator can only take the form of under the assumption that $z(n)$ is identically zero beyond n, and this may not be true, in general. Therefore, $[A_D^H]$ and $[A_M]^H$ are not identical for this problem unless some additional assumptions are made! In the signal processing literature, a clear distinction has been made between Eqs. (1.204) and (1.206). In Eq. (1.204), $[A_M]^H[A_M]$ is called the covariance matrix of the data, whereas $[A_D^H][A_D]$ in Eq. (1.206) is called the autocorrelation matrix of the data. It is well known that Eqs. (1.204) and (1.206) yield different results. In the electromagnetics literature, several techniques have been developed to solve Eq. (1.206). These techniques depend on the assumptions that need to be made regarding the behavior of $\mathbf{x}$ and $\mathbf{z}$ for $t \geq n$. One possible assumption may be that the waveforms become almost zero for $t \geq n$. An alternative numerical implementation has been considered in [53], in which instead of assuming the nature of decay for $\mathbf{x}$ and $\mathbf{z}$ for $t \geq n$, a weighted inner product has been defined to minimize the error introduced in discretizing the continuous adjoint operator. However, for non-equally sampled data, no generalization can be made.

Thus, it is important to note that the distinction between correlation (1.211) and a convolution (1.207) disappears when the kernel is symmetric. Therefore, in the frequency domain, a distinction between correlation and convolution in an integral equation setting cannot be made as the Green's function contains terms that are the absolute value of the distance and, therefore, is symmetric. Hence, this distinction does not exist for problems defined in the frequency domain. However, this is not true for time domain problems in which the convolution operator is not similar to its adjoint, which is a correlation operator as shown in [47].

1.13   CONCLUSION

This chapter begins by describing some fundamental mathematical definitions related to the concepts of various spaces and defining the concept of convergence.

Next, the concept of an ill-posed and a well-posed problem is described to illustrate that the concept of a solution to a problem may not be clearly defined.

Then an attempt has been made to bring out the mathematical differences between the variational method (Rayleigh-Ritz), Galerkin's method, and the method of least squares. It is seen that, for a variational method, the expansion functions need not be in the domain of the operator, whereas for Galerkin's and the method of least squares, they are required to be in the domain of the operator. Moreover, for Galerkin's method, the expansion functions have to span both the domain and range of the operator, whereas for the method of least squares, the operator operating on the expansion functions must span the range of the operator. For a variational method, however, the expansion functions have to be complete with respect to the inner product $\langle \mathcal{A}\mathbf{u}_i, \mathbf{u}_j \rangle$. Both Galerkin's method and the method of least squares are applicable to non-self-adjoint operators, whereas for the Rayleigh-Ritz method, one has to formulate the variational principle not only for the original operator but also for its adjoint. For a non-self-adjoint operator, either Galerkin's method or the variational method may not give monotonic convergence for the solution.

Only the method of least squares guarantees the convergence of the residuals as the order of the approximation gets higher. This advantage, is however, offset by the fact that the method of least squares requires considerably more computations than the other two methods. For Galerkin's method, the convergence of the residual is guaranteed only under special circumstances. For a variational method, the problem of setting up the functional $F(y)$ is a very tedious process, and often, the functional may not be defined. Finally, it is shown that the variational method converges as fast as, if not better than, the method of least squares.

It is important to note that all inner products in a numerical method need to be evaluated numerically. In this respect, it is essential that one uses a larger number of points to evaluate the integrals than the number of unknowns; otherwise one will obtain a point-matched solution irrespective of which method one started out with in solving the problem. However, from a purely computational point of view, all techniques essentially yield a solution corresponding to a weighted point-matched technique if the inner products are evaluated using a numerical quadrature. If the inner products are evaluated analytically, then this criterion does not hold.

The concept of iterative methods, particularly the conjugate direction method, is introduced to illustrate how the round-off and truncation error can be compensated in a computational procedure, and the choice of the basis functions can even be implicit. Two versions of the conjugate gradient method are described in which one can minimize at each iteration the error between the exact solution and the estimated one, even though the exact solution is not known. In the second version of the conjugate gradient method, the equation error is minimized at each iteration.

Depending on the functional form of the Green's function, the application of the conjugate gradient method to the solution of a matrix equation can yield different results, as opposed to the application of the conjugate gradient method to the direct solution of an operator equation. The difference becomes obvious for the solution of integral equations in the time domain and for nonequally sampled functionals. However in the frequency domain, there is no difference between the application of an iterative method to the solution of the operator equation directly and to a matrix equation as the kernel, namely the Green's function is symmetric in this case. However, because the kernel is not symmetric in the solution of a time domain integral equation, the results will be different when an iterative method is applied directly to the solution of an operator equation than applying it to the solution of a matricized version of the operator equation because the adjoint of a convolutional operator is the correlation operator.

REFERENCES

[1]   F. Reza, *Linear Spaces In Engineering*, Ginn and Company, Waltham, Mass., 1971.

[2]   R. F. Harrington, *Methods of System Analysis*, Lecture Notes, Department of Electrical and Computer Engineering, Syracuse University, Syracuse, N.Y., 1971.

[3]   I. Stakgold, *Green's Functions and Boundary Value Problems*, John Wiley & Sons, New York, 1979.

[4]   C. Dorny, *A Vector Space Approach to Models and Optimization*, John Wiley & Sons, New York, 1975.

[5]   T. K. Sarkar, D. D. Wiener, and V. K. Jain, "Some mathematical considerations in dealing with the inverse problem," *IEEE Trans. Antennas Prop.*, Mar. 1981, pp. 373–379.

[6]   M. Z. Nashed, "Approximate regularized solutions to improperly posed linear integral and operator equations," in D. Colton and R. Gilbert, (eds.), *Constructive and Computational Methods for Differential and Integral Equations*, Springer, New York, 1974, pp. 289–332.

[7]   M. M. Lavrentiev, "Some improperly posed problems of mathematical physics," in *Tracts in Natural Philosophy*, Vol. II., Springer-Verlag, Berlin, Germany, 1967.

[8]   A. N. Tykhonov and V. Y. Arsenin, *Solutions of Ill-Posed Problems*, Wiley, New York, 1997.

[9]   D. L. Phillips, "A technique for the numerical solution of certain integral equations of the first kind," *J. Assoc. Comput. Mach.*, Vol. 9, pp. 84–97, 1962.

[10] S. Twomey, "On numerical solution of Fredholm integral equations of the first kind by the inversion of the linear system produced by quadrature," *J. Assoc. Comput. Mach.*, Vol. 10, pp. 97–101, 1963.

[11] S. Twomey, "The application of numerical filtering to the solution of integral equations encountered in indirect sensing measurements," *J. Franklin Inst.*, pp. 95–109, 1965.

[12] W. L. Perry, "Approximate solution of inverse problems with piecewise continuous solutions," *Radio Sci.*, Vol. 12, pp. 637–642, 1977.

[13] A. E. Hoerl and R. W. Kennard, "Ridge regression—Past, present, and future," *Int. Symp. Ill-Posed Problems: Theory and Practice*, University of Delaware, Newark, Oct. 1979.

[14] M. Z. Nashed, Ed., *Generalized Inverses and Applications*, Academic, New York, 1976.

[15] R. M. Lewis, "Physical optics inverse diffraction," *IEEE Trans. Antennas Prop.*, Vol. AP-17, pp. 308–314, 1969.

[16] N. N. Bojarski, *Three-Dimensional Electromagnetic Short Pulse Inverse Scattering*, Syracuse University Res. Corp., Syracuse, N.Y., 1967.

[17] W. L. Perry, "On the Bojarski-Lewis inverse scattering method," *IEEE Trans. Antennas Prop.*, Vol. AP-22, pp. 826–829, 1974.

[18] V. F. Turchin, V. P. Kozlov, and M. S. Malkevich, "The use of mathematical statistics methods in the solution of incorrectly posed problems," *Soviet Phys. Uspekhi*, Vol. 13, pp. 681–703, 1971.

[19] J. H. Franklin, "On Tikhonov's method for ill-posed problems," *Math. Comput.*, Vol. 28, pp. 889–907, 1974.

[20] J. H. Franklin, "Well-posed stochastic extensions of ill-posed linear problems," *J. Math. Anal. Appl.*, Vol. 31, pp. 682–716, 1970.

[21] O. N. Strand and E. R. Westwater, "Statistical estimation of the numerical solution of a Fredholm integral equation of the first kind," *J. Assoc. Comput. Mach.*, Vol. 15, pp. 100–114, 1968.

[22] R. F. Harrington, *Field Computation by Moment Methods*, Macmillan, London, U.K. 1968.

[23] K. Rektorys, *Variational Methods in Mathematical Sciences and Engineering*, D. Reidel, Hingham, Mass., 1972.

[24] S. G. Mikhlin, *Variational Methods in Mathematical Physics*, Pergamon Press, New York, 1964.

[25] S. G. Mikhlin, *The Problem of the Minimum of a Quadratic Functional*, Holden-Day, San Francisco, Calif., 1965.

[26] M. A. Krasnoselskii, G. M. Vainikko, P. P. Zabreiko, Y. B. Rutitskii, and V. Y. Stetwenko, *Approximate Solution of Operator Equations*, Walters-Noordhoff, Leyden, Groningen, 1972.

[27] T. K. Sarkar, "A note on the choice of weighting functions in the method of moments," *IEEE Trans. Antennas Prop.*, Vol. AP-33, pp. 436–441, 1985.

[28] T. K. Sarkar, A. R. Djordjević, and E. Arvas, "On the choice of expansion and weighting functions in the method of moments," *IEEE Trans. Antennas Prop.*, Vol. AP-33, pp. 988–996, 1985.

[29] T. K. Sarkar, "A study of the various methods for computing electromagnetic field utilizing the thin wire integral equations," *Radio Sci.*, Vol. 18, pp. 29–38, Jan. 1983.

[30] S. G. Mikhlin, *The Numerical Performance of Variational Methods*, Watters-Noordhoff, Leyden, Groningen, 1971.

[31] B. A. Finlayson and L. E. Scriven, "The method of weighted residuals—A review," *Appl. Mech. Rev.*, Vol. 19, pp. 735–748, 1966.

[32] S. G. Mikhlin, "Stability of some computing processes," *Sov. Math.*, Engl. Transl., Vol. 6, pp. 931–933, 1965.

[33] T. K. Sarkar, "A note on the variational method (Rayleigh-Ritz), Galerkin's method, and the method of least squares," *Radio Sci.*, Vol. 18, pp. 1207–1224, 1983.

[34] J. Singer, "On the equivalence of the Galerkin and the Rayleigh-Ritz methods," *J. R. Aeronaut. Soc.*, Vol. 66, p. 592, 1960.

[35] A. B. Stephens, "The convergence of residuals for projective approximations," *SIAM J. Numer. Anal.*, Vol. 13, pp. 607–614, 1976.

[36] P. Linz, *Theoretical Numerical Analysis*, Wiley, New York, 1979.

[37] T. L. Phillips, "The use of collocation as a projection method for ends. This in turn could be interpreted as a large excitation solving linear operator equations," *SIAM J. Numer. Anal.*, Vol. 9, pp. 14–28, 1972.

[38] O. K. Bogarayan, "Convergence of the residual of the Bubnov-Galerkin and Ritz methods," *Sov. Math.* 2, Engl. Transl., pp. 1413–1415, 1961.

[39] A. R. Djordjevic and T. K. Sarkar, "A theorem on the moment methods," *IEEE Trans. Antennas and Prop.*, Vol. AP-35, pp. 353–355, 1987.

[40] G. Golub and C. F. van Loan, *Matrix Computations*, Johns Hopkins Univ. Press, Baltimore, Md., 1983.

[41] B. Noble, "Error analysis of collocation methods for solving Fredholm integral equations," in J. J. Miller, (ed.), *Topics in Numerical Analysis*, London: Academic, New York, 1973, pp. 211–232.

[42] R. F. Harrington, "Matrix methods for field problems," *Proc. IEEE*, Vol. 55, pp. 136–149, 1967.

[43] M. K. Hestenes, "Applications of the theory of quadratic forms in Hilbert space in the calculus of variations," *Pacific J. Math.*, Vol. 1, pp. 525–581, 1951.

[44] K. M. Hayes, "Iterative methods for solving linear problems in Hilbert space," in O. Taussky (ed.), *Contributions to Solution of Systems of Linear Equations and Determination of Eigenvalues*, Vol. 39, Nat. Bur. Standards, Appl. Math., 1954, pp. 71–104.

[45] M. K. Hestenes and B. Steifel, "Method of conjugate gradients for solving linear systems," *J. Res. Nat. Bur. Standards*, Vol. 49, pp. 409–436, 1952.

[46] T. K. Sarkar, "The application of the conjugate gradient method for the solution of operator equations arising in electromagnetic scattering from wire antennas," *Radio Sci.*, Vol. 19, pp. 1156–1172, 1984.

[47] T. K. Sarkar, S. M. Rao, and S. A. Dianat, "The application of the conjugate gradient method to the solution of transient electromagnetic scattering from thin wires," *Radio Sci.*, Vol. 19, pp. 1319–1326, 1984.

[48] T. K. Sarkar and E. Arvas, "On a class of finite step iterative methods (conjugate directions) for the solution of an operator equation arising in electromagnetics," *IEEE Trans. Antennas Prop.*, Vol. 33, pp. 1058–1066, 1985.

[49] T. K. Sarkar, E. Arvas, and S. M. Rao, "Application of FFT and the Conjugate Gradient Method for the Solution of Electromagnetic Radiation from Electrically Large and Small Conducting Bodies," *IEEE Transactions on Antennas and Propagation*, Vol. 34, No. 5, pp. 635–640, May 1986.

[50] T. K. Sarkar (ed.), *Application of the Conjugate Gradient Method in Electromagnetics and Signal Processing*, Elsevier, New York, 1991.

[51] M. Hestenes, *Conjugate Direction Methods in Optimization*, Springer Verlag, Berlin, Germany, 1978.

[52] Y. Zhang and T. K. Sarkar, *Parallel Solution of Integral Equation-Based EM Problems in the Frequency Domain*, John Wiley & Sons, Hoboken, N.J., 2009.

[53] F. I. Tseng and T. K. Sarkar, "Deconvolution of the impulse response of a conducting sphere by the conjugate gradient method," *IEEE Trans. Antennas Prop.*, Vol. 35, pp. 105–110, 1987.

2

ANALYSIS OF CONDUCTING STRUCTURES IN THE FREQUENCY DOMAIN

2.0 SUMMARY

In this chapter, the numerical procedures for analyzing the electromagnetic scattering from an arbitrarily-shaped conducting object in the frequency domain are presented. Based on the boundary condition of the electric or the magnetic fields, two different integral equations for analyzing the scattering problems for conducting objects are derived and termed as the *electric field integral equation* (EFIE) and the *magnetic field integral equation* (MFIE). In addition, these two integral equations can be combined to form a more general formulation called the *combined field integral equation* (CFIE). The EFIE can be used for objects with both closed and open surfaces, whereas the MFIE and CFIE are suitable only for closed structures. All these equations can be represented as a function of the induced surface current density. The scattered field can be calculated using the current density obtained by solving one of these equations. The integral equations (i.e., EFIE, MFIE, and CFIE), can be solved numerically using the method of moments through the triangular patch modeling scheme using the well-known triangular patch or the Rao–Wilton–Glisson (RWG) basis functions. Once the current distribution on the structure is obtained, the fields scattered from the structure of interest can be evaluated.

2.1 INTRODUCTION

Research in electromagnetics (EM) has been pursued at a fast pace. The major efforts have been placed in the development of the general purpose computational methodology, yielding a good enough approximation for the exact solution. An important breakthrough is that the solution to the intricate EM formulations can be obtained more efficiently by taking full advantage of the fast digital computers available today. The evolution of computer hardware and software further enables the electromagnetic analysis for large or complicated targets to be completed in an acceptable amount of time [1]. Analyzing electromagnetic radiation and scattering problems based on the integral equations is one of the most efficient solution procedures used in scientific research and commercial EM simulation software packages. This chapter, as an introductory section, reviews several types of integral equations to analyze the radiation or scattering problems, particularly for a conducting object of arbitrary shape. (The formulations for dielectric and conducting-dielectric composite objects will be given in the next two chapters). These integral equations include the EFIE, MFIE, and the CFIE. Here they are considered in the frequency domain, whereas the time domain formulations will be addressed in the later chapters. The field integral equation contains integral operators and transformed into a matrix equation that can be solved more easily by some known matrix techniques. This method is called the method of moments (MoM), as introduced in Chapter 1. Recall that in MoM, each unknown is expanded in a series of well-defined functions, called the expansion functions or basis functions. Then, by taking the inner product with a set of weighting functions (also called testing functions), the integral equation can be transformed into a matrix equation. The solution may be numerically accurate, depending on the choice of the basis and testing functions. Furthermore, Galerkin's method can be applied by choosing testing functions as the same as the basis functions. For numerical purposes, the objects are modeled using planar surface patches with the triangular shapes (i.e., many small triangular patches are used to approximately form the shape of the object). Once the formulation for the surface current over each patch is derived, the overall current as well as the scattered field from the object can be calculated. A set of basis functions and testing functions for the triangular patch model, which are called the RWG basis functions, has been proposed and has been demonstrated to obtain a good approximate solution to the integral equation. In this chapter, the procedures related to MoM, the triangular surface patch modeling, and use of the RWG basis functions for the analysis of scattering from conducting objects will be presented.

2.2   FREQUENCY DOMAIN ELECTRIC FIELD INTEGRAL EQUATION (FD-EFIE)

In this section, the EFIE in the frequency domain is used to describe how induced surface current is distributed on a conducting object when it is illuminated by an electromagnetic plane wave. As shown in Figure 2.1, the surface of a *perfectly-electric-conducting* (PEC) body is denoted by S, which may be either closed or open. This incident wave induces a surface current density $\mathbf{J}$ on S. Because the total tangential electric field must be zero on the conducting surface, it results in

$$\left(\mathbf{E}^i(\mathbf{r}) + \mathbf{E}^s(\mathbf{r}) \right)_{\tan} = 0 \ , \quad \mathbf{r} \in S \tag{2.1}$$

where $\mathbf{E}^i$ and $\mathbf{E}^s$, respectively, are the incident and the scattered electric fields. The subscript 'tan' denotes the tangential component. The scattered electric field $\mathbf{E}^s$ can be computed from the surface current as

$$\mathbf{E}^s(\mathbf{r}) = -j\omega\mathbf{A}(\mathbf{r}) - \nabla\Phi(\mathbf{r}) \tag{2.2}$$

where $\mathbf{A}$ is the magnetic vector potential and Φ is the electric scalar potential given by

$$\mathbf{A}(\mathbf{r}) = \frac{\mu}{4\pi}\int_S \mathbf{J}(\mathbf{r}')G(\mathbf{r},\mathbf{r}')\,dS' \tag{2.3}$$

$$\Phi(\mathbf{r}) = \frac{1}{4\pi\varepsilon}\int_S Q(\mathbf{r}')G(\mathbf{r},\mathbf{r}')\,dS' \tag{2.4}$$

$$G(\mathbf{r},\mathbf{r}') = \frac{e^{-jkR}}{R} \tag{2.5}$$

and $R = |\mathbf{r} - \mathbf{r}'|$ represents the distance between the observation point $\mathbf{r}$ and the source point $\mathbf{r}'$. The parameters μ and ε are the permeability and permittivity of the space, respectively, and k is the wave number defined as $k = \omega\sqrt{\mu\varepsilon}$. The term $G(\mathbf{r},\mathbf{r}')$ is the Green's function for a homogenous medium. In the previous equation, harmonic time dependence $e^{j\omega t}$ is assumed and has been suppressed for convenience.

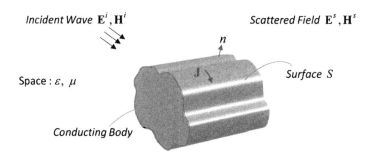

Figure 2.1. A conducting body illuminated by an electromagnetic plane wave.

The surface charge density Q, related to the surface divergence of $\mathbf{J}$ by the equation of continuity is given by

$$\nabla_S \cdot \mathbf{J}(\mathbf{r}) = -j\omega Q(\mathbf{r}) \tag{2.6}$$

Substituting Eq. (2.6) into Eq. (2.4) yields the following expression for the electric scalar potential:

$$\Phi(\mathbf{r}) = \frac{j}{4\pi\omega\varepsilon} \int_S \nabla'_S \cdot \mathbf{J}(\mathbf{r}') G(\mathbf{r},\mathbf{r}') \, dS' \tag{2.7}$$

Then, substituting Eq. (2.2) into Eq. (2.1) and grouped with Eqs. (2.3) and (2.7), the relation in the frequency domain can be obtained between the incident electric field and the induced surface current density. It is generally known as the frequency domain EFIE.

$$\left(j\omega\mathbf{A}(\mathbf{r}) + \nabla\Phi(\mathbf{r})\right)_{\text{tan}} = \left(\mathbf{E}^i(\mathbf{r})\right)_{\text{tan}} \tag{2.8a}$$

$$\mathbf{A}(\mathbf{r}) = \frac{\mu}{4\pi} \int_S \mathbf{J}(\mathbf{r}') G(\mathbf{r},\mathbf{r}') \, dS' \tag{2.8b}$$

$$\Phi(\mathbf{r}) = \frac{j}{4\pi\omega\varepsilon} \int_S \nabla'_S \cdot \mathbf{J}(\mathbf{r}') G(\mathbf{r},\mathbf{r}') \, dS' \tag{2.8c}$$

where $\mathbf{r} \in S$. This EFIE cannot be solved exactly using analytical methods only and thus a numerical solution is often used instead. One accurate and efficient method to numerically solve the unknown surface current in the previous EFIE is to apply the MoM along with the triangular surface patch model and the RWG basis functions. The basic concept of MoM, which has been given in Chapter 1, is to convert an operator equation into a matrix equation using a set of basis and testing functions. To determine the current, the surface of the object is usually modeled by the combination of many small triangular patches. The calculation will be carried out patch by patch, and the overall surface current can be obtained by summing up the contributions over all the patches. The accuracy of the MoM is indeed dependent on the selection of basis and testing functions. One common choice is the RWG function. The detailed procedures of the numerical method will be presented in the following sections.

2.3   TRIANGULAR SURFACE PATCH MODELING

To model an arbitrarily shaped surface, planar triangular patch modeling is particularly appropriate, as it is can accurately conform to any geometrical surface or boundary [2]. This patching scheme is easily specified as a possible computer input; a varying patch density can be used according to the resolution required in the surface geometry or induced current, and mathematically simple basis functions can be defined with no bothersome line charges along the boundary. Detailed considerations for patch modeling of arbitrary surfaces may be seen in [3]. Assuming that the structure to be studied is approximated by many planar triangular patches, consider a pair of triangles T_n^+ and T_n^- connected with the nth nonboundary edge, as illustrated in Figure 2.2.

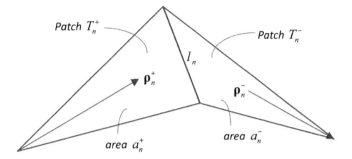

Figure 2.2. A pair of triangular patches associated with a nonboundary edge.

In Figure 2.2, l_n is the length of the nonboundary edge (edge common to both triangles) and a_n^+ and a_n^- are the areas of the triangles T_n^+ and T_n^-, respectively. Vectors $\boldsymbol{\rho}_n^+$ and $\boldsymbol{\rho}_n^-$ are the position vectors defined with respect to the free vertices (vertices not connected to the common edge) of T_n^+ and T_n^-, respectively. The position vector $\boldsymbol{\rho}_n^+$ is oriented from the free vertex of T_n^+ to any point inside the triangle. Similar remarks can be applied to the position vector $\boldsymbol{\rho}_n^-$, except its direction is toward the free vertex from any point in T_n^-. The plus or minus designation of the triangles is defined based on the assumption that the positive direction of the current flowing is from the triangle T_n^+ to T_n^-. From the properties of the RWG basis [2], the expansion function associated with the nth edge is defined on the pair of adjacent triangular patches as

$$\mathbf{f}_n(\mathbf{r}) = \mathbf{f}_n^+(\mathbf{r}) + \mathbf{f}_n^-(\mathbf{r}) \tag{2.9a}$$

$$\mathbf{f}_n^\pm(\mathbf{r}) = \begin{cases} \dfrac{l_n}{2a_n^\pm}\,\boldsymbol{\rho}_n^\pm, & \mathbf{r} \in T_n^\pm \\[2mm] 0, & \mathbf{r} \notin T_n^\pm \end{cases} \tag{2.9b}$$

Then the surface divergence of this basis function can be calculated [2] using $\left(\pm 1/\rho_n^\pm\right)\dfrac{\partial\left(\rho_n^\pm \mathbf{f}_n\right)}{\partial\boldsymbol{\rho}_n^\pm}$, which results in

$$\nabla_S \cdot \mathbf{f}_n(\mathbf{r}) = \nabla_S \cdot \mathbf{f}_n^+(\mathbf{r}) + \nabla_S \cdot \mathbf{f}_n^-(\mathbf{r}) \tag{2.10a}$$

$$\nabla_S \cdot \mathbf{f}_n^\pm(\mathbf{r}) = \begin{cases} \pm\dfrac{l_n}{a_n^\pm}, & \mathbf{r} \in T_n^\pm \\[2mm] 0, & \mathbf{r} \notin T_n^\pm \end{cases} \tag{2.10b}$$

With the defined expansion function, the solution of the EFIE can be found by the MoM.

2.4   SOLUTION OF THE EFIE

As has been introduced in Chapter 1, a general operator equation given by

$$\mathcal{L}\mathbf{x} = \mathbf{y} \tag{2.11}$$

can be solved by using the method of moment. In this section, the MoM steps are outlined to provide a more complete description of this solution procedure. The MoM starts with using a set of expansion functions or basis functions $\mathbf{f}_n$ to represent the unknown

$$\mathbf{x} = \sum_{i=1}^{N} \alpha_i \mathbf{f}_i \tag{2.12}$$

The operator equation in Eq. (2.11) is then expressed by

$$\mathcal{L}\mathbf{x} \approx \sum_{i=1}^{N} \alpha_i \mathcal{L}\mathbf{f}_i = \mathbf{y} \tag{2.13}$$

To weigh the approximation error to zero with respect to a certain set of weighting functions or testing functions $\mathbf{W}_j$, the inner product of Eq. (2.13) with $\mathbf{W}_j$ results in

$$\sum_{i=1}^{N} \alpha_i \left\langle \mathcal{L}\mathbf{f}_i, \mathbf{W}_j \right\rangle = \left\langle \mathbf{y}, \mathbf{W}_j \right\rangle \quad \text{for} \quad j = 1, 2, \ldots, N \tag{2.14}$$

This set of equations can be represented in a matrix form as

$$\left[L_{ij} \right] \left[\alpha_i \right] = \left[y_j \right] \tag{2.15}$$

In Eq. (2.15), $[\alpha]$ is the unknown to be solved for a given $[y]$ and, thus, leads to the solution of the original operator equation in Eq. (2.11). In the testing procedure, the complex inner products are used. Some important properties of the inner product are listed in section 1.3.5. From Eq. (1.9), the inner product for two vector functions $\mathbf{f}$ and $\mathbf{g}$ is defined as

$$\langle \mathbf{f}, \mathbf{g} \rangle = \int_S \mathbf{f} \cdot \mathbf{g}^* dS \tag{2.16}$$

If both the functions $\mathbf{f}$ and $\mathbf{g}$ contain two parts (i.e., $\mathbf{f} = \mathbf{f}^+ + \mathbf{f}^-$ and $\mathbf{g}^+ = \mathbf{g}^+ + \mathbf{g}^-$), then their inner product can be represented by

$$\langle \mathbf{g}, \mathbf{f} \rangle = \langle \mathbf{g}^+, \mathbf{f}^+ \rangle + \langle \mathbf{g}^+, \mathbf{f}^- \rangle + \langle \mathbf{g}^-, \mathbf{f}^+ \rangle + \langle \mathbf{g}^-, \mathbf{f}^- \rangle$$

$$= \sum_{p,q} \langle \mathbf{g}^p, \mathbf{f}^q \rangle \tag{2.17a}$$

$$\langle \mathbf{g}, \mathcal{A}\mathbf{f} \rangle = \langle \mathbf{g}^+, \mathcal{A}\mathbf{f}^+ \rangle + \langle \mathbf{g}^+, \mathcal{A}\mathbf{f}^- \rangle + \langle \mathbf{g}^-, \mathcal{A}\mathbf{f}^+ \rangle + \langle \mathbf{g}^-, \mathcal{A}\mathbf{f}^- \rangle$$

$$= \sum_{p,q} \langle \mathbf{g}^p, \mathcal{A}\mathbf{f}^q \rangle \tag{2.17b}$$

where the superscripts p and q can be $+$ or $-$. These definitions are useful to express the inner products in a more compact form as demonstrated in the following discussions.

Now consider solving the electric field integral equation of Eq. (2.8) using the MoM. The induced surface current $\mathbf{J}$ may be approximated by a set of expansion functions as

$$\mathbf{J}(\mathbf{r}) = \sum_{n=1}^{N} J_n \mathbf{f}_n(\mathbf{r}) \tag{2.18}$$

where N represents the number of common edges. Substituting the current $\mathbf{J}$ in Eq. (2.8) yields

$$\left(j\omega \mathbf{A}(\mathbf{r}) + \nabla \Phi(\mathbf{r}) \right)_{\text{tan}} = \left(\mathbf{E}^i(\mathbf{r}) \right)_{\text{tan}} \tag{2.19a}$$

$$\mathbf{A}(\mathbf{r}) = \sum_{n=1}^{N} \left[\frac{\mu}{4\pi} \int_S \mathbf{f}_n(\mathbf{r}') G(\mathbf{r}, \mathbf{r}') \, dS' \right] J_n \tag{2.19b}$$

$$\Phi(\mathbf{r}) = \sum_{n=1}^{N} \left[\frac{j}{4\pi\omega\varepsilon} \int_S \nabla'_S \cdot \mathbf{f}_n(\mathbf{r}') G(\mathbf{r}, \mathbf{r}') \, dS' \right] J_n \tag{2.19c}$$

Next apply Galerkin's method in the MoM context by selecting the testing function that is the same as the basis function. Applying the testing procedure, with the testing function $\mathbf{f}_m$, we obtain

$$\langle \mathbf{f}_m, j\omega \mathbf{A}(\mathbf{r}) \rangle + \langle \mathbf{f}_m, \nabla \Phi(\mathbf{r}) \rangle = \langle \mathbf{f}_m, \mathbf{E}^i(\mathbf{r}) \rangle \tag{2.20}$$

for $m = 1, 2, \dots, N$. The first term in Eq. (2.20) becomes

$$\langle \mathbf{f}_m, j\omega \mathbf{A}(\mathbf{r}) \rangle = \sum_{n=1}^{N} \left[\frac{j\omega\mu}{4\pi} \int_S \mathbf{f}_m(\mathbf{r}) \cdot \int_S \mathbf{f}_n(\mathbf{r}') \, G(\mathbf{r},\mathbf{r}') \, dS' \, dS \right] J_n \qquad (2.21)$$

or equivalently

$$\langle \mathbf{f}_m, j\omega \mathbf{A}(\mathbf{r}) \rangle = j\omega\mu \sum_{n=1}^{N} A_{mn} J_n \qquad (2.22a)$$

$$A_{mn} = \frac{1}{4\pi} \int_S \mathbf{f}_m(\mathbf{r}) \cdot \int_S \mathbf{f}_n(\mathbf{r}') G(\mathbf{r},\mathbf{r}') \, dS' \, dS \qquad (2.22b)$$

The second term in Eq. (2.20) contains the testing of the gradient of the electric scalar potential. This task can be accomplished by using the vector identity $\nabla \cdot (\phi \mathbf{A}) = \mathbf{A} \cdot (\nabla \phi) + \phi (\nabla \cdot \mathbf{A})$ or $\mathbf{A} \cdot (\nabla \phi) = \nabla \cdot (\phi \mathbf{A}) - \phi(\nabla \cdot \mathbf{A})$.

$$\langle \mathbf{f}_m, \nabla \Phi(\mathbf{r}) \rangle = \int_S \nabla_S \cdot \left(\mathbf{f}_m \, \Phi(\mathbf{r}) \right) dS - \int_S \left(\nabla_S \cdot \mathbf{f}_m \right) \Phi(\mathbf{r}) \, dS \qquad (2.23)$$

The first term in Eq. (2.23) can be converted into a contour integral using the Stoke's theorem. Because the normal component of the basis function is zero along the contour [3], we have

$$\int_S \nabla_S \cdot \left(\mathbf{f}_m \, \Phi(\mathbf{r}) \right) dS = 0 \qquad (2.24)$$

Therefore, the second term in Eq. (2.20) can be written as

$$\langle \mathbf{f}_m, \nabla \Phi(\mathbf{r}) \rangle = - \int_S \left(\nabla_S \cdot \mathbf{f}_m \right) \Phi(\mathbf{r}) \, dS \qquad (2.25)$$

Now substitute the scalar potential Φ given by Eq. (2.19c) into Eq. (2.25)

$$\langle \mathbf{f}_m, \nabla \Phi(\mathbf{r}) \rangle = - \sum_{n=1}^{N} \left[\frac{j}{4\pi\omega\varepsilon} \int_S \nabla_S \cdot \mathbf{f}_m(\mathbf{r}) \int_S \nabla_S' \cdot \mathbf{f}_n(\mathbf{r}') G(\mathbf{r},\mathbf{r}') \, dS' \, dS \right] J_n \qquad (2.26)$$

resulting in

$$\langle \mathbf{f}_m, \nabla \Phi(\mathbf{r}) \rangle = - \frac{j}{\omega\varepsilon} \sum_{n=1}^{N} B_{mn} J_n \qquad (2.27a)$$

$$B_{mn} = \frac{1}{4\pi} \int_S \nabla_S \cdot \mathbf{f}_m(\mathbf{r}) \int_S \nabla'_S \cdot \mathbf{f}_n(\mathbf{r}') G(\mathbf{r},\mathbf{r}') \, dS' \, dS \tag{2.27b}$$

Similarly, the term on the right side of Eq. (2.20), which relates to the testing of the incident field, becomes

$$\langle \mathbf{f}_m, \mathbf{E}^i(\mathbf{r}) \rangle = \int_S \mathbf{f}_m(\mathbf{r}) \cdot \mathbf{E}^i(\mathbf{r}) \, dS = V_m \tag{2.28}$$

As a result of employing the MoM into the EFIE given in Eq. (2.8), we have

$$j\omega\mu \sum_{n=1}^N A_{mn} J_n - \frac{j}{\omega\varepsilon} \sum_{n=1}^N B_{mn} J_n = V_m \tag{2.29a}$$

where

$$A_{mn} = \frac{1}{4\pi} \int_S \mathbf{f}_m(\mathbf{r}) \cdot \int_S \mathbf{f}_n(\mathbf{r}') G(\mathbf{r},\mathbf{r}') dS' \, dS \tag{2.29b}$$

$$B_{mn} = \frac{1}{4\pi} \int_S \nabla_S \cdot \mathbf{f}_m(\mathbf{r}) \int_S \nabla'_S \cdot \mathbf{f}_n(\mathbf{r}') G(\mathbf{r},\mathbf{r}') dS' \, dS \tag{2.29c}$$

$$V_m = \int_S \mathbf{f}_m(\mathbf{r}) \cdot \mathbf{E}^i(\mathbf{r}) dS \tag{2.29d}$$

When the triangular patch modeling is used, the basis function $\mathbf{f}_n$ and the testing function $\mathbf{f}_m$ have both the + and − components. Eq. (2.29) is simplified by using the summation operator given in Eq. (2.17), resulting in

$$j\omega\mu \sum_{n=1}^N A_{mn} J_n - \frac{j}{\omega\varepsilon} \sum_{n=1}^N B_{mn} J_n = V_m \tag{2.30a}$$

where

$$A_{mn} = \sum_{p,q} A_{mn}^{pq} = \sum_{p,q} \left(\frac{1}{4\pi} \int_S \mathbf{f}_m^p(\mathbf{r}) \cdot \int_S \mathbf{f}_n^q(\mathbf{r}') G(\mathbf{r},\mathbf{r}') dS' \, dS \right) \tag{2.30b}$$

$$B_{mn} = \sum_{p,q} B_{mn}^{pq} = \sum_{p,q} \left(\frac{1}{4\pi} \int_S \nabla_S \cdot \mathbf{f}_m^p(\mathbf{r}) \int_S \nabla'_S \cdot \mathbf{f}_n^q(\mathbf{r}') G(\mathbf{r},\mathbf{r}') \, dS' \, dS \right) \tag{2.30c}$$

$$V_m = \sum_p V_m^p = \sum_p \left(\int_S \mathbf{f}_m^p(\mathbf{r}) \cdot \mathbf{E}^i(\mathbf{r}) dS \right) \tag{2.30d}$$

Substituting the basis and testing functions on the triangular patches defined in Eq. (2.9b) and their divergence given by Eq. (2.10b) into Eq. (2.30) and approximating the average charge density by the value of ρ at the centroid of the triangular patch and, in addition, approximating the distance by r_m^{cp} is equivalent to replacing the integral over the testing triangle by its value at the centroid as seen in Figure 2.3 [2]. In addition, one obtains

$$(a_n^+ + a_n^-)\mathbf{f}_n^{avg} \equiv \int_{T_n^+ + T_n^-} \mathbf{f}_n \, dS = \frac{l_n}{2}(\boldsymbol{\rho}_n^{c+} + \boldsymbol{\rho}_n^{c-}) = l_n(\mathbf{r}_n^{c+} - \mathbf{r}_n^{c-}) \tag{2.31a}$$

resulting in

$$j\omega\mu\sum_{n=1}^{N} A_{mn} J_n - \frac{j}{\omega\varepsilon}\sum_{n=1}^{N} B_{mn} J_n = V_m \tag{2.31b}$$

where

$$A_{mn} = \sum_{p,q} A_{mn}^{pq} = \sum_{p,q} \left(\frac{l_m l_n}{16\pi} \boldsymbol{\rho}_m^{cp} \cdot \frac{1}{a_n^q} \int_{T_n^q} \boldsymbol{\rho}_n^q \frac{e^{-jkR_m^p}}{R_m^p} \, dS' \right) \tag{2.31c}$$

$$B_{mn} = \sum_{p,q} B_{mn}^{pq} = \pm\sum_{p,q} \left(\frac{l_m l_n}{4\pi} \frac{1}{a_n^q} \int_{T_n^q} \frac{e^{-jkR_m^p}}{R_m^p} \, dS' \right) \tag{2.31d}$$

$$V_m = \sum_{p} V_m^p = \sum_{p} \frac{l_m}{2} \boldsymbol{\rho}_m^{cp} \cdot \mathbf{E}^i(\mathbf{r}_m^{cp}) \tag{2.31e}$$

and $R_m^p = |\mathbf{r}_m^{cp} - \mathbf{r}'|$. The sign in Eq. (2.31$d$) is positive, when p and q are the same and negative otherwise.

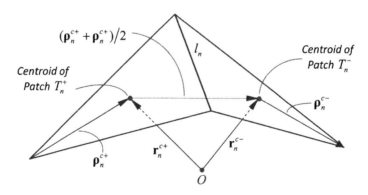

Figure 2.3. Centroid approximations on the triangular patches.

The integrals in Eqs. (2.31c) and (2.31d) may be evaluated by the method described in [4]. Furthermore, Eq. (2.31b) can be expressed as

$$\sum_{n=1}^{N} Z_{mn} J_n = V_m \tag{2.32}$$

where

$$Z_{mn} = j\omega \mu A_{mn} - \frac{j}{\omega \varepsilon} B_{mn} = jk\eta \left(A_{mn} - \frac{B_{mn}}{k^2} \right) \tag{2.33}$$

In Eq. (2.33), the parameter k is the wave number and parameter η is the intrinsic wave impedance defined as $\eta = \sqrt{\mu / \varepsilon}$. Equation (2.32) is associated with each common edge, for $m = 1, 2, ..., N$, and it finally may be written in a matrix form as

$$[Z_{mn}][J_n] = [V_m] \tag{2.34}$$

where $[Z_{mn}]$ is a $N \times N$ matrix and $[J_n]$ and $[V_m]$ are column vectors of length N. The solution of Eq. (2.34) is a column vector of the coefficients of current expansion $[J_n]$, and thus, it can lead to the solution of current density as well as the scattered electric field.

2.5   FREQUENCY DOMAIN MAGNETIC FIELD INTEGRAL EQUATION (FD-MFIE)

Applying the boundary condition for the magnetic field, a MFIE for a closed conducting body may be derived as follows. Let S represent the surface of a closed conducting body that is illuminated by an electromagnetic wave. Also, let $\mathbf{H}^i$ represent the incident magnetic field and $\mathbf{H}^s$ represent the scattered magnetic field caused by the induced current $\mathbf{J}$. Then we have,

$$\mathbf{n} \times \left(\mathbf{H}^i(\mathbf{r}) + \mathbf{H}^s(\mathbf{r}) \right) = \mathbf{J}(\mathbf{r}), \ \mathbf{r} \in S \tag{2.35}$$

where $\mathbf{n}$ represents an outward-directed unit vector normal to the surface S at a field point $\mathbf{r}$. Note that the body must be a closed body to define a unique outward normal vector. The scattered field can be written in terms of the magnetic vector potential function, given by

$$\mathbf{H}^s(\mathbf{r}) = \frac{1}{\mu} \nabla \times \mathbf{A}(\mathbf{r}) \tag{2.36}$$

It is noted that the magnetic vector potential $\mathbf{A}$ is the same as in Eq. (2.3). Applying the Cauchy principal value to the curl term, Eq. (2.36) can be rewritten as [5]

$$\mathbf{n} \times \mathbf{H}^s(\mathbf{r}) = \frac{\mathbf{J}(\mathbf{r})}{2} + \mathbf{n} \times \frac{1}{\mu} \nabla \times \tilde{\mathbf{A}}(\mathbf{r}) \tag{2.37}$$

where

$$\tilde{\mathbf{A}}(\mathbf{r}) = \frac{\mu}{4\pi} \int_{S_0} \mathbf{J}(\mathbf{r}') G(\mathbf{r},\mathbf{r}') \, dS' \tag{2.38}$$

In Eq. (2.38), S_0 denotes the surface with the removal of the singularity at $\mathbf{r} = \mathbf{r}'$ (or $R = 0$) from the surface S. Note that Eq. (2.35) along with Eq. (2.37) represents the magnetic field integral equation. Now substituting Eq. (2.37) into Eq. (2.35), we obtain the relation between the incident magnetic field and the induced surface current density, and it is generally called the frequency domain MFIE. It is given by

$$\frac{\mathbf{J}(\mathbf{r})}{2} - \mathbf{n} \times \frac{1}{\mu} \nabla \times \tilde{\mathbf{A}}(\mathbf{r}) = \mathbf{n} \times \mathbf{H}^i(\mathbf{r}) \tag{2.39}$$

2.6   SOLUTION OF THE MFIE

Employing the testing procedure of the MoM to Eq. (2.39) results in the matrix equation

$$\left\langle \mathbf{f}_m, \frac{\mathbf{J}(\mathbf{r})}{2} \right\rangle - \left\langle \mathbf{f}_m, \mathbf{n} \times \frac{1}{\mu} \nabla \times \tilde{\mathbf{A}}(\mathbf{r}) \right\rangle = \left\langle \mathbf{f}_m, \mathbf{n} \times \mathbf{H}^i(\mathbf{r}) \right\rangle \tag{2.40}$$

The surface current is approximated by the current expansion functions,

$$\mathbf{J}(\mathbf{r}) = \sum_{n=1}^{N} J_n \mathbf{f}_n(\mathbf{r})$$

as described in Eq. (2.9). The first term of the inner product in Eq. (2.40) yields

$$\left\langle \mathbf{f}_m, \frac{\mathbf{J}(\mathbf{r})}{2} \right\rangle = \frac{1}{2} \sum_{n=1}^{N} \left(\int_S \mathbf{f}_m(\mathbf{r}) \cdot \mathbf{f}_n(\mathbf{r}) \, dS \right) J_n \tag{2.41}$$

or equivalently

$$\left\langle \mathbf{f}_m, \frac{\mathbf{J}(\mathbf{r})}{2} \right\rangle = \sum_{n=1}^{N} C_{mn} J_n \tag{2.42a}$$

$$C_{mn} = \frac{1}{2} \int_S \mathbf{f}_m(\mathbf{r}) \cdot \mathbf{f}_n(\mathbf{r}) \, dS = \frac{1}{2} \langle \mathbf{f}_m, \mathbf{f}_n \rangle \qquad (2.42b)$$

The second term of Eq. (2.40) can be rewritten by substituting the magnetic vector potential $\tilde{\mathbf{A}}(\mathbf{r})$ given in Eq. (2.39b) as

$$\left\langle \mathbf{f}_m, \frac{\mathbf{n} \times \nabla \times \tilde{\mathbf{A}}(\mathbf{r})}{\mu} \right\rangle = \frac{1}{4\pi} \int_S \mathbf{f}_m(\mathbf{r}) \cdot \mathbf{n} \times \int_{S_0} \nabla \times \left(\mathbf{J}(\mathbf{r}') G(\mathbf{r}, \mathbf{r}') \right) dS' \, dS \qquad (2.43)$$

where the curl operator is taken inside the integral for the case $\mathbf{r}$ not equal to $\mathbf{r}'$ only. Using a vector identity, the curl term inside the integral is given by

$$\nabla \times \left(\mathbf{J}(\mathbf{r}') \, G(\mathbf{r}, \mathbf{r}') \right) = \mathbf{J}(\mathbf{r}') \times \nabla' G(\mathbf{r}, \mathbf{r}') \qquad (2.44)$$

Substituting Eq. (2.44) into Eq. (2.43) and using the expansion for the current, one obtains

$$\left\langle \mathbf{f}_m, \frac{\mathbf{n} \times \nabla \times \tilde{\mathbf{A}}(\mathbf{r})}{\mu} \right\rangle = \sum_{n=1}^{N} \left[\frac{1}{4\pi} \int_S \mathbf{f}_m(\mathbf{r}) \cdot \mathbf{n} \times \int_S \mathbf{f}_n(\mathbf{r}') \times \nabla' G(\mathbf{r}, \mathbf{r}') \, dS' \, dS \right] J_n \qquad (2.45)$$

which results in

$$\left\langle \mathbf{f}_m, \frac{\mathbf{n} \times \nabla \times \tilde{\mathbf{A}}(\mathbf{r})}{\mu} \right\rangle = \sum_{n=1}^{N} D_{mn} J_n \qquad (2.46a)$$

$$D_{mn} = \frac{1}{4\pi} \int_S \mathbf{f}_m(\mathbf{r}) \cdot \mathbf{n} \times \int_S \mathbf{f}_n(\mathbf{r}') \times \nabla' G(\mathbf{r}, \mathbf{r}') \, dS' \, dS \,. \qquad (2.46b)$$

Finally, the term on the right side of Eq. (2.40), which relates to the testing of the incident field, yields

$$\left\langle \mathbf{f}_m, \mathbf{n} \times \mathbf{H}^i(\mathbf{r}) \right\rangle = \int_S \mathbf{f}_m(\mathbf{r}) \cdot \mathbf{n} \times \mathbf{H}^i(\mathbf{r}) \, dS = V_m \qquad (2.47)$$

As a result of employing the MoM into the MFIE given in Eq. (2.40), we have

$$\sum_{n=1}^{N} C_{mn} J_n - \sum_{n=1}^{N} D_{mn} J_n = V_m \qquad (2.48a)$$

where

$$C_{mn} = \frac{1}{2}\int_S \mathbf{f}_m(\mathbf{r})\cdot\mathbf{f}_n(\mathbf{r})\,dS = \frac{1}{2}\langle \mathbf{f}_m, \mathbf{f}_n\rangle \qquad (2.48b)$$

$$D_{mn} = \frac{1}{4\pi}\int_S \mathbf{f}_m(\mathbf{r})\cdot\mathbf{n}\times\int_S \mathbf{f}_n(\mathbf{r}')\times\nabla'G(\mathbf{r},\mathbf{r}')\,dS'\,dS \qquad (2.48c)$$

$$V_m = \int_S \mathbf{f}_m(\mathbf{r})\cdot\mathbf{n}\times\mathbf{H}^i(\mathbf{r})\,dS \qquad (2.48d)$$

When the RWG modeling is used, the basis function $\mathbf{f}_n$ and testing function $\mathbf{f}_m$ have both $+$ and $-$ components. Thus, Eq. (2.48) results in

$$\sum_{n=1}^N C_{mn}J_n - \sum_{n=1}^N D_{mn}J_n = V_m \qquad (2.49a)$$

where

$$C_{mn} = \sum_{p,q} C_{mn}^{pq} = \sum_{p,q}\left(\frac{1}{2}\int_S \mathbf{f}_m^p\cdot\mathbf{f}_n^q\,dS\right) \qquad (2.49b)$$

$$D_{mn} = \sum_{p,q} D_{mn}^{pq} = \sum_{p,q}\left(\frac{1}{4\pi}\int_S \mathbf{f}_m^p(\mathbf{r})\cdot\mathbf{n}\times\int_S \mathbf{f}_n^q(\mathbf{r}')\times\nabla'G(\mathbf{r},\mathbf{r}')\,dS'\,dS\right) \qquad (2.49c)$$

$$V_m = \sum_p V_m^p = \sum_p\left(\int_S \mathbf{f}_m^p(\mathbf{r})\cdot\mathbf{n}\times\mathbf{H}^i(\mathbf{r})\,dS\right) \qquad (2.49d)$$

Substituting the basis and testing functions on the triangular patches in Eq. (2.9b) into Eq. (2.49) and approximating the average charge density by the value of ρ at the centroid of the triangular patch, and in addition approximating the distance by $\mathbf{r}_m^{cp}$, one obtains using the procedure described in [2]

$$\sum_{n=1}^N C_{mn}J_n - \sum_{n=1}^N D_{mn}J_n = V_m \qquad (2.50a)$$

where

$$C_{mn}^{pq} = \pm\frac{l_m l_n}{8a_n^q}\left(\begin{array}{c}\dfrac{3}{4}\left|\mathbf{r}_m^{cp}\right|^2 + \dfrac{1}{12}\left(\left|\mathbf{r}_{m1}\right|^2 + \left|\mathbf{r}_{m2}\right|^2 + \left|\mathbf{r}_{m3}\right|^2\right) \\ -\mathbf{r}_m^{cp}\left(\mathbf{r}_{m1}+\mathbf{r}_{n1}\right)+\mathbf{r}_{m1}\mathbf{r}_{n1}\end{array}\right) \qquad (2.50b)$$

$$D_{mn}^{pq} = \frac{l_m l_n}{16\pi a_m^p a_n^q}\int_S \boldsymbol{\rho}_m^p\cdot\mathbf{n}\times\int_S \boldsymbol{\rho}_n^q\times\mathbf{R}(1+jkR)\frac{e^{-jkR}}{R^3}\,dS'\,dS \qquad (2.50c)$$

$$V_m^p = \frac{l_m}{2}\,\boldsymbol{\rho}_m^{cp}\cdot\mathbf{n}\times\mathbf{H}^i(\mathbf{r}_m^{cp}) \qquad (2.50d)$$

where $\mathbf{r}_{m1}$, $\mathbf{r}_{m2}$, and $\mathbf{r}_{m3}$ are the position vectors from the three vertices of the triangle T_m^P, and $\mathbf{r}_m^{cp}$ is the centroid of triangle T_m^P. $\mathbf{r}_{m1}$ and $\mathbf{r}_{n1}$ are position vectors of the free vertex of triangle T_m^P and T_n^q, respectively. Note that if the field point does not lie on the triangle T_n^q (i.e., $\mathbf{r} \notin T_n^q$) the result is $C_{mn}^{pq} = 0$. In Eq. (2.49), the sign is positive when p and q are the same and is negative otherwise. The expression in Eq. (2.50b) is zero if m and n do not share a common triangle, and it is twice the value if $m = n$. The integral in Eq. (2.50c) may be computed numerically using the Gaussian quadrature. In Eq. (2.50d), $\mathbf{n}$ is associated with the source triangle. Furthermore, Eq. (2.50a) can be expressed as

$$\sum_{n=1}^{N} Z_{mn} J_n = V_m \tag{2.51}$$

where

$$Z_{mn} = C_{mn} - D_{mn} \tag{2.52}$$

Equation (2.51) is associated with each common edge, for $m = 1, 2, ..., N$. Finally, it may be written in a matrix form as

$$[Z_{mn}][J_n] = [V_m] \tag{2.53}$$

where $[Z_{mn}]$ is a $N \times N$ matrix and $[J_n]$ and $[V_m]$ are column vectors of length N. The solution of Eq. (2.53) is a column vector of the coefficients of the current expansion $[J_n]$, and thus, it can provide a solution to the current density and leads to the analysis of the scattered field. Comparing Eq. (2.53) with Eq. (2.34), it is noted that both the MFIE and EFIE have a similar form.

2.7   FREQUENCY DOMAIN COMBINED FIELD INTEGRAL EQUATION (FD-CFIE)

In the analysis of closed conducting bodies at the frequencies corresponding to the internal resonance of the structure, spurious solutions may be produced when either the EFIE or MFIE is used. This is because the operator is singular at an internal resonant frequency of the closed structure. One possible way to obtain a unique solution for closed bodies at this internal resonant frequency is to combine the EFIE and MFIE in a linear fashion. This combination results in the CFIE. The frequency domain CFIE is obtained by a linear combination of the EFIE with the MFIE [5] through

$$(1 - \kappa)\left(-\mathbf{E}^s(\mathbf{r})\right)_{\tan} + \kappa\eta\left(\mathbf{J}(\mathbf{r}) - \mathbf{n} \times \mathbf{H}^s(\mathbf{r})\right)$$
$$= (1 - \kappa)\left(\mathbf{E}^i(\mathbf{r})\right)_{\tan} + \kappa\eta\left(\mathbf{n} \times \mathbf{H}^i(\mathbf{r})\right) \tag{2.54}$$

where κ is the parameter of linear combination, which is between 0 (EFIE) and 1 (MFIE), and η is the wave impedance of the space.

The matrix equations of EFIE and MFIE, which are given by Eq. (2.34) and Eq. (2.53), respectively, can be rewritten as

$$\left[Z_{mn}^E\right]\left[J_n\right]=\left[V_m^E\right] \tag{2.55}$$

$$\left[Z_{mn}^H\right]\left[J_n\right]=\left[V_m^H\right] \tag{2.56}$$

Thus, a matrix equation for the frequency domain CFIE can be obtained directly from Eqs. (2.55) and (2.56) as

$$\left[Z_{mn}\right]\left[J_n\right]=\left[V_m\right] \tag{2.57a}$$

where

$$Z_{mn}=\left(1-\kappa\right)Z_{mn}^E+\kappa\eta\,Z_{mn}^H \tag{2.57b}$$

$$V_m=\left(1-\kappa\right)V_m^E+\kappa\eta\,V_m^H \tag{2.57c}$$

2.8   FAR FIELD COMPUTATION IN THE FREQUENCY DOMAIN

Once the current density induced on the scatterer **J** has been obtained the scattered field at any point in space can easily be computed. However, to compute the far field, the term with the gradient of the electric scalar potential Φ in Eq. (2.2) is neglected, resulting in

$$\mathbf{E}(\mathbf{r})\approx-j\omega\mathbf{A}=-\frac{j\omega\mu}{4\pi}\int_S\mathbf{J}(\mathbf{r}')\frac{e^{-jkR}}{R}\,dS' \tag{2.58}$$

Substituting the current expansion Eq. (2.18) into Eq. (2.58), one obtains

$$\mathbf{E}(\mathbf{r})=-\frac{j\omega\mu}{4\pi}\sum_{n=1}^N\left(\int_S\mathbf{f}_n(\mathbf{r}')\frac{e^{-jkR}}{R}\,dS'\right)J_n \tag{2.59}$$

In the far field, the following approximations are made:

$$R=\left|\mathbf{r}-\mathbf{r}'\right|\approx r \quad\text{for the amplitude term } 1/R$$

$$R=\left|\mathbf{r}-\mathbf{r}'\right|\approx r-\hat{\mathbf{r}}\bullet\mathbf{r}' \quad\text{for phase term } e^{-jkR}$$

where $\hat{\mathbf{r}} = \mathbf{r}/r$ is a unit vector in the direction of observation given $\hat{\mathbf{r}} = \hat{\mathbf{x}} \sin\theta \cos\phi + \hat{\mathbf{y}} \sin\theta \sin\phi + \hat{\mathbf{z}} \cos\theta$. θ and ϕ are the angles of radiation in the general spherical coordinates. With these approximations, we have

$$\mathbf{E}(\mathbf{r}) = -\frac{j\omega\mu}{4\pi} \frac{e^{-jkr}}{r} \sum_{n=1}^{N} \left(\int_S \mathbf{f}_n(\mathbf{r}') e^{jk\hat{\mathbf{r}}\cdot\mathbf{r}'} dS' \right) J_n \qquad (2.60)$$

Using the summation operator for $\mathbf{f}_n$ defined in Eq. (2.17), the integral in Eq. (2.60) can be expressed as

$$\int_S \mathbf{f}_n(\mathbf{r}') e^{jk\hat{\mathbf{r}}\cdot\mathbf{r}'} dS' = \sum_q \int_S \mathbf{f}_n^q(\mathbf{r}') e^{jk\hat{\mathbf{r}}\cdot\mathbf{r}'} dS'$$
$$= \sum_q \frac{l_n}{2a_n^q} \int_{T_n^q} \boldsymbol{\rho}_n^q e^{jk\hat{\mathbf{r}}\cdot\mathbf{r}'} dS' \qquad (2.61)$$

This integral in Eq. (2.61) is evaluated by approximating the integrand by the value at the centroid of the source triangle T_n^q. Approximating $\boldsymbol{\rho}_n^q \approx \boldsymbol{\rho}_n^{cq}$ and $\mathbf{r}' \approx \mathbf{r}_n^{cq}$. One obtains

$$\int_S \mathbf{f}_n(\mathbf{r}') e^{jk\hat{\mathbf{r}}\cdot\mathbf{r}'} dS' \approx \frac{l_n}{2} \sum_q \boldsymbol{\rho}_n^{cq} e^{jk\hat{\mathbf{r}}\cdot\mathbf{r}_n^{cq}} \qquad (2.62)$$

Substituting Eq. (2.62) into Eq. (2.60), results in

$$\mathbf{E}(\mathbf{r}) = -\frac{j\omega\mu}{8\pi} \frac{e^{-jkr}}{r} \sum_{n=1}^{N} \left(l_n \sum_q \boldsymbol{\rho}_n^{cq} e^{jk\hat{\mathbf{r}}\cdot\mathbf{r}_n^{cq}} \right) J_n \qquad (2.63)$$

or equivalently

$$\mathbf{E}(\mathbf{r}) = -\frac{j\omega\mu}{8\pi} \frac{e^{-jkr}}{r} \sum_{n=1}^{N} \mathbf{W}_n J_n \qquad (2.64a)$$

$$\mathbf{W}_n = l_n \sum_q \boldsymbol{\rho}_n^{cq} e^{jk\hat{\mathbf{r}}\cdot\mathbf{r}_n^{cq}} \qquad (2.64b)$$

The normalized far field is defined as

$$r\mathbf{E}(\mathbf{r}) e^{jkr} = -\frac{jk\eta}{8\pi} \sum_{n=1}^{N} \mathbf{W}_n J_n \qquad (2.65)$$

where the parameter k is the wave number and the parameter η is the intrinsic impedance of the medium surrounding the scatterer.

Numerical results using these formulations will be presented in Chapter 4.

2.9   CONCLUSION

Analysis of the electromagnetic scattering by arbitrarily shaped conducting objects in the frequency domain is presented in this chapter Method of Moment, triangular surface patch modeling and the RWG basis functions have been used to solve the electric field integral equation, the magnetic field integral equation, and the combined field integral equation. Of them, the EFIE can be used for both open and closed objects, whereas the MFIE and the CFIE is used only for closed surfaces. Numerical examples using these formulations for the analysis of composite conducting and dielectric structures will be presented in Chapter 4.

REFERENCES

[1]   Y. Zhang, and T. K. Sarkar, *Parallel Solution of Integral Equation-based EM Problems in the Frequency Domain*, John Wiley & Sons, New York, 2009.

[2]   S. M. Rao, D. R. Wilton, and A. W. Glisson, "Electromagnetic scattering by surfaces of arbitrary shape," *IEEE Trans. Antennas Prop.*, Vol. 30, pp. 409–418, 1982.

[3]   S. M. Rao, *Electromagnetic Scattering and Radiation of Arbitrarily-Shaped Surfaces by Triangular Patch Modeling*, PhD Dissertation, University Mississippi, August, 1980.

[4]   D. R. Wilton, S. M. Rao, A. W. Glisson, D. H. Schaubert, O. M. Al-Bundak, and C. M. Butler, "Potential integrals for uniform and linear source distributions on polygonal and polyhedral domains," *IEEE Trans. Antennas Prop.*, Vol. 32, pp. 276–281, 1984.

[5]   A. J. Poggio and E. K. Miller, "Integral equation solutions of three dimensional scattering problems," *Computer Techniques for Electromagnetics*, Pergamon Press, New York, 1973.

3

ANALYSIS OF DIELECTRIC OBJECTS IN THE FREQUENCY DOMAIN

3.0   SUMMARY

In this chapter, numerical methods used to analyze the electromagnetic scattering from three-dimensional (3-D) homogeneous dielectric bodies of arbitrary shapes in the frequency domain are addressed. Five different integral equation formulations are described here, viz. the electric field integral equation (EFIE), the magnetic field integral equation (MFIE), the combined field integral equation (CFIE), the *Poggio–Miller–Chang–Harrington–Wu integral equation* (PMCHW), and the *Muller integral equation*. The EFIE and MFIE are derived using the boundary conditions of the electric or the magnetic field, respectively, whereas the other equations consider both boundary conditions. In addition, a thorough study on different combinations of the testing functions in the CFIE and the Muller equation are presented. It is demonstrated that not all CFIE formulations are accurate enough in removing defects that occur at frequencies corresponding to an internal resonance of the structure. At the end of this chapter, the equivalent electric current, magnetic current, far scattered field, and *radar cross-section* (RCS) are computed based on the described formulations for three dielectric scatterers of different shapes, viz. a sphere, a cube, and a finite circular cylinder. Through the comparison of these results, the effectiveness of different formulations is illustrated.

3.1   INTRODUCTION

This section deals with the scattering from arbitrarily shaped 3-D homogeneous dielectric bodies in the frequency domain. Furthermore, the simpler EFIE or MFIE formulations fail at certain discrete frequencies that correspond to the internal resonances of the structure. One possible way to obtain a unique solution at an internal resonant frequency of the structure under analysis is to combine a weighted linear sum of the EFIE with MFIE and thereby minimize the effects of spurious solutions [1]. The CFIE, as the name implies, is a combination of EFIE and MFIE. The use of a combined field formulation does not eliminate the problem of the internal resonances; it merely pushes them off the real frequency axis into the complex s-plane, thereby minimizing the effects of the spurious modes when observed along the real frequency axis. Although an integral equation formulation has been used for analyzing 3-D dielectric bodies in the frequency domain, only a few researchers have applied it to the analysis of scattering by arbitrarily shaped 3-D objects with triangular patch modeling [2–9].

The PMCHW formulation was proposed [2,5], in which Rao–Wilton–Glisson (RWG) functions have been used as both the basis and the testing functions, and also to approximate both the electric and the magnetic currents. However, a similar scheme, when used to solve the EFIE or MFIE, yielded ill-conditioned matrices and inaccurate results. To overcome these problems for the EFIE formulation, the electric current is expanded using the RWG functions, whereas the magnetic current is expanded using another set of basis functions given by $\mathbf{n} \times$ RWG, which are pointwise spatially orthogonal to the RWG set [3]. Here, $\mathbf{n}$ is the unit normal pointing outward from the surface. In addition, the RWG basis functions are also used as the testing functions. Next, Rao and Wilton proposed the CFIE along with EFIE and MFIE for the analysis of scattering by arbitrarily shaped 3-D dielectric bodies for the first time [4]. In their work, RWG functions are used to approximate the electric current, but the magnetic current is approximated by $\mathbf{n} \times$ RWG [3], and a line testing is used. In a recent paper [6], Sheng et al. proposed a CFIE formulation for the same problem. In Sheng's work, the RWG functions are used as basis functions to approximate both the electric and magnetic currents, and a linear combination of RWG and $\mathbf{n} \times$ RWG functions are used as testing functions. This yields a well-conditioned matrix. Sheng also presented a set of four CFIE formulations by dropping one of the testing terms in $\langle$ RWG, EFIE $\rangle + \langle \mathbf{n} \times$ RWG, EFIE $\rangle + \langle$ RWG, MFIE $\rangle + \langle \mathbf{n} \times$ RWG, MFIE $\rangle$.

For the Muller's formulation, the equivalent currents are expanded using the RWG functions, and a combination of the RWG functions and its orthogonal component is used for testing. All possible cases of the testing procedure are investigated using the four parameters in conjunction with the testing functions. Further, a new set of pulse basis functions on triangular patches was used to solve the EFIE, MFIE, and CFIE formulations [10]. In this chapter, various integral formulations are investigated and several combinations of CFIE are proposed with different choices of testing functions. The results illustrate that not all CFIE formulations are stable.

This chapter is organized as follows. In section 3.2, the integral equations, EFIE, MFIE, CFIE, PMCHW, and the Muller equation are introduced and the different testing and basis functions are discussed in section 3.3. Numerical procedures to solve them are given in sections 3.4–3.8. The evaluation of various integrals appearing in these numerical solutions is given in the appendix at the end of this chapter. Section 3.9 presents the accuracy comparison of using several formulations of CFIE, and section 3.10 provides the Muller's formulations. In section 3.11, more than 30 numerical results for the analysis of a dielectric sphere, a cube, and a finite circular cylinder using different formulations are presented and compared with the Mie series solution and the available software—HOBBIES (*higher order basis function based integral equation solver*) [11]. Finally, some conclusions are presented in section 3.12

3.2   INTEGRAL EQUATIONS

In this section, the mathematical steps are described to obtain the integral equations to analyze the electromagnetic scattering by arbitrarily-shaped 3-D homogeneous dielectric bodies. To simplify the presentation, the formulations are only presented for a single dielectric body, and it is possible to extend this formulation for handling multiple dielectric bodies.

Consider a homogeneous dielectric body with permittivity ε_2 and permeability μ_2 placed in an infinite homogeneous medium with permittivity ε_1 and permeability μ_1 as shown in Figure 3.1. A lossy material body can be handled by considering ε_2, μ_2 to be complex quantities. Throughout this chapter, the subscripts 1 and 2 are used to represent the region outside and inside the dielectric body to be studied, respectively. The structure is now illuminated by an incident plane wave denoted by $(\mathbf{E}^i, \mathbf{H}^i)$. It may be noted that the incident field is defined to exist if the structure were not present. By invoking the equivalence principle [12], the following two relations corresponding to the exterior and interior regions are formed in terms of the equivalent electric current $\mathbf{J}$ and the magnetic current $\mathbf{M}$ on the surface S of the dielectric body.

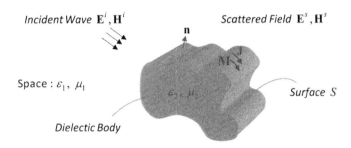

Incident Wave $\mathbf{E}^i, \mathbf{H}^i$ **n**   *Scattered Field* $\mathbf{E}^s, \mathbf{H}^s$

Space : ε_1, μ_1 *Surface S*

Dielectic Body

Figure 3.1. Homogeneous dielectric body illuminated by an electromagnetic plane wave.

While applying the equivalence principle for the region external to the body, the field in the interior region is set to zero and the entire interior space is filled by the homogeneous medium (ε_1, μ_1) which was originally external to the body only. As a result, the whole problem space is filled with a homogeneous medium allowing us to use the free space Green's function to calculate the fields. Equivalent electric and magnetic currents, $\mathbf{J}$ and $\mathbf{M}$, are then placed on the surface of the body to preserve the continuity of the fields across the dielectric boundary. By enforcing the continuity of the tangential electric and magnetic field at S_-, where S_- is selected to be slightly interior to S, we obtain

$$\left(-\mathbf{E}_1^s(\mathbf{J},\mathbf{M})\right)_{\text{tan}} = \left(\mathbf{E}^i\right)_{\text{tan}} \tag{3.1}$$

$$\left(-\mathbf{H}_1^s(\mathbf{J},\mathbf{M})\right)_{\text{tan}} = \left(\mathbf{H}^i\right)_{\text{tan}} \tag{3.2}$$

where the subscript 1 indicates that the scattered field $(\mathbf{E}_1^s, \mathbf{H}_1^s)$ is computed in the external region. Using similar procedures for the interior region, the fields in the region external to the body are set to zero and thus, the external region is replaced by the material (ε_2, μ_2) so that the currents are entirely located in a homogeneous medium, which was originally inside the dielectric body only. By enforcing the continuity of the tangential electric and magnetic fields at S_+, where S_+ is the surface slightly exterior to S, the following equations are derived:

$$\left(-\mathbf{E}_2^s(\mathbf{J},\mathbf{M})\right)_{\text{tan}} = 0 \tag{3.3}$$

$$\left(-\mathbf{H}_2^s(\mathbf{J},\mathbf{M})\right)_{\text{tan}} = 0 \tag{3.4}$$

where the subscript 2 indicates that the scattered field $(\mathbf{E}_2^s, \mathbf{H}_2^s)$ is computed in the internal region. In Eqs. (3.1)–(3.4), the subscript 'tan' refers to the tangential component of the fields only. Because both the internal and external problems are formulated in terms of the electric current $\mathbf{J}$ and the magnetic current $\mathbf{M}$ radiating in respective homogeneous mediums, the scattered electric and magnetic fields caused by these currents can be expressed analytically as

$$\mathbf{E}_v^s(\mathbf{J}) = -j\omega\mathbf{A}_v - \nabla\Phi_v \tag{3.5}$$

$$\mathbf{E}_v^s(\mathbf{M}) = -\frac{1}{\varepsilon_v}\nabla\times\mathbf{F}_v \tag{3.6}$$

$$\mathbf{H}_v^s(\mathbf{J}) = \frac{1}{\mu_v}\nabla\times\mathbf{A}_v \tag{3.7}$$

$$\mathbf{H}_v^s(\mathbf{M}) = -j\omega\mathbf{F}_v - \nabla\Psi_v \tag{3.8}$$

In Eqs. (3.5)–(3.8), the symbol v is used to represent the region in which the currents $\mathbf{J}$ or $\mathbf{M}$ are radiating and can be assigned as either 1 for the exterior region or 2 for the interior region. The parameters $\mathbf{A}_v$, $\mathbf{F}_v$, Φ_v, and Ψ_v are the magnetic vector potential, electric vector potential, electric scalar potential and magnetic scalar potential, respectively, which are explicitly given by

$$\mathbf{A}_v(\mathbf{r}) = \frac{\mu_v}{4\pi}\int_S \mathbf{J}(\mathbf{r}')G_v(\mathbf{r},\mathbf{r}')dS' \tag{3.9}$$

$$\mathbf{F}_v(\mathbf{r}) = \frac{\varepsilon_v}{4\pi}\int_S \mathbf{M}(\mathbf{r}')G_v(\mathbf{r},\mathbf{r}')dS' \tag{3.10}$$

$$\Phi_v(\mathbf{r}) = \frac{j}{4\pi\omega\varepsilon_v} \int_S \nabla'_S \cdot \mathbf{J}(\mathbf{r}')G_v(\mathbf{r},\mathbf{r}')dS' \tag{3.11}$$

$$\Psi_v(\mathbf{r}) = \frac{j}{4\pi\omega\mu_v} \int_S \nabla'_S \cdot \mathbf{M}(\mathbf{r}')G_v(\mathbf{r},\mathbf{r}')dS' \tag{3.12}$$

$$G_v(\mathbf{r},\mathbf{r}') = \frac{e^{-jk_vR}}{R}, \quad R = |\mathbf{r}-\mathbf{r}'|. \tag{3.13}$$

In Eqs. (3.9)–(3.12), a $e^{j\omega t}$ time dependence has been assumed and suppressed for convenience. R represents the distance between the observation point $\mathbf{r}$ and the source point $\mathbf{r}'$ with respect to a global coordinate origin. $k_v = \omega\sqrt{\mu_v\varepsilon_v}$ is the wave number in medium v. ω is the angular frequency in rad/sec. The term $G_v(\mathbf{r},\mathbf{r}')$ represents the free-space Green's function for a homogeneous medium.

Examining Eqs. (3.1)–(3.4), we note that there are only two unknowns $\mathbf{J}$ and $\mathbf{M}$ to be obtained from these four equations. Based on various combinations of part or all of these four equations, it is possible to obtain several integral equations of which the following five integral equations are popular:

- EFIE
- MFIE
- CFIE
- PMCHW
- Muller

Note that the two Eqs. (3.1) and (3.3) are purely from the electric field boundary conditions and can be combined to form the EFIE formulation. It gives the relation in the frequency domain between the incident electric field and the scattered electric field caused by the induced electric and magnetic currents as

$$\left(-\mathbf{E}^s_v(\mathbf{J},\mathbf{M})\right)_{\tan} = \begin{cases} \left(\mathbf{E}^i\right)_{\tan}, & v=1 \\ 0, & v=2 \end{cases} \tag{3.14}$$

Dual to the EFIE formulation, the MFIE formulation is obtained by choosing the two magnetic field conditions only, that is, Eq. (3.2) and Eq. (3.4), as

$$\left(-\mathbf{H}^s_v(\mathbf{J},\mathbf{M})\right)_{\tan} = \begin{cases} \left(\mathbf{H}^i\right)_{\tan}, & v=1 \\ 0, & v=2 \end{cases} \tag{3.15}$$

However, both the EFIE and MFIE formulations fail at the frequencies when the surface S is covered by a perfect electric conductor and forms a resonant cavity, irrespective of the material filling the cavity. This situation of loading the cavity only changes its internal resonant frequency. For the CFIE formulation, a set of two integral equations are formed from the linear combination of EFIE in Eq. (3.14) and MFIE in Eq. (3.15) by using the following form:

$$\kappa \left(-\mathbf{E}_\nu^s (\mathbf{J}, \mathbf{M}) \right)_{\tan} + (1 - \kappa) \eta_1 \left(-\mathbf{H}_\nu^s (\mathbf{J}, \mathbf{M}) \right)_{\tan}$$
$$= \begin{cases} \kappa \left(\mathbf{E}^i \right)_{\tan} + (1 - \kappa) \eta_1 \left(\mathbf{H}^i \right)_{\tan}, & \nu = 1 \\ 0, & \nu = 2 \end{cases} \qquad (3.16)$$

where η_1 is the wave impedance of the exterior region and κ is a scale factor between 0 and 1. An alternative way to combine the four equations, Eq. (3.1) to (3.4), is the PMCHW formulation. In this formulation, the set of four equations is reduced to two by adding Eq. (3.1) to Eq. (3.3) and Eq. (3.2) to Eq. (3.4), giving

$$\left(-\mathbf{E}_1^s (\mathbf{J}, \mathbf{M}) - \mathbf{E}_2^s (\mathbf{J}, \mathbf{M}) \right)_{\tan} = \left(\mathbf{E}^i \right)_{\tan} \qquad (3.17a)$$

$$\left(-\mathbf{H}_1^s (\mathbf{J}, \mathbf{M}) - \mathbf{H}_2^s (\mathbf{J}, \mathbf{M}) \right)_{\tan} = \left(\mathbf{H}^i \right)_{\tan} \qquad (3.17b)$$

In addition, the PMCHW equations can be extended in Eq. (3.17) by introducing two more parameters α and β as

$$\left(-\mathbf{E}_1^s (\mathbf{J}, \mathbf{M}) - \alpha \mathbf{E}_2^s (\mathbf{J}, \mathbf{M}) \right)_{\tan} = \left(\mathbf{E}^i \right)_{\tan} \qquad (3.18a)$$

$$\left(-\mathbf{H}_1^s (\mathbf{J}, \mathbf{M}) - \beta \mathbf{H}_2^s (\mathbf{J}, \mathbf{M}) \right)_{\tan} = \left(\mathbf{H}^i \right)_{\tan} \qquad (3.18b)$$

where $\alpha = \varepsilon_1 / \varepsilon_2$ and $\beta = \mu_1 / \mu_2$. This is called the Muller formulation [1]. The Muller equation is same as the PMCHW formulation when $\alpha = \beta = 1$.

3.3   EXPANSION AND TESTING FUNCTIONS

For any one of the integral equations mentioned previously, note that there are the two unknowns, $\mathbf{J}$ and $\mathbf{M}$, and they can be determined by using similar MoM numerical procedures introduced in Chapter 2. First, the dielectric object is assumed to be a closed body and modeled by the planar triangular patches. Next, the unknown currents $\mathbf{J}$ and $\mathbf{M}$ need to be expanded with suitable basis functions, and tested with the suitable testing functions. In the following, we define two sets of functions that may be useful. The first is the RWG functions defined in Chapter 2. As noted earlier, the RWG vector function associated with the nth edge is given by

$$\mathbf{f}_n (\mathbf{r}) = \mathbf{f}_n^+ (\mathbf{r}) + \mathbf{f}_n^- (\mathbf{r}) \qquad (3.19a)$$

$$\mathbf{f}_n^{\pm}(\mathbf{r}) = \begin{cases} \dfrac{l_n}{2a_n^{\pm}}\, \boldsymbol{\rho}_n^{\pm}, & \mathbf{r} \in T_n^{\pm} \\[2ex] 0, & \mathbf{r} \notin T_n^{\pm} \end{cases} \tag{3.19b}$$

and the surface divergence of Eq. (3.19) is

$$\nabla_S \bullet \mathbf{f}_n(\mathbf{r}) = \nabla_S \bullet \mathbf{f}_n^{+}(\mathbf{r}) + \nabla_S \bullet \mathbf{f}_n^{-}(\mathbf{r}) \tag{3.20a}$$

$$\nabla_S \bullet \mathbf{f}_n^{\pm}(\mathbf{r}) = \begin{cases} \pm\dfrac{l_n}{a_n^{\pm}}, & \mathbf{r} \in T_n^{\pm} \\[2ex] 0, & \mathbf{r} \notin T_n^{\pm} \end{cases}. \tag{3.20b}$$

In Eqs. (3.19) and (3.20), l_n is the length of the nonboundary edge and a_n^{+} and a_n^{-} are the areas of the triangles T_n^{+} and T_n^{-}, respectively. Vectors $\boldsymbol{\rho}_n^{+}$ and $\boldsymbol{\rho}_n^{-}$ are the position vectors defined with respect to the free vertices of T_n^{+} and T_n^{-}.

Next, we define another vector basis function $\mathbf{g}_n$ that is point-wise orthogonal to the RWG function $\mathbf{f}_n$ in the same triangle pair, given by

$$\mathbf{g}_n(\mathbf{r}) = \mathbf{n} \times \mathbf{f}_n(\mathbf{r}) \tag{3.21}$$

where $\mathbf{n}$ is the unit normal pointing outward from the surface [3]. These two sets of functions are used in several different ways to solve five integral equations defined earlier. The various choices of the expansion and testing functions for the electric and magnetic currents in the five formulations, EFIE, MFIE, CFIE, PMCHW, and the Muller equation, are summarized in Table 3.1. The reason for a specific choice will be discussed in the following sections.

Table 3.1. Expansion and Testing Functions in Different Formulations

Formulation	Testing function	Current expansion function J	M
EFIE	$\mathbf{f}_m$	$\mathbf{f}_n$	$\mathbf{g}_n$
MFIE	$\mathbf{f}_m$	$\mathbf{g}_n$	$\mathbf{f}_n$
CFIE	$\pm\mathbf{f}_m \pm \mathbf{g}_n$	$\mathbf{f}_n$	$\mathbf{f}_n$
PMCHW	$\mathbf{f}_m$	$\mathbf{f}_n$	$\mathbf{f}_n$
Muller	$\pm\mathbf{f}_m \pm \mathbf{g}_n$	$\mathbf{f}_n$	$\mathbf{f}_n$

Next, the numerical procedures to solve the various integral equations using the method of moments (MoM) methodology are outlined.

3.4   SOLUTION OF THE ELECTRIC FIELD INTEGRAL EQUATION (EFIE)

To solve the EFIE for a dielectric object in Eq. (3.14), the unknown electric current $\mathbf{J}$ and the magnetic current $\mathbf{M}$ are first approximated in terms of the vector basis functions $\mathbf{f}_n$ in Eq. (3.19) and $\mathbf{g}_n$ in Eq. (3.21), respectively, as

$$\mathbf{J}(\mathbf{r}) = \sum_{n=1}^{N} J_n \, \mathbf{f}_n(\mathbf{r}) \tag{3.22a}$$

$$\mathbf{M}(\mathbf{r}) = \sum_{n=1}^{N} M_n \, \mathbf{g}_n(\mathbf{r}) \tag{3.22b}$$

where J_n and M_n are constants to be determined and N is the number of edges on the dielectric surface for the triangulated model approximating the surface of the dielectric body. The next step in the application of MoM is to select a suitable testing procedure. As mentioned in Table 3.1, the testing function is the RWG function $\mathbf{f}_m$. This choice results in a simple numerical procedure as we will observe.

The complex inner product is defined as

$$\langle \mathbf{f} \cdot \mathbf{g} \rangle = \int_S \mathbf{f} \cdot \mathbf{g}^* \, dS \tag{3.23}$$

As the testing functions are real, the conjugate can be eliminated. However, as noted in Chapter 1, a symmetric product does not define a norm; because, for an element $j1$, its norm is -1, which is incorrect, as a norm can never be a negative number. Testing the electric field integral equation Eq. (3.14) with $\mathbf{f}_m$ yields

$$\langle \mathbf{f}_m, -\mathbf{E}_v^s(\mathbf{J}) \rangle + \langle \mathbf{f}_m, -\mathbf{E}_v^s(\mathbf{M}) \rangle = \begin{cases} \langle \mathbf{f}_m, \mathbf{E}^i \rangle, & v = 1 \\ 0, & v = 2 \end{cases} \tag{3.24}$$

for $m = 1, 2, \ldots, N$. The first term in Eq. (3.24) is the testing of the scattered electric field as a result of the electric current, whereas the second term is the testing as a result of the magnetic current. For the term related to the electric current, it can be represented by the magnetic vector potential $\mathbf{A}$ and electric scalar potential Φ from Eq. (3.5) as

$$\langle \mathbf{f}_m, -\mathbf{E}_v^s(\mathbf{J}) \rangle = \langle \mathbf{f}_m, j\omega \mathbf{A}_v \rangle + \langle \mathbf{f}_m, \nabla \Phi_v \rangle \tag{3.25}$$

Note that the right-hand side of Eq. (3.25) is indeed same as the left-hand side of Eq. (2.20) in Chapter 2, which dealt with the analysis of the electric scattered field from a conducting object, except a subscript v that can be assigned as either 1 or

2 is added to indicate the region where the fields are considered. Meanwhile, the unknown electric current $\mathbf{J}$ is expanded by the basis function $\mathbf{f}_n$ in both Eqs. (3.25) and (2.20). The detailed steps of the derivation are available in [4,5]. When the testing of Eq. (2.29) is applied, one obtains

$$\left\langle \mathbf{f}_m, -\mathbf{E}_v^s(\mathbf{J}) \right\rangle = j\omega\mu_v \sum_{n=1}^{N} A_{mn,v} J_n - \frac{j}{\omega\varepsilon_v} \sum_{n=1}^{N} B_{mn,v} J_n \qquad (3.26a)$$

where

$$A_{mn,v} = \frac{1}{4\pi} \int_S \mathbf{f}_m(\mathbf{r}) \cdot \int_S \mathbf{f}_n(\mathbf{r}') G_v(\mathbf{r},\mathbf{r}') \, dS' dS \qquad (3.26b)$$

$$B_{mn,v} = \frac{1}{4\pi} \int_S \nabla_S \cdot \mathbf{f}_m(\mathbf{r}) \int_S \nabla_S' \cdot \mathbf{f}_n(\mathbf{r}') G_v(\mathbf{r},\mathbf{r}') \, dS' \, dS \qquad (3.26c)$$

Now examine the second term in Eq. (3.24) that describes the scattered electric field caused by the magnetic current. Extracting the Cauchy principal value from the curl term on a planar surface in Eq. (3.6) and using Eq. (3.10), the following can be written:

$$-\mathbf{E}_v^s(\mathbf{M}) = \frac{1}{\varepsilon_v} \nabla \times \mathbf{F}_v(\mathbf{r}) = \pm \frac{1}{2} \mathbf{n} \times \mathbf{M}(\mathbf{r}) + \frac{1}{\varepsilon_v} \nabla \times \tilde{\mathbf{F}}_v(\mathbf{r}) \qquad (3.27)$$

where $\tilde{\mathbf{F}}_v$ is defined by Eq. (3.10) with the removal of the singularity at $\mathbf{r} = \mathbf{r}'$ (or $R = 0$) from the integration. In Eq. (3.27), the positive sign is used when $v = 1$ and negative sign is used when $v = 2$. Therefore, by substituting Eq. (3.27), the second term of Eq. (3.24) is now rewritten as

$$\left\langle \mathbf{f}_m, -\mathbf{E}_v^s(\mathbf{M}) \right\rangle = \left\langle \mathbf{f}_m, \pm \frac{1}{2} \mathbf{n} \times \mathbf{M} \right\rangle + \left\langle \mathbf{f}_m, \frac{1}{\varepsilon_v} \nabla \times \tilde{\mathbf{F}}_v \right\rangle \qquad (3.28)$$

The magnetic current $\mathbf{M}$ in the first term on the right-hand side of Eq. (3.28) can be replaced by the expansion function defined in Eq. (3.22b)

$$\left\langle \mathbf{f}_m, \pm \frac{1}{2} \mathbf{n} \times \mathbf{M} \right\rangle = \sum_{n=1}^{N} C_{mn,v} M_n \qquad (3.29a)$$

where

$$C_{mn,v} = \begin{cases} \dfrac{1}{2} \int_S \mathbf{f}_m(\mathbf{r}) \cdot \mathbf{n} \times \mathbf{g}_n(\mathbf{r}) dS, & v = 1 \\[2mm] -\dfrac{1}{2} \int_S \mathbf{f}_m(\mathbf{r}) \cdot \mathbf{n} \times \mathbf{g}_n(\mathbf{r}) dS, & v = 2 \end{cases} \qquad (3.29b)$$

Then, in the second term on the right-hand side of Eq. (3.28), the electric vector potential $\tilde{\mathbf{F}}_v$ can be substituted using Eq. (3.10) to yield

$$\left\langle \mathbf{f}_m, \frac{1}{\varepsilon_v} \nabla \times \tilde{\mathbf{F}}_v \right\rangle = \sum_{n=1}^{N} D_{mn,v} M_n \tag{3.30a}$$

where

$$D_{mn,v} = \frac{1}{4\pi} \int_S \mathbf{f}_m(\mathbf{r}) \cdot \int_S \mathbf{g}_n(\mathbf{r}') \times \nabla' G_v(\mathbf{r},\mathbf{r}') dS' dS \tag{3.30b}$$

So far, all the terms in the left-hand side of Eq. (3.24) are expanded and given by Eqs. (3.26), (3.29), and (3.30). Finally, the term on the right-hand side of Eq. (3.24), which is the testing of the incident electric field, can be expressed as

$$V_{m,v}^E = \begin{cases} \int_S \mathbf{f}_m(\mathbf{r}) \cdot \mathbf{E}^i(\mathbf{r}) dS, & v = 1 \\ 0, & v = 2 \end{cases} \tag{3.31}$$

As an overall result of employing MoM in the EFIE given by Eq. (3.14) for dielectric objects, Eqs. (3.26), (3.29), (3.30), and (3.31) can be integrated to result in

$$\sum_{n=1}^{N} jk_v \eta_v \left(A_{mn,v} - \frac{B_{mn,v}}{k_v^2} \right) J_n + \sum_{n=1}^{N} \left(C_{mn,v} + D_{mn,v} \right) M_n = V_{m,v}^E \tag{3.32a}$$

where

$$A_{mn,v} = \frac{1}{4\pi} \int_S \mathbf{f}_m(\mathbf{r}) \cdot \int_S \mathbf{f}_n(\mathbf{r}') G_v(\mathbf{r},\mathbf{r}') dS' dS \tag{3.32b}$$

$$B_{mn,v} = \frac{1}{4\pi} \int_S \nabla_S \cdot \mathbf{f}_m(\mathbf{r}) \int_S \nabla_S' \cdot \mathbf{f}_n(\mathbf{r}') G_v(\mathbf{r},\mathbf{r}') dS' dS \tag{3.32c}$$

$$C_{mn,v} = \begin{cases} \dfrac{1}{2} \int_S \mathbf{f}_m(\mathbf{r}) \cdot \mathbf{n} \times \mathbf{g}_n(\mathbf{r}) dS, & v = 1 \\ -\dfrac{1}{2} \int_S \mathbf{f}_m(\mathbf{r}) \cdot \mathbf{n} \times \mathbf{g}_n(\mathbf{r}) dS, & v = 2 \end{cases} \tag{3.32d}$$

$$D_{mn,v} = \frac{1}{4\pi} \int_S \mathbf{f}_m(\mathbf{r}) \cdot \int_S \mathbf{g}_n(\mathbf{r}') \times \nabla' G_v(\mathbf{r},\mathbf{r}') dS' dS \tag{3.32e}$$

$$V_{m,v}^E = \begin{cases} \int_S \mathbf{f}_m(\mathbf{r}) \cdot \mathbf{E}^i(\mathbf{r}) dS, & v = 1 \\ 0, & v = 2 \end{cases} \tag{3.32f}$$

for $m = 1, 2, \ldots, N$. The parameters k_v and η_v are given by $k_v = \omega\sqrt{\mu_v \varepsilon_v}$ and $\eta_v = \sqrt{\mu_v / \varepsilon_v}$ where k_v and η_v are the wave number and the intrinsic impedance of the medium v, respectively. Finally, Eq. (3.32) may be written in a $2N \times 2N$ linear matrix equations form as

$$\begin{bmatrix} \left[jk_1\eta_1\left(A_{mn,1} - B_{mn,1}/k_1^2\right)\right] & \left[C_{mn,1} + D_{mn,1}\right] \\ \left[jk_2\eta_2\left(A_{mn,2} - B_{mn,2}/k_2^2\right)\right] & \left[C_{mn,2} + D_{mn,2}\right] \end{bmatrix} \begin{bmatrix} J_n \\ M_n \end{bmatrix} = \begin{bmatrix} \left[V_{m,1}^E\right] \\ \left[V_{m,2}^E\right] \end{bmatrix}. \tag{3.33}$$

To solve Eq. (3.33), the integrals A_{mn}, B_{mn}, C_{mn}, and D_{mn} must be evaluated first. The detailed expressions for the above terms will be given in the appendix at the end of this chapter.

3.5 SOLUTION OF THE MAGNETIC FIELD INTEGRAL EQUATION (MFIE)

The MFIE has been defined in Eq. (3.15), which is the dual of the EFIE. As listed in Table 3.1, the electric current $\mathbf{J}$ and the magnetic current $\mathbf{M}$ of the MFIE are expanded by the functions $\mathbf{g}_n$ and $\mathbf{f}_n$, respectively, and then tested with the function $\mathbf{f}_m$. Therefore, the approximation for the currents are given by

$$\mathbf{J(r)} = \sum_{n=1}^{N} J_n \, \mathbf{g}_n(\mathbf{r}) \tag{3.34a}$$

$$\mathbf{M(r)} = \sum_{n=1}^{N} M_n \, \mathbf{f}_n(\mathbf{r}) \tag{3.34b}$$

and the testing of the MFIE with $\mathbf{f}_m$ yields

$$\left\langle \mathbf{f}_m, -\mathbf{H}_v^s(\mathbf{J})\right\rangle + \left\langle \mathbf{f}_m, -\mathbf{H}_v^s(\mathbf{M})\right\rangle = \begin{cases} \left\langle \mathbf{f}_m, \mathbf{H}^i\right\rangle, & v = 1 \\ 0, & v = 2 \end{cases} \tag{3.35}$$

The procedure of applying MoM to solve the MFIE for dielectric objects is the same as presented for the EFIE. The procedure of the derivation is not repeated here in detail, and the results are found directly as

$$-\sum_{n=1}^{N}\left(C_{mn,v} + D_{mn,v}\right)J_n + \sum_{n=1}^{N} j\frac{k_v}{\eta_v}\left(A_{mn,v} - \frac{B_{mn,v}}{k_v^2}\right)M_n = V_{m,v}^H \tag{3.36a}$$

where

$$A_{mn,v} = \frac{1}{4\pi}\int_S \mathbf{f}_m(\mathbf{r}) \cdot \int_S \mathbf{f}_n(\mathbf{r}')G_v(\mathbf{r},\mathbf{r}')\,dS'\,dS \tag{3.36b}$$

$$B_{mn,v} = \frac{1}{4\pi} \int_S \nabla_S \cdot \mathbf{f}_m(\mathbf{r}) \int_S \nabla'_S \cdot \mathbf{f}_n(\mathbf{r}') G_v(\mathbf{r},\mathbf{r}') \, dS' \, dS \qquad (3.36c)$$

$$C_{mn,v} = \begin{cases} \dfrac{1}{2} \int_S \mathbf{f}_m(\mathbf{r}) \cdot \mathbf{n} \times \mathbf{g}_n(\mathbf{r}) \, dS, & v = 1 \\[2mm] -\dfrac{1}{2} \int_S \mathbf{f}_m(\mathbf{r}) \cdot \mathbf{n} \times \mathbf{g}_n(\mathbf{r}) \, dS, & v = 2 \end{cases} \qquad (3.36d)$$

$$D_{mn,v} = \frac{1}{4\pi} \int_S \mathbf{f}_m(\mathbf{r}) \cdot \int_S \mathbf{g}_n(\mathbf{r}') \times \nabla' G_v(\mathbf{r},\mathbf{r}') \, dS' \, dS \qquad (3.36e)$$

$$V_{m,v}^H = \begin{cases} \int_S \mathbf{f}_m(\mathbf{r}) \cdot \mathbf{H}^i(\mathbf{r}) \, dS, & v = 1 \\[2mm] 0, & v = 2 \end{cases} \qquad (3.36f)$$

The integrals $A_{mn,v}$, $B_{mn,v}$, $C_{mn,v}$, and $D_{mn,v}$ are the same as those in Eq. (3.32). Note that Eq. (3.36) can also be obtained directly from EFIE in Eq. (3.32) by using the principle of duality [8]. Similarly, Eq. (3.36) is associated with each edge, and it may be written in a matrix form for $m = 1, 2, \ldots, N$ as

$$\begin{bmatrix} [-C_{mn,1} - D_{mn,1}] & \left[jk_1/\eta_1 \left(A_{mn,1} - B_{mn,1}/k_1^2 \right) \right] \\ [-C_{mn,2} - D_{mn,2}] & \left[jk_2/\eta_2 \left(A_{mn,2} - B_{mn,2}/k_2^2 \right) \right] \end{bmatrix} \begin{bmatrix} [J_n] \\ [M_n] \end{bmatrix} = \begin{bmatrix} [V_{m,1}^H] \\ [V_{m,2}^H] \end{bmatrix} \qquad (3.37)$$

Eq. (3.37) is a $2N \times 2N$ linear equations in matrix form. Solving the currents J_n and M_n in this matrix equation can lead to the determination of the scattered field.

3.6   SOLUTION OF THE COMBINED FIELD INTEGRAL EQUATION (CFIE)

For the CFIE formulation, we note that it is not possible to simply combine the EFIE and the MFIE solutions described in the previous two sections. The primary difficulty is the expansion functions for $\mathbf{J}$ and $\mathbf{M}$ are not the same in both formulations. Hence, we adapt the following procedure.

Here, the basis function $\mathbf{f}_n$ is used to expand both $\mathbf{J}$ and $\mathbf{M}$ as

$$\mathbf{J}(\mathbf{r}) = \sum_{n=1}^{N} J_n \mathbf{f}_n(\mathbf{r}) \qquad (3.38a)$$

$$\mathbf{M}(\mathbf{r}) = \sum_{n=1}^{N} M_n \mathbf{f}_n(\mathbf{r}) \qquad (3.38b)$$

However, a combination of the RWG basis function $\mathbf{f}_n$ and its pointwise orthogonal function $\mathbf{g}_n$ is used for testing which converts the integral equation into a matrix equation. We also note that this special choice of testing scheme is required to make the numerical solution well conditioned and free of instabilities. When $\mathbf{f}_m + \mathbf{g}_m$ is used for testing the CFIE in Eq. (3.16), one obtains

$$\kappa \left\langle \mathbf{f}_m + \mathbf{g}_m, -\mathbf{E}_\nu^s \right\rangle + (1-\kappa)\eta_1 \left\langle \mathbf{f}_m + \mathbf{g}_m, -\mathbf{H}_\nu^s \right\rangle$$

$$= \begin{cases} \kappa \left\langle \mathbf{f}_m + \mathbf{g}_m, \mathbf{E}^i \right\rangle + (1-\kappa)\eta_1 \left\langle \mathbf{f}_m + \mathbf{g}_m, \mathbf{H}^i \right\rangle, & \nu = 1 \\ 0, & \nu = 2 \end{cases} \qquad (3.39)$$

Next, we consider yet another situation. Note that the set of four boundary integral equations in Eqs. (3.1)–(3.4) can also be represented as

$$-\mathbf{n} \times \mathbf{E}_1^s (\mathbf{J}, \mathbf{M}) = \mathbf{n} \times \mathbf{E}^i \qquad (3.40)$$

$$-\mathbf{n} \times \mathbf{H}_1^s (\mathbf{J}, \mathbf{M}) = \mathbf{n} \times \mathbf{H}^i \qquad (3.41)$$

$$-\mathbf{n} \times \mathbf{E}_2^s (\mathbf{J}, \mathbf{M}) = 0 \qquad (3.42)$$

$$-\mathbf{n} \times \mathbf{H}_2^s (\mathbf{J}, \mathbf{M}) = 0 \qquad (3.43)$$

Adding Eqs. (3.1)–(3.4) to Eqs. (3.40)–(3.43), respectively, another formulation of CFIE is obtained, and it is similar to Eq. (3.16). Applying the testing procedure with $\mathbf{f}_m$ as testing functions, one finds that

$$\left\langle \mathbf{f}_m, -\mathbf{E}_\nu^s - \mathbf{n} \times \mathbf{E}_\nu^s \right\rangle + \eta_1 \left\langle \mathbf{f}_m, -\mathbf{H}_\nu^s - \mathbf{n} \times \mathbf{H}_\nu^s \right\rangle$$

$$= \begin{cases} \left\langle \mathbf{f}_m, \mathbf{E}^i + \mathbf{n} \times \mathbf{E}^i \right\rangle + \eta_1 \left\langle \mathbf{f}_m, \mathbf{H}^i + \mathbf{n} \times \mathbf{H}^i \right\rangle, & \nu = 1 \\ 0, & \nu = 2 \end{cases} \qquad (3.44)$$

By using the vector identity $\mathbf{A} \cdot \mathbf{B} \times \mathbf{C} = \mathbf{C} \cdot \mathbf{A} \times \mathbf{B}$, the following relations are obtained:

$$\left\langle \mathbf{f}_m, \mathbf{n} \times \mathbf{E} \right\rangle = \left\langle -\mathbf{n} \times \mathbf{f}_m, \mathbf{E} \right\rangle = \left\langle -\mathbf{g}_m, \mathbf{E} \right\rangle \qquad (3.45a)$$

$$\left\langle \mathbf{f}_m, \mathbf{n} \times \mathbf{H} \right\rangle = \left\langle -\mathbf{n} \times \mathbf{f}_m, \mathbf{H} \right\rangle = \left\langle -\mathbf{g}_m, \mathbf{H} \right\rangle \qquad (3.45b)$$

Using the equalities in Eq. (3.45), Eq. (3.44) can be written as

$$\kappa \left\langle \mathbf{f}_m - \mathbf{g}_m, -\mathbf{E}_\nu^s \right\rangle + (1-\kappa)\eta_1 \left\langle \mathbf{f}_m - \mathbf{g}_m, -\mathbf{H}_\nu^s \right\rangle$$

$$= \begin{cases} \kappa \left\langle \mathbf{f}_m - \mathbf{g}_m, \mathbf{E}^i \right\rangle + (1-\kappa)\eta_1 \left\langle \mathbf{f}_m - \mathbf{g}_m, \mathbf{H}^i \right\rangle, & \nu = 1 \\ 0, & \nu = 2 \end{cases} \tag{3.46}$$

It is important to note that the testing function is $\mathbf{f}_m + \mathbf{g}_m$ in Eq. (3.39) and $\mathbf{f}_m - \mathbf{g}_m$ in Eq. (3.46) for the same CFIE. This implies that $\mathbf{f}_m + \mathbf{g}_m$ or $\mathbf{f}_m - \mathbf{g}_m$ can be used to test either the part of EFIE or MFIE, and this may result in eight different CFIE combinations as $\left\langle \mathbf{f}_m \pm \mathbf{g}_m, \mathbf{E} \right\rangle \pm \left\langle \mathbf{f}_m \pm \mathbf{g}_m, \mathbf{H} \right\rangle$. For this, the expression for CFIE is generalized by involving four parameters, f_E, g_E, f_H, and g_H, in conjunction with the testing functions as

$$(1-\kappa)\left\langle f_E \mathbf{f}_m + g_E \mathbf{g}_m, -\mathbf{E}_\nu^s \right\rangle + \kappa\eta_1 \left\langle f_H \mathbf{f}_m + g_H \mathbf{g}_m, -\mathbf{H}_\nu^s \right\rangle$$

$$= \begin{cases} (1-\kappa)\left\langle f_E \mathbf{f}_m + g_E \mathbf{g}_m, \mathbf{E}^i \right\rangle + \kappa\eta_1 \left\langle f_H \mathbf{f}_m + g_H \mathbf{g}_m, \mathbf{H}^i \right\rangle, & \nu = 1 \\ 0, & \nu = 2 \end{cases} \tag{3.47}$$

where κ is a parameter for linear combination that can have any value between 0 and 1. The testing coefficients f_E, g_E, f_H, and g_H may be either $+1$ or -1. If $f_E = g_E = f_H = g_H = 1$, then Eq. (3.47) is the same as Eq. (3.39), or it becomes Eq. (3.46) when $f_E = f_H = 1$ and $g_E = g_H = -1$.

To transform Eq. (3.47) into a matrix equation, this equation is divided into two parts: the electric field and the magnetic field parts. For the part related to the electric field only, we have

$$\left\langle f_E \mathbf{f}_m + g_E \mathbf{g}_m, -\mathbf{E}_\nu^s (\mathbf{J}, \mathbf{M}) \right\rangle = \begin{cases} \left\langle f_E \mathbf{f}_m + g_E \mathbf{g}_m, \mathbf{E}^i \right\rangle, & \nu = 1 \\ 0, & \nu = 2 \end{cases} \tag{3.48}$$

This equation is called the TENE formulation [6] in which TE stands for the testing of the electric field with the tangential function $\mathbf{f}_m$ and NE denotes that with the normal function $\mathbf{g}_m$. Thus, by using a similar procedure as in EFIE, we have

$$\sum_{n=1}^{N} jk_\nu \eta_\nu \left(A_{mn,\nu}^E - \frac{B_{mn,\nu}^E}{k_\nu^2} \right) J_n + \sum_{n=1}^{N} \left(C_{mn,\nu}^E + D_{mn,\nu}^E \right) M_n = V_{m,\nu}^E \tag{3.49a}$$

where

$$A_{mn,\nu}^E = f_E A_{mn,\nu}^f + g_E A_{mn,\nu}^g \tag{3.49b}$$

$$B_{mn,\nu}^E = f_E B_{mn,\nu}^f + g_E B_{mn,\nu}^g \tag{3.49c}$$

$$C_{mn,v}^E = \begin{cases} +f_E C_{mn}^f + g_E C_{mn}^g, & v = 1 \\ -f_E C_{mn}^f - g_E C_{mn}^g, & v = 2 \end{cases} \tag{3.49d}$$

$$D_{mn,v}^E = f_E D_{mn,v}^f + g_E D_{mn,v}^g \tag{3.49e}$$

$$V_{m,v}^E = \begin{cases} \int_S (f_E \mathbf{f}_m + g_E \mathbf{g}_m) \cdot \mathbf{E}^i dS, & v = 1 \\ 0, & v = 2 \end{cases} \tag{3.49f}$$

In Eqs. (3.49b)–(3.49e), the elements with the superscript f indicate that these elements take the inner products with $\mathbf{f}_m$, whereas those with superscript g denote that the inner products are taken with $\mathbf{g}_m$. The elements are given by

$$A_{mn,v}^f = \frac{1}{4\pi} \int_S \mathbf{f}_m(\mathbf{r}) \cdot \int_S \mathbf{f}_n(\mathbf{r}') G_v(\mathbf{r},\mathbf{r}') dS' dS \tag{3.50a}$$

$$A_{mn,v}^g = \frac{1}{4\pi} \int_S \mathbf{g}_m(\mathbf{r}) \cdot \int_S \mathbf{f}_n(\mathbf{r}') G_v(\mathbf{r},\mathbf{r}') dS' dS \tag{3.50b}$$

$$B_{mn,v}^f = \frac{1}{4\pi} \int_S \nabla_S \cdot \mathbf{f}_m(\mathbf{r}) \int_S \nabla_S' \cdot \mathbf{f}_n(\mathbf{r}') G_v(\mathbf{r},\mathbf{r}') dS' dS \tag{3.50c}$$

$$B_{mn,v}^g = \frac{1}{4\pi} \int_S \nabla_S \cdot \mathbf{g}_m(\mathbf{r}) \int_S \nabla_S' \cdot \mathbf{f}_n(\mathbf{r}') G_v(\mathbf{r},\mathbf{r}') dS' dS \tag{3.50d}$$

$$C_{mn}^f = \frac{1}{2} \int_S \mathbf{f}_m(\mathbf{r}) \cdot \mathbf{n} \times \mathbf{f}_n(\mathbf{r}) dS \tag{3.50e}$$

$$C_{mn}^g = \frac{1}{2} \int_S \mathbf{g}_m(\mathbf{r}) \cdot \mathbf{n} \times \mathbf{f}_n(\mathbf{r}) dS \tag{3.50f}$$

$$D_{mn,v}^f = \frac{1}{4\pi} \int_S \mathbf{f}_m(\mathbf{r}) \cdot \int_S \mathbf{f}_n(\mathbf{r}') \times \nabla' G_v(\mathbf{r},\mathbf{r}') dS' dS \tag{3.50g}$$

$$D_{mn,v}^g = \frac{1}{4\pi} \int_S \mathbf{g}_m(\mathbf{r}) \cdot \int_S \mathbf{f}_n(\mathbf{r}') \times \nabla' G_v(\mathbf{r},\mathbf{r}') dS' dS \tag{3.50h}$$

Therefore, from Eq. (3.49), a matrix equation can be obtained for the testing related only to the electric field as

$$\begin{bmatrix} \left[jk_1\eta_1 \left(A_{mn,1}^E - B_{mn,1}^E/k_1^2 \right) \right] & \left[C_{mn,1}^E + D_{mn,1}^E \right] \\ \left[jk_2\eta_2 \left(A_{mn,2}^E - B_{mn,2}^E/k_2^2 \right) \right] & \left[C_{mn,2}^E + D_{mn,2}^E \right] \end{bmatrix} \begin{bmatrix} [J_n] \\ [M_n] \end{bmatrix} = \begin{bmatrix} [V_{m,1}^E] \\ [V_{m,2}^E] \end{bmatrix} \tag{3.51}$$

Next, we write the testing equation corresponding to the magnetic field only from Eq. (3.47) as

$$\left\langle f_H \mathbf{f}_m + g_H \mathbf{g}_m, -\mathbf{H}_v^s(\mathbf{J}, \mathbf{M}) \right\rangle = \begin{cases} \left\langle f_H \mathbf{f}_m + g_H \mathbf{g}_m, \mathbf{H}^i \right\rangle, & v = 1 \\ 0, & v = 2 \end{cases} \tag{3.52}$$

This equation is called the THNH formulation [6] in which TH stands for the testing to the magnetic field with the tangential function $\mathbf{f}_n$ and NH denotes that with the normal function $\mathbf{g}_m$. Thus, by using a similar procedure as in the MFIE, the following is obtained

$$-\sum_{n=1}^{N} \left(C_{mn,v}^H + D_{mn,v}^H \right) J_n + \sum_{n=1}^{N} j \frac{k_v}{\eta_v} \left(A_{mn,v}^H - \frac{B_{mn,v}^H}{k_v^2} \right) M_n = V_{m,v}^H \tag{3.53a}$$

where

$$A_{mn,v}^H = f_H A_{mn,v}^f + g_H A_{mn,v}^g \tag{3.53b}$$

$$B_{mn,v}^H = f_H B_{mn,v}^f + g_H B_{mn,v}^g \tag{3.53c}$$

$$C_{mn,v}^H = \begin{cases} +f_H C_{mn}^f + g_H C_{mn}^g, & v = 1 \\ -f_H C_{mn}^f - g_H C_{mn}^g, & v = 2 \end{cases} \tag{3.53d}$$

$$D_{mn,v}^H = f_H D_{mn,v}^f + g_H D_{mn,v}^g \tag{3.53e}$$

$$V_{m,v}^H = \begin{cases} \int_S (f_H \mathbf{f}_m + g_H \mathbf{g}_m) \cdot \mathbf{H}^i dS, & v = 1 \\ 0, & v = 2 \end{cases} \tag{3.53f}$$

In Eqs. (3.53b)–(3.53e), the elements with superscripts f and g are the same as those in Eqs. (3.50). Note that Eq. (3.53) can be obtained directly from Eq. (3.49) by using duality. The matrix equation corresponding to Eq. (3.53) is given by

$$\begin{bmatrix} \left[-(C_{mn,1}^H + D_{mn,1}^H) \right] & \left[jk_1/\eta_1 \left(A_{mn,1}^H - B_{mn,1}^H/k_1^2 \right) \right] \\ \left[-(C_{mn,2}^H + D_{mn,2}^H) \right] & \left[jk_2/\eta_2 \left(A_{mn,2}^H - B_{mn,2}^H/k_2^2 \right) \right] \end{bmatrix} \begin{bmatrix} [J_n] \\ [M_n] \end{bmatrix} = \begin{bmatrix} [V_{m,1}^H] \\ [V_{m,2}^H] \end{bmatrix}. \tag{3.54}$$

The matrix equations Eq. (3.51) and Eq. (3.54) are rewritten, respectively, as

$$\left[Z_{mn}^E \right] [C_n] = \left[V_m^E \right] \tag{3.55a}$$

$$\left[Z_{mn}^H \right] [C_n] = \left[V_m^H \right] \tag{3.55b}$$

where $C_n = J_n$ and $C_{(N+n)} = M_n$ for $n = 1, 2, ..., N$. Finally, by combining TENE and THNH, as given in Eqs. (3.55a) and (3.55b), a matrix equation is formed for the general CFIE testing equation given by Eq. (3.47), which is termed TENE-THNH, as

$$[Z_{mn}][C_n] = [V_m] \qquad (3.56a)$$

where the matrix elements are given by

$$Z_{mn} = (1-\kappa)Z_{mn}^E + \kappa\eta_1 Z_{mn}^H \qquad (3.56b)$$

$$V_m = (1-\kappa)V_m^E + \kappa\eta_1 V_m^H \qquad (3.56c)$$

for $m = 1, 2, ..., 2N$ and $n = 1, 2, ..., 2N$. To solve the previous matrix equation, it is required that the integrals A_{mn}^f, B_{mn}^f, C_{mn}^f, D_{mn}^f, A_{mn}^g, B_{mn}^g, C_{mn}^g, and D_{mn}^g are determined first. The formulations of applying the RWG basis function on the triangular patches for approximating these integrals is given in the appendix at the end of this chapter.

3.7   SOLUTION OF THE PMCHW INTEGRAL EQUATION

For the PMCHW formulation of Eq. (3.17), the expansion functions for both the electric current $\mathbf{J}$ and the magnetic current $\mathbf{M}$ are $\mathbf{f}_n$ as that in the CFIE given by Eq. (3.38); and the testing function is $\mathbf{f}_m$. Applying the testing procedure to Eq. (3.17) yields

$$\left\langle \mathbf{f}_m, -\mathbf{E}_1^s(\mathbf{J},\mathbf{M}) \right\rangle + \left\langle \mathbf{f}_m, -\mathbf{E}_2^s(\mathbf{J},\mathbf{M}) \right\rangle = \left\langle \mathbf{f}_m, \mathbf{E}^i \right\rangle \qquad (3.57a)$$

$$\left\langle \mathbf{f}_m, -\mathbf{H}_1^s(\mathbf{J},\mathbf{M}) \right\rangle + \left\langle \mathbf{f}_m, -\mathbf{H}_2^s(\mathbf{J},\mathbf{M}) \right\rangle = \left\langle \mathbf{f}_m, \mathbf{H}^i \right\rangle \qquad (3.57b)$$

Eq. (3.57) contains two testing equations in which Eq. (3.57a) is related only to the electric field and Eq. (3.57b) is for the magnetic field. First, let us look at Eq. (3.57a) and compare it with the testing of EFIE in Eq. (3.24). It can be observed that they have a similar form, except Eq. (3.57a) includes the testing of both the scattered electric fields $\mathbf{E}_1^s$ and $\mathbf{E}_2^s$. Therefore, the result from applying MoM can be derived by adding the EFIE results in Eq. (3.33) for these two fields. Furthermore, the expansion function for $\mathbf{M}$ is now $\mathbf{f}_n$ and is different from $\mathbf{g}_n$ in the EFIE; thus, the formulation for the PMCHW has to be modified, mainly in the integrals $C_{mn,v}$ and $D_{mn,v}$. The integral $C_{mn,v}$ becomes zero in the PMCHW formulation because, for regions 1 and 2, the terms are negative of each other and when combined, yield zero value. Also, for the integral $D_{mn,v}$, one needs to replace the term $\mathbf{g}_n$ with $\mathbf{f}_n$. The result is given by

$$\sum_{n=1}^{N}\sum_{v=1}^{2} jk_v\eta_v\left(A_{mn,v} - \frac{B_{mn,v}}{k_v^2}\right)J_n + \sum_{n=1}^{N}\sum_{v=1}^{2} D_{mn,v}\, M_n = V_m^E \qquad (3.58a)$$

$$A_{mn,v} = \frac{1}{4\pi}\int_S \mathbf{f}_m(\mathbf{r})\cdot\int_S \mathbf{f}_n(\mathbf{r}')G_v(\mathbf{r},\mathbf{r}')\,dS'\,dS \qquad (3.58b)$$

$$B_{mn,v} = \frac{1}{4\pi}\int_S \nabla_S\cdot\mathbf{f}_m(\mathbf{r})\int_S \nabla_S'\cdot\mathbf{f}_n(\mathbf{r}')G_v(\mathbf{r},\mathbf{r}')\,dS'\,dS \qquad (3.58c)$$

$$D_{mn,v} = \frac{1}{4\pi}\int_S \mathbf{f}_m(\mathbf{r})\cdot\int_S \mathbf{f}_n(\mathbf{r}')\times\nabla' G_v(\mathbf{r},\mathbf{r}')\,dS'\,dS \qquad (3.58d)$$

$$V_m^E = \int_S \mathbf{f}_m(\mathbf{r})\cdot\mathbf{E}^i(\mathbf{r})\,dS \qquad (3.58e)$$

In Eq. (3.58), $A_{mn,v}$ and $B_{mn,v}$ are the same as those used for the EFIE in Eq. (3.32).

Consider the testing in Eq. (3.57b) and compare it with the testing of MFIE in Eq. (3.35); one can find that they have the same form, except Eq. (3.57b) includes both the testing of the scattered magnetic fields $\mathbf{H}_1^s$ and $\mathbf{H}_2^s$. Therefore, one can directly apply the result of MFIE in Eq. (3.36) for these two fields and sum them together. Also the integral $C_{mn,v}$ becomes zero, and the vector function $\mathbf{g}_n$ in the integral $D_{mn,v}$ has to be replaced by $\mathbf{f}_n$. The result can be expressed as

$$-\sum_{n=1}^{N}\sum_{v=1}^{2} D_{mn,v}J_n + \sum_{n=1}^{N}\sum_{v=1}^{2} j\frac{k_v}{\eta_v}\left(A_{mn,v} - \frac{B_{mn,v}}{k_v^2}\right)M_n = V_m^H \qquad (3.59a)$$

where

$$A_{mn,v} = \frac{1}{4\pi}\int_S \mathbf{f}_m(\mathbf{r})\cdot\int_S \mathbf{f}_n(\mathbf{r}')G_v(\mathbf{r},\mathbf{r}')\,dS'\,dS \qquad (3.59b)$$

$$B_{mn,v} = \frac{1}{4\pi}\int_S \nabla_S\cdot\mathbf{f}_m(\mathbf{r})\int_S \nabla_S'\cdot\mathbf{f}_n(\mathbf{r}')G_v(\mathbf{r},\mathbf{r}')\,dS'\,dS \qquad (3.59c)$$

$$D_{mn,v} = \frac{1}{4\pi}\int_S \mathbf{f}_m(\mathbf{r})\cdot\int_S \mathbf{f}_n(\mathbf{r}')\times\nabla' G_v(\mathbf{r},\mathbf{r}')\,dS'\,dS \qquad (3.59d)$$

$$V_m^H = \int_S \mathbf{f}_m(\mathbf{r})\cdot\mathbf{H}^i(\mathbf{r})\,dS \qquad (3.59e)$$

The integrals $A_{mn,v}$, $B_{mn,v}$, and $D_{mn,v}$ are the same as those in Eq. (3.58), respectively. Note that Eq. (3.59) can be obtained directly from Eq. (3.58) by using duality without any additional effort. Equation (3.58) and Eq. (3.59) are associated with each edge, where $m = 1, 2, ..., N$. Therefore, a matrix equation can be obtained as

$$\begin{bmatrix} \left[\displaystyle\sum_{v=1}^{2} jk_v\eta_v\left(A_{mn,v} - \frac{B_{mn,v}}{k_v^2}\right) \right] & \left[\displaystyle\sum_{v=1}^{2} D_{mn,v} \right] \\[1em] \left[-\displaystyle\sum_{v=1}^{2} D_{mn,v} \right] & \left[\displaystyle\sum_{v=1}^{2} j\frac{k_v}{\eta_v}\left(A_{mn,2} - \frac{B_{mn,v}}{k_v^2}\right) \right] \end{bmatrix} \begin{bmatrix} [J_n] \\[0.5em] [M_n] \end{bmatrix}$$

$$= \begin{bmatrix} [V_m^E] \\[0.5em] [V_m^H] \end{bmatrix} \qquad (3.60)$$

3.8   SOLUTION OF THE MULLER INTEGRAL EQUATION

The last formulation to be introduced is the Muller integral equation. It was first introduced in Eq. (3.18). Note that it becomes the PMCHW formulation when $\alpha = \beta = 1$. Similar to CFIE and PMCHW, the RWG function $\mathbf{f}_n$ is chosen as the expansion functions for both the electric current $\mathbf{J}$ and the magnetic current $\mathbf{M}$, as given by Eq. (3.38). The testing function selection follows the same procedure as CFIE. In the CFIE formulation, Eq. (3.47), a combination of RWG function $\mathbf{f}_n$ and its orthogonal component $\mathbf{g}_n = \mathbf{n} \times \mathbf{f}_n$, is used as the testing function to convert the integral equation into a matrix equation.

Now consider the testing procedure for the Muller equation. By substituting Eqs. (3.5)–(3.8) into Eqs. (3.18) and extracting the Cauchy principal value from the curl term, Eq. (3.18) can be rewritten as

$$\left. \begin{pmatrix} j\omega(\mathbf{A}_1 + \alpha\mathbf{A}_2) + \nabla(\Phi_1 + \alpha\Phi_2) + (1-\alpha)\frac{1}{2}\mathbf{n}\times\mathbf{M} \\[0.5em] + PV\left(\dfrac{1}{\varepsilon_1}\nabla\times\mathbf{F}_1 + \alpha\dfrac{1}{\varepsilon_2}\nabla\times\mathbf{F}_2\right) \end{pmatrix} \right|_{\text{tan}} = \left(\mathbf{E}^i\right)_{\text{tan}} \qquad (3.61a)$$

$$\left. \begin{pmatrix} j\omega(\mathbf{F}_1 + \beta\mathbf{F}_2) + \nabla(\Psi_1 + \beta\Psi_2) - (1-\beta)\frac{1}{2}\mathbf{n}\times\mathbf{J} \\[0.5em] - PV\left(\dfrac{1}{\mu_1}\nabla\times\mathbf{A}_1 + \beta\dfrac{1}{\mu_2}\nabla\times\mathbf{A}_2\right) \end{pmatrix} \right|_{\text{tan}} = \left(\mathbf{H}^i\right)_{\text{tan}} \qquad (3.61b)$$

where PV denotes the principal value. In the PMCHW formulation, when $\alpha = \beta = 1$, the third terms in Eqs. (3.61a) and (3.61b) are dropped out. In this PMCHW integral equation, an accurate and stable solution can be obtained by using the RWG function $\mathbf{f}_n$ as the testing functions [2,6,13]. To test the third terms $\mathbf{n}\times\mathbf{M}$ and $\mathbf{n}\times\mathbf{J}$ in Eq. (3.61), the spatially orthogonal component $\mathbf{g}_m$ is applied to the Muller equation. Now, a general expression is presented for testing the Muller equations using the following four parameters in conjunction with the testing functions as:

$$\left\langle f_E \mathbf{f}_m + g_E \mathbf{g}_m, - \mathbf{E}_1^s(\mathbf{J},\mathbf{M}) - \alpha \mathbf{E}_2^s(\mathbf{J},\mathbf{M}) \right\rangle = \left\langle f_E \mathbf{f}_m + g_E \mathbf{g}_m, \mathbf{E}^i \right\rangle \qquad (3.62a)$$

$$\left\langle f_H \mathbf{f}_m + g_H \mathbf{g}_m, - \mathbf{H}_1^s(\mathbf{J},\mathbf{M}) - \beta \mathbf{H}_2^s(\mathbf{J},\mathbf{M}) \right\rangle = \left\langle f_H \mathbf{f}_m + g_H \mathbf{g}_m, \mathbf{H}^i \right\rangle \qquad (3.62b)$$

where the testing coefficients, f_E, g_E, f_H, and g_H, may be +1 or −1. This equation is referred to as the TENE-THNH formulation of the Muller equation [13].

3.9   STUDY OF VARIOUS CFIE FORMULATIONS

In this section, the general CFIE formulation (TENE-THNH) is studied, which is described by Eq. (3.47) and results in the matrix given by Eq. (3.56). As discussed in Section 3.6, there are eight different formulations for the general CFIE formulation by setting the testing coefficients $f_E = 1$ and g_E, f_H, and g_H to either +1 or −1. All eight combinations are listed in Table 3.2, and they are simply called CFIE-1 to CFIE-8. The value $\kappa = 0.5$ is used for all cases.

TABLE 3.2. CFIE Formulations with Different Combinations of Testing Coefficients and the Averaged Difference of Monostatic RCS Between the Mie Series and the CFIE Solutions for the Dielectric Sphere in Figure 3.2

Formulation	Testing coefficients				Testing method		$\Delta\sigma$
	f_E	g_E	f_H	g_H	E-Field	H-Field	(dBm^2)
CFIE-1	+1	+1	+1	+1	$\left\langle \mathbf{f}_m + \mathbf{g}_m, \mathbf{E} \right\rangle$	$\left\langle \mathbf{f}_m + \mathbf{g}_m, \mathbf{H} \right\rangle$	1.02
CFIE-2	+1	+1	+1	−1	$\left\langle \mathbf{f}_m + \mathbf{g}_m, \mathbf{E} \right\rangle$	$\left\langle \mathbf{f}_m - \mathbf{g}_m, \mathbf{H} \right\rangle$	4.61
CFIE-3	+1	+1	−1	+1	$\left\langle \mathbf{f}_m + \mathbf{g}_m, \mathbf{E} \right\rangle$	$\left\langle -\mathbf{f}_m + \mathbf{g}_m, \mathbf{H} \right\rangle$	0.40
CFIE-4	+1	+1	−1	−1	$\left\langle \mathbf{f}_m + \mathbf{g}_m, \mathbf{E} \right\rangle$	$\left\langle -\mathbf{f}_m - \mathbf{g}_m, \mathbf{H} \right\rangle$	0.73
CFIE-5	+1	−1	+1	+1	$\left\langle \mathbf{f}_m - \mathbf{g}_m, \mathbf{E} \right\rangle$	$\left\langle \mathbf{f}_m + \mathbf{g}_m, \mathbf{H} \right\rangle$	0.41
CFIE-6	+1	−1	+1	−1	$\left\langle \mathbf{f}_m - \mathbf{g}_m, \mathbf{E} \right\rangle$	$\left\langle \mathbf{f}_m - \mathbf{g}_m, \mathbf{H} \right\rangle$	0.85
CFIE-7	+1	−1	−1	+1	$\left\langle \mathbf{f}_m - \mathbf{g}_m, \mathbf{E} \right\rangle$	$\left\langle -\mathbf{f}_m + \mathbf{g}_m, \mathbf{H} \right\rangle$	1.10
CFIE-8	+1	−1	−1	−1	$\left\langle \mathbf{f}_m - \mathbf{g}_m, \mathbf{E} \right\rangle$	$\left\langle -\mathbf{f}_m - \mathbf{g}_m, \mathbf{H} \right\rangle$	4.61

To verify the efficacy of these formulations, scattering from a dielectric sphere is computed. The sphere has a diameter of 1 m, a relative permittivity $\varepsilon_r = 2$ and centered at the origin, as shown in Figure 3.2. By selecting 12 and 24 divisions along θ and ϕ directions, the sphere is discretized by 528 patches and 792 edges. In the numerical calculation, the sphere is illuminated from the top by an incident x-polarized plane wave with the propagation vector $\hat{\mathbf{k}} = -\hat{\mathbf{z}}$. The analysis is to be done over a frequency range of $0 < f \leq 400$ MHz at an interval of 4 MHz with

100 frequency samples. In addition, the numerical results are compared with the analytical Mie series solution.

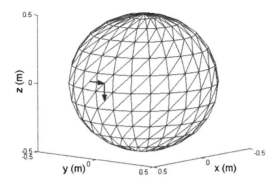

Figure 3.2. Triangle surface patch model of a dielectric sphere (diameter of 1 m and a relative permittivity $\varepsilon_r = 2$). θ- and ϕ-directed arrows represent position and direction of sampled currents **J** and **M**, respectively.

Figure 3.3 shows the monostatic RCS of the sphere obtained from the eight CFIE formulations in Table 3.2 and the Mie solution, in which CFIE-1–CFIE-4 are shown in Figure 3.3(*a*), whereas CFIE-5–CFIE-8 are given in Figure 3.3(*b*). As shown in Figure 3.3(*a*), CFIE-3 agrees with the Mie solution well, whereas CFIE-2 has an obvious discrepancy for the frequencies higher than 200MHz. CFIE-1 and CFIE-4 have two breakpoints in the proximity of 262 MHz and 369 MHz that should be the internal resonant frequencies of the sphere. In Figure 3.3(*b*), only CFIE-5 shows good agreement with the Mie solution, but the others do not. The CFIE-6 and CFIE-7 also suffer from the breakpoints near the internal resonant frequencies.

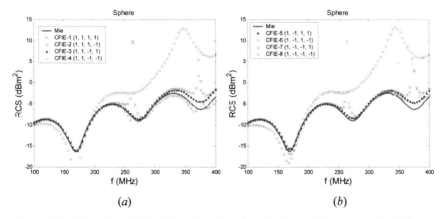

(*a*)   (*b*)

Figure 3.3. Monostatic RCS of the dielectric sphere in Figure 3.2 computed by different CFIE formulations.

The discrepancy between the CFIE solutions and the Mie series solution in the studied frequency range can be evaluated by defining an averaged difference of the monostatic RCS as

$$\Delta\sigma = \left[\sum_1^M \left| \sigma_{\text{Mie}} - \sigma_{\text{numerical}} \right| \right] / M \tag{3.63}$$

where σ denotes the RCS. M is the number of samples in the whole frequency range, and it is equal to 100 in this example. These average differences $\Delta\sigma$ for the eight CFIE formulations are given also in Table 3.2. CFIE-3 and CFIE-5 have the lowest errors as 0.40 and 0.41, respectively, which is much lower than all the others.

In the previous CFIE formulation, it is termed TENE-THNH, as there are four terms to form the CFIE (i.e., CFIE = EFIE + $\mathbf{n}\times$EFIE + MFIE + $\mathbf{n}\times$MFIE with $\mathbf{f}_m$ as the testing functions or CFIE = EFIE + MFIE with $\mathbf{f}_m + \mathbf{g}_m$ as the testing functions). One suggestion from [6] is to drop one of these four terms and thus the CFIE formulations can further be extended to TENE-TH, TENE-NH, TE-THNH, or NE-THNH, depending on which term is neglected. For example, by setting g_H to zero, the testing of the magnetic field contains only the tangential function $\mathbf{f}_m$ and thus is referred to as TENE-TH. Applying this scheme to the eight different CFIE formulations of Table 3.2 results in 16 possible formulations for the CFIE with different testing coefficients, which are listed in Table 3.3.

TABLE 3.3. CFIE Formulations with Different Combinations of Testing Coefficients and the Averaged Difference of Far Field and Monostatic RCS between the Mie Series and the CFIE Solution for the Dielectric Sphere depicted in Figure 3.2

Formulation		Testing coefficients				Δe_θ	Δe_ϕ	$\Delta\sigma$
		f_E	g_E	f_H	g_H	(mV)	(mV)	(dBm^2)
TENE-TH	(1)	+1	+1	+1	0	12.0	16.7	0.79
	(2)	+1	+1	−1	0	4.1	6.4	0.30
	(3)	+1	−1	+1	0	4.7	6.0	0.34
	(4)	+1	−1	−1	0	4.5	24.0	0.51
TENE-NH	(1)	+1	+1	0	+1	3.5	12.9	0.28
	(2)	+1	+1	0	−1	7.8	32.1	0.22
	(3)	+1	−1	0	+1	2.4	20.3	0.37
	(4)	+1	−1	0	−1	8.2	21.4	0.48
TE-THNH	(1)	+1	0	+1	+1	9.4	5.5	0.61
	(2)	+1	0	+1	−1	8.1	9.8	0.59
	(3)	+1	0	−1	+1	8.7	7.2	0.55
	(4)	+1	0	−1	−1	6.9	16.4	0.49
NE-THNH	(1)	0	+1	+1	+1	5.1	12.9	0.42
	(2)	0	+1	+1	−1	5.7	16.4	0.48
	(3)	0	+1	−1	+1	5.7	21.5	0.52
	(4)	0	+1	−1	−1	5.3	15.6	0.45

The computation results of these 16 formulations for the dielectric sphere of Figure 3.2 are shown in Figure 3.4. The results from TENE-TH, TENE-NH, TE-THNH, and NE-THNH are given in Figure 3.4 (a), (b), (c), and (d), respectively, along with the Mie series solution represented by the solid line.

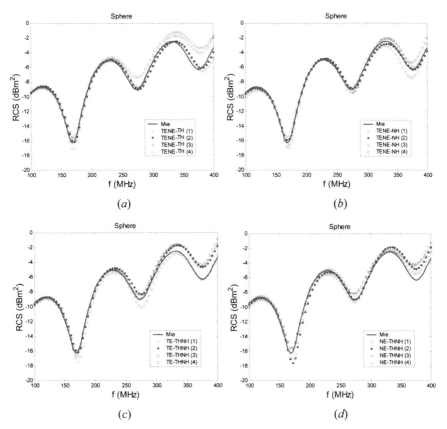

Figure 3.4. Monostatic RCS of the dielectric sphere in Figure 3.2 computed by the 16 CFIE formulations of Table 3.3: (a) TENE-TH. (b) TENE-NH. (c) TE-THNH. (d) NE-THNH.

It is noted that the internal resonance problem does not occur in all 16 CFIE formulations. All the numerical solutions match with the Mie series solution well. Only a small difference can be observed in the high-frequency region. The discrepancies between the numerical and the Mie solution can be evaluated by calculating the averaged difference in the monostatic RCS using Eq. (3.63). By using the same definition as in Eq. (3.63), the averaged difference in the far field θ- and ϕ-components can also be computed. The average differences in the monostatic RCS and the far field are summarized in Table 3.3. Among the 16 cases, the smallest averaged difference in the RCS is 0.22 dBm^2 from the TENE-NH(2) case, whereas TENE-TH(2) not only has a small difference in RCS as 0.30

dBm^2, but also it is the most accurate for both the θ- and ϕ-components in the far field ($\Delta e_\theta = 4.1$ and $\Delta e_\phi = 6.4$). Figure 3.5 shows the far field for TENE-TH(2) and TENE-NH(2) against the Mie solution. The TENE-TH(2) clearly shows a better agreement to the Mie solution than the TENE-NH(2). It can be confirmed that the TENE-TH(2) formulation, which is obtained by setting $f_E = 1$, $g_E = 1$, $f_H = -1$, and $g_H = 0$ in Eq. (3.47), can offer a better approximation to the exact solution in both the far field and the monostatic RCS computations.

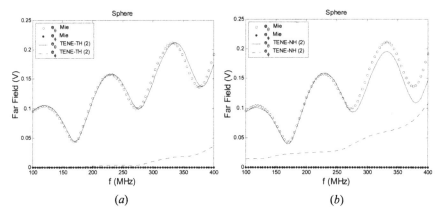

Figure 3.5. Comparison of the far field for the dielectric sphere computed by Mie and CFIE solutions: (a) TENE-TH(2) ($f_E = 1$, $g_E = 1$, $f_H = -1$, $g_H = 0$). (b) TENE-NH(2) ($f_E = 1$, $g_E = 1$, $f_H = 0$, $g_H = -1$).

In the CFIE formulation, there is a parameter κ representing the linear combination of the electric and the magnetic fields. Now the effect of the combination parameter κ is investigated and is shown in Figure 3.6, where only TENE-TH (2) formulation is considered.

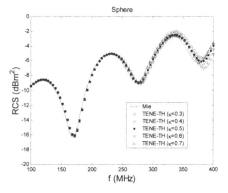

Figure 3.6. Monostatic RCS of the dielectric sphere computed by TENE-TH(2) ($f_E = 1$, $g_E = 1$, $f_H = -1$, $g_H = 0$) as varying CFIE combining parameter κ.

The solutions of TENE-TH(2) are plotted against the Mie series solution in Figure 3.6 for different values 0.3, 0.4, 0.5, 0.6, or 0.7 of κ. All the results show an excellent agreement with the Mie solution at the lower frequency region and very small discrepancies at the higher frequencies. Table 3.4 summarizes the averaged difference in the far field and monostatic RCS. It is evident from Figure 3.6 and Table 3.4 that the solution of CFIE is not so sensitive to the choice of the value of κ over a wide range.

Table 3.4. Average Difference of the Far field and Monostatic RCS for the Dielectric Sphere Computed by Using TENE-TH(2) Formulation and Different Values of κ

κ	Δe_θ (mV)	Δe_ϕ (mV)	$\Delta \sigma$ (dBm^2)
0.3	6.0	5.0	0.41
0.4	4.8	5.7	0.34
0.5	4.1	6.4	0.30
0.6	4.0	6.8	0.29
0.7	4.5	6.9	0.31

3.10   STUDY OF DIFFERENT MULLER FORMULATIONS

The testing of the Muller integral equation is given by Eq. (3.62). It contains the four testing coefficients f_E, g_E, f_H, and g_H. Similar to the CFIE, we may set these coefficients to +1 or –1 and obtain four different cases for the Muller formulation: TENE-THNH(1) to TENE-THNH(4). Also, one of the terms can be ignored, and thus, four additional types of formulations are obtained and termed as TENE-TH ($g_H = 0$), TENE-NH ($f_H = 0$), TE-THNH ($g_E = 0$), and NE-THNH ($f_E = 0$), depending on which term is neglected. By selecting different signs of the testing coefficients, one may have a total of eight possible cases for these four types. All 12 formulations are listed in Table 3.5.

To evaluate the validity of the 12 different formulations of the Muller integral equation, the bistatic RCS of a dielectric sphere having a diameter of 1 m, a relative permittivity of $\varepsilon_r = 4$, and centered at the origin, as shown in Figure 3.7, is computed and compared with the Mie series solution. Figure 3.8 shows the bistatic RCS solution computed from the four cases of TENE-THNH formulations against the Mie solution. All numerical results using the Muller TENE-THNH formulations do not exhibit the spurious resonance and agree well with the Mie solution. The average differences between the Muller formulations and the Mie solution are calculated by using Eq. (3.63) and shown in Table 3.5. The values of the average differences are from about 0.10 to 0.16. In fact, the error is smaller than that of the PMCHW and CFIE formulations for the same example, which are 0.44 and 0.19, respectively. However, when using the two formulations of TENE-THNH(2) and TENE-THNH(3) to solve for the equivalent currents, the results are unstable in the low-frequency region (0–100 MHz).

The bistatic RCS computed using the scheme with dropping one of the testing terms are shown in Figure 3.9 and compared with the Mie solution. It is observed that the solutions from TENE-TH in Figure 3.9(*a*) and from TE-THNH in Figure 3.9(*c*) match with the Mie solution, whereas the solutions from TENE-NH in Figure 3.9(*b*) and the NE-THNH in Figure 3.9(*d*) are actually inaccurate, when the Muller integral equation given by Eq. (3.18) is tested with $\mathbf{g}_m$ only. The average differences for the bistatic RCS between these formulations and the Mie solution are also given in Table 3.5. Compared with the four cases of TENE-THNH, the scheme with the removal of one of the testing terms causes a higher discrepancy to the exact solution.

Table 3.5. **Muller Formulation with Different Combinations of the Testing Coefficients and the Average Difference of Bistatic RCS Between the Mie Series and the Muller Solutions for the Dielectric Sphere**

Formulation	Testing coefficients				Testing method		$\Delta\sigma$
	f_E	g_E	f_H	g_H	E-field	H-field	(dBm^2)
TENE-THNH(1)	+1	+1	+1	+1	$\langle \mathbf{f}_m + \mathbf{g}_m, \mathbf{E} \rangle$	$\langle \mathbf{f}_m + \mathbf{g}_m, \mathbf{H} \rangle$	0.12
TENE-THNH(2)	+1	+1	+1	−1	$\langle \mathbf{f}_m + \mathbf{g}_m, \mathbf{E} \rangle$	$\langle \mathbf{f}_m - \mathbf{g}_m, \mathbf{H} \rangle$	0.14
TENE-THNH(3)	+1	−1	+1	+1	$\langle \mathbf{f}_m - \mathbf{g}_m, \mathbf{E} \rangle$	$\langle \mathbf{f}_m + \mathbf{g}_m, \mathbf{H} \rangle$	0.10
TENE-THNH(4)	+1	−1	+1	−1	$\langle \mathbf{f}_m - \mathbf{g}_m, \mathbf{E} \rangle$	$\langle \mathbf{f}_m - \mathbf{g}_m, \mathbf{H} \rangle$	0.16
TENE-TH(1)	+1	+1	+1	0	$\langle \mathbf{f}_m + \mathbf{g}_m, \mathbf{E} \rangle$	$\langle \mathbf{f}_m, \mathbf{H} \rangle$	0.31
TENE-TH(2)	+1	−1	+1	0	$\langle \mathbf{f}_m - \mathbf{g}_m, \mathbf{E} \rangle$	$\langle \mathbf{f}_m, \mathbf{H} \rangle$	0.34
TENE-NH(1)	+1	+1	0	+1	$\langle \mathbf{f}_m + \mathbf{g}_m, \mathbf{E} \rangle$	$\langle \mathbf{g}_m, \mathbf{H} \rangle$	2.38
TENE-NH(2)	+1	−1	0	+1	$\langle \mathbf{f}_m - \mathbf{g}_m, \mathbf{E} \rangle$	$\langle \mathbf{g}_m, \mathbf{H} \rangle$	2.15
TE-THNH(1)	+1	0	+1	+1	$\langle \mathbf{f}_m, \mathbf{E} \rangle$	$\langle \mathbf{f}_m + \mathbf{g}_m, \mathbf{H} \rangle$	0.14
TE-THNH(2)	+1	0	+1	−1	$\langle \mathbf{f}_m, \mathbf{E} \rangle$	$\langle \mathbf{f}_m - \mathbf{g}_m, \mathbf{H} \rangle$	0.19
NE-THNH(1)	0	+1	+1	+1	$\langle \mathbf{g}_m, \mathbf{E} \rangle$	$\langle \mathbf{f}_m + \mathbf{g}_m, \mathbf{H} \rangle$	1.62
NE-THNH(2)	0	+1	+1	−1	$\langle \mathbf{g}_m, \mathbf{E} \rangle$	$\langle \mathbf{f}_m - \mathbf{g}_m, \mathbf{H} \rangle$	2.07

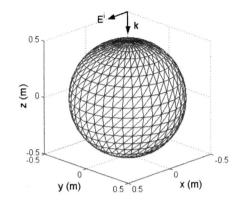

Figure 3.7. Triangular surface patching of a dielectric sphere with diameter of 1 m and a relative permittivity $\varepsilon_r = 4$. There are 1,224 patches and 1,836 edges.

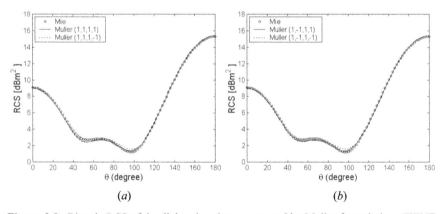

Figure 3.8. Bistatic RCS of the dielectric sphere computed by Muller formulations (TENE-THNH): (a) $\langle \mathbf{f}_m + \mathbf{g}_m, \mathbf{E} \rangle$ and $\langle \mathbf{f}_m \pm \mathbf{g}_m, \mathbf{H} \rangle$. (b) $\langle \mathbf{f}_m - \mathbf{g}_m, \mathbf{E} \rangle$ and $\langle \mathbf{f}_m \pm \mathbf{g}_m, \mathbf{H} \rangle$.

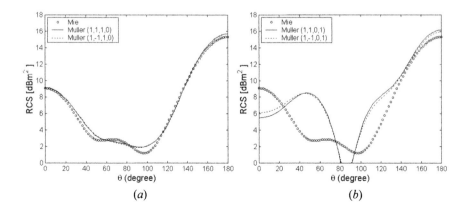

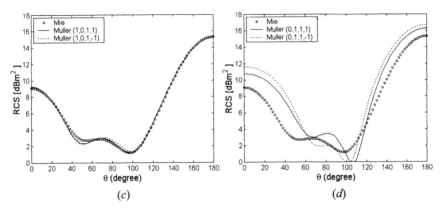

Figure 3.9. Bistatic RCS of the dielectric sphere computed by Muller formulations: (a) TENE-TH, $\langle \mathbf{f}_m \pm \mathbf{g}_m, \mathbf{E} \rangle$ and $\langle \mathbf{f}_m, \mathbf{H} \rangle$, (b) TENE-NH, $\langle \mathbf{f}_m \pm \mathbf{g}_m, \mathbf{E} \rangle$ and $\langle \mathbf{g}_m, \mathbf{H} \rangle$. (c) TE-THNH, $\langle \mathbf{f}_m, \mathbf{E} \rangle$ and $\langle \mathbf{f}_m \pm \mathbf{g}_m, \mathbf{H} \rangle$, (d) NE-THNH, $\langle \mathbf{g}_m, \mathbf{E} \rangle$ and $\langle \mathbf{f}_m \pm \mathbf{g}_m, \mathbf{H} \rangle$.

3.11 NUMERICAL EXAMPLES

In this section, the numerical results using the following ten different formulations

1. PMCHW
2. EFIE
3. MFIE
4. CFIE: TENE with $f_E = 1$, $g_E = 1$, $f_H = 0$, $g_H = 0$
5. CFIE: THNH with $f_E = 0$, $g_E = 0$, $f_H = -1$, $g_H = 1$
6. CFIE: TENE-THNH with $f_E = 1$, $g_E = 1$, $f_H = -1$, $g_H = 1$
7. CFIE: TENE-TH with $f_E = 1$, $g_E = 1$, $f_H = -1$, $g_H = 0$
8. CFIE: TENE-NH with $f_E = 1$, $g_E = 1$, $f_H = 0$, $g_H = 1$
9. CFIE: TE-THNH with $f_E = 1$, $g_E = 0$, $f_H = -1$, $g_H = 1$
10. CFIE: NE-THNH with $f_E = 0$, $g_E = 1$, $f_H = -1$, $g_H = 1$

for three differently shaped dielectric objects are presented. Different versions of the CFIE are selected to compare with the EFIE, MFIE, and PMCHW methods. The sixth method in the list is the general form of CFIE, called TENE-THNH, which contains both the tangential and normal testing to the electric and magnetic fields. As introduced in Section 3.9, peaks and discontinuities may be exhibited in some TENE-THNH formulations. Here the testing coefficients are chosen as $f_E = 1$, $g_E = 1$, $f_H = -1$, and $g_H = 1$, which is the CFIE-3 in Table 3.2. It offers the best approximation to the Mie solution and does not suffer from the internal resonance problem among all TENE-THNH formulations for the example in Section 3.9. Methods 7–10 can be referred to as the TENE-TH(2), TENE-NH(1), TE-THNH(3), and NE-THNH(3) formulations of the CFIE in Table 3.3, respectively. In each of them, one of the testing terms is removed. The TENE or THNH methods consist of either an electric or magnetic field only; this is considered to be a special case of CFIE, with $\kappa = 0$ or $\kappa = 1$ to differentiate from

EFIE or MFIE as previously described in Sections 3.4 and 3.5. The value $\kappa = 0.5$ is chosen for all CFIE formulations.

The object to be analyzed is a dielectric sphere, a cube, or a cylinder with a relative permittivity $\varepsilon_r = 2$. In the numerical calculation, the scatterers are illuminated from the top by an incident x-polarized plane wave with a propagation vector $\hat{\mathbf{k}} = -\hat{\mathbf{z}}$, as used in the previous section. The frequency range over which the results are calculated is $0 < f \leq 400$ MHz at an interval of 4 MHz. The equivalent electric and magnetic currents, the far field, and the monostatic RCS are computed. The equivalent currents computed by the PMCHW formulation are used to compare with the results from the other formulations. In the analysis of the sphere, the computed far fields and RCS are compared with the analytical Mie series solution. Because the analytical solution of the far field or the monostatic RCS for a cube or cylinder are not known, the commercially available higher order basis-function-based integral equation solver (HOBBIES) [11] solution is used instead. Both the Mie and HOBBIES solutions are calculated at the same frequency interval of 4 MHz.

3.11.1 Dielectric Sphere

The first example is the dielectric sphere as shown in Figure 3.2. The θ-directed electric current and the ϕ-directed magnetic current, as indicated by arrows in Figure 3.2, are observed.

3.11.1.1 Analysis of a Sphere by PMCHW. Figure 3.10 shows the computed far field and monostatic RCS using the PMCHW formulation. The far field and RCS agree well with the Mie solution except for a small difference in the high-frequency region. The currents are used to compare with other numerical results and thus will be shown later.

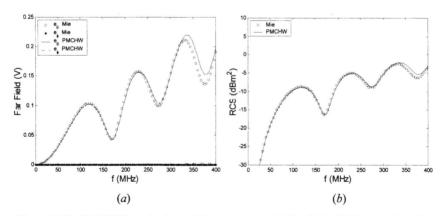

(a) (b)

Figure 3.10. PMCHW results for a dielectric sphere: (a) Far field. (b) Monostatic RCS.

3.11.1.2 Analysis of a Sphere by EFIE. Figure 3.11 shows the electric and magnetic currents, the far field, and the monostatic RCS computed by the EFIE formulation in Eq. (3.33). The electric current of the EFIE as shown in Figure 3.11(*a*) exhibits a null at about 369 MHz whereas that of the PMCHW does not. The basis function for the magnetic current in the EFIE is $\mathbf{g}_n$ and it is different from $\mathbf{f}_n$ in PMCHW. They cannot be compared so that only the result of EFIE is shown in Figure 3.11(*b*). From Figure 3.11 (*c*) and (*d*), it is found that there are peaks and discontinuities near the internal resonant frequencies at 262 MHz and 369 MHz in the far field and RCS results, although they match well with the analytical Mie solutions in the low-frequency region.

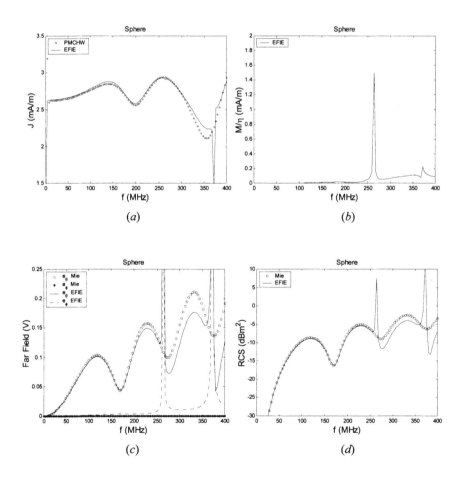

Figure 3.11. EFIE results for a dielectric sphere: (*a*) Electric current. (*b*) Magnetic current. (*c*) Far field. (*d*) Monostatic RCS.

3.11.1.3 Analysis of a Sphere by MFIE. Figure 3.12 shows the results of MFIE obtained using Eq. (3.37). Because the basis function for the electric current in the MFIE is $\mathbf{g}_n$ and it is different from $\mathbf{f}_n$ in the PMCHW, they are not compared in Figure 3.12(a). The magnetic current in Figure 3.12(b) agrees well with those from PMCHW except at the internal resonant frequencies. From Figure 3.12(c) and (d), it is shown clearly that there are peaks and discontinuities near the resonant frequencies of 262 MHz and 369 MHz in the far field and RCS results, despite the good agreement with the Mie solution at the low-frequency region.

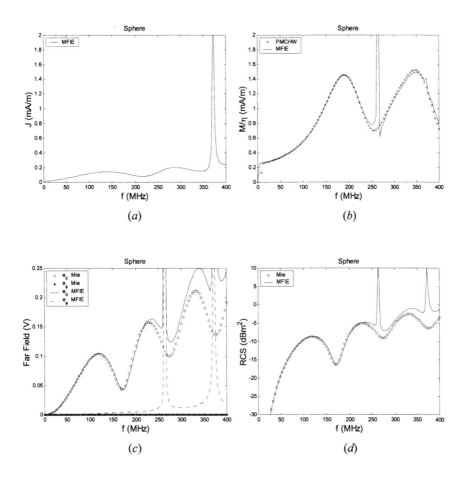

Figure 3.12. MFIE results for a dielectric sphere: (a) Electric current. (b) Magnetic current. (c) Far field. (d) Monostatic RCS.

3.11.1.4 Analysis of a Sphere by TENE of CFIE. Figure 3.13 shows the computed numerical results using the TENE formulation. Similar to the EFIE, the results from TENE is related only to the electric field. Because of this, peaks at the internal resonant frequencies are observed in the currents, the far field, and the RCS. In contrast to the EFIE results in Figure 3.11, the electric current in the EFIE formulation shows a better agreement with the PMCHW solution than that from TENE, even though $\mathbf{g}_m$ is used in EFIE as the expansion function for the magnetic current.

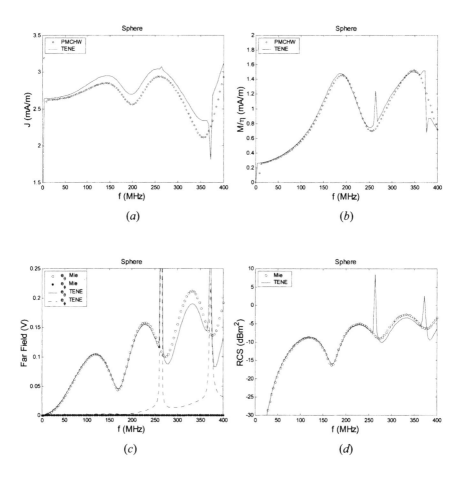

Figure 3.13.   TENE ($f_E = 1$, $g_E = 1$, $f_H = 0$, $g_H = 0$) results for a dielectric sphere: (*a*) Electric current. (*b*) Magnetic current. (*c*) Far field. (*d*) Monostatic RCS.

3.11.1.5 Analysis of a Sphere by THNH of CFIE. Figure 3.14 shows the numerical results for THNH. Similar to the MFIE, the THNH formulation is based on only the magnetic field. For this reason, peaks at the internal resonant frequencies are observed in the figures. It is interesting to note that the magnetic current in the MFIE formulation agrees better with the PMCHW solution than those from THNH, even though $\mathbf{g}_m$ is used in MFIE as the expansion function for electric current.

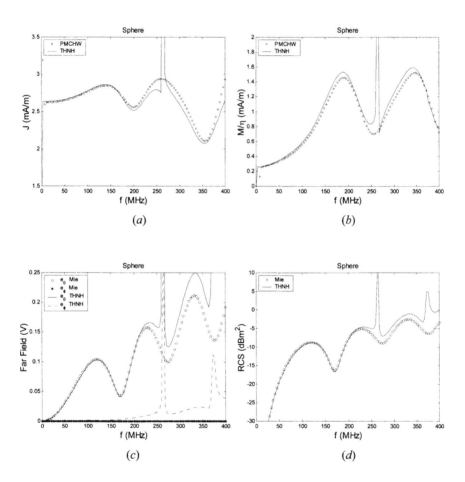

Figure 3.14. THNH ($f_E = 0$, $g_E = 0$, $f_H = -1$, $g_H = 1$) results for a dielectric sphere: (*a*) Electric current. (*b*) Magnetic current. (*c*) Far field. (*d*) Monostatic RCS.

3.11.1.6 Analysis of a Sphere by TENE-THNH of CFIE. Figure 3.15 gives the
results computed by the general form of CFIE, called TENE-THNH, which
combines the TENE and THNH formulations. In this version of CFIE, no resonant
peak or discontinuity is observed in the plots. As shown in Figure 3.15, the TENE-
THNH results agree well with the PMCHW for the equivalent currents and with
the Mie solution for the far fields and RCS, respectively.

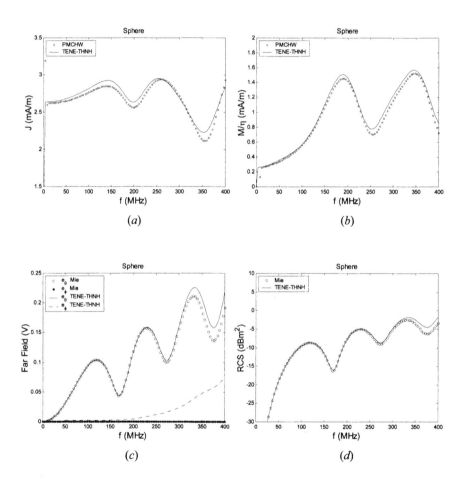

Figure 3.15. TENE-THNH ($f_E = 1$, $g_E = 1$, $f_H = -1$, $g_H = 1$) results for a dielectric
sphere: (*a*) Electric current. (*b*) Magnetic current. (*c*) Far field. (*d*) Monostatic RCS.

3.11.1.7 Analysis of a Sphere by TENE-TH of CFIE.

The TENE-TH formulation is obtained when the term $\langle \mathbf{g}_m, \mathbf{H} \rangle$ is removed from the general form of CFIE, as discussed in Section 3.9. The TENE-TH results are shown in Figure 3.16. All numerical results given by TENE-TH do not show the internal resonant problem. The results of the magnetic current, the far field, and RCS agree excellently with the Mie solution, whereas the electric current exhibits a higher value than PMCHW in the entire frequency range.

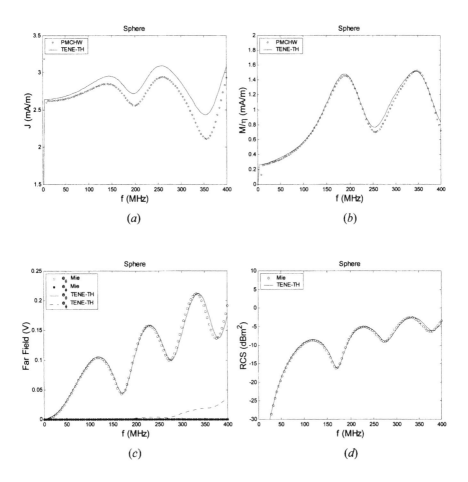

Figure 3.16. TENE-TH ($f_E = 1$, $g_E = 1$, $f_H = -1$, $g_H = 0$) results for a dielectric sphere: (a) Electric current. (b) Magnetic current. (c) Far field. (d) Monostatic RCS.

3.11.1.8 Analysis of a Sphere by TENE-NH of CFIE. The TENE-NH formulation
is obtained when the term $\langle \mathbf{f}_m, \mathbf{H} \rangle$ is removed from the general form of CFIE, as
discussed in Section 3.9. The TENE-NH results are shown in Figure 3.17. All
numerical results do not show the internal resonant problem. The results of the
RCS agree well with the Mie solution, whereas the currents and the far field show
a little difference from the PMCHW and the Mie solutions.

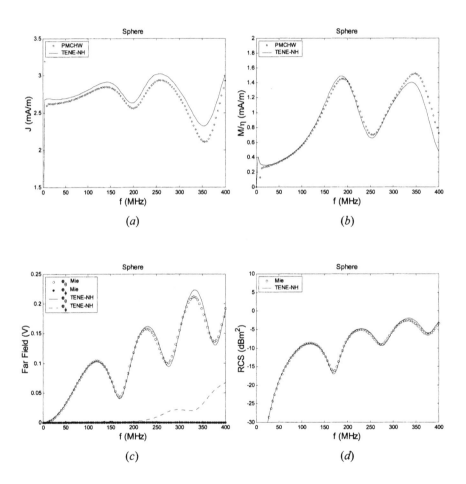

Figure 3.17. TENE-NH $(f_E = 1, g_E = 1, f_H = 0, g_H = 1)$ results for a dielectric
sphere: (*a*) Electric current. (*b*) Magnetic current. (*c*) Far field. (*d*) Monostatic RCS.

3.11.1.9 Analysis of a Sphere by TE-THNH of CFIE. The TE-THNH formulation is obtained when the term $\langle \mathbf{g}_m, \mathbf{E} \rangle$ is removed from the general form of CFIE, as discussed in Section 3.9. The TE-THNH results are shown in Figure 3.18. All numerical results do not show the internal resonant problem. The results of the electric and magnetic currents, the far field, and RCS generally agree with the PMCHW and the Mie solutions but little errors appear in the high-frequency region.

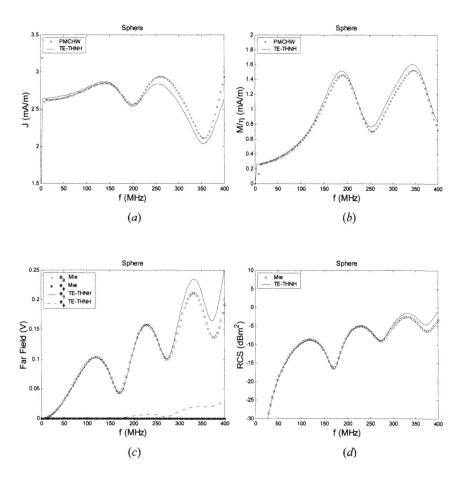

Figure 3.18. TE-THNH ($f_E = 1$, $g_E = 0$, $f_H = -1$, $g_H = 1$) results for a dielectric sphere: (*a*) Electric current. (*b*) Magnetic current. (*c*) Far field. (*d*) Monostatic RCS.

3.11.1.10 Analysis of a Sphere by NE-THNH of CFIE. The NE-THNH formulation is obtained when the term $\langle \mathbf{f}_m, \mathbf{E} \rangle$ is removed from the general form of CFIE, as discussed in Section 3.9. The NE-THNH results are shown in Figure 3.19. All numerical results do not show the internal resonant problem. The results of the electric and magnetic currents, the far field, and RCS agree generally with the PMCHW and the Mie solutions, but little errors appear in the high-frequency region.

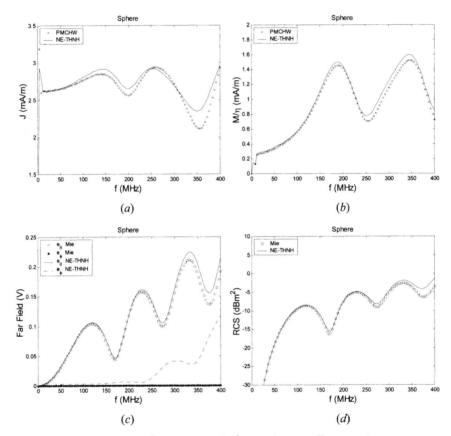

Figure 3.19. NE-THNH ($f_E = 0$, $g_E = 1$, $f_H = -1$, $g_H = 1$) results for a dielectric sphere: (*a*) Electric current. (*b*) Magnetic current. (*c*) Far field. (*d*) Monostatic RCS.

To summarize these ten cases, the results given by the EFIE, MFIE, TENE, and THNH exhibit peaks or discontinuities at the proximity of the internal resonant frequencies. For those cases that do not exhibit the resonant problem, the averaged differences of the far field and monostatic RCS are evaluated and summarized in Table 3.6. It is found that the TENE-NH results has the smallest error in the θ-component of the far field and the RCS, but the difference in the ϕ-component of the far field is relatively large.

TABLE 3.6. **The Averaged Difference of the Far Field and Monostatic RCS Between Mie and the Numerical Solution for the Dielectric Sphere in Figure 3.2**

Formulation	Δe_θ (mV)	Δe_ϕ (mV)	$\Delta \sigma$ (dBm^2)
PMCHW	4.7	0.8	0.34
TENE-THNH	5.3	16.1	0.40
TENE-TH	4.1	6.4	0.30
TENE-NH	3.5	12.9	0.28
TE-THNH	8.7	7.2	0.55
NE-THNH	5.7	21.5	0.52

3.11.2 Dielectric Cube

As a second example, consider the dielectric cube with 1 m on each side and centered at the origin as shown in Figure 3.20. There are eight divisions along each direction, respectively. This results in a total of 768 patches and 1152 edges. The z-directed electric current and the y-directed magnetic current, as indicated by arrows on a side face of the cube in Figure 3.20, are observed.

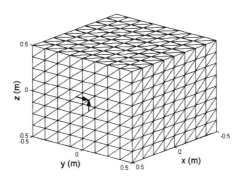

Figure 3.20. Triangle surface patching of a dielectric cube (side 1 m). z- and y-directed arrows represent position and direction of sampled currents **J** and **M**, respectively.

The computed currents of the dielectric cube are compared with the PMCHW solution. Because the Mie solution of the far field or the monostatic RCS for a cube are not known, the commercial EM software HOBBIES is used as a reference for comparison. To verify the validity of the HOBBIES solution, it is compared with the Mie solution for the sphere example in the previous subsection. The number of unknowns is 2400 in the computation using HOBBIES. More unknowns are necessary in HOBBIES to guarantee numerical convergence of the computations. Figure 3.21 compares the HOBBIES and Mie solutions for the sphere, and a good agreement is shown in the whole frequency range. The average difference between them is 0.42 for the RCS and 0.1 for either the θ- or ϕ-components of the far field.

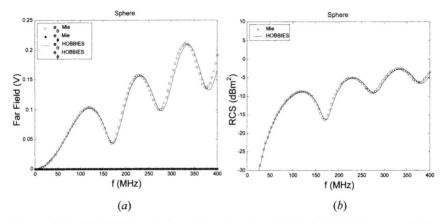

(a) (b)

Figure 3.21. HOBBIES results for a dielectric sphere: (a) Far field. (b) Monostatic RCS.

3.11.2.1 Analysis of a Cube by PMCHW. Figure 3.22 shows the computed far field and monostatic RCS for the cube defined in Figure 3.20 using the PMCHW formulation. The far field and RCS agree well with the HOBBIES solution except that a small difference in the high-frequency region is observed. The currents of PMCHW will be shown later for comparing with those given by other formulations.

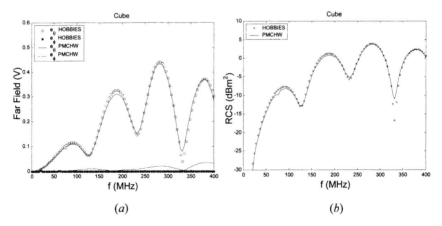

(a) (b)

Figure 3.22. PMCHW results for a dielectric cube: (a) Far field. (b) Monostatic RCS.

3.11.2.2 Analysis of a Cube by EFIE. Figure 3.23 shows the electric and the magnetic currents, the far field, and the monostatic RCS computed by the EFIE formulation in Eq. (3.33). The computed electric current from the EFIE as shown in Figure 3.23(*a*) exhibits a null at about 335 MHz, whereas that from the PMCHW does not. The basis function for the magnetic current in the EFIE is $\mathbf{g}_n$, and it is different from $\mathbf{f}_n$ in PMCHW. They cannot be compared so that only the result from the EFIE is shown in Figure 3.23(*b*). From Figure 3.23 (*b*), (*c*), and (*d*), it is found that there are peaks and discontinuities near the internal resonant frequencies at 212 MHz, 335 MHz, and 367 MHz in the far field and RCS results, although they match with the HOBBIES solution at the low frequencies.

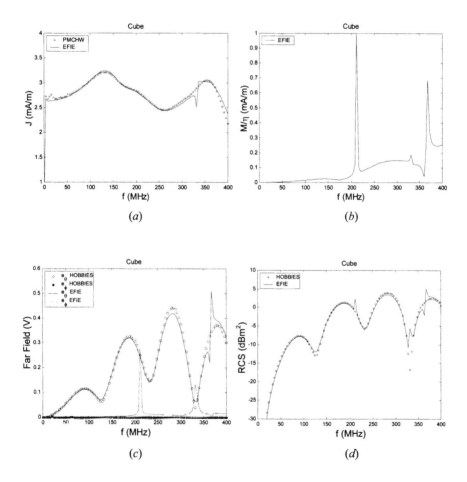

Figure 3.23. EFIE results for a dielectric cube: (*a*) Electric current. (*b*) Magnetic current. (*c*) Far field. (*d*) Monostatic RCS.

3.11.2.3 Analysis of a Cube by MFIE. Figure 3.24 shows the results of MFIE obtained by using Eq. (3.37). Because the basis function for the electric current in the MFIE is $\mathbf{g}_n$, and it is different from $\mathbf{f}_n$ in the PMCHW; they are not compared in Figure 3.24(a). The magnetic current in Figure 3.24(b) agrees well with those of PMCHW except that peaks are exhibited in the in the proximity of the internal resonant frequencies 212 MHz and 367 MHz. From Figures 3.24(c) and (d), it is clearly shown that there are peaks and discontinuities at the resonant frequencies of 212 MHz, 335 MHz, and 367 MHz in the far field and RCS results, despite the good agreement with the HOBBIES solution at the low-frequency region.

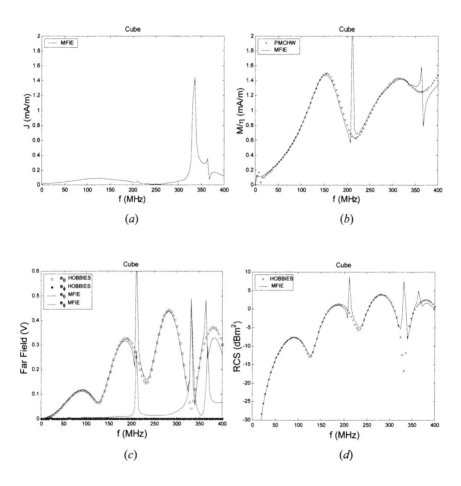

Figure 3.24. MFIE results for a dielectric cube: (a) Electric current. (b) Magnetic current. (c) Far field. (d) Monostatic RCS.

3.11.2.4 Analysis of a Cube by TENE of CFIE. Figure 3.25 shows the numerical results from the TENE formulation. Similar to the EFIE, the formulation from TENE is related to only the electric field. Because of this, peaks at the internal resonant frequencies are observed in the currents, the far field, and the RCS. In contrast to the EFIE results in Figure 3.23, spurious peaks in the TENE solution are much larger and should lead to a larger average error.

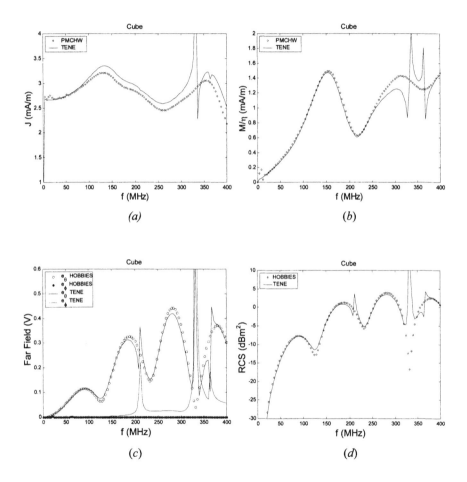

Figure 3.25. TENE ($f_E = 1$, $g_E = 1$, $f_H = 0$, $g_H = 0$) results for a dielectric cube: (*a*) Electric current. (*b*) Magnetic current. (*c*) Far field. (*d*) Monostatic RCS.

3.11.2.5 Analysis of a Cube by THNH of CFIE. Figure 3.26 shows the numerical results from THNH. Similar to the MFIE, the formulation from THNH is based on only the magnetic field. For this reason, peaks at the internal resonant frequencies are observed in the results of the currents, the far field, and the RCS.

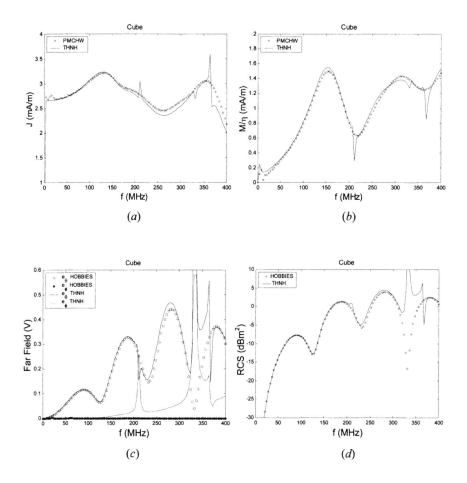

Figure 3.26. THNH ($f_E = 0$, $g_E = 0$, $f_H = -1$, $g_H = 1$) results for a dielectric cube: (*a*) Electric current. (*b*) Magnetic current. (*c*) Far field. (*d*) Monostatic RCS.

3.11.2.6 Analysis of a Cube by TENE-THNH of CFIE. Figure 3.27 gives the results computed by the general form of CFIE, called TENE-THNH, which combines the TENE and THNH formulations. In this version of CFIE, no resonant peak or discontinuity is observed in the plots. As shown in Figure 3.27, the TENE-THNH results agree well with the PMCHW for the equivalent currents and with the HOBBIES solution for the far fields and RCS, respectively, except for some small error in the high-frequency region.

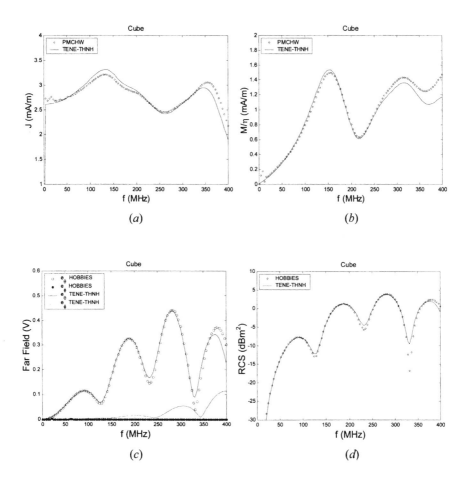

Figure 3.27. TENE-THNH ($f_E = 1$, $g_E = 1$, $f_H = -1$, $g_H = 1$) results for a dielectric cube: (*a*) Electric current. (*b*) Magnetic current. (*c*) Far field. (*d*) Monostatic RCS.

3.11.2.7 Analysis of a Cube by TENE-TH of CFIE. The TENE-TH formulation is obtained when the term $\langle \mathbf{g}_m, \mathbf{H} \rangle$ is removed from the general form of CFIE, as discussed in Section 3.9. The TENE-TH results for a dielectric cube are shown in Figure 3.28. All numerical results given by TENE-TH do not show the internal resonant problem. The results for the magnetic current, the far field, and RCS agree well with the HOBBIES solution, whereas the electric current exhibits a higher value than that of the PMCHW in the entire frequency range.

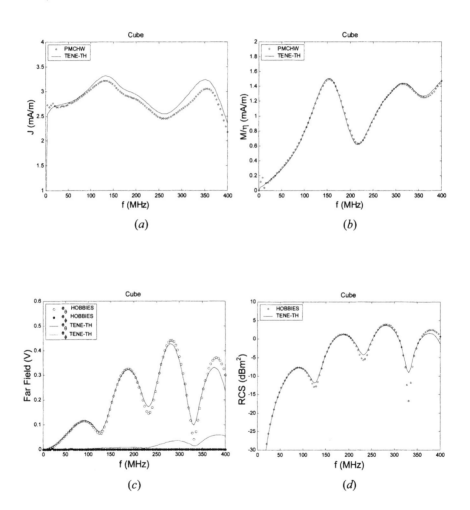

Figure 3.28. TENE-TH ($f_E = 1$, $g_E = 1$, $f_H = -1$, $g_H = 0$) results for a dielectric cube: (*a*) Electric current. (*b*) Magnetic current. (*c*) Far field. (*d*) Monostatic RCS.

3.11.2.8 Analysis of a Cube by TENE-NH of CFIE. The TENE-NH formulation is obtained when the term $\langle \mathbf{f}_m, \mathbf{H} \rangle$ is removed from the general form of CFIE, as discussed in Section 3.9. The TENE-NH results for a dielectric cube are shown in Figure 3.29. All numerical results do not show the internal resonant problem. The results for the far field and RCS agree very well with the HOBBIES solutions, whereas the currents show some differences from the PMCHW solutions.

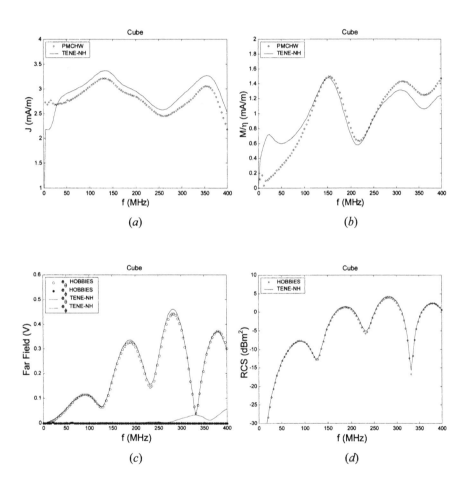

Figure 3.29. TENE-NH ($f_E = 1$, $g_E = 1$, $f_H = 0$, $g_H = 1$) results for a dielectric cube: (*a*) Electric current. (*b*) Magnetic current. (*c*) Far field. (*d*) Monostatic RCS.

3.11.2.9 Analysis of a Cube by TE-THNH of CFIE. The TE-THNH formulation is obtained when the term $\langle \mathbf{g}_m, \mathbf{E} \rangle$ is removed from the general form of CFIE, as discussed in Section 3.9. The TE-THNH results for a dielectric cube are shown in Figure 3.30. All numerical results do not show the internal resonant problem. The results for the electric and magnetic currents, the far field, and RCS generally agree with the PMCHW and the Mie solutions, but little errors are generated in the solution corresponding to the high-frequency region.

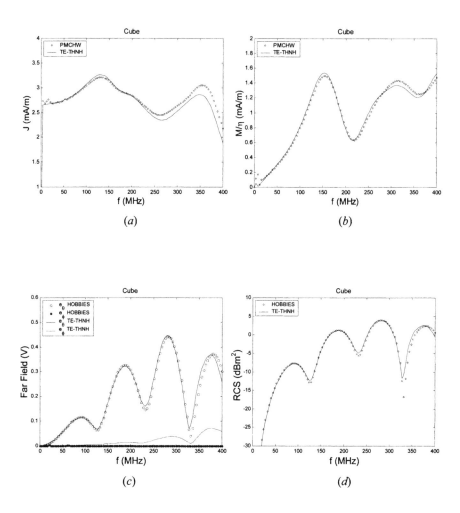

Figure 3.30. TE-THNH ($f_E = 1$, $g_E = 0$, $f_H = -1$, $g_H = 1$) results for a dielectric cube: (a) Electric current. (b) Magnetic current. (c) Far field. (d) Monostatic RCS.

3.11.2.10 Analysis of a Cube by NE-THNH of CFIE. The NE-THNH formulation is obtained when the term $\langle \mathbf{f}_m, \mathbf{E} \rangle$ is removed from the general form of CFIE, as discussed in Section 3.9. The NE-THNH results for the dielectric cube are shown in Figure 3.31. All numerical results do not show the internal resonant problem. The results for the RCS agree well with the HOBBIES solutions, but little errors appear for the far field in the high-frequency region. The computed electric and magnetic currents are of similar shapes as the PMCHW current but obvious discrepancies are observed particularly in the low-frequency region of the magnetic current and in the high-frequency region of the electric current.

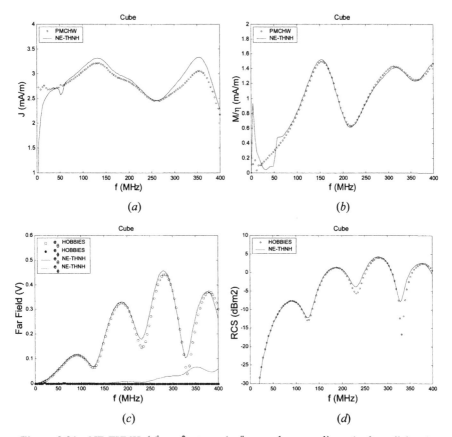

Figure 3.31. NE-THNH ($f_E = 0$, $g_E = 1$, $f_H = -1$, $g_H = 1$) results for a dielectric cube: (*a*) Electric current. (*b*) Magnetic current. (*c*) Far field. (*d*) Monostatic RCS.

To summarize the ten analyses for the cube, the results given by the EFIE, MFIE, TENE, and THNH exhibit peaks or discontinuities at the proximity of the internal resonant frequencies. For those cases that do not exhibit the resonant problem, the averaged differences of the far field and monostatic RCS are evaluated and summarized in Table 3.7. As in the analysis of the sphere, the TENE-NH offers the best numerical approximation because it has the smallest error in the θ- and ϕ- components of the far field and the RCS.

TABLE 3.7. Averaged Difference of the Far Field and Monostatic RCS Between HOBBIES and the Numerical Solution for the Dielectric Cube in Figure 3.18

Formulation	Δe_θ (mV)	Δe_ϕ (mV)	$\Delta\sigma$ (dBm^2)
PMCHW	8.2	10.9	0.65
TENE-THNH	9.5	23.7	0.50
TENE-TH	11.6	16.3	0.59
TENE-NH	6.4	9.5	0.35
TE-THNH	9.0	20.1	0.52
NE-THNH	12.0	18.1	0.68

3.11.3 Dielectric Cylinder

As a final example, the numerical results are presented for a finite dielectric cylinder with a radius of 0.5 m and a height of 1 m and centered at the origin, as shown in Figure 3.32. The cylinder is subdivided into 4, 24, and 8 divisions along r, ϕ, and z directions, respectively. This represents a total of 720 patches with 1080 edges. The z-directed electric current and the ϕ-directed magnetic current are observed at a location indicated by the arrows in Figure 3.32. The computed currents are compared with those of the PMCHW and the far fields, and the monostatic RCS are compared with the HOBBIES solution. The number of unknowns used is 2,688 in the computation using HOBBIES.

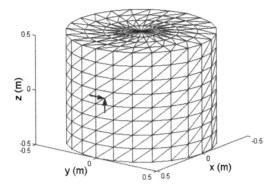

Figure 3.32. Triangle surface patching of a dielectric cylinder (radius 0.5 m and height 1 m). z- and ϕ-directed arrows represent position and direction of sampled currents **J** and **M**, respectively.

3.11.3.1 Analysis of a Cylinder by PMCHW. Figure 3.33 shows the results using the PMCHW. The far field and RCS agree well with the HOBBIES solution in the entire frequency region.

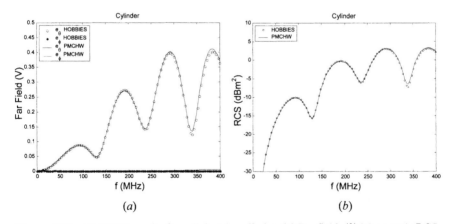

Figure 3.33. PMCHW results for a dielectric cylinder: (*a*) Far field. (*b*) Monostatic RCS.

3.11.3.2 Analysis of a Cylinder by EFIE. Figure 3.34 shows the electric and magnetic currents, the far field, and the monostatic RCS computed by the EFIE formulation in Eq. (3.33). The electric current of the EFIE as shown in Figure 3.34(*a*) exhibits peaks at about 328 MHz and 366 MHz, whereas that of the PMCHW does not. The basis function for the magnetic current in the EFIE is $\mathbf{g}_n$ and it is different from $\mathbf{f}_n$ in PMCHW. They cannot be compared so that only the result of EFIE is shown in Figures 3.34(*b*). From Figure 3.34 (*b*), (*c*), and (*d*), it is found that there are peaks and discontinuities near the internal resonant frequencies at 230 MHz, 328 MHz, and 366 MHz and 367 MHz in the far field and RCS results, although they match with the HOBBIES solutions at the low frequencies.

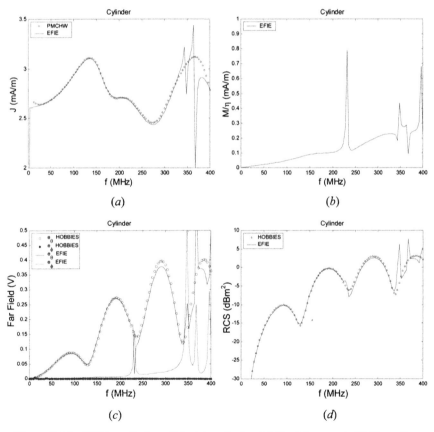

Figure 3.34. EFIE results for a dielectric cylinder: (*a*) Electric current. (*b*) Magnetic current. (*c*) Far field. (*d*) Monostatic RCS.

3.11.3.3 Analysis of a Cylinder by MFIE. Figure 3.35 shows the results of MFIE obtained by using Eq. (3.37). Because the basis function for the electric current in the MFIE is $\mathbf{g}_n$, and it is different from $\mathbf{f}_n$ in the PMCHW, they are not compared in Figure 3.35(a). The magnetic current in Figure 3.35(b) agrees well with those of PMCHW, except that peaks appear in the in the proximity of the internal resonant frequencies 212 MHz and 367 MHz. From Figures 3.35(c) and (d), it is clear that there are peaks and discontinuities at the resonant frequencies of 230 MHz, 328 MHz, and 366 MHz in the far field and RCS results, despite the good agreement with the HOBBIES solution at the low-frequency region.

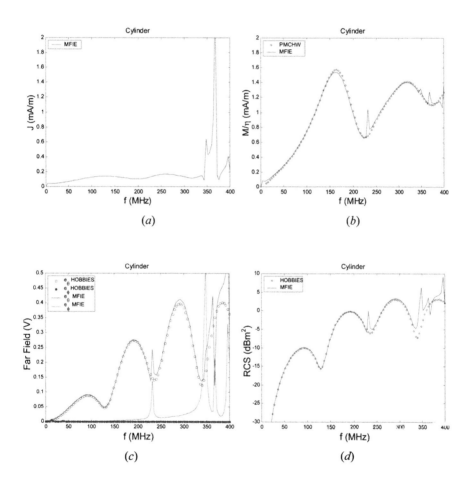

Figure 3.35. MFIE results for a dielectric cylinder: (a) Electric current. (b) Magnetic current. (c) Far field. (d) Monostatic RCS.

3.11.3.4 Analysis of a Cylinder by TENE of CFIE. Figure 3.36 shows the numerical results from the TENE formulation. Similar to the EFIE, the formulation of TENE is related to only the electric field. Because of this, peaks at the internal resonant frequencies are observed in the currents, the far field, and the RCS. In contrast to the EFIE results in Figure 3.24, spurious peaks in the TENE solution are relatively larger and should lead to a larger average error.

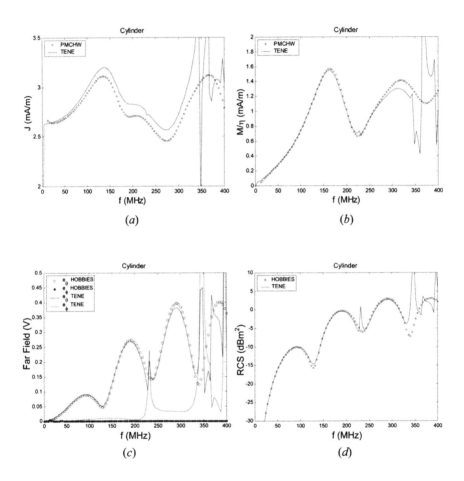

Figure 3.36. TENE ($f_E = 1$, $g_E = 1$, $f_H = 0$, $g_H = 0$) results for a dielectric cylinder: (*a*) Electric current. (*b*) Magnetic current. (*c*) Far field. (*d*) Monostatic RCS.

3.11.3.5 Analysis of a Cylinder by THNH of CFIE. Figure 3.37 shows the numerical results for THNH. Similar to the MFIE, the formulation of THNH is based on only the magnetic field. For this reason, peaks at the internal resonant frequencies are observed in the results for the currents, the far field, and the RCS.

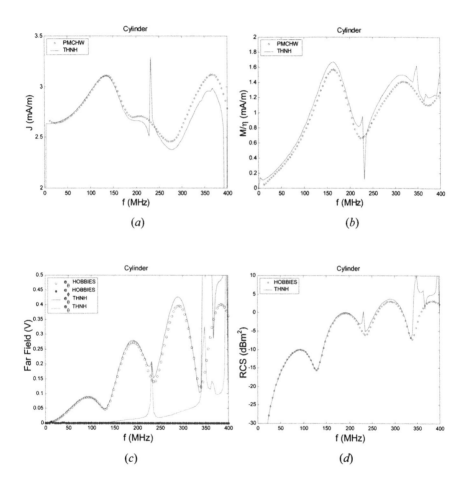

Figure 3.37. THNH $(f_E - 0,\ g_E = 0,\ f_H = -1,\ g_H = 1)$ results for a dielectric cylinder: (a) Electric current. (b) Magnetic current. (c) Far field. (d) Monostatic RCS.

3.11.3.6 Analysis of a Cylinder by TENE-THNH of CFIE. Figure 3.38 gives the
results computed for a cylinder by the general form of CFIE, called TENE-THNH,
which combines the TENE and THNH formulations. In this version of CFIE, no
resonant peak or discontinuity is observed in the plots. As shown in Figure 3.38,
the TENE-THNH results agree well with the PMCHW for the equivalent currents
and with the HOBBIES solution for the far fields and RCS, respectively, except
for some small error in the high-frequency region.

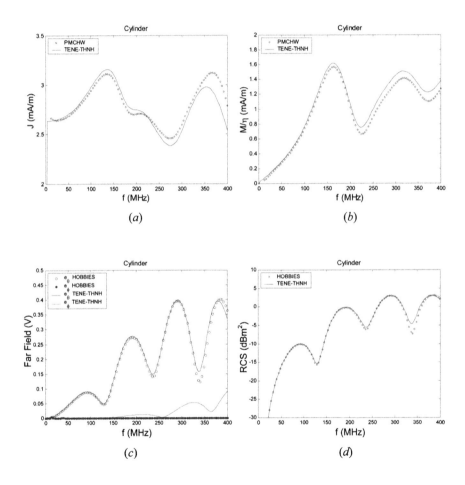

Figure 3.38. TENE-THNH ($f_E = 1$, $g_E = 1$, $f_H = -1$, $g_H = 1$) results for a dielectric
cylinder: (*a*) Electric current. (*b*) Magnetic current. (c) Far field. (*d*) Monostatic RCS.

3.11.3.7 Analysis of a Cylinder by TENE-TH of CFIE. The TENE-TH formulation is obtained when the term $\langle \mathbf{g}_m, \mathbf{H} \rangle$ is removed from the general form of CFIE, as discussed in Section 3.9. The TENE-TH results for a dielectric cylinder are shown in Figure 3.39. All numerical results given by TENE-TH do not show the internal resonant problem. The results for the magnetic current, the far field, and RCS agree well with the HOBBIES solution, whereas the electric current exhibits a higher value than that of the PMCHW in the entire frequency range.

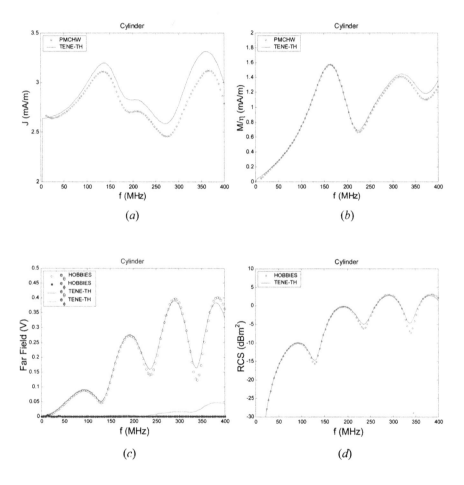

Figure 3.39. TENE-TH ($f_E = 1$, $g_E = 1$, $f_H = -1$, $g_H = 0$) results for a dielectric cylinder: (*a*) Electric current. (*b*) Magnetic current. (*c*) Far field. (*d*) Monostatic RCS.

3.11.3.8 Analysis of a Cylinder by TENE-NH of CFIE. The TENE-NH formulation is obtained when the term $\langle \mathbf{f}_m, \mathbf{H} \rangle$ is removed from the general form of CFIE, as discussed in Section 3.9. The TENE-NH results for a dielectric cube are shown in Figure 3.40. All numerical results do not show the internal resonant problem. The results for the currents, far field, and RCS agree with the PMCHW and HOBBIES solutions, whereas some differences are observed in the high-frequency region.

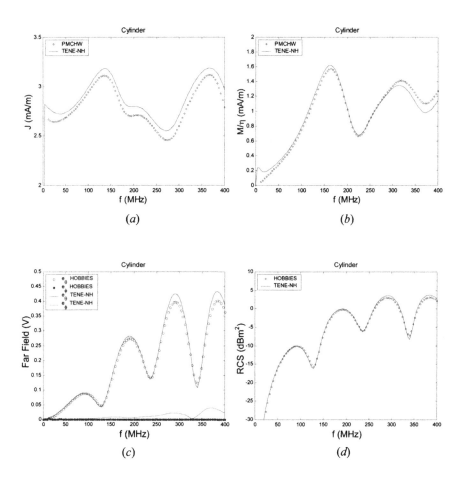

Figure 3.40. TENE-NH ($f_E = 1$, $g_E = 1$, $f_H = 0$, $g_H = 1$) results for a dielectric cylinder: (*a*) Electric current. (*b*) Magnetic current. (*c*) Far field. (*d*) Monostatic RCS.

3.11.3.9 Analysis of a Cylinder by TE-THNH of CFIE.

The TE-THNH formulation is obtained when the term $\langle \mathbf{g}_m, \mathbf{E} \rangle$ is removed from the general form of CFIE, as discussed in Section 3.9. The TE-THNH results for a dielectric cube are shown in Figure 3.41. All numerical results do not show the internal resonant problem. The results for the electric and magnetic currents, the far field, and RCS generally agree with the PMCHW and the Mie solutions, but little errors are shown in the high-frequency region.

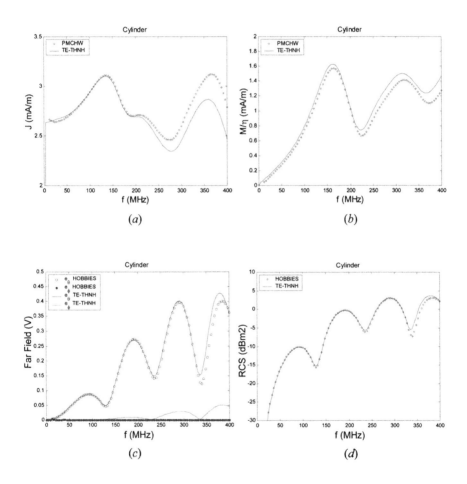

Figure 3.41. TE-THNH ($f_E = 1$, $g_E = 0$, $f_H = -1$, $g_H = 1$) results for a dielectric cylinder: (*a*) Electric current. (*b*) Magnetic current. (*c*) Far field. (*d*) Monostatic RCS.

3.11.3.10 Analysis of a Cylinder by NE-THNH of CFIE. The NE-THNH formulation is obtained when the term $\langle \mathbf{f}_m, \mathbf{E} \rangle$ is removed from the general form of CFIE, as discussed in Section 3.9. The NE-THNH results for the dielectric cube are shown in Figure 3.42. All numerical results do not show the internal resonant problem. The results for the RCS agree well with the HOBBIES solutions, but little errors are displayed for the far field in the high-frequency region. The computed currents and far field match with the PMCHW and HOBBIES solutions but with larger values in the entire frequency range.

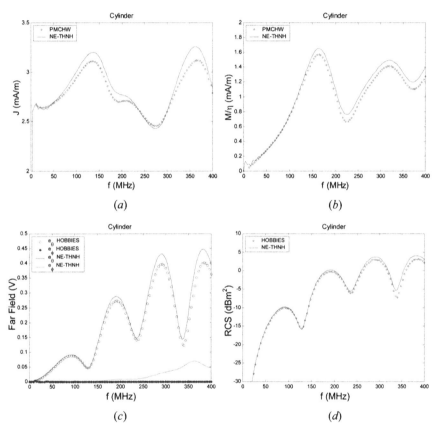

Figure 3.42. NE-THNH ($f_E = 0$, $g_E = 1$, $f_H = -1$, $g_H = 1$) results for a dielectric cylinder: (*a*) Electric current. (*b*) Magnetic current. (*c*) Far field. (*d*) Monostatic RCS.

From these ten analyses for the cylinder, similar observations are obtained as that for a sphere and a cube. The results given by the EFIE, MFIE, TENE, and THNH exhibit peaks or discontinuities at the proximity of the internal resonant frequencies. For those cases that do not exhibit the resonant problem, the averaged differences of the far field and monostatic RCS are evaluated and summarized in Table 3.8. The PMCHW method offers the minimum error than the other schemes. In addition, one can conclude that the TENE-THNH results produce a small difference in the θ-component of the far field, but the difference in the ϕ-component of the far field is relatively large. When the averaged difference for both the far field and RCS is considered, TENE-NH or TENE-TH formulations can also be chosen.

TABLE 3.8. Averaged Difference of the Far Field and Monostatic RCS Between HOBBIES and Numerical Solution for the Dielectric Cylinder in Figure 3.29

Formulation	Δe_θ (mV)	Δe_ϕ (mV)	$\Delta\sigma$ (dBm^2)
PMCHW	3.7	1.5	0.35
TENE-THNH	5.8	16.9	0.47
TENE-TH	7.4	10.6	0.55
TENE-NH	9.7	11.5	0.47
TE-THNH	7.5	13.0	0.55
NE-THNH	15.7	18.5	0.78

3.11.4 Comparison among CFIE, PMCHW, and Muller Equations

Based on the same example in Section 3.10, Figure 3.43 shows the bistatic RCS solutions obtained from CFIE, PMCHW and Muller (TETH and NENH) formulations compared with those from the Mie series solution. The CFIE solution is computed by choosing the testing coefficients as $f_E = 1$, $g_E = 1$, $f_H = -1$, and $g_H = 1$ and the combination parameter $\kappa = 0.5$. From Figure 3.43, it is observed that there is an agreement between Muller (TETH) and PMCHW solutions. The reason for this is that the RWG function does not test the term $\mathbf{n} \times \mathbf{M}$ and $\mathbf{n} \times \mathbf{J}$ in Eq. (3.61) well, resulting in a solution similar to the PMCHW. All numerical results match with the Mie solution except for the NENH of the Muller method, which exhibits spurious resonances.

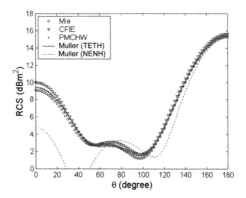

Figure 3.43. Bistatic RCS of the dielectric sphere computed by CFIE, PMCHW, and Muller (TETH and NENH) formulations.

3.12   CONCLUSION

A set of coupled integral equations is used to analyze the scattering from 3-D dielectric objects. The integral equation formulations are derived using the equivalence principle and using the continuity conditions of the fields. To obtain a numerical solution, MoM is employed in conjunction with the planar triangular patch basis function. The EFIE and MFIE formulations give relatively valid solutions, except at frequencies that correspond to an internal resonant frequency of the scatterer, when they are compared with the TENE and THNH, which also exhibit resonance problems. To obtain the CFIE, four testing coefficients are introduced in the combination of the TENE and THNH. As a result, there are eight different cases of the CFIE formulation of which only two formulations are stable and not affected by the presence of internal resonances. Also presented are 16 possible cases of the CFIE by dropping one of terms in the testing procedure and verifying whether all 16 formulations give valid solutions. The important point to note is that not all possible CFIE formulations eliminate the internal resonance problem. When the CFIE is used with other formulations, for example, to analyze the scattering from composite structures with conductors and dielectrics, TENE-THNH, TENE-TH, or TENE-NH may be used.

Also, the numerical scheme for solving the Muller integral equation is investigated to analyze scattering from 3-D arbitrarily shaped dielectric objects using the triangular patch model. The Muller formulation gives accurate and valid solutions at an internal resonant frequency of the scatterer only when testing both the electric and magnetic fields with the combination of the RWG function and its orthogonal component at the same time. All solutions obtained using the TENE-TH and TE-THNH formulations by dropping one of the testing terms are less accurate. The NE-THNH and TENE-NH formulations, testing one of the Muller equations with $\mathbf{g}_m$ only, do not give valid solutions. Finally, two testing methods are proposed for the Muller equation that are not affected by the internal resonance problem and yield a stable solution even at the low frequency by choosing either RWG $+\mathbf{n} \times$ RWG or RWG $-\mathbf{n} \times$ RWG for testing the electric and magnetic fields at the same time.

3.13   APPENDIX

To solve the matrix equations of (3.33),(3.37),(3.54), and (3.60), the integrals A_{mn}, B_{mn}, C_{mn}, and D_{mn} are required to be computed first. In this section, the formulations of applying the RWG basis function over the triangular patches for approximating these integrals are shown. The subscript ν represents the medium only, so it can be ignored here for a simpler presentation. Note that while computing the elements of the matrix, the appropriate material parameters should be included in evaluating the integrals.

3.13.1 Evaluation of the Integrals A_{mn}, A_{mn}^f, and A_{mn}^g

The EFIE and MFIE formulations are given by Eqs. (3.32) and (3.36). Considering the term A_{mn} and substituting $\mathbf{f}_n$ by the RWG basis function given by Eq. (3.19), we obtain four terms. The summation operator is defined as

$$A_{mn} = A_{mn}^{++} + A_{mn}^{+-} + A_{mn}^{-+} + A_{mn}^{--} \equiv \sum_{p,q} A_{mn}^{pq} \tag{3.A1}$$

where

$$A_{mn}^{pq} = \frac{1}{4\pi} \int_S \mathbf{f}_m^p(\mathbf{r}) \cdot \int_S \mathbf{f}_n^q(\mathbf{r}') G(\mathbf{r},\mathbf{r}') \, dS' \, dS \tag{3.A2}$$

and p and q can be either $+$ or $-$. If the integral on the unprimed variable is evaluated by approximating the integrand by their respective values at the centroid of the testing triangle T_m^p, Eq. (3.A2) becomes

$$A_{mn}^{pq} = \frac{l_m l_n}{16\pi} \boldsymbol{\rho}_m^{cp} \cdot \frac{1}{a_n^q} \int_{T_n^q} \boldsymbol{\rho}_n^q \frac{e^{-jkR_m^p}}{R_m^p} \, dS' \tag{3.A3}$$

where

$$R_m^p = \left| \mathbf{r}_m^{cp} - \mathbf{r}' \right| \tag{3.A4}$$

and $\mathbf{r}_m^{cp}$ is the position vector of the centroid of the triangle T_m^p. Equation (3.50a) is the same as Eq. (3.32b) or Eq. (3.A1), and Eq. (3.50b) may be evaluated by replacing $\boldsymbol{\rho}_m^{cp}$ with $\mathbf{n} \times \boldsymbol{\rho}_m^{cp}$ in Eq. (3.A3).

3.13.2 Evaluation of the Integrals B_{mn}, B_{mn}^f, and B_{mn}^g

By substituting Eq. (3.20) into Eq. (3.36c), one obtains

$$B_{mn} = B_{mn}^{++} + B_{mn}^{+-} + B_{mn}^{-+} + B_{mn}^{--} \equiv \sum_{p,q} B_{mn}^{pq} \tag{3.A5}$$

where

$$B_{mn}^{pq} = \frac{1}{4\pi} \int_S \nabla_S \cdot \mathbf{f}_m^p(\mathbf{r}) \int_S \nabla_S' \cdot \mathbf{f}_n^q(\mathbf{r}') G(\mathbf{r},\mathbf{r}') \, dS' \, dS \tag{3.A6}$$

Approximating the integrand by their respective values at the centroid of the testing triangle T_m^p, Eq. (3.A6) becomes

$$B_{mn}^{pq} = \frac{l_m l_n}{4\pi} \frac{1}{a_n^q} \int_{T_n^q} \frac{e^{-jkR_m^p}}{R_m^p} \, dS' \tag{3.A7}$$

where R_m^p is given by Eq. (3.A4). Equation (3.50c) is the same as Eq. (3.36c) or Eq. (3.A5). Equation (3.50d) is evaluated through

$$B_{mn}^g = \sum_{p,q} B_{mn}^{pq,g} \tag{3.A8}$$

where

$$B_{mn}^{pq,g} = \frac{1}{4\pi} \int_S \mathbf{n} \times \mathbf{f}_m^p(\mathbf{r}) \cdot \int_S \nabla_S' \cdot \mathbf{f}_n^q(\mathbf{r}') \nabla' G(\mathbf{r},\mathbf{r}') dS' dS \tag{3.A9}$$

Approximating the integrand by the irrespective values at the centroid of the testing triangle T_m^p, Eq. (3.A9) becomes

$$B_{mn}^{pq,g} = \frac{l_m l_n}{8\pi} \mathbf{n} \times \mathbf{\rho}_m^p \cdot \frac{1}{a_n^q} \int_{T_n^q} \nabla' \frac{e^{-jkR_m^p}}{R_m^p} dS' \tag{3.A10}$$

The evaluation of the potential integrals in Eqs. (3.A3), (3.A7), and (3.A10) may be carried out by using the numerical methods specially developed for triangular regions in [14–17].

3.13.3   Evaluation of the Integrals C_{mn}, C_{mn}^f, and C_{mn}^g

Using a vector identity [12], Eq. (3.32d) is given by

$$C_{mn} = \frac{1}{2} \int_S \mathbf{f}_m(\mathbf{r}) \cdot \mathbf{n} \times (\mathbf{n} \times \mathbf{f}_n(\mathbf{r})) dS = -\sum_{p,q} \frac{1}{2} \int_S \mathbf{f}_m^p \cdot \mathbf{f}_n^q \, dS \tag{3.A11}$$

The integral of Eq. (3.A11) can be computed analytically, and the result is given by [18]

$$\frac{1}{2} \int_S \mathbf{f}_m^p \cdot \mathbf{f}_n^q \, dS$$

$$= \pm \frac{l_m l_n}{8a} \left[\frac{3}{4} |\mathbf{r}^c|^2 + \frac{1}{12} \left(|\mathbf{r}_1|^2 + |\mathbf{r}_2|^2 + |\mathbf{r}_3|^2 \right) - \mathbf{r}^c \cdot (\mathbf{r}_m + \mathbf{r}_n) + \mathbf{r}_m \cdot \mathbf{r}_n \right] \tag{3.A12}$$

where $\mathbf{r}_1$, $\mathbf{r}_2$, and $\mathbf{r}_3$ are the position vectors for the three vertices of the triangle T_m^p or T_n^q, $\mathbf{r}^c$ is the centroid of the triangle T_m^p or T_n^q, and A is the area of T_m^p. The vectors $\mathbf{r}_m$ and $\mathbf{r}_n$ are the position vectors for the free vertex of the triangles T_m^p and T_n^q, respectively. Note that if the field point does not lie on the triangle T_n^q, (i.e. $\mathbf{r} \notin T_n^q$), which results in $C_{mn}^{pq} = 0$. In Eq. (3.A12), the sign is positive when p and q are same and negative otherwise.

Using the summation operator, Eq. (3.50e) is given by

$$C_{mn}^f = \frac{1}{2}\int_S \mathbf{f}_m(\mathbf{r}) \cdot \mathbf{n} \times \mathbf{f}_n(\mathbf{r})\,dS = \sum_{p,q}\frac{1}{2}\int_S \mathbf{f}_m^p \cdot \mathbf{n} \times \mathbf{f}_n^q\,dS \qquad (3.A13)$$

By evaluating the element of Eq. (3.A13) analytically, the result is given by

$$\frac{1}{2}\int_S \mathbf{f}_m^p \cdot \mathbf{n} \times \mathbf{f}_n^q\,dS = \pm\frac{l_m l_n}{8a}\mathbf{n}\cdot\big((\mathbf{r}_m - \mathbf{r}_n)\times\mathbf{r}^c - (\mathbf{r}_m\times\mathbf{r}_n)\big) \qquad (3.A14)$$

By using a vector identity [18], Eq. (3.50f) may be written as

$$C_{mn}^g = \frac{1}{2}\int_S \mathbf{n}\times\mathbf{f}_m(\mathbf{r})\cdot\mathbf{n}\times\mathbf{f}_n(\mathbf{r})\,dS = \sum_{p,q}\frac{1}{2}\int_S \mathbf{f}_m^p(\mathbf{r})\cdot\mathbf{f}_n^q(\mathbf{r})\,dS \qquad (3.A15)$$

where the integral is same as in Eq. (3.A12).

Note that, if m and n does not belong to the same triangle, then $C_{mn} = 0$. And it is double of whatever expression one uses for it when $m = n$.

3.13.4   Evaluation of the Integrals D_{mn}, D_{mn}^f, and D_{mn}^g

By using the summation operator, the terms of Eq. (3.32e) are written as

$$D_{mn}^{pq} = \frac{1}{4\pi}\int_S \mathbf{f}_m^p(\mathbf{r})\cdot\int_S \mathbf{n}'\times\mathbf{f}_n^q(\mathbf{r})\times\nabla'G(\mathbf{r},\mathbf{r}')\,dS'\,dS \qquad (3.A16)$$

where $\mathbf{n}'$ denotes the unit vector normal to the triangle T_n^q. Thus, substituting Eq. (3.19b) into Eq. (3.A16), obtains

$$D_{mn}^{pq} = \frac{l_m l_n}{16\pi a_m^p a_n^q}\int_{T_m^p}\boldsymbol{\rho}_m^p\cdot\int_{T_n^q}(\mathbf{n}'\times\boldsymbol{\rho}_n^q)\times\mathbf{R}(1+jkR)\frac{e^{-jkR}}{R^3}\,dS'\,dS \qquad (3.A17)$$

This integral may be computed using a Gaussian quadrature scheme over the unprimed and primed coordinates numerically. Other integrals of Eqs. (3.50g), (3.50h), (3.58d), and (3.59d) may be evaluated in a similar fashion.

3.13.5   Evaluation of the Integrals V_{mn}^E, V_{mn}^H, V_m^E, V_m^H, $V_{m,v}^E$, and $V_{m,v}^H$

Consider Eq. (3.32f) for $v = 1$ and apply a centroid testing, yielding

$$\int_S \mathbf{f}_m(\mathbf{r})\cdot\mathbf{E}^i(\mathbf{r})\,dS = \int_S \mathbf{f}_m^+(\mathbf{r})\cdot\mathbf{E}^i(\mathbf{r})\,dS + \int_S \mathbf{f}_m^-(\mathbf{r})\cdot\mathbf{E}^i(\mathbf{r})\,dS$$

$$= \frac{l_m}{2} \left(\boldsymbol{\rho}_m^{c+} \cdot \mathbf{E}^i (\mathbf{r}_m^{c+}) + \boldsymbol{\rho}_m^{c-} \cdot \mathbf{E}^i (\mathbf{r}_m^{c-}) \right) \qquad (3.A18)$$

Other integrals may be evaluated in a similar fashion.

REFERENCES

[1]   R. F. Harrington, "Boundary integral formulations for homogeneous material bodies," *Jour. Electromagn. Waves Appl.*, Vol. 3, pp. 1–15, 1989.

[2]   K. Umashankar, A. Taflove, and S. M. Rao, "Electromagnetic scattering by arbitrary shaped three-dimensional homogeneous lossy dielectric bodies," *IEEE Trans. Antennas Prop.*, Vol. 34, pp. 758–766, 1986.

[3]   T. K. Sarkar, S. M. Rao, and A. R. Djordjevic, "Electromagnetic scattering and radiation from finite microstrip structures," *IEEE Trans. Microwave Theory Tech.*, Vol. 38, pp. 1568–1575, 1990.

[4]   S. M. Rao and D. R. Wilton, "*E*-field, *H*-field, and combined field solution for arbitrarily shaped three-dimensional dielectric bodies," *Electromagnetics*, Vol. 10, pp. 407–421, 1990.

[5]   S. M. Rao, C. C. Cha, R. L. Cravey, and D. L. Wilkes, "Electromagnetic scattering from arbitrary shaped conducting bodies coated with lossy materials of arbitrary thickness," *IEEE Trans. Antennas Prop.*, Vol. 39, pp. 627–631, 1991.

[6]   X. Q. Sheng, J. M. Jin, J. M. Song, W. C. Chew, and C. C. Lu, "Solution of combined-field integral equation using multilevel fast multipole algorithm for scattering by homogeneous bodies," *IEEE Trans. Antennas Prop.*, Vol. 46, pp. 1718–1726, 1998.

[7]   A. I. Mackenzie, S. M. Rao, and M. E. Baginski, "Electromagnetic scattering from arbitrarily shaped dielectric bodies using paired pulse vector basis functions and method of moments," *IEEE Trans. Antennas Prop.*, Vol. 57, pp. 2076–2083, 1999.

[8]   S. M. Rao, D. R. Wilton, and A. W. Glisson, "Electromagnetic scattering by surfaces of arbitrary shape," *IEEE Trans. Antennas Prop.*, Vol. 30, pp. 409–418, 1982.

[9]   J. R. Mautz and R. F. Harrington, "Electromagnetic scattering from a homogeneous material body of revolution," *Arch. Elektron. Ubertragungstech.*, Vol. 33, pp. 71–80, 1979.

[10] M. S. Yeung, "Single integral equation for electromagnetic scattering by three-dimensional homogeneous dielectric objects," *IEEE Trans. Antennas Prop.*, Vol. 47, pp. 1615–1622, 1999.

[11] Y. Zhang, and T. K. Sarkar, *Parallel Solution of Integral Equation-based EM Problems in the Frequency Domain*, John Wiley & Sons, New York, 2009.

[12] R. F. Harrington, *Time Harmonic Electromagnetics*, McGraw-Hill, New York, N.Y., 1961.

[13] B. H. Jung, T. K. Sarkar, and Y.-S. Chung, "A survey of various frequency domain integral equations for the analysis of scattering from three-dimensional dielectric objects," *J. Electromagn. Waves Appl.*, Vol. 16. pp. 1419–1421, 2002.

[14] D. R. Wilton, S. M. Rao, A. W. Glisson, D. H. Schaubert, O. M. Al-Bundak, and C. M. Butler, "Potential integrals for uniform and linear source distributions on polygonal and polyhedral domains," *IEEE Trans. Antennas Prop.*, Vol. 32, pp. 276–281, 1984.

[15]   S. Caorsi, D. Moreno, and F. Sidoti, "Theoretical and numerical treatment of surface integrals involving the free-space green's function," *IEEE Trans. Antennas Prop.*, Vol. 41, pp. 1296–1301, 1993.

[16]   R. D. Graglia, "On the numerical integration of the linear shape functions times the 3-D green's function or its gradient on a plane triangle," *IEEE Trans. Antennas Prop.*, Vol. 41, pp. 1448–1455, 1993.

[17]   T. F. Eibert and V. Hansen, "On the calculation of potential integrals for linear source distributions on triangular domains," *IEEE Trans. Antennas Prop.*, Vol. 43, pp. 1499–1502, 1995.

[18]   S. M. Rao, *Electromagnetic Scattering and Radiation of Arbitrarily-Shaped Surfaces by Triangular Patch Modeling*, PhD Dissertation, University Mississippi, August, 1980.

4

ANALYSIS OF COMPOSITE STRUCTURES IN THE FREQUENCY DOMAIN

4.0 SUMMARY

In this chapter, formulations are presented for the analysis of electromagnetic scattering from an arbitrarily shaped three-dimensional (3-D) composite structure, such as a perfectly conducting and piecewise homogeneous dielectric composite body and the conducting object coated with dielectric material. The formulations include the electric field integral equation (EFIE), the magnetic field integral equation (MFIE), the combined field integral equation (CFIE), and the Poggio–Miller–Chang– Harrington–Wu equation (PMCHW). To evaluate the efficacies of different formulations, these integral equations are used to analyze four different 3-D composite scatterers, and their results are compared with each other as well as a commercial software package, higher order basis function based integral equations (HOBBIES). In the analysis, the equations are solved by using MoM along with the triangular patch model and Rao–Wilton–Glisson (RWG) functions. From the comparison, it is found that CFIE can approximate the HOBBIES solution better than the EFIE and MFIE, particularly for preventing from spurious solutions, which are found at the frequencies corresponding to internal resonances of the structure.

4.1 INTRODUCTION

Many publications have reported the analysis of a composite structure that is a combination of conducting and dielectric bodies. Some earlier techniques have been used to solve two-dimensional problems and bodies of revolution [1–5]. In addition, various methods using integral equations have been developed for 3-D composite bodies [6–9]. There are four sets of coupled integral equations: the EFIE, MFIE, CFIE, and the PMCHW equation. The EFIE, MFIE, and CFIE are formed by considering only the electric fields, only the magnetic fields, or only the electric and magnetic combined fields for the continuity conditions at both the conducting and dielectric surface boundaries. In addition, different continuity conditions for the conducting and dielectric surfaces can be mixed to extend into other forms of integral equations; for example, choosing the CFIE equation for the conducting surface and the PMCHW equation for the dielectric surface, another formulation is obtained and is called C-PMCHW. By changing the equations for the conducting surface from CFIE to EFIE or to MFIE, new formulations called E-PMCHW or H-PMCHW can be obtained.

Similar to the analysis of the 3-D conducting or dielectric objects using the surface integral equations in Chapter 2 and 3, the integral equations can be solved numerically by using method of moments (MoM) with the triangular patch modeling and the RWG basis function [6]. Sarkar et al. proposed the EFIE for the analysis of scattering by arbitrarily shaped 3-D complex bodies for the first time [7]. In their work, the electric current is expanded using the RWG functions, whereas the magnetic current is expanded using its point-wise spatially orthogonal function, $\mathbf{n} \times$ RWG. Here $\mathbf{n}$ is the unit normal pointing outward from the surface. In the testing procedure, the same RWG function is used.

In the analysis of conducting-dielectric composite bodies, spurious solutions at the frequencies corresponding to an internal resonance of the structure are found when only the EFIE or the MFIE is used. One possible remedy is to use the combined field formulations. A CFIE formulation has been proposed by Sheng et al. for a homogeneous dielectric problem [10]. In this work, the RWG functions are used as basis functions to approximate both the electric and magnetic currents, and RWG and $\mathbf{n} \times$ RWG functions are used as testing functions. Based on Sheng's work, Jung et al. presented a set of CFIE formulations by choosing a combination of testing functions and dropping one of the testing terms in CFIE [11]. Using these testing techniques, very good approximation to the exact solution has been achieved without any internal resonant problems for the analysis of 3-D arbitrarily shaped composite structures.

The problem of electromagnetic scattering from a coated structure that consists of conducting objects enclosed by the dielectric material has been studied extensively. Some techniques were proposed to solve two–dimensional and three-dimensional structures [2,4,5,12]. The formulation EFIE-PMCHW [9,13], which is the combination of EFIE for conducting boundaries and PMCHW formulation for dielectric bodies, was proposed for the analysis of an arbitrarily shaped 3-D coated structures. In the numerical solution procedures, the RWG function has been used for both the basis and testing functions.

In this chapter, some particular choices of expansion and testing functions for CFIE and PMCHW are introduced. Section 4.2 describes various integral equations for analyzing composite structure and the suitable basis and testing functions for them. Some numerical examples are given in section 4.3 for composite structures to demonstrate the efficacies of the introduced formulations. Sections 4.4 and 4.5 provide the equations and numerical examples for the analysis of the scattering from coated structures. All examples here are compared with the solutions from a commercial software package—HOBBIES [12]. Finally, some conclusions are drawn in section 4.6.

4.2 INTEGRAL EQUATIONS FOR COMPOSITE STRUCTURES

The integral equations for analyzing the composite structures are presented in this section. The central idea of this analysis is to treat the composite object as many individual conducting and dielectric bodies immersed in free space. To simplify the presentation, it is assumed here that the composite structure consists of only one conductor and one dielectric body, as shown in Fig. 4.1. The described formulation for this case can be extended to apply for multiple complex structures. Although the bodies shown in Figure 4.1 are separated, it generally is not necessary to separate them. If the conducting and dielectric bodies are joined together, then they are treated as two bodies with a zero-thickness layer of free space separating them.

The composite structure in Figure 4.1 is illuminated by an incident plane wave in which the electric and magnetic fields are denoted by $\mathbf{E}^i$ and $\mathbf{H}^i$. The incident field is defined as that would exist in space if the structure were not present. The regions external and internal to the dielectric body are characterized by the medium parameters (ε_1, μ_1) and (ε_2, μ_2), respectively. The surfaces of the conducting and dielectric scatterers are denoted by S_c and S_d. Furthermore, it is assumed that the dielectric body is a closed body so that a unique outward normal vector can be defined unambiguously.

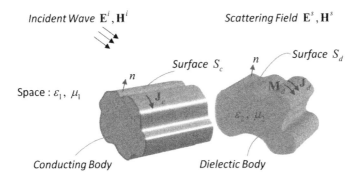

Figure 4.1. A conducting-dielectric composite body illuminated by an electromagnetic wave.

By using the equivalence principle [14], the conducting body is first replaced by an equivalent surface current $\mathbf{J}_c$ radiating in the space (ε_1, μ_1). On the conducting surface S_c, the total tangential electric field must approach zero. The equivalent principle is employed next to split the original problem into two parts, which are equivalent to the exterior and the interior region of the dielectric body. In the equivalent problem for the exterior region, there is a restriction for the fields external to the dielectric body that it should remain as the original, whereas the fields inside the surface S_d can be chosen freely. For simplicity, the interior fields are set to zero, and the entire interior region of the dielectric body is assumed to be replaced by the material of the external region (i.e. the medium with ε_1 and μ_1). Because the tangential components of the fields are not continuous across the dielectric interface S_d, an equivalent electric current $\mathbf{J}_d$ and a magnetic current $\mathbf{M}_d$ on the surface S_d are required to make up for this discontinuity. Because these currents now radiate in a homogenous unbounded medium, the Maxwell's equation and free space Green's function of that medium can be used for the computation of the scattered fields.

By enforcing the boundary conditions on the electric and magnetic fields across the conducting and dielectric interfaces, a set of coupled integral equations may be written in terms of the incident fields and the scattered fields given by the equivalent electric current $\mathbf{J}_c$ on the conducting surface S_c, and electric current $\mathbf{J}_d$ and magnetic current $\mathbf{M}_d$ on the dielectric surface S_d, as

$$\left(\mathbf{E}_{1c}^s(\mathbf{J}_c)+\mathbf{E}_{1c}^s(\mathbf{J}_d)+\mathbf{E}_{1c}^s(\mathbf{M}_d)+\mathbf{E}^i\right)_{\tan}=0, \qquad \text{for } \mathbf{r}\in S_c \qquad (4.1a)$$

$$\left(\mathbf{E}_{1d}^s(\mathbf{J}_c)+\mathbf{E}_{1d}^s(\mathbf{J}_d)+\mathbf{E}_{1d}^s(\mathbf{M}_d)+\mathbf{E}^i\right)_{\tan}=0, \qquad \text{for } \mathbf{r}\in S_d \qquad (4.1b)$$

$$\mathbf{J}_c-\mathbf{n}\times\left(\mathbf{H}_{1c}^s(\mathbf{J}_c)+\mathbf{H}_{1c}^s(\mathbf{J}_d)+\mathbf{H}_{1c}^s(\mathbf{M}_d)\right)=\mathbf{n}\times\mathbf{H}^i, \quad \text{for } \mathbf{r}\in S_c, \qquad (4.1c)$$

$$\left(\mathbf{H}_{1d}^s(\mathbf{J}_c)+\mathbf{H}_{1d}^s(\mathbf{J}_d)+\mathbf{H}_{1d}^s(\mathbf{M}_d)+\mathbf{H}^i\right)_{\tan}=0, \qquad \text{for } \mathbf{r}\in S_d \qquad (4.1d)$$

where $\mathbf{E}^i$ and $\mathbf{E}^s$ are the incident and scattered electric fields, respectively, and $\mathbf{H}^i$ and $\mathbf{H}^s$ are the incident and scattered magnetic fields, respectively. The subscript "tan" denotes the tangential component of the fields. The subscript "1" associated with the scattered fields represents the fields to be computed in the region external to the dielectric body, and the subscripts "c" and "d" denote that these boundary conditions are enforced at the boundary of the conducting and dielectric surfaces, respectively.

In the equivalent problem for the interior region, the material parameters inside the dielectric body must be kept as the original, whereas the exterior region has no restrictions. The fields external to the surface S_d are set to zero for convenience, and the entire exterior material parameters are replaced and set to be the same as the interior material of the dielectric body (i.e., the medium with ε_2 and μ_2). Similar to the external equivalent situation, equivalent currents on the surface S_d are required to satisfy the boundary continuity for the fields. It turns out that the equivalent electric and magnetic currents on S_d are the negative of those for the exterior equivalent problem (i.e., $-\mathbf{J}_d$ and $-\mathbf{M}_d$, which is from the direction of the unit normal vector pointing into the external medium). Enforcing the continuity of the tangential fields across the dielectric interface yields

$$\left(\mathbf{E}_{2d}^s(-\mathbf{J}_d)+\mathbf{E}_{2d}^s(-\mathbf{M}_d)\right)_{\tan}=0, \quad \text{for } \mathbf{r}\in S_d \qquad (4.2a)$$

$$\left(\mathbf{H}_{2d}^{s}(-\mathbf{J}_{d})+\mathbf{H}_{2d}^{s}(-\mathbf{M}_{d})\right)_{\text{tan}}=0, \quad \text{for } \mathbf{r}\in S_{d} \quad (4.2b)$$

where the subscript "2" associated with the scattered fields, represents the fields to be computed in the region internal to the dielectric body

From Eqs. (4.1) and (4.2), when only the electric field boundary conditions are considered, the electric field integral equation for the composite object can be formed by combining Eqs (4.1a), (4.1b), and (4.2a). This gives the relation in the frequency domain between the incident and the scattered electric fields for a conducting-dielectric composite body caused by the induced electric and magnetic currents on the conducting and dielectric surfaces as

$$\left(-\mathbf{E}_{1c}^{s}(\mathbf{J}_{c},\mathbf{J}_{d},\mathbf{M}_{d})\right)_{\text{tan}}=\left(\mathbf{E}^{i}\right)_{\text{tan}} \quad \text{for } \mathbf{r}\in S_{c} \quad (4.3a)$$

$$\left(-\mathbf{E}_{1d}^{s}(\mathbf{J}_{c},\mathbf{J}_{d},\mathbf{M}_{d})\right)_{\text{tan}}=\left(\mathbf{E}^{i}\right)_{\text{tan}} \quad \text{for } \mathbf{r}\in S_{d} \quad (4.3b)$$

$$\left(-\mathbf{E}_{2d}^{s}(\mathbf{J}_{d},\mathbf{M}_{d})\right)_{\text{tan}}=0 \quad \text{for } \mathbf{r}\in S_{d} \quad (4.3c)$$

Equation (4.3) is called EFIE because only the electric field relations are introduced in the continuity conditions for both the conducting and the dielectric surfaces.

Similarly, by combining Eqs. (4.1c), (4.1d), and (4.2b), the magnetic field integral equation for the composite object can be formed. It describes the relationship in the frequency domain between the incident and the scattered magnetic fields for a conducting-dielectric composite body caused by the induced electric and magnetic currents on the conducting and dielectric surfaces as

$$\mathbf{J}_{c}-\mathbf{n}\times\mathbf{H}_{1c}^{s}(\mathbf{J}_{c},\mathbf{J}_{d},\mathbf{M}_{d})=\mathbf{n}\times\mathbf{H}^{i} \quad \text{for } \mathbf{r}\in S_{c} \quad (4.4a)$$

$$\left(-\mathbf{H}_{1d}^{s}(\mathbf{J}_{c},\mathbf{J}_{d},\mathbf{M}_{d})\right)_{\text{tan}}=\left(\mathbf{H}^{i}\right)_{\text{tan}} \quad \text{for } \mathbf{r}\in S_{d} \quad (4.4b)$$

$$\left(-\mathbf{H}_{2d}^{s}(\mathbf{J}_{d},\mathbf{M}_{d})\right)_{\text{tan}}=0 \quad \text{for } \mathbf{r}\in S_{d} \quad (4.4c)$$

Equation (4.4) is called MFIE because the continuity conditions are enforced on the magnetic electric fields only.

In the analysis of composite structures at a certain frequency, which corresponds to an internal resonance of the closed body, spurious solutions may be obtained for either the EFIE or MFIE. One possible way of obtaining a unique solution at an internal resonant frequency of the structures under analysis is to combine the EFIE and MFIE through a linear summation. By linearly combining Eq. (4.3a) with Eq. (4.4a), Eq. (4.3b) with Eq. (4.4b), and Eq. (4.3c) with Eq. (4.4c), results in the CFIE as

$$(1-\alpha_c)\left(-\mathbf{E}_{1c}^s(\mathbf{J}_c,\mathbf{J}_d,\mathbf{M}_d)\right)_{\tan} + \alpha_c\eta_1\left(\mathbf{J}_c - \mathbf{n}\times\mathbf{H}_{1c}^s(\mathbf{J}_c,\mathbf{J}_d,\mathbf{M}_d)\right)$$
$$= (1-\alpha_c)\left(\mathbf{E}^i\right)_{\tan} + \alpha_c\eta_1\left(\mathbf{n}\times\mathbf{H}^i\right) \qquad \text{for } \mathbf{r} \in S_c \qquad (4.5a)$$

$$(1-\alpha_d)\left(-\mathbf{E}_{1d}^s(\mathbf{J}_c,\mathbf{J}_d,\mathbf{M}_d)\right)_{\tan} + \alpha_d\eta_1\left(-\mathbf{H}_{1d}^s(\mathbf{J}_c,\mathbf{J}_d,\mathbf{M}_d)\right)_{\tan}$$
$$= (1-\alpha_d)\left(\mathbf{E}^i\right)_{\tan} + \alpha_d\eta_1\left(\mathbf{H}^i\right)_{\tan} \qquad \text{for } \mathbf{r} \in S_d \qquad (4.5b)$$

$$(1-\alpha_d)\left(-\mathbf{E}_{2d}^s(\mathbf{J}_d,\mathbf{M}_d)\right)_{\tan} + \alpha_d\eta_2\left(-\mathbf{H}_{2d}^s(\mathbf{J}_d,\mathbf{M}_d)\right)_{\tan}$$
$$= 0 \qquad\qquad \text{for } \mathbf{r} \in S_d \qquad (4.5c)$$

where η_1 and η_2 are the wave impedance of the exterior and interior region, respectively. The parameters α_c and α_d are the usual combination parameters for the boundary conditions enforced on the conducting and dielectric surfaces and they can have any value between 0 and 1. When the conductor is open, the parameter α_c is set to zero and the formulation becomes EFIE.

An alternate formulation called C-PMCHW can be obtained by adding Eq. (4.3b) to Eq. (4.3c) and Eq. (4.4b) to Eq. (4.4c), and using them in place of Eq. (4.5b) and Eq. (4.5c), to yield

$$(1-\alpha_c)\left(-\mathbf{E}_{1c}^s(\mathbf{J}_c,\mathbf{J}_d,\mathbf{M}_d)\right)_{\tan} + \alpha_c\eta_1\left(\mathbf{J}_c - \mathbf{n}\times\mathbf{H}_{1c}^s(\mathbf{J}_c,\mathbf{J}_d,\mathbf{M}_d)\right)$$
$$= (1-\alpha_c)\left(\mathbf{E}^i\right)_{\tan} + \alpha_c\eta_1\left(\mathbf{n}\times\mathbf{H}^i\right) \qquad \text{for } \mathbf{r} \in S_c \qquad (4.6a)$$

$$\left(-\mathbf{E}_{1d}^s(\mathbf{J}_c,\mathbf{J}_d,\mathbf{M}_d) - \mathbf{E}_{2d}^s(\mathbf{J}_d,\mathbf{M}_d)\right)_{\tan} = \left(\mathbf{E}^i\right)_{\tan} \qquad \text{for } \mathbf{r} \in S_d \qquad (4.6b)$$

$$\left(-\mathbf{H}_{1d}^s(\mathbf{J}_c,\mathbf{J}_d,\mathbf{M}_d) - \mathbf{H}_{2d}^s(\mathbf{J}_d,\mathbf{M}_d)\right)_{\tan} = \left(\mathbf{H}^i\right)_{\tan} \qquad \text{for } \mathbf{r} \in S_d \qquad (4.6c)$$

Note that Eq. (4.6a), which is obtained from the boundary condition on the conducting surface S_c, is the same as Eq. (4.5a). Setting the combination parameter α_c in Eq. (4.6a) to extreme values leads to two special formulations [5]:

- when $\alpha_c = 0$, Eq. (4.6) is called E-PMCHW
- when $\alpha_c = 1$, Eq. (4.6) is called H-PMCHW

To solve the described integral equations for composite objects, the method of moment (MoM) along with a planar triangular patch model and RWG functions are used, which have been applied for the pure conducting structure in Chapter 2 and the pure dielectric structure in Chapter 3. To approximate the electric or magnetic currents on the conducting and dielectric surfaces, the RWG basis function $\mathbf{f}_n$ and its point-wise orthogonal vector basis function $\mathbf{g}_n = \mathbf{n}\times\mathbf{f}_n$

can be used on each triangular patch, where $\mathbf{n}$ is the vector normal to the surface. The electric currents $\mathbf{J}_c$ and $\mathbf{J}_d$ and the magnetic current $\mathbf{M}_d$ on the scatterers may be approximated in terms of these two basis functions as

$$\mathbf{J}_c(\mathbf{r}) = \sum_{n=1}^{N_c} J_{c,n} \mathbf{f}_n(\mathbf{r}) \tag{4.7a}$$

$$\mathbf{J}_d(\mathbf{r}) = \sum_{n=1}^{N_d} J_{d,n} \mathbf{f}_{n+N_c}(\mathbf{r}) \tag{4.7b}$$

$$\mathbf{M}_d(\mathbf{r}) = \sum_{n=1}^{N_d} M_n \mathbf{g}_n(\mathbf{r}) \tag{4.7c}$$

where N_c and N_d are the numbers of common edges on the conducting and dielectric surfaces in the triangular patch model, respectively. By choosing the basis functions for the currents as Eq. (4.7) and the RWG function $\mathbf{f}_m$ as the testing function, the EFIE given by Eqs. (4.3) yields

$$\left\langle \mathbf{f}_m, -\mathbf{E}_{1c}^s \left(\mathbf{J}_c(\mathbf{f}_n), \mathbf{J}_d(\mathbf{f}_n), \mathbf{M}_d(\mathbf{g}_n) \right) \right\rangle = \left\langle \mathbf{f}_m, \mathbf{E}^i \right\rangle \tag{4.8a}$$

$$\left\langle \mathbf{f}_m, -\mathbf{E}_{1d}^s \left(\mathbf{J}_c(\mathbf{f}_n), \mathbf{J}_d(\mathbf{f}_n), \mathbf{M}_d(\mathbf{g}_n) \right) \right\rangle = \left\langle \mathbf{f}_m, \mathbf{E}^i \right\rangle \tag{4.8b}$$

$$\left\langle \mathbf{f}_m, -\mathbf{E}_{2d}^s \left(\mathbf{J}_d(\mathbf{f}_n), \mathbf{M}_d(\mathbf{g}_n) \right) \right\rangle = 0 \tag{4.8c}$$

for $m = 1, 2, \ldots, N_c + N_d$. For the MFIE formulation, the electric current $\mathbf{J}_c$ on the conductor and the magnetic current $\mathbf{M}_d$ on the dielectric structure are expanded by the function $\mathbf{f}_n$, whereas the electric current $\mathbf{J}_d$ is approximated by $\mathbf{g}_n$ as

$$\mathbf{J}_c(\mathbf{r}) = \sum_{n=1}^{N_c} J_{c,n} \mathbf{f}_n(\mathbf{r}) \tag{4.9a}$$

$$\mathbf{J}_d(\mathbf{r}) = \sum_{n=1}^{N_d} J_{d,n} \mathbf{g}_n(\mathbf{r}) \tag{4.9b}$$

$$\mathbf{M}_d(\mathbf{r}) = \sum_{n=1}^{N_d} M_n \mathbf{f}_{n+N_c}(\mathbf{r}) \tag{4.9c}$$

The expansions of the currents on the dielectric surface in Eq. (4.9) are dual to those in Eq. (4.7) for the EFIE formulation. The MFIE given by Eq. (4.4) is expanded by using Eq. (4.9) and is tested with the RWG function $\mathbf{f}_m$, yielding

$$\left\langle \mathbf{f}_m, \mathbf{J}_c(\mathbf{f}_n) - \mathbf{n} \times \mathbf{H}_{1c}^s \left(\mathbf{J}_c(\mathbf{f}_n), \mathbf{J}_d(\mathbf{g}_n), \mathbf{M}_d(\mathbf{f}_n) \right) \right\rangle = \left\langle \mathbf{f}_m, \mathbf{n} \times \mathbf{H}^i \right\rangle \tag{4.10a}$$

$$\left\langle \mathbf{f}_m, -\mathbf{H}_{1d}^s \left(\mathbf{J}_c(\mathbf{f}_n), \mathbf{J}_d(\mathbf{g}_n), \mathbf{M}_d(\mathbf{f}_n) \right) \right\rangle = \left\langle \mathbf{f}_m, \mathbf{H}^i \right\rangle \tag{4.10b}$$

$$\left\langle \mathbf{f}_m, -\mathbf{H}_{2d}^s \left(\mathbf{J}_d(\mathbf{g}_n), \mathbf{M}_d(\mathbf{f}_n) \right) \right\rangle = 0 \tag{4.10c}$$

Now consider the CFIE formulation. All the currents $\mathbf{J}_c$, $\mathbf{J}_d$, and $\mathbf{M}_d$ are expanded by the RWG basis function $\mathbf{f}_n$ as

$$\mathbf{J}_c(\mathbf{r}) = \sum_{n=1}^{N_c} J_{c,n} \mathbf{f}_n(\mathbf{r}) \tag{4.11a}$$

$$\mathbf{J}_d(\mathbf{r}) = \sum_{n=1}^{N_d} J_{d,n} \mathbf{f}_{n+N_c}(\mathbf{r}) \tag{4.11b}$$

$$\mathbf{M}_d(\mathbf{r}) = \sum_{n=1}^{N_d} M_n \mathbf{f}_n(\mathbf{r}) \tag{4.11c}$$

In the CFIE formulation, Eq. (4.5a) is tested with $\mathbf{f}_m$, and a combination of $\mathbf{f}_m$ and $\mathbf{g}_m$ are used as the testing functions for Eqs. (4.5b) and (4.5c). By choosing the testing functions for the electric field part as $\mathbf{f}_m + \mathbf{g}_m$ and for the magnetic field part as $-\mathbf{f}_m + \mathbf{g}_m$ [11], we obtain

$$\left(1 - \alpha_c\right) \left\langle \mathbf{f}_m, -\mathbf{E}_{1c}^s \left(\mathbf{J}_c(\mathbf{f}_n), \mathbf{J}_d(\mathbf{f}_n), \mathbf{M}_d(\mathbf{f}_n) \right) \right\rangle$$
$$+ \alpha_c \eta_1 \left\langle \mathbf{f}_m, \mathbf{J}_c(\mathbf{f}_n) - \mathbf{n} \times \mathbf{H}_{1c}^s \left(\mathbf{J}_c(\mathbf{f}_n), \mathbf{J}_d(\mathbf{f}_n), \mathbf{M}_d(\mathbf{f}_n) \right) \right\rangle$$
$$= \alpha_c \eta_1 \left\langle \mathbf{f}_m, \mathbf{n} \times \mathbf{H}^i \right\rangle + \left(1 - \alpha_c\right) \left\langle \mathbf{f}_m, \mathbf{E}^i \right\rangle \tag{4.12a}$$

$$\left(1 - \alpha_d\right) \left\langle \mathbf{f}_m + \mathbf{g}_m, -\mathbf{E}_{1d}^s \left(\mathbf{J}_c(\mathbf{f}_n), \mathbf{J}_d(\mathbf{f}_n), \mathbf{M}_d(\mathbf{f}_n) \right) \right\rangle$$
$$+ \alpha_d \eta_1 \left\langle -\mathbf{f}_m + \mathbf{g}_m, -\mathbf{H}_{1d}^s \left(\mathbf{J}_c(\mathbf{f}_n), \mathbf{J}_d(\mathbf{f}_n), \mathbf{M}_d(\mathbf{f}_n) \right) \right\rangle$$
$$= \alpha_d \eta_1 \left\langle -\mathbf{f}_m + \mathbf{g}_m, \mathbf{H}^i \right\rangle + \left(1 - \alpha_d\right) \left\langle \mathbf{f}_m + \mathbf{g}_m, \mathbf{E}^i \right\rangle \tag{4.12b}$$

$$\left(1 - \alpha_d\right) \left\langle \mathbf{f}_m + \mathbf{g}_m, -\mathbf{E}_{2d}^s \left(\mathbf{J}_d(\mathbf{f}_n), \mathbf{M}_d(\mathbf{f}_n) \right) \right\rangle$$
$$+ \alpha_d \eta_2 \left\langle -\mathbf{f}_m + \mathbf{g}_m, -\mathbf{H}_{2d}^s \left(\mathbf{J}_d(\mathbf{f}_n), \mathbf{M}_d(\mathbf{f}_n) \right) \right\rangle$$
$$= 0 \tag{4.12c}$$

Note that it is also possible to choose the testing functions as $\mathbf{f}_m - \mathbf{g}_m$ for the electric field part and $\mathbf{f}_m + \mathbf{g}_m$ for the magnetic field part [11]. Setting the combination parameters α_c and α_d in Eq. (4.12) to extreme values leads to the following special formulations [10,11]:

- When $\alpha_d = 0$, Eq. (4.12) is called TENE
- When $\alpha_d = 1$, Eq. (4.12) is called THNH
- When $\alpha_c = \alpha_d = 0$, Eq. (4.12) is called EFIE-TENE,
- When $\alpha_c = \alpha_d = 1$, Eq. (4.12) is called MFIE-THNH

The EFIE-TENE formulation is obtained when both α_c and α_d are zero. In this case, the formulation only considers the electric field boundary conditions. Similarly, when both α_c and α_d are one, the MFIE-THNH formulation only deals with the magnetic field conditions. Note that even though EFIE-TENE and MFIE-THNH consists of only equations for the electric and the magnetic fields, respectively, they are the special cases of CFIE and should be differentiated from the EFIE or MFIE, as described in Eqs. (4.8) and (4.10).

In the C-PMCHW formulation, the RWG functions $\mathbf{f}_n$ are used to expand the currents $\mathbf{J}_c$, $\mathbf{J}_d$, and $\mathbf{M}_d$ as in Eq. (4.11) for the CFIE. Also, the function $\mathbf{f}_m$ is used as the testing function. Applying the testing procedure to Eq. (4.6), gives

$$
\begin{aligned}
&\left(1-\alpha_c\right)\left\langle \mathbf{f}_m, -\mathbf{E}_{1c}^s\left(\mathbf{J}_c(\mathbf{f}_n), \mathbf{J}_d(\mathbf{f}_n), \mathbf{M}_d(\mathbf{f}_n)\right)\right\rangle \\
&+\alpha_c\eta_1\left\langle \mathbf{f}_m, \mathbf{J}_c(\mathbf{f}_n) - \mathbf{n}\times\mathbf{H}_{1c}^s\left(\mathbf{J}_c(\mathbf{f}_n), \mathbf{J}_d(\mathbf{f}_n), \mathbf{M}_d(\mathbf{f}_n)\right)\right\rangle \\
&= \alpha_c\eta_1\left\langle \mathbf{f}_m, \mathbf{n}\times\mathbf{H}^i\right\rangle + \left(1-\alpha_c\right)\left\langle \mathbf{f}_m, \mathbf{E}^i\right\rangle
\end{aligned}
\tag{4.13a}
$$

$$
\left\langle \mathbf{f}_m, -\mathbf{E}_{1d}^s\left(\mathbf{J}_c(\mathbf{f}_n), \mathbf{J}_d(\mathbf{f}_n), \mathbf{M}_d(\mathbf{f}_n)\right) - \mathbf{E}_{2d}^s\left(\mathbf{J}_d(\mathbf{f}_n), \mathbf{M}_d(\mathbf{f}_n)\right)\right\rangle = \left\langle \mathbf{f}_m, \mathbf{E}^i\right\rangle
\tag{4.13b}
$$

$$
\left\langle \mathbf{f}_m, -\mathbf{H}_{1d}^s\left(\mathbf{J}_c(\mathbf{f}_n), \mathbf{J}_d(\mathbf{f}_n), \mathbf{M}_d(\mathbf{f}_n)\right) - \mathbf{H}_{2d}^s\left(\mathbf{J}_d(\mathbf{f}_n), \mathbf{M}_d(\mathbf{f}_n)\right)\right\rangle = \left\langle \mathbf{f}_m, \mathbf{E}^i\right\rangle
\tag{4.13c}
$$

The expansion and testing functions used in the described four formulations, viz., EFIE, MFIE, CFIE, and C-PMCHW, are summarized in Table 4.1. The detailed procedures for converting the integral equations into matrix equations for these four formulations can be found in Chapter 2 and Chapter 3.

Table 4.1. Testing and Expansion Functions used in the Four Formulations

Formulations	Expansion functions			Testing functions	
	$\mathbf{J}_c$	$\mathbf{J}_d$	$\mathbf{M}_d$	PEC	Dielectric
EFIE	$\mathbf{f}_n$	$\mathbf{f}_n$	$\mathbf{n}\times\mathbf{f}_n$	$\mathbf{f}_m$	$\mathbf{f}_m$
MFIE	$\mathbf{f}_n$	$\mathbf{n}\times\mathbf{f}_n$	$\mathbf{f}_n$	$\mathbf{f}_m$	$\mathbf{f}_m$
CFIE	$\mathbf{f}_n$	$\mathbf{f}_n$	$\mathbf{f}_n$	$\mathbf{f}_m$	$\pm\mathbf{f}_m + \mathbf{n}\times\mathbf{f}_m$
C-PMCHW	$\mathbf{f}_n$	$\mathbf{f}_n$	$\mathbf{f}_n$	$\mathbf{f}_m$	$\mathbf{f}_m$

To summarize this section, the EFIE, MFIE, CFIE, and C-PMCHW (including E-PMCHW and H-PMCHW) formulations are given by Eqs. (4.3), (4.4), (4.5), and (4.6), Using the testing functions as listed in Table 4.1, the testing of these integral equations for these four cases result in Eqs. (4.8), (4.10), (4.12), and (4.13), respectively.

4.3   NUMERICAL EXAMPLES FOR COMPOSITE OBJECTS

In this section, the numerical results obtained from the integral equations for composite objects described in the previous section are presented and compared. In the numerical computations, the scatterers are illuminated from the top ($-z$ direction) by an incident x-polarized plane wave. The frequency range, over which the results are calculated, is between 0 MHz to 500 MHz at an interval of 4 MHz. All computed results are compared with the solution obtained from a commercial software package, HOBBIES [12], at the same interval of 4 MHz.

The first example is a composite sphere structure as shown in Figure 4.2, which consists of a perfectly-electric-conducting (PEC) sphere and a dielectric sphere. The conducting sphere has a diameter of 1 m and is centered at the origin of the coordinates. The dielectric has a diameter of 0.8 m and a relative permittivity of $\varepsilon_r = 2$. The separation between their centers is 1 m. The surface of each sphere is modeled by 528 triangular patches, which results in 792 edges. The total number of unknowns for the whole structure is $N = N_c + 2N_d = 2376$. In HOBBIES, a similarly number of unknowns, 2304, is used for computation.

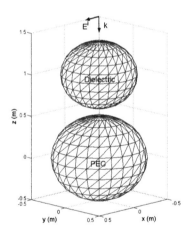

Figure 4.2.   A composite sphere structure consisting of a PEC sphere with a diameter of 1 m and a dielectric sphere with a diameter of 0.8 m and $\varepsilon_r = 2$.

Figure 4.3 shows the comparison of the radar cross-section (RCS) solution obtained from the EFIE and MFIE, with the solution from HOBBIES. The figure shows that there are several discontinuities in both the EFIE and MFIE results, which exhibit at the internal resonant frequencies of the two spheres. Besides these discontinuities, both the EFIE and MFIE agree with the HOBBIES solution.

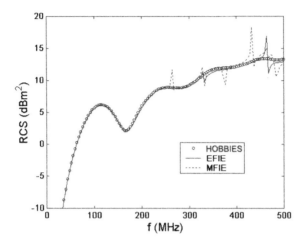

Figure 4.3. Comparison of monostatic RCS for the composite sphere structure computed by the EFIE and MFIE formulations along with the HOBBIES solution.

Figure 4.4 compares the results from HOBBIES solution with the special cases of the CFIE that relates only to either the electric or the magnetic field (i.e., EFIE-TENE [when $\alpha_c = \alpha_d = 0$] and MFIE-THNH [when $\alpha_c = \alpha_d = 1$]), in comparison to the. For the latter formulations, discontinuities are still observed at the internal resonant frequencies. At other frequencies, the results from EFIE-TENE and MFIE-THNH are in good agreement with the HOBBIES solution.

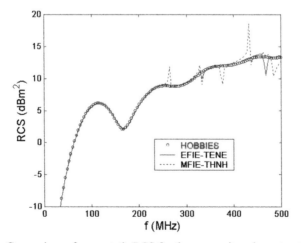

Figure 4.4. Comparison of monostatic RCS for the composite sphere structure computed by EFIE-TENE (CFIE with $\alpha_c = \alpha_d = 0$) and MFIE-THNH (CFIE with $\alpha_c = \alpha_d = 1$) formulations along with the HOBBIES solution.

From Figures 4.3 and 4.4, we observe that all the four formulations, which only make use of either electric or magnetic field boundary conditions (i.e., EFIE, MFIE, EFIE-TENE, and MFIE-THNH), suffer from internal resonant problems. In addition, more discontinuities are obtained in the results of the MFIE and MFIE-THNH, which consider only the magnetic field, than in the results of the EFIE and EFIE-TENE, which consider only the electric field.

Figure 4.5 shows the numerical results for C-PMCHW with $\alpha_c = 0.5$ and CFIE with $\alpha_c = \alpha_d = 0.5$. All the numerical results do not exhibit any spurious resonant peak and agree excellently with the HOBBIES solution. However, when the formulations of E-PMCHW (when $\alpha_c = 0$) and H-PMCHW (when $\alpha_c = 1$) are used, discontinuities resulting from the internal resonance are displayed in the solutions and therefore are not presented.

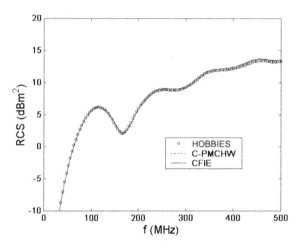

Figure 4.5. Comparison of monostatic RCS for the composite sphere structure computed by C-PMCHW ($\alpha_c = 0.5$) and CFIE ($\alpha_c = \alpha_d = 0.5$) formulations along with the HOBBIES solution.

As a second example, consider the composite structure consisting of a conducting and a dielectric cube, as shown in Figure 4.6. The conducting cube has a length of 1 m for each side and is centered at the origin of the coordinates. The dielectric cube has a length of 0.8 m and a relative permittivity of $\varepsilon_r = 2$. The separation between their centers is 1 m. The surface of each cube is modeled by 768 triangular patches and results in 1152 edges. The total number of unknowns for the whole structure is $N = N_c + 2N_d = 3456$. In HOBBIES, 3600 unknowns are used for computation.

Figure 4.7 shows the RCS computed using the EFIE and MFIE and compares those with the HOBBIES solution. Figure 4.8 shows the results of EFIE-TENE and MFIE-THNH. It is shown that there are several discontinuities near the resonant frequencies for the two cubes in each figure. It can be observed from Figures 4.7 and 4.8 that, except for the discontinuities at or near the resonant frequencies, there is still good agreement with the HOBBIES solution. Meanwhile, the formulations given by the magnetic boundary conditions (MFIE and MFIE-THNH) have more discontinuities than those formulations using electric field boundary conditions (EFIE and EFIE-TENE).

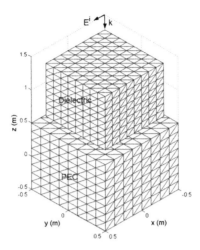

Figure 4.6. Composite cube structure consisting of a perfectly conducting cube with a length of 1 m and a dielectric cube with a length of 0.8 m and $\varepsilon_r = 2$.

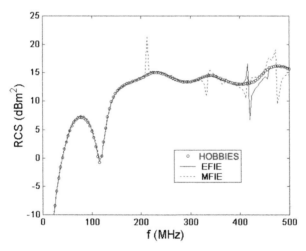

Figure 4.7. Comparison of monostatic RCS for the composite cube structure computed by EFIE and MFIE formulations along with the HOBBIES result.

Figure 4.9 shows the numerical results of C-PMCHW with $\alpha_c = 0.5$ and CFIE with $\alpha_c = \alpha_d = 0.5$. All numerical results do not exhibit the spurious resonant peaks and agree perfectly with the HOBBIES solution. It is difficult to distinguish between the C-PMCHW and the CFIE solutions. When E-PMCHW with $\alpha_c = 0$ or H-PMCHW with $\alpha_c = 1$ is used, the results exhibit discontinuities.

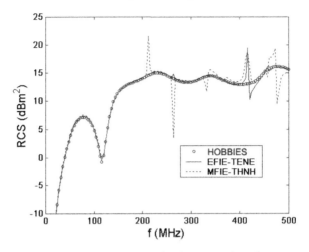

Figure 4.8. Comparison of monostatic RCS for the composite cube structure computed by EFIE-TENE (CFIE with $\alpha_c = \alpha_d = 0$) and MFIE-THNH (CFIE with $\alpha_c = \alpha_d = 1$) formulations along with the HOBBIES result.

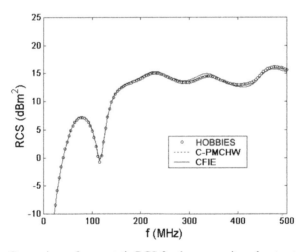

Figure 4.9. Comparison of monostatic RCS for the composite cube structure computed by C-PMCHW ($\alpha_c = 0.5$) and CFIE ($\alpha_c = \alpha_d = 0.5$) formulations along with the HOBBIES result.

4.4   INTEGRAL EQUATIONS FOR COATED OBJECTS

In this section, several formulations are presented to analyze the coated composite structures using surface integral equations. To simplify the presentation, it is assumed that there is only one conducting body enclosed by a homogeneous dielectric material, as shown in Figure 4.10. The formulation derived for this case can be extended to analysis of structures with multiple conducting objects. This coated structure is illuminated by an incident plane wave with the electric and magnetic fields denoted by $\mathbf{E}^i$ and $\mathbf{H}^i$. The incident field is defined to exist in space if the structure were not present. The regions external and internal to the dielectric body are characterized by the medium parameters (ε_1, μ_1) and (ε_2, μ_2), respectively. The surfaces of the conducting and dielectric bodies are denoted by S_c and S_d.

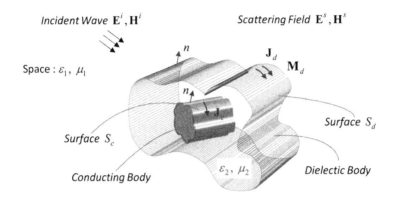

Figure 4.10. A conducting body enclosed by dielectric material and illuminated by an electromagnetic wave.

The equivalent principle is employed to split the original problem into two parts, which are equivalent to the exterior and the interior region of the dielectric body, respectively. In the equivalent problem for the exterior region, the surface of the dielectric body S_d is replaced by a fictitious mathematical surface, and the entire interior region of the dielectric body is assumed to be replaced by the material of the external region (i.e. the medium with ε_1 and μ_1). Because the equivalence is now for the exterior region, the field inside the surface S_d can be set to zero for convenience. To ensure the continuity of the tangential electric and magnetic fields, there must be equivalent electric and magnetic currents, $\mathbf{J}_d$ and $\mathbf{M}_d$, flowing on the surface S_d. Since these currents radiate in a homogeneous unbounded medium, the scattered fields can be computed by using the Maxwell's equations. By enforcing the boundary conditions on the electric and magnetic fields across the interface, a set of coupled equations are derived

$$\left(\mathbf{E}_{1d}^s(\mathbf{J}_d) + \mathbf{E}_{1d}^s(\mathbf{M}_d) + \mathbf{E}^i\right)_{\tan} = 0 \quad \text{for } \mathbf{r} \in S_d \tag{4.14a}$$

$$\left(\mathbf{H}_{1d}^s(\mathbf{J}_d) + \mathbf{H}_{1d}^s(\mathbf{M}_d) + \mathbf{H}^i\right)_{\tan} = 0 \quad \text{for } \mathbf{r} \in S_d \tag{4.14b}$$

where $\mathbf{E}^i$ and $\mathbf{E}^s$ are the incident and scattered electric fields, respectively, and $\mathbf{H}^i$ and $\mathbf{H}^s$ are the incident and scattered magnetic fields. The subscript "tan" denotes the tangential component. The subscript "1", associated with the scattered fields, represents the fields to be computed in the region external to the dielectric body, and "d" denotes the equations that are enforced at the boundary of the dielectric surface.

In the equivalent problem for the interior region, the material parameters inside the dielectric body must be kept as the original, whereas the exterior region has no restrictions. The fields external to the fictitious mathematical surface S_d are set to zero and the corresponding exterior space is replaced by the original interior material (i.e. the medium with ε_2 and μ_2). Moreover, the conducting surface, which is located inside the dielectric body, is replaced by a fictitious mathematical surface so that the present equivalent situation is again materially homogeneous. As before, equivalent currents on the surface S_d are set up to satisfy the continuity of the fields. It turns out that these equivalent currents are the negative of the currents for the exterior equivalent problem (i.e., $-\mathbf{J}_d$ and $-\mathbf{M}_d$), which is a result of the direction of the unit normal vector pointing into the external medium. In addition, an equivalent electric current $\mathbf{J}_c$ is introduced on the conducting surface S_c. Here, only the electric current is required because the tangential electric field is zero on the conducting surface. By enforcing the continuity of the tangential fields across both the conducting and dielectric interfaces, we have

$$\left(\mathbf{E}^s_{2d}(\mathbf{J}_c)+\mathbf{E}^s_{2d}(-\mathbf{J}_d)+\mathbf{E}^s_{2d}(-\mathbf{M}_d)\right)_{\tan}=0 \quad \text{for } \mathbf{r} \in S_d \tag{4.15a}$$

$$\left(\mathbf{H}^s_{2d}(\mathbf{J}_c)+\mathbf{H}^s_{2d}(-\mathbf{J}_d)+\mathbf{H}^s_{2d}(-\mathbf{M}_d)\right)_{\tan}=0 \quad \text{for } \mathbf{r} \in S_d \tag{4.15b}$$

$$\left(\mathbf{E}^s_{2c}(\mathbf{J}_c)+\mathbf{E}^s_{2c}(-\mathbf{J}_d)+\mathbf{E}^s_{2c}(-\mathbf{M}_d)\right)_{\tan}=0 \quad \text{for } \mathbf{r} \in S_c \tag{4.15c}$$

$$\mathbf{J}_c-\mathbf{n}\times\left(\mathbf{H}^s_{2c}(\mathbf{J}_c)+\mathbf{H}^s_{2c}(-\mathbf{J}_d)+\mathbf{H}^s_{2c}(-\mathbf{M}_d)\right)=0 \quad \text{for } \mathbf{r} \in S_c \tag{4.15d}$$

where the subscript "2" associated with the scattered fields represents the fields to be computed in the region internal to the dielectric body and "c" denotes the equations that are enforced on the conducting surface.

When only the boundary conditions for the electric field in Eqs. (4.14) and (4.15) are considered, the EFIE for the coated object can be formed by grouping Eqs (4.14a), (4.15a), and (4.15c). It gives the relation in the frequency domain between the incident and the scattered electric fields for a conducting body enclosed by a homogeneous dielectric material. The fields from the induced electric and magnetic currents on the conducting and dielectric surfaces are

$$\left(-\mathbf{E}^s_{1d}(\mathbf{J}_d,\mathbf{M}_d)\right)_{\tan}=\left(\mathbf{E}^i\right)_{\tan} \qquad \text{for } \mathbf{r} \in S_d \tag{4.16a}$$

$$\left(\mathbf{E}_{2d}^s(\mathbf{J}_c) - \mathbf{E}_{2d}^s(\mathbf{J}_d,\mathbf{M}_d)\right)_{\text{tan}} = 0 \qquad \text{for } \mathbf{r} \in S_d \qquad (4.16b)$$

$$\left(\mathbf{E}_{2c}^s(\mathbf{J}_c) - \mathbf{E}_{2c}^s(\mathbf{J}_d,\mathbf{M}_d)\right)_{\text{tan}} = 0 \qquad \text{for } \mathbf{r} \in S_c \qquad (4.16c)$$

In this section, Eq. (4.16) is called EFIE-EFIE because the continuity conditions of the electric field are used for both conducting and dielectric surfaces.

Similarly, by grouping Eqs (4.14b), (4.15b), and (4.15d), the magnetic field integral equation for the coated object can be formed. It gives the relationship in the frequency domain between the incident and the scattered magnetic fields for a conducting body coated with homogeneous dielectric material from the induced electric and magnetic currents on the conducting and dielectric surfaces as

$$\left(-\mathbf{H}_{1d}^s(\mathbf{J}_d,\mathbf{M}_d)\right)_{\text{tan}} = \left(\mathbf{H}^i\right)_{\text{tan}} \qquad \text{for } \mathbf{r} \in S_d \qquad (4.17a)$$

$$\left(\mathbf{H}_{2d}^s(\mathbf{J}_c) - \mathbf{H}_{2d}^s(\mathbf{J}_d,\mathbf{M}_d)\right)_{\text{tan}} = 0 \qquad \text{for } \mathbf{r} \in S_d \qquad (4.17b)$$

$$\mathbf{J}_c - \mathbf{n} \times \left(\mathbf{H}_{2c}^s(\mathbf{J}_c) - \mathbf{H}_{2c}^s(\mathbf{J}_d,\mathbf{M}_d)\right) = 0 \qquad \text{for } \mathbf{r} \in S_c \qquad (4.17c)$$

This formulation is called MFIE-MFIE, because the continuity conditions of the magnetic field are used for both conducting and dielectric surfaces.

In the analysis of composite structures at or near the frequency that corresponds to an internal resonance of the closed body, spurious solutions may be obtained when using either the EFIE-EFIE or MFIE-MFIE. To overcome this problem, the electric fields in EFIE-EFIE and magnetic fields in MFIE-MFIE for the same region and the same boundary are added linearly to form the combined field integral equation. Linearly combining Eq. (4.16a) with Eq. (4.17a), Eq. (4.16b) with Eq. (4.17b), and Eq. (4.16c) with Eq. (4.17c), gives

$$(1-\alpha_d)\left(-\mathbf{E}_{1d}^s(\mathbf{J}_d,\mathbf{M}_d)\right)_{\text{tan}} + \alpha_d\eta_1\left(-\mathbf{H}_{1d}^s(\mathbf{J}_d,\mathbf{M}_d)\right)_{\text{tan}}$$
$$= (1-\alpha_d)\left(\mathbf{E}^i\right)_{\text{tan}} + \alpha_d\eta_1\left(\mathbf{H}^i\right)_{\text{tan}} \qquad \text{for } \mathbf{r} \in S_d \qquad (4.18a)$$

$$(1-\alpha_d)\left(\mathbf{E}_{2d}^s(\mathbf{J}_c) - \mathbf{E}_{2d}^s(\mathbf{J}_d,\mathbf{M}_d)\right)_{\text{tan}}$$
$$+ \alpha_d\eta_2\left(\mathbf{H}_{2d}^s(\mathbf{J}_c) - \mathbf{H}_{2d}^s(\mathbf{J}_d,\mathbf{M}_d)\right)_{\text{tan}} = 0 \qquad \text{for } \mathbf{r} \in S_d \qquad (4.18b)$$

$$(1-\alpha_c)\left(\mathbf{E}_{2c}^s(\mathbf{J}_c) - \mathbf{E}_{2c}^s(\mathbf{J}_d,\mathbf{M}_d)\right)_{\text{tan}}$$
$$+ \alpha_c\eta_2\left(\mathbf{J}_c - \mathbf{n} \times \left(\mathbf{H}_{2c}^s(\mathbf{J}_c) - \mathbf{H}_{2c}^s(\mathbf{J}_d,\mathbf{M}_d)\right)\right) = 0 \qquad \text{for } \mathbf{r} \in S_c \qquad (4.18c)$$

where η_1 and η_2 are the wave impedance of the exterior and interior regions. The parameter α_c and α_d are the usual combination parameters that can be assigned to any value between 0 and 1. This formulation is called CFIE-CFIE because the continuity conditions of the combined electric and magnetic fields are used for both the conducting and dielectric surfaces. For the open conducting structure, α_c needs to be set to zero and this results in EFIE-CFIE formulation.

An alternative way to combine the electric and magnetic field formulations, the electric fields in EFIE-EFIE evaluated at the boundary of the dielectric surface S_d are added linearly. Similar summation is also performed for the magnetic fields in MFIE-MFIE. By linearly adding Eq. (4.16a) to Eq. (4.16b) and Eq. (4.17a) to Eq. (4.17b), the CFIE-PMCHW formulation is obtained

$$\left(-\mathbf{E}_{1d}^s(\mathbf{J}_d,\mathbf{M}_d)-\mathbf{E}_{2d}^s(\mathbf{J}_d,\mathbf{M}_d)+\mathbf{E}_{2d}^s(\mathbf{J}_c)\right)_{\tan}=\left(\mathbf{E}^i\right)_{\tan} \qquad \text{for } \mathbf{r}\in S_d \qquad (4.19a)$$

$$\left(-\mathbf{H}_{1d}^s(\mathbf{J}_d,\mathbf{M}_d)-\mathbf{H}_{2d}^s(\mathbf{J}_d,\mathbf{M}_d)+\mathbf{H}_{2d}^s(\mathbf{J}_c)\right)_{\tan}=\left(\mathbf{H}^i\right)_{\tan} \qquad \text{for } \mathbf{r}\in S_d \qquad (4.19b)$$

$$(1-\alpha_c)\left(\mathbf{E}_{2c}^s(\mathbf{J}_c)-\mathbf{E}_{2c}^s(\mathbf{J}_d,\mathbf{M}_d)\right)_{\tan}$$
$$+\alpha_c\eta_2\left(\mathbf{J}_c-\mathbf{n}\times\left(\mathbf{H}_{2c}^s(\mathbf{J}_c)-\mathbf{H}_{2c}^s(\mathbf{J}_d,\mathbf{M}_d)\right)\right)=0 \qquad \text{for } \mathbf{r}\in S_c \qquad (4.19c)$$

Note that Eq. (4.19c), which is derived at the boundary of the conducting surface S_c, is kept the same as Eq. (4.18c). The formulation given by Eq. (4.19) is called E-PMCHW when the parameter α_c in Eq. (4.19c) is zero, or it is called H-PMCHW when α_c is equal to one [5].

To solve the resulting equations obtained for coated objects, the method of moment along with the triangular surface patch model and the RWG basis functions are used. The expansion and testing functions for these formulations are chosen the same as those for the composite objects, and one can refer to Table 5.1 for more information.

Similar to the EFIE formulation for the composite objects, the electric current $\mathbf{J}_c$ and $\mathbf{J}_d$ and the magnetic current $\mathbf{M}_d$ in EFIE-EFIE formulation given by Eq. (4.16) are approximated by Eq. (4.7) and tested with $\mathbf{f}_m$ and yield

$$\left\langle \mathbf{f}_m, -\mathbf{E}_{1d}^s\left(\mathbf{J}_d(\mathbf{f}_n),\mathbf{M}_d(\mathbf{g}_n)\right)\right\rangle=\left\langle \mathbf{f}_m,\mathbf{E}^i\right\rangle \qquad (4.20a)$$

$$\left\langle \mathbf{f}_m, \mathbf{E}_{2d}^s\left(\mathbf{J}_c(\mathbf{f}_n)\right)-\mathbf{E}_{2d}^s\left(\mathbf{J}_d(\mathbf{f}_n),\mathbf{M}_d(\mathbf{g}_n)\right)\right\rangle=0 \qquad (4.20b)$$

$$\left\langle \mathbf{f}_m, \mathbf{E}_{2c}^s\left(\mathbf{J}_c(\mathbf{f}_n)\right)-\mathbf{E}_{2c}^s\left(\mathbf{J}_d(\mathbf{f}_n),\mathbf{M}_d(\mathbf{g}_n)\right)\right\rangle=0 \qquad (4.20c)$$

For the MFIE-MFIE formulation given by Eq. (4.17), the equivalent electric current $\mathbf{J}_c$ on the conductor, the electric current $\mathbf{J}_d$, and the magnetic current $\mathbf{M}_d$ may be expanded by Eq. (4.9) and tested with $\mathbf{f}_m$, yielding

$$\left\langle \mathbf{f}_m, -\mathbf{H}_{1d}^s \left(\mathbf{J}_d(\mathbf{g}_n), \mathbf{M}_d(\mathbf{f}_n) \right) \right\rangle = \left\langle \mathbf{f}_m, \mathbf{H}^i \right\rangle \tag{4.21a}$$

$$\left\langle \mathbf{f}_m, \mathbf{H}_{2d}^s \left(\mathbf{J}_c(\mathbf{f}_n) \right) - \mathbf{H}_{2d}^s \left(\mathbf{J}_d(\mathbf{g}_n), \mathbf{M}_d(\mathbf{f}_n) \right) \right\rangle = 0 \tag{4.21b}$$

$$\left\langle \mathbf{f}_m, \mathbf{J}_c(\mathbf{f}_n) - \mathbf{n} \times \left(\mathbf{H}_{2c}^s \left(\mathbf{J}_c(\mathbf{f}_n) \right) - \mathbf{H}_{2c}^s \left(\mathbf{J}_d(\mathbf{g}_n), \mathbf{M}_d(\mathbf{f}_n) \right) \right) \right\rangle = 0 \tag{4.21c}$$

Consider the CFIE-CFIE formulation now. The equivalent currents $\mathbf{J}_c$, $\mathbf{J}_d$, and $\mathbf{M}_d$ are all approximated by the RWG basis function $\mathbf{f}_n$ as given by Eq. (4.11). Two vector basis functions are used in the testing for the fields at the dielectric surface. By choosing the testing functions for electric field part as $\mathbf{f}_m + \mathbf{g}_m$ and for the magnetic field part as $-\mathbf{f}_m + \mathbf{g}_m$ [11], the CFIE-CFIE formulation given by Eq. (4.18) results in

$$(1 - \alpha_d) \left\langle \mathbf{f}_m + \mathbf{g}_m, -\mathbf{E}_{1d}^s \left(\mathbf{J}_d(\mathbf{f}_n), \mathbf{M}_d(\mathbf{f}_n) \right) \right\rangle$$
$$+ \alpha_d \eta_1 \left\langle -\mathbf{f}_m + \mathbf{g}_m, -\mathbf{H}_{1d}^s \left(\mathbf{J}_d(\mathbf{f}_n), \mathbf{M}_d(\mathbf{f}_n) \right) \right\rangle \tag{4.22a}$$
$$= (1 - \alpha_d) \left\langle \mathbf{f}_m + \mathbf{g}_m, \mathbf{E}^i \right\rangle + \alpha_d \eta_1 \left\langle -\mathbf{f}_m + \mathbf{g}_m, \mathbf{H}^i \right\rangle$$

$$(1 - \alpha_d) \left\langle \mathbf{f}_m + \mathbf{g}_m, \mathbf{E}_{2d}^s \left(\mathbf{J}_c(\mathbf{f}_n) \right) - \mathbf{E}_{2d}^s \left(\mathbf{J}_d(\mathbf{f}_n), \mathbf{M}_d(\mathbf{f}_n) \right) \right\rangle$$
$$+ \alpha_d \eta_2 \left\langle -\mathbf{f}_m + \mathbf{g}_m, \mathbf{H}_{2d}^s \left(\mathbf{J}_c(\mathbf{f}_n) \right) - \mathbf{H}_{2d}^s \left(\mathbf{J}_d(\mathbf{f}_n), \mathbf{M}_d(\mathbf{f}_n) \right) \right\rangle = 0 \tag{4.22b}$$

$$(1 - \alpha_c) \left\langle \mathbf{f}_m, \mathbf{E}_{2c}^s \left(\mathbf{J}_c(\mathbf{f}_n) \right) - \mathbf{E}_{2c}^s \left(\mathbf{J}_d(\mathbf{f}_n), \mathbf{M}_d(\mathbf{f}_n) \right) \right\rangle$$
$$+ \alpha_c \eta_2 \left\langle \mathbf{f}_m, \mathbf{J}_c(\mathbf{f}_n) - \mathbf{n} \times \left(\mathbf{H}_{2c}^s \left(\mathbf{J}_c(\mathbf{f}_n) \right) - \mathbf{H}_{2c}^s \left(\mathbf{J}_d(\mathbf{f}_n), \mathbf{M}_d(\mathbf{f}_n) \right) \right) \right\rangle = 0 \tag{4.22c}$$

Note that Eq. (4.22c) is tested with only RWG function for the conducting surface.

In the CFIE-PMCHW formulation, the procedure is the same as CFIE-CFIE; the RWG functions $\mathbf{f}_n$ are used to expand all the equivalent currents $\mathbf{J}_c$ $\mathbf{J}_d$, and $\mathbf{M}_d$, and $\mathbf{f}_m$ is used as the testing function. Applying the testing procedure to Eq. (4.19), we obtain

$$\left\langle \mathbf{f}_m, -\mathbf{E}_{1d}^s \left(\mathbf{J}_d(\mathbf{f}_n), \mathbf{M}_d(\mathbf{f}_n) \right) \right.$$
$$\left. -\mathbf{E}_{2d}^s \left(\mathbf{J}_d(\mathbf{f}_n), \mathbf{M}_d(\mathbf{f}_n) \right) + \mathbf{E}_{2d}^s \left(\mathbf{J}_c(\mathbf{f}_n) \right) \right\rangle = \left\langle \mathbf{f}_m, \mathbf{E}^i \right\rangle \tag{4.23a}$$

$$\left\langle \mathbf{f}_m, -\mathbf{H}_{1d}^s \left(\mathbf{J}_d(\mathbf{f}_n), \mathbf{M}_d(\mathbf{f}_n) \right) \right.$$
$$\left. -\mathbf{H}_{2d}^s \left(\mathbf{J}_d(\mathbf{f}_n), \mathbf{M}_d(\mathbf{f}_n) \right) + \mathbf{H}_{2d}^s \left(\mathbf{J}_c(\mathbf{f}_n) \right) \right\rangle = \left\langle \mathbf{f}_m, \mathbf{H}^i \right\rangle \tag{4.23b}$$

$$(1-\alpha_c)\langle \mathbf{f}_m, \mathbf{E}_{2c}^s(\mathbf{J}_c(\mathbf{f}_n)) - \mathbf{E}_{2c}^s(\mathbf{J}_d(\mathbf{f}_n), \mathbf{M}_d(\mathbf{f}_n)) \rangle + \alpha_c \eta_2 \langle \mathbf{f}_m, \mathbf{J}_c(\mathbf{f}_n)$$

$$-\mathbf{n} \times (\mathbf{H}_{2c}^s(\mathbf{J}_c(\mathbf{f}_n)) - \mathbf{H}_{2c}^s(\mathbf{J}_d(\mathbf{f}_n), \mathbf{M}_d(\mathbf{f}_n))) \rangle = 0 \qquad (4.23c)$$

Note that Eq. (4.23) can be converted into a matrix equation for the CFIE-PMCHW formulation. A detail procedure to convert the integral equations into a matrix equation can be found in [11]. In this section, the formulation for analyzing the scattering of a coated structure was presented. The EFIE, MFIE, CFIE, and C-PMCHW formulation are given by Eqs. (4.16), (4.17), (4.18), and (4.19), respectively. Using the expansion and testing functions listed in Table 5.1, the testing of the integral equations for these four cases result in Eqs. (4.20), (4.21), (4.22), and (4.23), respectively.

4.5   NUMERICAL EXAMPLES FOR COATED OBJECTS

In this section, the numerical results obtained from the formulations for coated objects described in the previous section are presented and compared. In the numerical computations, the scatterers are illuminated from the top ($-z$ direction) by an incident x-polarized plane wave. The frequency range, over which the RCS is calculated, is between 0 MHz to 500 MHz at an interval of 4 MHz. All computed results are compared with the solution obtained from a commercial software package, HOBBIES [12], at the same interval of 4 MHz.

The first example is a conducting disk enclosed by a dielectric cylinder, as shown in Figure 4.11. The conducting disk has a radius of 0.4 m and is centered at the origin of the coordinates. The dielectric cylinder has a radius of 0.5 m and a height of 0.5 m. The relative permittivity is $\varepsilon_r = 2$. The surface of the conducting disk is modeled with 160 triangular patches, which results in a total of 240 common edges. The outer surface of the dielectric cylinder is modeled with 528 triangular patches and has a total of 792 edges. The total number of unknowns for the whole structure is $N = N_c + 2N_d = 1824$. The number of unknowns is 2134 in the computation using HOBBIES.

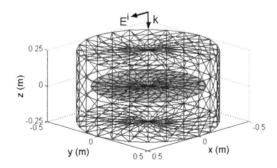

Figure 4.11. A conducting disk with a radius of 0.4 m enclosed by a dielectric cylinder ($\varepsilon_r = 2$) with a radius of 0.5 m and a height of 0.5 m.

Figure 4.12 shows the comparison of the RCS obtained from EFIE-EFIE and EFIE-CFIE (CFIE-CFIE with $\alpha_c = 0$ and $\alpha_d = 0.5$), and EFIE-PMCHW (CFIE-PMCHW with $\alpha_c = 0$) formulations along with the HOBBIES solution. It is shown that there are discontinuities in the solution of EFIE-EFIE formulation because of the internal resonances of the structure. The numerical results of EFIE-CFIE and EFIE-PMCHW do not exhibit spurious resonant peaks. Their results are almost identical, and both of them agree well with the HOBBIES solution, except for a little discrepancy of having lower RCS values at frequencies higher than 400 MHz.

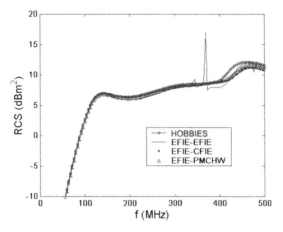

Figure 4.12. Comparison of monostatic RCS for the coated structure in Figure 4.11 computed by EFIE-EFIE, EFIE-CFIE (CFIE-CFIE $\alpha_c = 0$ and $\alpha_d = 0.5$), and EFIE-PMCHW (CFIE-PMCHW with $\alpha_c = 0$) along with the HOBBIES solution.

As a second example, consider a conducting sphere coated with a dielectric material shown in Figure 4.13.

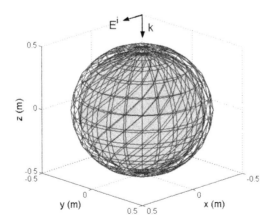

Figure 4.13. A conducting sphere coated with a dielectric material ($\varepsilon_r = 2$). The conducting sphere has a radius of 0.45 m, and the coating thickness is 0.05 m.

The conducting sphere has a radius of 0.45 m and is centered at the origin of the coordinates. The thickness of the dielectric material is 0.05 m, and its relative permittivity is $\varepsilon_r = 2$. Both the surfaces of the conducting sphere and the outer dielectric sphere are modeled with 528 triangular patches and each generates 792 edges. The total number of unknowns for this structure is $N = 2376$. The number of unknowns is 2304 in the computation using HOBBIES.

Figure 4.14 shows the RCS computed by using EFIE-EFIE and MFIE-MFIE formulations against the analytical Mie series solution. It can be found that there are many discontinuities resulting from the internal resonances of the structure. Particularly, the discontinuities in the MFIE-MFIE result are more serious than the EFIE-EFIE one. These two formulations can only provide good approximations to the Mie solution in the band below the resonant frequencies.

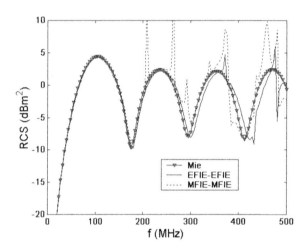

Figure 4.14. Comparison of monostatic RCS for the coated sphere in Figure 4.13 computed by EFIE-EFIE and MFIE-MFIE formulations along with the Mie solution.

Figure 4.15 shows the numerical results of the CFIE-CFIE formulation with $\alpha_c = \alpha_d = 0.5$ and CFIE-PMCHW formulation with $\alpha_c = 0.5$, against the HOBBIES and the analytical Mie solutions. The numerical results from these two formulations do not exhibit the spurious resonant peaks and agree excellently with the HOBBIES solution. But it should be noted that there is a discrepancy between these results and the Mie solution, as at high frequency, the structure was not finely meshed. For the special cases of the CFIE-PMCHW formulation, both the results from EFIE-PMCHW ($\alpha_c = 0$) and MFIE-PMCHW ($\alpha_c = 1$) had the discontinuity problems, and their results are therefore not shown here.

As a third example, consider a conducting sphere enclosed by a dielectric cube as shown in Figure 4.16. The conducting sphere has a radius of 0.4 m and is centered at the origin of the coordinates. The dielectric cube has a side length of 1 m. The relative permittivity is $\varepsilon_r = 2$. The surface of the conducting sphere is modeled with 528 triangular patches and, thus, has 792 edges. The surface of the dielectric cube is modeled with 768 triangular patches and has 1152 edges. The total number of unknowns for the whole structure is $N = 3096$. The number of unknowns is 3168 in the computation using HOBBIES.

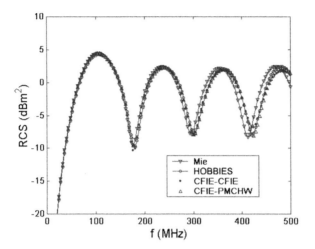

Figure 4.15. Comparison of monostatic RCS for the coated structure in Figure 4.13 computed by CFIE-CFIE ($\alpha_c = \alpha_d = 0.5$) and CFIE-PMCHW ($\alpha_c = 0.5$) formulations with the exact solution and the HOBBIES result.

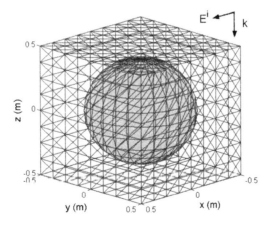

Figure 4.16.   A perfectly conducting sphere enclosed by a dielectric cube ($\varepsilon_r = 2$). The conducting sphere has a radius of 0.4 m, and the dielectric cube has a side of 1 m.

Figure 4.17 shows the RCS computed using EFIE-EFIE and MFIE-MFIE formulations, and compares with the HOBBIES solution. It is shown in the figure that there are several discontinuities resulting from the internal resonances of the structure, although there is good agreement between the solutions in the low frequency region. The peaks in the MFIE-MFIE solution are more serious than those in the EFIE-EFIE solution.

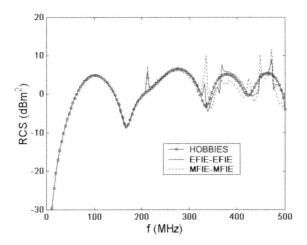

Figure 4.17. Comparison of monostatic RCS for the coated structure in Figure 4.16 computed by EFIE-EFIE and MFIE-MFIE formulations along with the HOBBIES result.

Figure 4.18 shows the comparison of the numerical results of the CFIE-CFIE with $\alpha_c = \alpha_d = 0.5$, the CFIE-PMCHW formulation with $\alpha_c = 0.5$, and the HOBBIES solution. All numerical results do not exhibit the spurious resonant peaks and agree well. Again, there is some discrepancy in the results at high frequencies, as more triangular patches are necessary to discretize the structure.

4.6   CONCLUSION

Various sets of coupled integral equations are considered to analyze the scattering from 3-D arbitrarily shaped conducting-dielectric composite or conductor coated with dielectric structures. The integral equations are derived using the equivalence principle and using the continuity conditions for the fields at the boundary surfaces. Numerical solutions for these equations are obtained through using MoM in conjunction with the planar triangular patch basis function. To validate the efficacies of the described formulations, the monostatic radar cross-sections of two composite and three coated structures are computed and compared with a commercial software package—HOBBIES. In the composite scatterer examples, the EFIE and MFIE formulations cannot give valid solutions at the internal resonant frequencies of the scatterers; particularly, the discontinuities in the MFIE results are relatively larger. Although stable solutions can be obtained by using the CFIE and C-PMCHW formulations and they have excellent agreements with the HOBBIES solutions. From the three examples of the coated structure, it is found that EFIE-EFIE and MFIE-MFIE suffer from the discontinuities resulting from the internal resonances of the structures. This internal resonance problem can be solved by using either the CFIE-CFIE or CFIE-PMCHW formulation, and their results match very well with the HOBBIES solutions except for small discrepancies at the high frequency region because of an insufficient number of triangular patches used to discretize the structure.

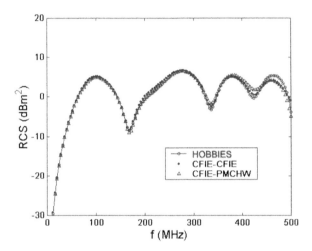

Figure 4.18. Comparison of monostatic radar cross-sections for the coated structure in Fig. 4.16 computed by CFIE-CFIE ($\alpha_c = \alpha_d = 0.5$) and CFIE-PMCHW ($\alpha_c = 0.5$) formulations with the HOBBIES result.

REFERENCES

[1]   S. M. Rao, E. Arvas, and T. K. Sarkar, "Combined field solution for TM scattering from multiple conducting and dielectric cylinders of arbitrary cross section," *IEEE Trans. Antennas Prop.,* Vol. 35, pp. 447–451, 1987.

[2]   E. Arvas and T. K. Sarkar, "RCS of two-dimensional structures consisting of both dielectric and conductors of arbitrary cross section," *IEEE Trans. Antennas Prop.,* Vol. 37, pp. 546–554, 1989.

[3]   P. M. Goggans and T. H. Shumpert, "CFIE MM solution for TE and TM incidence on a 2-D conducting body with dielectric filled cavity," *IEEE Trans. Antennas Prop.,* Vol. 38, pp. 1645–1649, 1990.

[4]   J. M. Putman and L. N. Medgyesi-Mitschang, "Combined field integral equation formulation for inhomogeneous two- and three-dimensional bodies: the junction problem," *IEEE Trans. Antennas Prop.,* Vol. 39, pp. 667–672, 1991.

[5]   A. A. Kishk and L. Shafai, "Different formulations for numerical solution of single or multibodies of revolution with mixed boundary conditions," *IEEE Trans. Antennas Prop.*, Vol. 34, pp. 666–673, 1986.

[6]   S. M. Rao, D. R. Wilton, and A. W. Glisson, "Electromagnetic scattering by surfaces of arbitrary shape," *IEEE Trans. Antennas Prop.*, Vol. 30, pp. 409–418, 1982.

[7]   T. K. Sarkar, S. M. Rao, and A. R. Djordjevic, "Electromagnetic scattering and radiation from finite microstrip structures," *IEEE Trans. Antennas Prop.*, Vol. 38, pp. 1568–1575, 1990.

[8]   S. M. Rao, T. K. Sarkar, and A. R. Djordevic, "Electromagnetic radiation and scattering from finite conducting and dielectric structures: surface/surface formulation," *IEEE Trans. Antennas Prop.*, Vol. 39, pp. 1034–1037, 1991.

[9]   S. M. Rao, C. C. Cha, R. L. Cravey, and D. L. Wilkes, "Electromagnetic scattering from arbitrary shaped conducting bodies coated with lossy materials of arbitrary thickness," *IEEE Trans. Antennas Prop.*, Vol. 39, pp. 627–631, 1991.

[10]   X. Q. Sheng, J. M. Jin, J. M. Song, W. C. Chew, and C. C. Lu, "Solution of combined-field integral equation using multilevel fast multipole algorithm for scattering by homogeneous bodies," *IEEE Trans. Antennas Prop.*, Vol. 46, pp. 1718–1726, 1998.

[11]   B. H. Jung, T. K. Sarkar, and Y.-S. Chung, "A survey of various frequency domain integral equations for the analysis of scattering from three-dimensional dielectric objects," *Electromagn. Waves Appl.*, Vol. 16. pp. 1419–1421, 2002.

[12]   Y. Zhang and T. K. Sarkar, *Parallel Solution of Integral Equation-based EM Problems in the Frequency Domain*, John Wiley & Sons, New York, 2009.

[13]   A. A. Kishk, A. W. Glisson, and P. M. Goggans, "Scattering from conductors coated with materials of arbitrary thickness," *IEEE Trans. Antennas Prop.*, Vol. 40, pp. 108–112, 1992.

[14]   R. F. Harrington, *Time Harmonic Electromagnetics*, McGraw-Hill, New York, 1961.

5

ANALYSIS OF CONDUCTING WIRES IN THE TIME DOMAIN

5.0 SUMMARY

In this chapter, the transient electromagnetic response from a conducting wire will be developed using the *marching-on-in-time* (MOT) and the *marching-on-in-degree* (MOD) methods. Here, we present an implicit MOT scheme with the central finite-difference for the solution of the time-domain electric field integral equation (TD-EFIE). Although the implicit scheme is accurate and stable for many cases, the computational accuracy and stability are dependent on the choice of the time step, which limits its applications to complex structures. Recently, the MOD technique has been used instead of the conventional MOT technique. The advantage of the MOD technique is that it does not have a Courant–Friedrich-Levy sampling criteria relating the spatial to the temporal sampling. The procedure for solving the integral equations using the MOD method is based on the Galerkin's method, which involves separate spatial and temporal testing procedures. Piecewise triangular basis functions are used for both spatial expansion and testing. The time variation in the MOD scheme is approximated by an orthogonal temporal basis function set—*associated Laguerre functions*, which is derived from the *Laguerre polynomials*. These basis functions are also used for temporal testing. Usually the temporal testing will be carried out after the spatial testing is performed. However, a reverse of this testing sequence was recently proposed to analytically handle the retarded-time term and thus to obtain more stable results. In this chapter, only the MOD with this novel scheme is addressed. Several numerical examples using MOT and MOD techniques are presented to illustrate the validity of these methodologies. Another advantage of the MOD technique is that one can use any type of incident wave including a rectangular pulse as excitations.

In section 5.1 the time-domain integral equation for conducting wires based on the electric field (TD-EFIE) is presented. Section 5.2 briefly describes the marching-on-in-time (MOT) procedure for solving the TD-EFIE. The steps of the solution only are outlined, as this technique will be described in detail in section 6.4 for the general case of transient scattering from three-dimensional (3-D) arbitrarily-shaped conducting objects. A new method called the MOD method for solving the TD-EFIE is given in sections 5.3. (The MOD techniques for 3-D conducting objects using two different testing sequences are given in Chapter 6 and Chapter 8, respectively). In section 5.4, numerical results obtained by these two methods are shown for scattering and radiation from wire structures and compared with the inverse discrete Fourier transform (IDFT) of the frequency domain EFIE solution followed by some conclusions in section 5.5. The appendix 5.6 provides some properties of the Laguerre polynomials used here.

5.1   TIME DOMAIN ELECTRIC FIELD INTEGRAL EQUATION (TD-EFIE) FOR CONDUCTING WIRES

The objective of this section is to obtain the current distribution on a thin wire as a function of time when the wire is illuminated by an incident electromagnetic pulse as shown in Figure 5.1. Thus, the induced currents on a conducting thin-wire structure can be obtained by enforcing that the total tangential electric field is zero on the surface of the wire structure. On an arbitrarily-shaped conducting wire illuminated by a transient electromagnetic wave, the total tangential electric field on the wire surface S is equated to zero for all times. Therefore, we have

$$\left(\mathbf{E}^i(\mathbf{r},t)+\mathbf{E}^s(\mathbf{r},t)\right)_{\tan}=0 \quad \text{for } \mathbf{r}\in S \tag{5.1}$$

where $\mathbf{E}^i$ is the incident electric field defined in the absence of the scatterer and $\mathbf{E}^s$ is the scattered electric field caused by the induced current $\mathbf{J}$. The subscript "tan" denotes the tangential component. The scattered electric field is expressed in terms of the scalar and vector potentials as

$$\mathbf{E}^s(\mathbf{r},t)=-\frac{\partial}{\partial t}\mathbf{A}(\mathbf{r},t)-\nabla\varPhi(\mathbf{r},t) \tag{5.2}$$

where $\mathbf{A}$ and $\varPhi$ are the magnetic vector and the electric scalar potential, respectively. They are given by

$$\mathbf{A}(\mathbf{r},t)=\frac{\mu}{4\pi}\int_S \frac{\mathbf{J}(\mathbf{r}',\tau)}{R}dS' \tag{5.3}$$

$$\varPhi(\mathbf{r},t)=\frac{1}{4\pi\varepsilon}\int_S \frac{Q(\mathbf{r}',\tau)}{R}dS' \tag{5.4}$$

and $R=\sqrt{|\mathbf{r}-\mathbf{r}'|^2+a^2}$ represents the distance between the observation point $\mathbf{r}$ arbitrarily located on the surface of the wire, and the source point $\mathbf{r}'$ located along the axis of the wire. a is the radius of the wire antenna. The retarded time τ is defined as $\tau=t-R/c$, where c is the velocity of the electromagnetic wave propagation in that space. The parameters μ and ε are the permeability and permittivity of free space. The surface-charge density Q is related to the surface divergence of $\mathbf{J}$ based on the equation of continuity, as

$$\nabla \cdot \mathbf{J}(\mathbf{r},t) = -\frac{\partial}{\partial t} Q(\mathbf{r},t) \tag{5.5}$$

Substituting Eq. (5.5) into Eq. (5.4) gives the expression for the electric scalar potential as

$$\Phi(\mathbf{r},t) = -\frac{1}{4\pi\varepsilon} \int_S \int_0^\tau \frac{\nabla' \cdot \mathbf{J}(\mathbf{r}',t')}{R} \, dt' \, dS' \tag{5.6}$$

Then, substituting Eq. (5.2) into Eq. (5.1) along with Eqs. (5.3) and (5.6) constitutes the TD–EFIE

$$\left(\frac{\partial}{\partial t} \mathbf{A}(\mathbf{r},t) + \nabla \Phi(\mathbf{r},t) \right)_{\text{tan}} = \left(\mathbf{E}^i(\mathbf{r},t) \right)_{\text{tan}} \tag{5.7a}$$

$$\mathbf{A}(\mathbf{r},t) = \frac{\mu}{4\pi} \int_S \frac{\mathbf{J}(\mathbf{r}',\tau)}{R} \, dS' \tag{5.7b}$$

$$\Phi(\mathbf{r},t) = -\frac{1}{4\pi\varepsilon} \int_S \int_0^\tau \frac{\nabla' \cdot \mathbf{J}(\mathbf{r}',t')}{R} \, dt' \, dS' \tag{5.7c}$$

from which the unknown current $\mathbf{J}$ is to be determined.

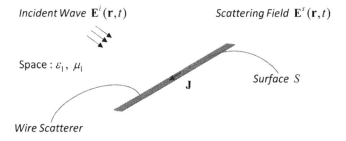

Figure 5.1. A thin wire structure illuminated by an electromagnetic wave.

5.2   SOLUTION OF THE TD-EFIE USING THE MARCHING-ON-IN-TIME (MOT) METHOD

Classically, the integral equation of Eq. (5.7) generally is solved by the MOT procedure. For the transient analysis from wire antennas, the MOT solution procedure first was applied by Sayre and Harrington [1]. However, the first working computer program for transient analysis from wire antennas is contained in the works of Bennett [2]. Other researchers [3–7] also have applied the MOT method with much success. The MOT method will not be presented in detail in this section, as it has been discussed at length for the solution of a general problem of electromagnetic scattering by 3-D arbitrarily-shaped conducting objects in section 6.4. In that section, it has been applied to obtain solutions from the TD-EFIE, TD-MFIE, and TD-CFIE equations. However, we provide a short summary in this section for completeness to illustrate the basic philosophy of this methodology.

The solution procedure for MOT starts by factoring the unknown current $\mathbf{J}$ in terms of the scalar temporal and vectorial spatial variables. This can be written as

$$\mathbf{J}(\mathbf{r},t) = \sum_{n=1}^{N} J_n(t)\mathbf{f}_n(\mathbf{r}) \tag{5.8}$$

where N is the number of spatial unknown approximating the spatial variation of the current on the conducting wire structure.

For finding the numerical solution of the TD-EFIE, the time axis is divided into segments of equal interval Δt and a discrete time instant t_i is defined as $t_i = i\Delta t$, where $\Delta t = t_i - t_{i-1}$. Next, assume that all current coefficients are zero for $t < 0$, which implies that a causal solution is going to be sought. It is also assumed that when calculating the unknown coefficients $J_n(t)$ at $t = t_i$, all coefficients at previous time instants are known. To approximate the time derivative in the TD-EFIE at the time instant $t = t_i$ using the backward finite difference [8], Eq. (5.7a) can be rewritten as follows:

$$\left(\frac{\mathbf{A}(\mathbf{r},t_i) - \mathbf{A}(\mathbf{r},t_{i-1})}{\Delta t} + \nabla \Phi(\mathbf{r},t_i) \right)_{\tan} = \left(\mathbf{E}^i(\mathbf{r},t_i) \right)_{\tan} \tag{5.9}$$

or it can be approximated by using the central finite difference [9] at time $t = t_{i-1/2}$ to obtain

$$\left(\frac{\mathbf{A}(\mathbf{r},t_i) - \mathbf{A}(\mathbf{r},t_{i-1})}{\Delta t} + \frac{\nabla \Phi(\mathbf{r},t_i) + \nabla \Phi(\mathbf{r},t_{i-1})}{2} \right)_{\tan} = \left(\mathbf{E}^i(\mathbf{r},t_{i-1/2}) \right)_{\tan} \tag{5.10}$$

where central finite difference is used for the time-derivative of the vector potential and time averaging is used for the scalar potential term.

Equations (5.9) and (5.10) are now solved by applying Galerkin's method in the MoM context, and hence, the testing function is the same as the expansion function. By choosing the expansion function $\mathbf{f}_m$ also as the testing function and defining the inner product between two functions by Eq. (1.9), Eq. (5.9) becomes

$$\frac{1}{\Delta t}\left\langle \mathbf{f}_m, \mathbf{A}(\mathbf{r},t_i)\right\rangle + \left\langle \mathbf{f}_m, \nabla \Phi(\mathbf{r},t_i)\right\rangle = \left\langle \mathbf{f}_m, \mathbf{E}^i(\mathbf{r},t_i)\right\rangle + \frac{1}{\Delta t}\left\langle \mathbf{f}_m, \mathbf{A}(\mathbf{r},t_{i-1})\right\rangle \qquad (5.11)$$

and Eq. (5.10) is rewritten as

$$\frac{1}{\Delta t}\left\langle \mathbf{f}_m, \mathbf{A}(\mathbf{r},t_i)\right\rangle + \frac{1}{2}\left\langle \mathbf{f}_m, \nabla \Phi(\mathbf{r},t_i)\right\rangle = \left\langle \mathbf{f}_m, \mathbf{E}^i(\mathbf{r},t_{i-1/2})\right\rangle$$

$$+ \frac{1}{\Delta t}\left\langle \mathbf{f}_m, \mathbf{A}(\mathbf{r},t_{i-1})\right\rangle - \frac{1}{2}\left\langle \mathbf{f}_m, \nabla \Phi(\mathbf{r},t_{i-1})\right\rangle \qquad (5.12)$$

The left-hand sides of Eqs. (5.11) and (5.12) contain the terms at $t = t_i$, and the right-hand side consists of known quantities, that is, potentials at $t = t_{i-1}$ and the incident field. Note that both sides of Eq. (5.12) include the scalar potential terms, and the incident field is considered at the center of the time step. Thus, from either Eq. (5.11) or Eq. (5.12), a matrix equation can be obtained as follows:

$$[\alpha_{mn}][J_n(t_i)] = [\beta_m(t_i)] \qquad (5.13)$$

The detailed mathematical step to obtain a matrix equation will be described in section 6.4. The matrix $[\alpha_{mn}]$ is not a function of time, and hence, the inverse of the matrix needs to be computed only once at the beginning of the computation step. Finally, by solving Eq. (5.13) at each time step, the time-domain current induced on the conducting surface may be obtained iteratively.

Some numerical simulations using this procedure will be presented in section 5.4 and will be compared with results obtained using the MOD solution and the inverse discrete Fourier transform (IDFT) of a frequency domain solution.

An important disadvantage with this procedure is the possible occurrence of rapidly growing spurious oscillations at later instants, which is apparently a result of the accumulation of errors during the calculations. These instabilities have been touched on by several authors [3,5–7]. Although it may be possible to eliminate these instabilities, in many cases by a smoothing procedure or by an implicit implementation as discussed in section 6.4; the accumulation of errors during calculation puts a limitation on the applicability of this technique [6]. Moreover, with the MOT procedure, the accuracy of the solution [6,7] cannot be verified easily, and usually there is no error estimation. That is why other researchers have discussed the use of alternate techniques [10–15], including iterative methods [6,7] in which the accumulation of errors in time does not take place. Also, for the iterative methods, the time and space discretizations are independent of one another, and so one does not have to worry about the connection between the space and the time discretizations because of Courant–Friedrich–Levy criteria. There are two iterative methods that have been used, namely the method of steepest descent [6,7] and the method of conjugate gradients [7]. However for arbitrarily-shaped structures the iterative methods may be difficult to apply.

In the next section, we describe a novel approach called the marching-on-in-degree (MOD) that will eliminate the instabilities of an MOT technique.

5.3 SOLUTION OF THE TD-EFIE USING THE MARCHING-ON-IN-DEGREE (MOD) METHOD

The electric current and the electric charge density described in Eq. (5.5) can be expressed in terms of a Hertz vector $\mathbf{u}(\mathbf{r}, t)$ defined by

$$\mathbf{J}(\mathbf{r}, t) = \frac{\partial}{\partial t} \mathbf{u}(\mathbf{r}, t) \tag{5.14}$$

$$Q(\mathbf{r}, t) = -\nabla \cdot \mathbf{u}(\mathbf{r}, t) \tag{5.15}$$

Substitution of Eqs. (5.14) and (5.15) into Eq. (5.7a) results in

$$\left(\frac{\mu}{4\pi} \frac{\partial^2}{\partial t^2} \int_L \frac{\mathbf{u}(\mathbf{r}', \tau)}{R} dl' - \frac{\nabla}{4\pi\varepsilon} \int_L \frac{\nabla' \cdot \mathbf{u}(\mathbf{r}', \tau)}{R} dl' \right)_{\text{tan}} = \left(\mathbf{E}^i(\mathbf{r}, t) \right)_{\text{tan}} \tag{5.16}$$

To solve the TD-EFIE of Eq. (5.16) using a Galerkin's method, it is necessary to first define the basis functions.

5.3.1 Spatial Basis Functions

The thin-wire structure to be analyzed is approximated by straight wire segments, as illustrated in Figure 5.2. We consider the segment pair $L_n^\pm$, which is assigned to the element from $\mathbf{r}_n$ to $\mathbf{r}_n^\pm$. The wire is divided into N segment pairs. The piecewise triangular basis function associated with the Hertz potential at the nth common node at $\mathbf{r}_n$ is defined by

$$\mathbf{f}_n(\mathbf{r}) = \mathbf{f}_n^+(\mathbf{r}) + \mathbf{f}_n^-(\mathbf{r}) \tag{5.17a}$$

where

$$\mathbf{f}_n^\pm(\mathbf{r}) = \begin{cases} \pm \dfrac{\mathbf{l}_n^\pm}{\Delta l_n^\pm}, & \mathbf{r} \in L_n^\pm \\ 0, & \mathbf{r} \notin L_n^\pm \end{cases} \tag{5.17b}$$

$$\Delta l_n^\pm = \sqrt{\left| \mathbf{r}_n - \mathbf{r}_n^\pm \right|^2 + a^2} \tag{5.17c}$$

$$\mathbf{l}_n^\pm = \mathbf{r} - \mathbf{r}_n^\pm \tag{5.17d}$$

In Eq. (5.17), $\Delta l_n^{\pm}$ is the length of the segment $L_n^{\pm}$, and $\mathbf{l}_n^{\pm}$ is the local position vector.

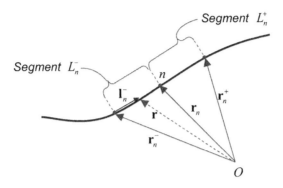

Figure 5.2. An arbitrarily-shaped conducting wire with a segmentation scheme.

5.3.2   Temporal Basis Functions

We can now express the Hertz vector in terms of the triangular vector basis function for the spatial variable and use the associated Laguerre functions for the expansion of the time variable. Hence,

$$\mathbf{u}(\mathbf{r}, t) = \sum_{n=1}^{N} u_n(t)\, \mathbf{f}_n(\mathbf{r}) \tag{5.18}$$

The associated Laguerre function set is defined by

$$\phi_j(t) = e^{-t/2} L_j(t) \tag{5.19}$$

where $L_j(t)$ is the Laguerre polynomial of order j. It is used to represent the causal temporal basis functions, and the transient variation introduced in Eq. (5.18) can be expanded as,

$$u_n(t) = \sum_{j=0}^{\infty} u_{n,j}\, \phi_j(st) \tag{5.20}$$

where $u_{n,j}$ are the unknown coefficients, and s is a scaling factor. By controlling the factor s, the support provided by the expansion functions can be increased or decreased. The expression of the first and second derivatives of the transient variations for the solution can be given analytically as (more detail can be obtained from the appendix 5.6)

$$\frac{d}{dt}u_n(t) = s \sum_{j=0}^{\infty} \left(\frac{1}{2}u_{n,j} + \sum_{k=0}^{j-1} u_{n,k} \right) \phi_j(st) \qquad (5.21a)$$

$$\frac{d^2}{dt^2}u_n(t) = s^2 \sum_{j=0}^{\infty} \left(\frac{1}{4}u_{n,j} + \sum_{k=0}^{j-1} (j-k)u_{n,k} \right) \phi_j(st) \qquad (5.21b)$$

This assumes that the functions $u_n(0) = 0$ and $du_n(0)/dt = 0$ as the time response has not started at $t = 0$ because of causality.

In summary, the following five characteristic properties of the associated Laguerre functions derived from Laguerre polynomials [16,17] have been used in this new formulation:

1. *Causality*: The Laguerre polynomials are defined over $0 \leq t < +\infty$ for our applications. Therefore, they are suitable to represent any natural time domain responses as they are always causal.
2. *Recursive computation*: The Laguerre polynomials of higher orders can be generated recursively using the lower orders through a stable computational process.
3. *Orthogonality*: With respect to a weighting function, the Laguerre polynomials are orthogonal with respect to each other. One can construct a set of orthonormal basis functions, which we call the weighted Laguerre polynomials or associated Laguerre functions. Physical quantities that are functions of time can be spanned in terms of these orthonormal basis functions.
4. *Convergence*: The associated Laguerre functions decay to zero as time goes to infinity and therefore the solution does not blow up for late times. Also, because the associated Laguerre functions form an orthonormal set, any arbitrary time function can be spanned by these basis functions and the approximation converges.
5. *Separability of the space and time variables*: Because of the additivity property of the weighted Laguerre functions, the spatial and the temporal variables can be separated completely and the time variable can be eliminated completely from all computations except the calculation of the excitation coefficient, which is determined by the excitation waveform only. This eliminates the interpolation that is necessary to estimate values of the current or the charge at time instances that do not correspond to a sampled time instance. Therefore, the values of the current can be obtained at any time accurately.

Substituting Eqs. (5.18)–(5.21) into Eq. (5.16) results in

$$\left(\frac{\mu s^2}{4\pi} \sum_{n=1}^{N} \sum_{j=0}^{\infty} \left(\frac{1}{4}u_{n,j} + \sum_{k=0}^{j-1}(j-k)u_{n,k} \right) \int_L \frac{\phi_j(s\tau)\, \mathbf{f}_n(\mathbf{r}')}{R}\, dl'$$

$$-\frac{1}{4\pi\varepsilon}\sum_{n=1}^{N}\sum_{j=0}^{\infty}u_{n,j}\int_{L}\nabla\left(\frac{\phi_{j}(s\tau)\,\nabla'\boldsymbol{\cdot}\,\mathbf{f}_{n}(\mathbf{r}')}{R}\right)dl'\bigg|_{\text{tan}} = \left(\mathbf{E}^{i}(\mathbf{r},t)\right)_{\text{tan}} \tag{5.22}$$

Next we employ the Galerkin's method to solve this integral equation.

5.3.3   Testing of the Integral Equation

Through the Galerkin's method and taking a temporal testing to Eq. (5.22) with $\phi_{i}(st)$ $(i=0,1,2,\dots,M)$, where M is the maximum order of the Laguerre functions to be evaluated from the time-bandwidth product of the waveform, we have

$$\frac{\mu s^{2}}{4\pi}\sum_{n=1}^{N}\sum_{j=0}^{\infty}\left(\frac{1}{4}u_{n,j}+\sum_{k=0}^{j-1}(j-k)u_{n,k}\right)\int_{L}\frac{1}{R}I_{ij}(sR/c)\mathbf{f}_{n}(\mathbf{r}')dl'$$

$$-\frac{1}{4\pi\varepsilon}\sum_{n=1}^{N}\sum_{j=0}^{\infty}u_{n,j}\int_{L}\nabla\left(\frac{1}{R}I_{ij}(sR/c)\nabla'\boldsymbol{\cdot}\,\mathbf{f}_{n}(\mathbf{r}')\right)dl'=V_{i}(\mathbf{r}) \tag{5.23a}$$

where

$$I_{ij}(sR/c)=\int_{sR/c}^{\infty}\phi_{i}(st)\phi_{j}(st-sR/c)d(st)$$

$$=\begin{cases} e^{(-sR/(2c))}\left(L_{i-j}(sR/c)-L_{i-j-1}(sR/c)\right) & j<i \\[2mm] e^{(-sR/(2c))} & j=i \\[2mm] 0 & j>i \end{cases} \tag{5.23b}$$

$$V_{i}(\mathbf{r})=\int_{0}^{\infty}\phi_{i}(st)\,\mathbf{E}^{i}(\mathbf{r},t)\,d(st)\,. \tag{5.23c}$$

Because of this orthogonality condition, we can change the upper limit of the sum in Eq. (5.23a) from ∞ to i. Then performing the spatial testing with $\mathbf{f}_{m}(\mathbf{r})$ $(m=1,2,\dots,N)$, we obtain

$$\sum_{n=1}^{N}\left(\frac{\mu s^{2}}{4}\alpha_{mn}+\frac{1}{\varepsilon}\beta_{mn}\right)u_{n,i}=\Omega_{mi}-\mu s^{2}\sum_{n=1}^{N}\sum_{k=0}^{i-1}(i-k)u_{n,k}\,\alpha_{mn}$$

$$-\mu s^{2}\sum_{n=1}^{N}\sum_{j=0}^{i-1}\sum_{k=0}^{j-1}(j-k)u_{n,k}\,A_{mnij}\,-\sum_{n=1}^{N}\sum_{j=0}^{i-1}\left(\frac{\mu s^{2}}{4}A_{mnij}+\frac{1}{\varepsilon}B_{mnij}\right)u_{n,j} \tag{5.24a}$$

where

$$\alpha_{mn} = \int_L \mathbf{f}_m(\mathbf{r}) \cdot \int_{L'} \frac{e^{(-sR/(2c))}}{4\pi R} \mathbf{f}_n(\mathbf{r}') dl' \, dl \tag{5.24b}$$

$$\beta_{mn} = \int_L \nabla \cdot \mathbf{f}_m(\mathbf{r}) \int_{L'} \frac{e^{(-sR/(2c))}}{4\pi R} \nabla' \cdot \mathbf{f}_n(\mathbf{r}') dl' \, dl \tag{5.24c}$$

$$A_{mnij} = \int_L \mathbf{f}_m(\mathbf{r}) \cdot \int_{L'} \frac{I_{ij}(sR/c)}{4\pi R} \mathbf{f}_n(\mathbf{r}') dl' \, dl \tag{5.24d}$$

$$B_{mnij} = \int_L \nabla \cdot \mathbf{f}_m(\mathbf{r}) \int_{L'} \frac{I_{ij}(sR/c)}{4\pi R} \nabla' \cdot \mathbf{f}_n(\mathbf{r}') dl' \, dl \; . \tag{5.24e}$$

$$\Omega_{m,i} = \int_L \mathbf{f}_m(\mathbf{r}) \cdot \mathbf{V}_i(\mathbf{r}) dl \; . \tag{5.24f}$$

To evaluate the integrals numerically, we use the vector equation of a straight-line segment. Assuming the vectors of the two end-points of this segment is represented by $\mathbf{r}_n$ and $\mathbf{r}_n^{\pm}$ (as seen in Fig. 5.2), then any point on the segment is defined by

$$\mathbf{r} = \mathbf{r}_n + \alpha \, \Delta l \, \mathbf{e}_n, \qquad (0 \le \alpha \le 1) \tag{5.25a}$$

where

$$\mathbf{e}_n = \frac{\mathbf{r}_n^{\pm} - \mathbf{r}_n}{\Delta l} = \frac{1}{\Delta l} \left(\Delta x \, \mathbf{e}_x + \Delta y \, \mathbf{e}_y + \Delta z \, \mathbf{e}_z \right) \tag{5.25b}$$

where $\mathbf{e}_n$ is the unit vector associated with each segment, Δl is the length of the segment. Δx, Δy, and Δz represent the various coordinates between the two end points of the segment. So, the spatial basis function and Eqs. (5.24b)–(5.24f) can be expressed as,

$$\mathbf{f}_n^{\pm}(\mathbf{r}) = \begin{cases} \pm \alpha \mathbf{e}_n, & \mathbf{r} \in L_n^{\pm} \\ 0, & \mathbf{r} \notin L_n^{\pm} \end{cases} \tag{5.26}$$

$$\alpha_{mn} = \sum_{pq} \int_0^1 \int_0^1 \Delta l_m^p \, \Delta l_n^q \, \alpha \, \alpha' \frac{e^{(-sR/(2c))}}{4\pi R} (\mathbf{e}_m^p \cdot \mathbf{e}_n^q) d\alpha' \, d\alpha \tag{5.27}$$

$$\beta_{mn} = \sum_{pq} \int_0^1 \int_0^1 \frac{e^{(-sR/(2c))}}{4\pi R} d\alpha' \, d\alpha \tag{5.28}$$

$$A_{mnij} = \sum_{pq} \int_0^1 \int_0^1 \Delta l_m^p \, \Delta l_n^q \, \alpha \, \alpha' \frac{I_{ij}(sR/c)}{4\pi R} (\mathbf{e}_m^p \cdot \mathbf{e}_n^q) d\alpha' \, d\alpha \tag{5.29}$$

$$B_{mnij} = \sum_{pq} \int_0^1 \int_0^1 \frac{I_{ij}(sR/c)}{4\pi R} da' d\alpha \tag{5.30}$$

$$\Omega_{mi} = \sum_p \int_0^1 \alpha \, \Delta l_m^p \, \mathbf{e}_m^p \cdot \int_0^\infty \phi_i(st) \mathbf{E}^i(\mathbf{r},t) \, d(st) \, d\alpha \tag{5.31}$$

where $p,q = \pm$. It should be noted that because the observation point is located on the surface of the wire, and the current is assumed to be located along the wire axis; the distance R is always greater than or equal to the radius of the wire, and there is no singularity in the integrals that one needs to worry about.

Finally, Eq. (5.23a) can be written as

$$[Z_{mn}][u_{n,i}] = [\gamma_{m,i}]. \quad i = 1, 2, 3, \ldots, M \tag{5.32a}$$

where

$$Z_{mn} = \frac{\mu s^2}{4} \alpha_{mn} + \frac{1}{\varepsilon} \beta_{mn} \tag{5.32b}$$

$$\gamma_{m,i} = \Omega_{mi} - \mu s^2 \sum_{n=1}^{N} \sum_{k=0}^{j-1} (i-k) u_{n,k} \alpha_{mn}$$

$$-\mu s^2 \sum_{n=1}^{N} \sum_{j=0}^{i-1} \sum_{k=0}^{j-1} (j-k) u_{n,k} A_{mnij} - \sum_{n=1}^{N} \sum_{j=0}^{i-1} \left(\frac{\mu s^2}{4} A_{mnij} + \frac{1}{\varepsilon} B_{mnij} \right) u_r \tag{5.32c}$$

The matrix element Z_{mn} is not a function of the order of the temporal testing functions. By solving the matrix equation Eq. (5.32) in a marching-on-in-degree manner, the electric current is expressed using the relation Eqs. (5,14), (5,18), and (5.20).

We need to know the finite value of the number of temporal basis functions, $m = 1, 2, \ldots, M$ needed in computing Eq. (5.32). Based on the time–bandwidth product of the signals that we are trying to approximate, it is possible to exactly reproduce any arbitrarily-shaped transient waveform of duration T_f and with a total bandwidth $2B$ by a set of $2BT_f + 1$ associated Laguerre functions. Thus, the minimum number M of the temporal basis functions required to represent that function becomes [18]

$$M \geq 2BT_f + 1 \tag{5.33}$$

Note that in Eq. (5.31), the upper limit of the integral in $\Omega_{m,i}$ can be replaced by the time sT_f instead of infinity.

For analysis of radiation problems, a spatial delta function generator at the feed node is considered. The inner product given by Eq. (5.24f) with the impressed field and the associated Laguerre functions of order m will be given by

$$\Omega_{m,i} = \int_0^{sT_f} \phi_i(st) v_m(t) \, d(st) \,. \tag{5.34}$$

where $v_m(t)$ is the time-dependent source voltage at the common node of the segment pair m.

5.3.4   Choice of Excitations

Usually, a Gaussian pulse is the most popular excitation for calculating the transient response of objects. It is considered to be of finite duration in time and also to be practically band-limited in the frequency domain. A Gaussian pulse is represented mathematically by

$$\mathbf{E}(\mathbf{r},t) = \mathbf{E}_0 \frac{4}{T\sqrt{\pi}} e^{-\gamma^2} \tag{5.35a}$$

where

$$\gamma = \frac{4}{T}(ct - ct_0 - \mathbf{r}\cdot\hat{\mathbf{k}}) \tag{5.35b}$$

The parameter c is the velocity of light, $\hat{\mathbf{k}}$ is the unit vector along the direction of the wave propagation, t is the time variable, T is the width of the pulse and ct_0 is the time delay at which the pulse reaches its peak. We use $T = 4\,\text{lm}$ and $ct_0 = 6$ lm. Both of these quantities are defined in light meters (lm), which is the time taken by light to traverse 1 m, which is 3.33 ns. This is generally used as the incident pulse in most time-domain methodologies, as the function needs to be differentiable. In this chapter, we also consider other types of incident signals like a triangular-shaped pulse or a rectangular pulse. In this way, response to different waveshapes, can be evaluated without performing any deconvolution [19].

In this chapter, we use the following triangular pulse for excitation:

$$\mathbf{E}(\mathbf{r},t) = \begin{cases} \frac{1}{3}(t-5)+1, & 2<t<5 \\ -\frac{1}{3}(t-5)+1, & 5\le t\le 8 \\ 0, & \text{others} \end{cases} \tag{5.36}$$

along with a rectangular pulse of the following shape:

$$\mathbf{E}(\mathbf{r},t) = \begin{cases} 1, & 4\le t<10 \\ 0, & \text{others} \end{cases} \tag{5.37}$$

Figure 5.3 presents all three types of signals in the time domain by Figure 5.3(a) and in the frequency domain by Figure 5.3(b).

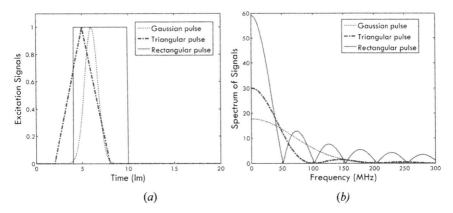

Figure 5.3. Different excitation signals in (*a*) time domain, and (*b*) frequency domain.

5.3.5   Treatment of Wire Junctions

Two conditions must be satisfied at a junction. The first one is the enforcement of the Kirchhoff's current law, which states that the sum of all the currents entering a junction equals the sum of the currents leaving the junction. The second condition is that the tangential component of the electric field must be continuous across the wire surfaces at the junction [19,20]. As shown in Figure 5.4, we replace the actual junction with a pair of wire segments that are not connected electrically but are overlapping with each other, as shown in Figure 5.4. Figure 5.4(*a*) displays the actual junction of a wire. We treat this junction electromagnetically by considering it to be two separate wires with overlapping segments, as shown in Figure 5.4(*b*), and the total current at the junction will be the sum of all currents at that physical location.

Next we present some numerical results.

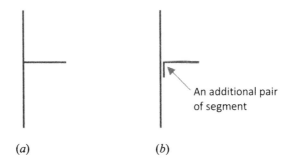

Figure 5.4. Treatment of a wire junction: (*a*) Actual junction. (*b*) Equivalent junction.

5.4   NUMERICAL EXAMPLES

Analysis of scattering and radiation from wire structures using the MOT and MOD methods are now discussed. The numerical results obtained by these two methods are compared with the inverse discrete Fourier transform (IDFT) of the frequency domain EFIE solution using the same basis functions.

5.4.1   Radiation from a Wire Antenna Solved by the MOT and MOD Methods

To evaluate the performance of the MOD method in contrast to the MOT approach, a straight-wire antenna of length $l = 1$ m located along the z-axis is considered. The straight-wire antenna is divided into 30 subsections. The radius of the wire is $a = 0.005$ m. The excitation is applied at the center of the wire, and thus, a dipole structure is formed. Figure 5.5 plots the input current at the dipole antenna using both MOT and the IDFT of a frequency domain solution. An implicit solution was generated using $c\Delta t = 2R_{\min}$, where $R_{\min}$ is the minimum distance between any two distinct segment centers of the triangular patches. It is important to note that the solution given by the MOT method using the backward finite difference has a lower accuracy than when using the central finite differencing scheme for early times but the latter suffers from late-time oscillations. The MOT (backward) is based on [21,22] and the MOT (central) is based on [19,23]. In addition, it is shown that the solution obtained from the MOD method is stable and has good agreement with the IDFT of the frequency domain solution.

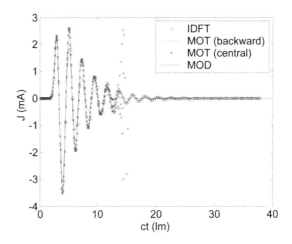

Figure 5.5. Transient input current at the center of a dipole antenna with a length $l = 1$ m and radius $a = 0.005$ m.

Figure 5.6 presents the transient response using the MOD method against the IDFT of the frequency domain solution for the radiated field from the dipole along the broadside direction. A very good agreement is obtained between these two responses.

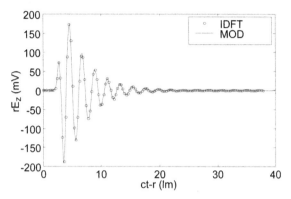

Figure 5.6. Transient far field from a dipole antenna with a length $l = 1$ m and radius $a = 0.005$ m.

When the straight wire is considered as a scatterer, an incident pulse of width $T = 2$ lm and $ct_0 = 3$ lm arriving from $\phi = 0°$ and $\theta = 90°$ with $\hat{\mathbf{k}} = -\hat{\mathbf{x}}$ and $\mathbf{E}_0 = \hat{\mathbf{z}}$ is considered. The pulse has a bandwidth of 500 MHz. Setting $s = 10^9$ and $M = 140$ is sufficient to obtain accurate solutions for the MOD method. For comparison, the results obtained by taking the IDFT of the solution from a frequency domain integral equation are presented. Figure 5.7 compares the transient response of the MOD method with the IDFT solution for the far field from the straight wire along the backward scattered direction. The two responses are almost identical, and it implies that the MOD method offers a very good approximation to the exact transient solution.

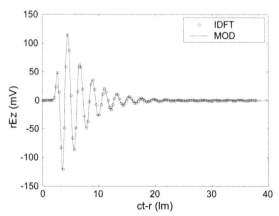

Figure 5.7. Transient far field from a straight wire scatterer with a length $l = 1$ m and radius $a = 0.005$ m.

For the same wire, the induced current at the center of the wire due to a field incident along the x-axis, which is polarized along the z-axis is plotted. Figure 5.8 shows the current at the center of the dipole with different space discretizations when the incident wave is a Gaussian pulse ($T = 4$ lm, $ct_0 = 6$ lm). In Figure 5.8,"irregular" means that the lengths of the segments have been chosen at random. Because the time variable is not present in the final equation, one does not have to worry about the Courant–Friedrich–Levy condition relating the spatial discretizations to the time discretizations. Here, we choose the maximum order of the Laguerre functions to be $M = 80$. From these results, we can see that the MOD method can obtain good results with different types of space discretizations. Figure 5.9 shows the current at the center of the wire when the incident wave is either a triangular or a rectangular pulse, respectively. Figure 5.10 plots the radiated far field when the excitation at the center of the dipoles is a Gaussian pulse, triangular pulse, and rectangular pulse, respectively. All the results are compared with the inverse discrete Fourier transform (IDFT) of the frequency domain result. They all agree well.

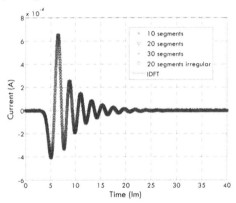

Figure 5.8. Current at the center of the dipole for different types of space discretizations.

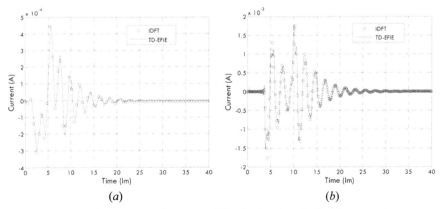

(a)   (b)

Figure 5.9. Current at the center of the dipole when the incident wave is
(a) a triangular pulse and (b) a rectangular pulse.

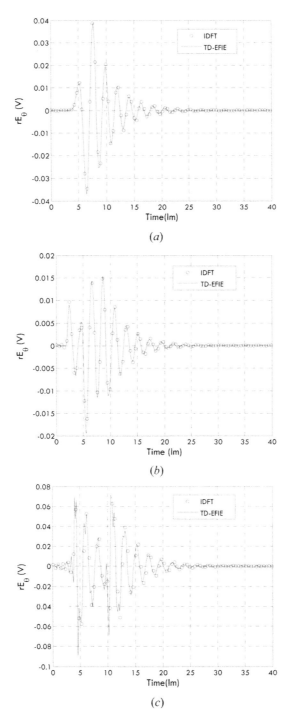

Figure 5.10. The radiated far field along $\theta = 90°$ and $\phi = 0°$ when the excitation is a (a) Gaussian pulse, (b) Triangular pulse, and (c) Rectangular pulse.

5.4.2   Radiation from a Circular Loop Antenna

Next, a loop antenna, whose radius is 0.5 m, is radiated by a Gaussian pulse ($T = 4 \, \text{lm}$, $ct_0 = 6 \, \text{lm}$). We check the current at the point (0.5, 0). Figure 5.11 plots the current at the right side of the node using different orders of the weighted Laguerre functions. If we choose M according to Eq. (5.31), then we can get accurate results, and they all agree with the IDFT result obtained from a frequency domain solution.

5.4.3   Radiation from a V-Shaped Antenna

We use a V-shaped antenna with length 0.5 m for each arm and the angle between the two arms is 60°. One wire is along the z-axis. The feed point is at the vertex. We again use a Gaussian pulse as an excitation. Figures 5.12 and 5.13 show the θ and ϕ components of the far field, respectively. It shows that the results for the radiation problem can be obtained with the same accuracy as solving a scattering problem.

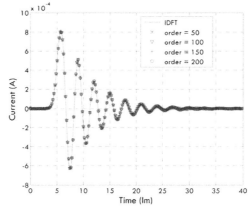

Figure 5.11. Current at a point on the loop using different orders of Laguerre polynomials.

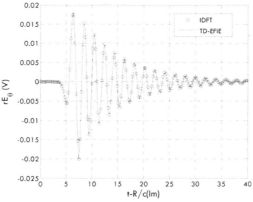

Figure 5.12. Far field of $\mathbf{E}_\theta$ along the direction $\theta = 90°$ and $\phi = 0°$.

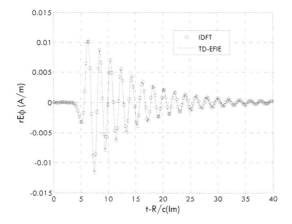

Figure 5.13. Far field of $\mathbf{E}_\phi$ along the direction $\theta = 90°$ and $\phi = 0°$.

5.4.4   Radiation from a Yagi Antenna

A Yagi antenna with six elements is shown in Figure 5.14. The incident wave is applied from $\theta = 90°$ and $\phi = 0°$, and the polarization of the electric field is along the ϕ direction. Figure 5.15 shows the current at the center of the driven element (point O) when using a Gaussian pulse as an incident wave. The results agree well with the IDFT of a frequency domain solution.

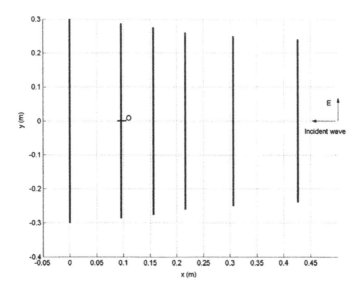

Figure 5.14. A Yagi antenna.

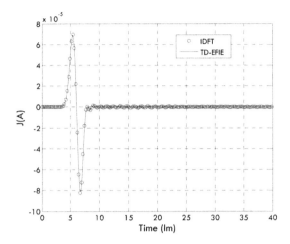

Figure 5.15. Current at the point O when the incident wave is a Gaussian pulse.

5.4.5   Radiation from an Equiangular Planar Tripole Antenna

Radiation from an equiangular planar tripole antenna with an arm length of 0.5 m as shown in Figure 5.16 is also simulated. In this problem, there is a junction at the origin. Figures 5.17, 5.18, and 5.19 show the current at the junction of the tripole antenna when it is excited using a Gaussian pulse, a triangular shaped pulse, and a rectangular pulse, respectively. All results are stable and agree well with the IDFT results. In Figure 5.19, there is small difference between the IDFT and the time domain results. This is because, in this case, the bandwidth of the rectangular pulse is very wide, and even use of a large number of associated Laguerre functions do not match the time-bandwidth product of the waveform.

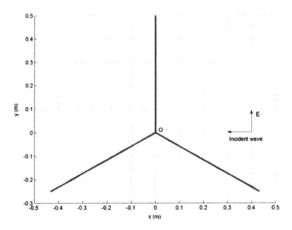

Figure 5.16. An equiangular planar tripole antenna.

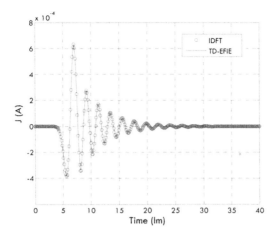

Figure 5.17. Current at the point O when using the Gaussian pulse as an incident wave.

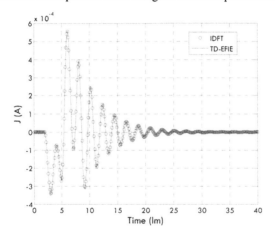

Figure 5.18. Current at the point O when using the triangular pulse as an incident wave.

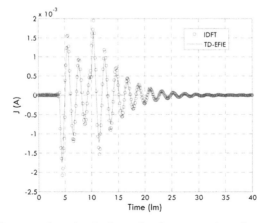

Figure 5.19. Current at the point O when using the rectangular pulse as an incident wave.

5.5   CONCLUSION

In this chapter, the procedures to analyze a wire scatterer and antenna by using the time-domain electric-field integral equation with the MOT and MOD methods have been presented. For the MOT method, because the size of the time step can be increased when the implicit scheme is used, the central finite difference approximates well the derivative of the magnetic vector potential and gives a more accurate result than the backward finite difference. As such, the marching-on-in-degree method can be used for solving the time-domain electric field integral equation for arbitrarily shaped conducting wire structures. The advantage of the MOD method is to guarantee the late time stability. The temporal derivatives can also be treated analytically.

This method eliminates the time variable completely from the computations, and hence, it is a recursion in the order of the Laguerre polynomials. The solution is thus independent of the time step, and a Courant condition is not necessary. In addition, any arbitrary waveshape can be used as an excitation, including discontinuous excitations like a rectangular pulse. Transient currents and far fields obtained by this method are accurate and stable when compared with the inverse Fourier transformed data of a frequency domain solution.

5.6   APPENDIX

The definition and properties of Laguerre polynomials are listed in this appendix.

Definition:

Consider the following set of functions [16]:

$$L_j(t) = \frac{e^t}{j!} \frac{d^j}{dt^j}\left(t^j e^{-t}\right), \quad 0 \le t < \infty, \quad j = 0, 1, 2, \dots \tag{5.A1}$$

These are the Laguerre polynomials of degree j. They are causal (i.e., they are defined for $t \ge 0$). They can be computed through a stable recursive procedure and obtained as

$$L_0(t) = 1 \tag{5.A2}$$

$$L_1(t) = 1 - t \tag{5.A3}$$

$$L_j(t) = \frac{1}{j}\left((2j - 1 - t)L_{j-1}(t) - (j - 1)L_{j-2}(t)\right) \quad j \ge 1 \tag{5.A4}$$

The main properties that will be used in the temporal procedures are described briefly in the following paragraphs.

Orthogonality:

The associated Laguerre functions are orthogonal as

$$\int_0^\infty e^{-t} L_i(t) L_j(t) \, dt = \delta_{ij} = \begin{cases} 1, & i = j \\ 0, & i \neq j \end{cases} \tag{5.A5}$$

Laguerre Transform:

A causal time-dependent function $f(t)$ for $t \geq 0$ can be expanded as

$$f(t) = \sum_{j=0}^\infty f_j \phi_j(t) \tag{5.A6}$$

where $\phi_j(t)$ is the associated Laguerre function defined in Eq. (5.19). Based on the orthogonal property given by Eq. (5.A5), the multiplication of the function $f(t)$ with $\phi_j(t)$ and integrating from zero to infinity yields

$$\int_0^\infty \phi_i(t) f(t) dt = f_i \tag{5.A7}$$

The expression in Eq. (5.A7) is called the Laguerre transform.

Derivative:

Using the Laguerre transform, an analytic representation for the time derivative of the function $f(t)$ can be obtained as

$$\int_0^\infty \phi_i(t) \frac{d}{dt} f(t) dt = \frac{1}{2} f_i + \sum_{k=0}^{i-1} f_k \tag{5.A8}$$

where $f(0) = 0$ is assumed, and $\phi_i(\infty) = 0$ is used.

  Using the relations (5.A6)–(5.A8), we can expand the derivative of the function $f(t)$ as

$$\frac{d}{dt} f(t) = \sum_{j=0}^\infty \left(\frac{1}{2} f_j + \sum_{k=0}^{j-1} f_k \right) \phi_j(t) \tag{5.A9}$$

Similarly, the result for the second derivative of the function $f(t)$ is given as

$$\frac{d^2}{dt^2}f(t) = \sum_{j=0}^{\infty}\left[\left(\frac{1}{4}f_j + \sum_{k=0}^{j-1}(j-k)f_k\right)\right]\phi_j(t) \tag{5.A10}$$

Integral:

Consider an integral given as

$$I_{ij}(y) = \int_0^{\infty}\phi_i(x)\phi_j(x-y)dx \tag{5.A11}$$

Through a change of variable $z = x - y$, and substituting the expression of $\phi_j(t)$ given by Eq. (5.19), Eq. (5.A11) yields

$$I_{ij}(y) = e^{-y/2}\int_{-y}^{\infty}e^{-z}L_i(z+y)L_j(z)dz \tag{5.A12}$$

Using the properties of Laguerre polynomials [16; Eqs. (8.971) and (8.974)], we obtain

$$L_i(z+y) = \sum_{k=0}^{i}L_k(z)\left(L_{i-k}(y) - L_{i-k-1}(y)\right) \tag{5.A13}$$

Substituting Eq. (5.A13) into Eq. (5.A12), one obtains

$$I_{ij}(y) = e^{-y/2}\sum_{k=0}^{i}\left(L_{i-k}(y) - L_{i-k-1}(y)\right)\int_{-y}^{\infty}e^{-z}L_k(z)L_j(z)dz \tag{5.A14}$$

Because the Laguerre polynomial is defined for $z \geq 0$, the lower limit of the integral in Eq. (5.A14) may be changed from $-y$ to zero. Moreover, Eq. (5.14) can be computed by using the orthogonal property given by Eq. (5.A5) and yields the integral of associated Laguerre functions as

$$\int_0^{\infty}\phi_i(x)\phi_j(x-y)dx = \begin{cases} e^{-y/2}\left(L_{i-j}(y) - L_{i-j-1}(y)\right) & j \leq i \\ 0 & j > i \end{cases} \tag{5.A15}$$

REFERENCES

[1]   E. Sayre and R. F. Harrington, "Time domain radiation and scattering by thin wires," *Appl. Sci. Res.*, Vol. 26, pp. 413–444, 1972.

[2] C. L. Bennett, *A Technique for Computing Approximate Electromagnetic Pulse Response of Conducting Bodies*, Ph.D. Dissertation, Purdue University, Lafayette, Ind., 1968.

[3] C. L. Bennett and W. L. Weeks, "Transient scattering from conducting cylinders," *IEEE Tran. Antennas Prop.*, Vol. 18, pp. 621–633, 1970.

[4] A. M. Auckenthaler and C. L Bennett, "Computer solution of transient time domain thin-wire antenna problems," *IEEE Trans.Micro. Theory and Tech.*, Vol. 19, pp. 892–893, 1971.

[5] E. K. Miller and J. A. Landt, "Direct time domain techniques for transient radiation and scattering from wires," *Proc. IEEE*, Vol. 68, pp. 1396–1423, 1980.

[6] G. C. Herman, *Scattering of Transient Acoustic Waves in Fluids and Solids*, Ph.D. Dissertation, Delft Inst. of Technology, Delft, The Netherlands, 1981.

[7] S. M. Rao, T. K. Sarkar and S. A. Dianat, "The application of the conjugate gradient method to the solution of transient electromagnetic scattering from thin wires," *Radio Sci.*, Vol. 19, pp. 1319–1326, 1984.

[8] S. M. Rao and T. K. Sarkar, "Transient analysis of electromagnetic scattering from wire structures utilizing an implicit time-domain integral-equation technique," *Micro. Opt. Tech. Lett.*, Vol. 17, pp. 66–69, 1998.

[9] B. H. Jung and T. K. Sarkar, "Time-domain electric-field integral equation with central finite difference," *Micro .Opt. Tech. Lett.*, Vol. 31, pp. 429–435, 2001.

[10] S. M. Booker, "Calculation of surface impedance effects on transient antenna radiation," *Radio Sci.*, Vol. 31, pp. 1663–1669, 1996.

[11] J. F. Callejon, A. R. Bretones, and R. G. Martin, "On the application of parametric models to the transient analysis of resonant and multiband antennas," *IEEE Trans. Antennas Prop.*, Vol. 46, pp. 312–317, 1998.

[12] K. Aygun, S. E. Fisher, A. A. Ergin, B. Shanker, and E. Michielssen, "Transient analysis of multi-element wire antennas mounted on arbitrarily shaped perfectly conducting bodies," *Radio Sci.*, Vol. 34, pp. 781–796, 1999.

[13] B. H. Jung, T. K. Sarkar, Y. S. Chung, and Z. Ji, "An accurate and stable implicit solution for transient scattering and radiation from wire structures," *Microw. Opt. Tech. Lett.*, Vol. 34, pp. 354–359, 2002.

[14] A. G. Tijhuis, P. Zhongqiu, and A. B. Bretones, "Transient excitation of a straight thin-wire segment: a new look at an old problem," *IEEE Trans. Antennas Prop.*, Vol. 40, pp. 1132–1146, 1992.

[15] F. Bost L. Nicolas, and G. Rojat, "A time-domain integral formulation for the scattering by thin wires," *IEEE Trans. Magn.*, Vol. 36, pp. 868–871, 2000.

[16] A. D. Poularikas, *The Transforms and Applications Handbook*, IEEE Press, New York, 1996.

[17] I. S. Gradshteyn and I. M. Ryzhik, *Table of Integrals, Series, and Products*, Academic Press, New York, 1980.

[18] Y. S. Chung, T. K. Sarkar, and B. H. Jung, "Solution of a time-domain magnetic-field integral equation for arbitrarily closed conducting bodies using an unconditionally stable methodology," *Microw. Opti. Tech. Lett.*, Vol. 35, pp. 493–499, 2002,

[19] M. R. Zunoubi, N. H. Younan, J. H. Beggs, and C. D. Taylor, "FDTD analysis of linear antennas driven from a discrete impulse excitation," *IEEE Trans. Electromag. Compat.*, Vol. 39, pp. 247–250, 1997.

[20] T. T. Wu, and R.W. P. King, "The tapered antenna and its application to the junction problem for thin wires," *IEEE Trans. Antennas Prop.*, Vol. 24, pp. 42–45, 1976.

[21] S. M. Rao and T. K. Sarkar, "Transient analysis of electromagnetic scattering from wire structures utilizing an implicit time-domain integral equation technique," *Micro. Opt. Tech. Lett.*, Vol. 17, pp. 66–69, 1998.

[22]   S. M. Rao and T. K. Sarkar, "An efficient method to evaluate the time-domain scattering from arbitrarily shaped conducting bodies," *Micro. Opt. Tech. Lett.*, Vol. 17, pp. 321–325, 1998.
[23]   B. H. Jung and T. K. Sarkar, "Time-domain electric-field integral equation with central finite difference," *Micro. Optical Tech. Lett.*, Vol. 31, pp. 429–435, 2001.

6

ANALYSIS OF CONDUCTING STRUCTURES IN THE TIME DOMAIN

6.0 SUMMARY

In this chapter, the time-domain integral equations (TDIEs) are used to analyze transient electromagnetic responses from three-dimensional (3-D) arbitrarily shaped conducting bodies. They include the time-domain electric field integral equation (TD-EFIE), the time-domain magnetic field integral equation (TD-MFIE), and the time-domain combined field integral equation (TD-CFIE). Two solution methods, which are called marching-on-in-time (MOT) and marching-on-in-degree (MOD) are presented for solving these integral equations.

In the MOT method, the TD-EFIE is solved based on an implicit scheme that approximates the time derivative of the magnetic vector potential by a central finite difference scheme and the value of the scalar potential by its time average. Both the explicit and the implicit solution methods are described for the solution of the TD-MFIE. The solution to the TD-CFIE can then be obtained as a linear combination of the TD-EFIE and the TD-MFIE. When solving the TD-CFIE, it is recognized that the spectrum of the incident field may contain frequencies that may correspond to the internal resonances of the structure.

The MOD technique is based on the implementation of Galerkin's method that involves separate spatial and temporal testing. The triangular patch vector functions often called the RWG functions are used for spatial expansion and testing in the analysis of conducting structures. The temporal variables are approximated by using an orthonormal set of basis functions called the associated Laguerre functions, which are derived from the Laguerre polynomials. These basis functions are also used for temporal testing. Using these associated Laguerre functions, it is possible to evaluate the time derivatives in the various expressions in an analytic fashion. An alternative set of formulations for the MOD technique is also presented to solve the TDIEs. Numerical examples related to the TD-EFIE, TD-MFIE, and TD-CFIE when using the MOT, MOD, and the alternative MOD methods are presented, and the results are compared with those obtained from the inverse discrete Fourier transform (IDFT) of the frequency domain solutions to demonstrate the accuracy of the methods.

6.1   INTRODUCTION

The analysis of electromagnetic scattering from arbitrarily shaped conducting bodies in the time domain has been of considerable interest recently, and many approaches have been reported in the published literature. Among them, several approaches have been presented in solving the TDIEs using the triangular patch RWG basis functions to represent the spatial variations [1–5]. In the expression of the TD-EFIE, there is a time derivative of the magnetic vector potential. An explicit solution has been presented by differentiating the TD-EFIE and using second-order finite difference approximation involving the derivative [1]. But the computed results become unstable for late times. The late-time oscillations could be eliminated by taking the average value for the current [2]. However, as the incident field is also differentiated, this method cannot deal with excitations that may involve an impulse or a step function. To overcome this problem of the explicit technique, a backward finite difference scheme is used to approximate the derivative on the magnetic vector potential [4]. Also, many numerical results using the explicit method with forward and backward difference schemes to approximate the derivative have been reported [4–7]. Recently, an implicit scheme has been proposed to solve two- and three-dimensional scattering problems with improved late-time stability [8–12]. When the explicit method is used, the time step becomes very small, and the computed transient response becomes unstable because of the accumulation of numerical errors at each step, and the overall computation time is long. When the implicit method is used, a larger time step can be used. But the numerical error in the approximation for the time derivative using a finite difference is increased. This implies that the stability and accuracy of a solution procedure are dependent on the choice of the time step. A central finite difference methodology has also been used for approximating the time derivative of the TD-EFIE to improve the stability and the accuracy [13,14]. In addition, numerical methods based on the TD-MFIE have also been presented for calculating the transient response from three-dimensional closed conducting bodies [9,15]. It has been observed that the numerical results obtained using the TD-MFIE is more stable than those obtained using the TD-EFIE.

It is well known that the numerical solutions obtained using either EFIE or MFIE for the analysis of closed conducting structures in the frequency domain, do not provide a unique result when the incident field contains frequencies at which the structure is resonant. Similar problems are encountered in the time domain. The methods based on the TD-EFIE and TD-MFIE may produce spurious solutions in the analysis of closed conducting bodies at frequencies that correspond to the internal resonance of the structure. One possible way of obtaining a unique solution is to use the TD-CFIE by linearly combining the TD-EFIE with TD-MFIE. The TD-CFIE has been applied to the analysis of transient scattering by two- and three-dimensional bodies [16–18]. This formulation eliminates the resonance problem and sometimes also the late-time oscillation encountered in the TD-EFIE and slow varying oscillation of the solution in the TD-MFIE. However, solutions obtained using the TD-CFIE scheme still may have an instability problem, and its accuracy is dependent on the size of the time step.

To solve TDIEs, the MOT method has been widely used [5] for many years. However, a serious drawback of this procedure is that late-time instabilities occur in the solution in the form of high-frequency oscillations. In Chapter 5, a method using the Laguerre polynomials as temporal basis functions has been introduced. The associated Laguerre functions derived from the Laguerre polynomials are defined only over the interval from zero to infinity and, hence, are considered to be more suited for the analysis of transient problems, as they naturally enforce causality [19]. Using the associated Laguerre functions, a set of orthonormal basis functions is constructed. Transient quantities that are functions of time can be spanned in terms of these orthogonal basis functions. Because these temporal basis functions are convergent to zero as time increases to infinity, the transient response spanned by these basis functions is also expected to be convergent to zero as time progresses. Using the Galerkin's method, a temporal testing procedure is introduced. By applying the temporal testing procedure to the TDIEs, the numerical instabilities can be improved on in contrast to the MOT method. Another advantage of the MOD procedure over the MOT technique is that the connection between the spatial and the temporal discretizations related by the Courant–Friedrich–Levy criteria can be relaxed. In the MOD method, the solution progresses by increasing the degree of the polynomial involved in the temporal testing functions. Therefore, the unknown coefficients can be obtained by solving a matrix equation recursively with a finite number of basis functions. The minimum number of basis functions is dependent on the product of the time duration and the frequency bandwidth of the incident wave. In this chapter, alternative formulations of the MOD method are also presented to solve the TD-EFIE, TD-MFIE, and TD-CFIE.

In section 6.2, the TD-EFIE, TD-MFIE, and TD-CFIE for analyzing transient electromagnetic responses from three-dimensional arbitrarily shaped conducting bodies are presented. Section 6.3 will briefly review the triangular patch vector function (i.e., the RWG vector functions [20–23]), which will be used in both the MOT and the MOD methods. The details of generating the matrix equation using the MOT method will be given in section 6.4, whereas that for the MOD method and its alternative formulations will be presented in sections 6.5 and 6.6, respectively. Numerical examples illustrating the application of the different methods are given in section 6.7 followed by some conclusions in section 6.8.

6.2   TIME DOMAIN INTEGRAL EQUATIONS (TDIEs)

6.2.1   Time Domain Electric Field Integral Equation (TD-EFIE)

In this section, the TD-EFIE for the analysis of conducting structures is presented. Let S denote the surface of a closed or open conducting body illuminated by a transient electromagnetic wave. Enforcing the boundary condition that the total tangential electric field is zero on the surface for all time leads to

$$\left(\mathbf{E}^i(\mathbf{r},t)+\mathbf{E}^s(\mathbf{r},t)\right)_{\tan}=0, \quad \text{for } \mathbf{r}\in S \tag{6.1}$$

where $\mathbf{E}^i$ is the incident field and $\mathbf{E}^s$ is the scattered field because of the induced current $\mathbf{J}$ on the conducting structure. The subscript 'tan' denotes the tangential component. The scattered field is given by

$$\mathbf{E}^s(\mathbf{r},t)=-\frac{\partial}{\partial t}\mathbf{A}(\mathbf{r},t)-\nabla\Phi(\mathbf{r},t) \tag{6.2}$$

The expressions for $\mathbf{A}$ and Φ, which represent the magnetic vector and electric scalar potentials, respectively, can be written as

$$\mathbf{A}(\mathbf{r},t)=\frac{\mu}{4\pi}\int_S \frac{\mathbf{J}(\mathbf{r}',\tau)}{R}dS' \tag{6.3}$$

$$\Phi(\mathbf{r},t)=\frac{1}{4\pi\varepsilon}\int_S \frac{Q(\mathbf{r}',\tau)}{R}dS' \tag{6.4}$$

In Eqs. (6.3) and (6.4), $R=|\mathbf{r}-\mathbf{r}'|$ represents the distance between the arbitrarily located observation point $\mathbf{r}$ and the source point $\mathbf{r}'$. $\tau=t-R/c$ is the retarded time. μ and ε are the permeability and the permittivity of the medium, and c is the velocity of the propagation of the electromagnetic wave in that space. The surface charge density Q is related to the surface current density $\mathbf{J}$ by the equation of continuity

$$\nabla\cdot\mathbf{J}(\mathbf{r},t)=-\frac{\partial}{\partial t}Q(\mathbf{r},t) \tag{6.5}$$

Substituting Eq. (6.5) into Eq. (6.4) expresses the electric scalar potential as

$$\Phi(\mathbf{r},t)=-\frac{1}{4\pi\varepsilon}\int_S \int_0^\tau \frac{\nabla'\cdot\mathbf{J}(\mathbf{r}',t')}{R}dt'dS' \tag{6.6}$$

In summary, the TD-EFIE can be written as

$$\left(\frac{\partial}{\partial t}\mathbf{A}(\mathbf{r},t)+\nabla\Phi(\mathbf{r},t)\right)_{\tan}=\left(\mathbf{E}^i(\mathbf{r},t)\right)_{\tan} \tag{6.7a}$$

$$\mathbf{A}(\mathbf{r},t)=\frac{\mu}{4\pi}\int_S \frac{\mathbf{J}(\mathbf{r}',\tau)}{R}dS' \tag{6.7b}$$

$$\Phi(\mathbf{r},t)=\frac{1}{4\pi\varepsilon}\int_S \int_0^\tau \frac{\nabla'\cdot\mathbf{J}(\mathbf{r}',t')}{R}dt'dS' \tag{6.7c}$$

The unknown current $\mathbf{J}$ can be obtained from the solution of Eq. (6.7).

6.2.2   Time Domain Magnetic Field Integral Equation (TD-MFIE)

Using the boundary condition for the magnetic field, the TD-MFIE can be obtained. Let S represent the surface of a closed conducting body illuminated by a transient electromagnetic plane wave. The TD-MFIE is given by

$$\mathbf{n} \times \left(\mathbf{H}^i(\mathbf{r},t) + \mathbf{H}^s(\mathbf{r},t) \right) = \mathbf{J}(\mathbf{r},t) , \quad \mathbf{r} \in S \tag{6.8}$$

where $\mathbf{n}$ represents an outward-directed unit vector normal to the surface S at a field point. $\mathbf{H}^i$ is the incident field, and $\mathbf{H}^s$ is the scattered magnetic field caused by the induced current $\mathbf{J}$. The scattered magnetic field can be written in terms of the magnetic vector potential as

$$\mathbf{H}^s(\mathbf{r},t) = \frac{1}{\mu} \nabla \times \mathbf{A}(\mathbf{r},t) \tag{6.9}$$

where $\mathbf{A}$ is the magnetic vector potential given by Eq. (6.3). Extracting the Cauchy principal value from the curl term in Eq. (6.9) yields

$$\mathbf{n} \times \mathbf{H}^s(\mathbf{r}, t) = \frac{\mathbf{J}(\mathbf{r}, t)}{2} + \mathbf{n} \times \frac{1}{4\pi} \int_{S_0} \nabla \times \frac{\mathbf{J}(\mathbf{r}', \tau)}{R} dS' \tag{6.10}$$

where S_0 denotes the surface S with the singularity at $\mathbf{r} = \mathbf{r}'$ or $R = 0$ removed. Now, by substituting Eq. (6.10) into Eq. (6.8), one obtains

$$\frac{\mathbf{J}(\mathbf{r},t)}{2} - \mathbf{n} \times \frac{1}{4\pi} \int_{S_0} \nabla \times \frac{\mathbf{J}(\mathbf{r}',\tau)}{R} dS' = \mathbf{n} \times \mathbf{H}^i(\mathbf{r},t) \tag{6.11}$$

Equation (6.11) is the TD-MFIE, and the unknown current $\mathbf{J}$ is to be determined.

6.2.3   Time Domain Combined Field Integral Equation (TD-CFIE)

The time domain CFIE is obtained through a linear combination of the EFIE with the MFIE [23] through

$$(1-\kappa)\left(-\mathbf{E}^s(\mathbf{r},t)\right)_{\text{tan}} + \kappa\eta\left(\mathbf{J} - \mathbf{n} \times \mathbf{H}^s(\mathbf{r},t)\right)$$

$$= (1-\kappa)\left(\mathbf{E}^i(\mathbf{r},t)\right)_{\text{tan}} + \kappa\eta\left(\mathbf{n} \times \mathbf{H}^i(\mathbf{r},t)\right) \tag{6.12}$$

where κ is the parameter of the linear combination, which is between 0 (EFIE) and 1 (MFIE), and η is the wave impedance of the space. The advantage of the TD-CFIE over the TD-EFIE and TD-MFIE are that both of the latter equations fail to give a unique solution when the frequency spectrum of the incident wave includes internal resonant frequencies of the object. By obtaining a linear combination of the TD-EFIE and TD-MFIE, the real poles on the frequency axis are pushed in to the complex frequency plane, thereby eliminating the internal resonant poles on the ω-axis.

6.3   TRIANGULAR PATCH (RWG) VECTOR BASIS FUNCTION

To numerically solve the TDIEs either by using the MOT or MOD techniques, the structure of interest is modeled by triangular patches, and the unknown surface current is approximated by the RWG basis function over the patch, which has been introduced in the previous chapters [20]. In summary, on the nth edge, the RWG vector basis function is given by

$$\mathbf{f}_n(\mathbf{r}) = \mathbf{f}_n^+(\mathbf{r}) + \mathbf{f}_n^-(\mathbf{r}) \qquad (6.13a)$$

$$\mathbf{f}_n^\pm(\mathbf{r}) = \begin{cases} \dfrac{l_n}{2a_n^\pm}\boldsymbol{\rho}_n^\pm, & \mathbf{r} \in T_n^\pm \\ 0, & \mathbf{r} \notin T_n^\pm \end{cases} \qquad (6.13b)$$

where l_n is the length of the edge and $a_n^\pm$ is the area of the triangle $T_n^\pm$, and $\boldsymbol{\rho}_n^\pm$ is the position vector defined with respect to the free vertex of $T_n^\pm$.

6.4   COMPUTED NUMERICAL SOLUTIONS USING THE MOT

The electric current $\mathbf{J}$ on the surface of the scatterer may be approximated by

$$\mathbf{J}(\mathbf{r},t) = \sum_{n=1}^{N} J_n(t)\mathbf{f}_n(\mathbf{r}) \qquad (6.14)$$

where N represents the number of common edges. By applying the Galerkin's method and using the same RWG functions as both expansion and the testing, the operator equation can be transformed into a matrix equation. This matrix equation is a function of time and can be solved iteratively as the time progresses.

6.4.1 Solving the TD-EFIE by the MOT Method

To numerically solve the TD-EFIE given by Eq. (6.7) using the MOT technique, the time derivative terms are first approximated by finite differences. The time axis is divided into segments of equal intervals of duration Δt with the discrete time instant defined as $t_i = i\Delta t$, where $\Delta t = t_i - t_{i-1}$. The time derivative in Eq. (6.7) can be approximated by the forward finite difference at $t = t_{i-1}$ as

$$\left(\frac{\mathbf{A}(\mathbf{r},t_i) - \mathbf{A}(\mathbf{r},t_{i-1})}{\Delta t} + \nabla \Phi(\mathbf{r},t_{i-1}) \right)_{\tan} = \left(\mathbf{E}^i(\mathbf{r},t_{i-1}) \right)_{\tan} \qquad (6.15)$$

The time derivative in Eq. (6.7) can also be approximated by the backward finite difference at $t = t_i$ as

$$\left(\frac{\mathbf{A}(\mathbf{r},t_i) - \mathbf{A}(\mathbf{r},t_{i-1})}{\Delta t} + \nabla \Phi(\mathbf{r},t_i) \right)_{\tan} = \left(\mathbf{E}^i(\mathbf{r},t_i) \right)_{\tan} \qquad (6.16)$$

or by using the central finite difference at time $t = t_{i-1/2}$ as

$$\left(\frac{\mathbf{A}(\mathbf{r},t_i) - \mathbf{A}(\mathbf{r},t_{i-1})}{\Delta t} + \frac{\nabla \Phi(\mathbf{r},t_i) + \nabla \Phi(\mathbf{r},t_{i-1})}{2} \right)_{\tan} = \left(\mathbf{E}^i(\mathbf{r},t_{i-1/2}) \right)_{\tan} \qquad (6.17)$$

The forward finite difference scheme is usually used in the explicit method for solving the TD-EFIE [1] whereas the backward finite difference is used for the explicit and implicit solution procedures [4,8]. The central finite difference was proposed to improve the accuracy and stability for the implicit solution. In this scheme, the time averaging is also applied for the scalar potential term when the central finite difference is used for computing the time derivative associated with the vector potential. Equations (6.15)–(6.17) can be combined into a unified form so that a single formulation can implement all these three schemes by

$$\left(\frac{\mathbf{A}(\mathbf{r},t_i) - \mathbf{A}(\mathbf{r},t_{i-1})}{\Delta t} + (1-v)\nabla \Phi(\mathbf{r},t_i) + v\nabla \Phi(\mathbf{r},t_{i-1}) \right)_{\tan}$$
$$= \left(\mathbf{E}^i(\mathbf{r},t_{i-v}) \right)_{\tan} \qquad (6.18)$$

where $v = 1$ is for the forward, $v = 0$ for the backward, and $v = 1/2$ for the central finite difference approximation.

Equation (6.18) is now solved by applying the Galerkin's method. By choosing the expansion functions $\mathbf{f}_m$ as the testing functions and defining the inner product by Eq. (1.9), one obtains

$$A_m(t_i) + \Delta t \ (1-\nu) \ \phi_m(t_i) = \Delta t \ V_m(t_{i-\nu}) + A_m(t_{i-1}) - \Delta t \ \nu \ \phi_m(t_{i-1}) \qquad (6.19a)$$

where

$$A_m(t_i) = \langle \mathbf{f}_m, \mathbf{A}(\mathbf{r}, t_i) \rangle \qquad (6.19b)$$

$$\phi_m(t_i) = \langle \mathbf{f}_m, \nabla \Phi(\mathbf{r}, t_i) \rangle \qquad (6.19c)$$

$$V_m(t_i) = \langle \mathbf{f}_m, \mathbf{E}^i(\mathbf{r}, t_i) \rangle \qquad (6.19d)$$

In Eq. (6.19), both sides include the vector and the scalar potential terms and the left-hand side contains all the terms at $t = t_i$. The next step in the procedure is to substitute the current expansion functions defined by Eqs. (6.13) and (6.14) into Eqs. (6.19b) and (6.19c), and this yields

$$A_m(t_i) = \sum_{n=1}^{N} \sum_{p,q} A_{mn}^{pq} J_n(\tau_{mn}^{pq})$$

$$= \sum_{n=1}^{N} \left(A_{mn}^{++} J_n(\tau_{mn}^{++}) + A_{mn}^{+-} J_n(\tau_{mn}^{+-}) + A_{mn}^{-+} J_n(\tau_{mn}^{-+}) + A_{mn}^{--} J_n(\tau_{mn}^{--}) \right) \qquad (6.20)$$

$$\phi_m(t_i) = \sum_{n=1}^{N} \sum_{p,q} \phi_{mn}^{pq} Q_n(\tau_{mn}^{pq})$$

$$= \sum_{n=1}^{N} \left(\phi_{mn}^{++} Q_n(\tau_{mn}^{++}) + \phi_{mn}^{+-} Q_n(\tau_{mn}^{+-}) + \phi_{mn}^{-+} Q_n(\tau_{mn}^{-+}) + \phi_{mn}^{--} Q_n(\tau_{mn}^{--}) \right) \qquad (6.21)$$

where

$$A_{mn}^{pq} = \frac{\mu}{4\pi} \int_S \mathbf{f}_m^p(\mathbf{r}) \cdot \int_S \frac{\mathbf{f}_n^q(\mathbf{r}')}{R} dS' dS \qquad (6.22)$$

$$\phi_{mn}^{pq} = \frac{1}{4\pi\varepsilon} \int_S \nabla \cdot \mathbf{f}_m^p(\mathbf{r}) \int_S \frac{\nabla' \cdot \mathbf{f}_n^q(\mathbf{r}')}{R} dS' dS \qquad (6.23)$$

$$J_n(\tau_{mn}^{pq}) = J_n(t_i - R_{mn}^{pq}/c) \qquad (6.24)$$

$$Q_n(\tau_{mn}^{pq}) = \int_0^{\tau_{mn}^{pq}} J_n(t') dt' \qquad (6.25)$$

$$\tau_{mn}^{pq} = t_i - R_{mn}^{pq}/c \qquad (6.26)$$

$$R_{mn}^{pq} = \left| \mathbf{r}_m^{cp} - \mathbf{r}_n^{cq} \right| \qquad (6.27)$$

where p and q are $+$ or $-$, and $\mathbf{r}_m^{cp}$ is the position vector of the centroid of the triangle T_m^p. In deriving Eqs. (6.20) and (6.21), the simplified retarded time

$t_i - R/c \approx t_i - R_{mn}^{pq}/c$ is assumed and R_{mn}^{pq} represents the distance between the centers of testing triangle T_m^p and source triangle T_n^q. This assumption implies that the current distribution on a triangular patch does not vary appreciably with time.

If the integrals involving the unprimed variables in Eqs. (6.22) and (6.23) are evaluated by approximating the integrand by their respective values at the centroids of the testing triangle, Eqs. (6.22) and (6.23) become

$$A_{mn}^{pq} = \frac{\mu \, l_m \, l_n}{16\pi} \, \rho_m^{cp} \cdot \frac{1}{a_n^q} \int_{T_n^q} \frac{\rho_n^q}{R_m^p} \, dS' \tag{6.28}$$

$$\phi_{mn}^{pq} = \frac{l_m l_n}{4\pi\varepsilon} \frac{1}{a_n^q} \int_{T_n^q} \frac{1}{R_m^p} \, dS' \tag{6.29}$$

where $R_m^p = \left| \mathbf{r}_m^{cp} - \mathbf{r}' \right|$, l_n is the length of the nonboundary edge and a_n^q is the area of the source triangle T_n^q. Vectors ρ_n^q are the position vector defined with respect to the free vertices of T_n^p. l_m is the length of the edge and a_m^p is the area of the testing triangle T_n^q. The average charge density is given by the value of ρ at the centroid of the triangular patch $\mathbf{r}_m^{cp}$ (i.e., $\rho_m^{cp} = \rho_m^p / a_m^p$). The integrals in Eqs. (6.28) and (6.29) may be computed by the methods described in [21].

Finally, Eq. (6.19d) can be computed by approximating the testing integral by the value at the centroid of the testing triangle as

$$V_m(t_i) = \sum_p V_m^p(t_i) = V_m^+(t_i) + V_m^-(t_i) \tag{6.30}$$

$$V_m^p(t_i) = \int_S \mathbf{f}_m^p(\mathbf{r}) \cdot \mathbf{E}^i(\mathbf{r}, t_i) \, dS \approx \frac{l_m}{2} \rho_m^{cp} \cdot \mathbf{E}^i(\mathbf{r}_m^{cp}, t_i) \tag{6.31}$$

where p can be either $+$ or $-$. Furthermore, when $R_{mn}^{pq} \geq c\Delta t$ or $t_{j-1} < \tau_{mn}^{pq} \leq t_j$, where $t_j \leq t_{i-1}$, the current coefficients $J_n(\tau_{mn}^{pq})$ in Eq. (6.24) and the integration of the current coefficient $Q_n(\tau_{mn}^{pq})$ in Eq. (6.25) can be rewritten as

$$J_n(\tau_{mn}^{pq}) = (1-\delta)J_n(t_{j-1}) + \delta J_n(t_j) \tag{6.32}$$

$$Q_n(\tau_{mn}^{pq}) = \Delta t \sum_{k=0}^{j-2} J_n(t_k) + \frac{\Delta t}{2}(1+2\delta-\delta^2)J_n(t_{j-1}) + \frac{\Delta t}{2}\delta^2 J_n(t_j) \tag{6.33}$$

where

$$\delta = \frac{\tau_{mn}^{pq} - t_{j-1}}{\Delta t} \tag{6.34}$$

Using Eqs. (6.32) and (6.33), Eq. (6.19) for testing of TD-EFIE can be computed at the time step $t = t_i$ except for the terms for which $R_{mn}^{pq} < c\Delta t$. Those terms for which $R_{mn}^{pq} < c\Delta t$ or $t_{i-1} < \tau_{mn}^{pq} \leq t_i$, Eqs. (6.24) and (6.25) can be expressed as

$$J_n(\tau_{mn}^{pq}) = S_{mn}^{pq} J_n(t_{i-1}) + \left(1 - S_{mn}^{pq}\right) J_n(t_i) \tag{6.35}$$

$$Q_n(\tau_{mn}^{pq}) = \Delta t \sum_{k=0}^{i-2} J_n(t_k) + \frac{\Delta t}{2}\left(2 - S_{mn}^{pq2}\right) J_n(t_{i-1}) + \frac{\Delta t}{2}\left(1 - S_{mn}^{pq}\right)^2 J_n(t_i) \tag{6.36}$$

where

$$S_{mn}^{pq} = \frac{R_{mn}^{pq}}{c\Delta t} \tag{6.37}$$

Using Eqs. (6.36) and (6.37), Eqs. (6.20) and (6.21) can be decomposed into the terms for which $R_{mn}^{pq} \geq c\Delta t$ or $R_{mn}^{pq} < c\Delta t$ as

$$A_m(t_i) = \sum_{n=1}^{N} \breve{X}_{mn} J_n(t_i) + \breve{P}_m(t_{i-1}) + \hat{A}_m(t_i) \tag{6.38}$$

$$\phi_m(t_i) = \sum_{n=1}^{N} \breve{Y}_{mn} J_n(t_i) + \breve{B}_m(t_{i-1}) + \breve{D}_m(t_{i-2}) + \hat{\phi}_m(t_i) \tag{6.39}$$

where

$$\breve{X}_{mn} = A_{mn}^{++}\left(1 - S_{mn}^{++}\right) + A_{mn}^{+-}\left(1 - S_{mn}^{+-}\right) + A_{mn}^{-+}\left(1 - S_{mn}^{-+}\right) + A_{mn}^{--}\left(1 - S_{mn}^{--}\right) \tag{6.40}$$

$$\breve{P}_m(t_{i-1}) = \sum_{n=1}^{N}\left(A_{mn}^{++} S_{mn}^{++} + A_{mn}^{+-} S_{mn}^{+-} + A_{mn}^{-+} S_{mn}^{-+} + A_{mn}^{--} S_{mn}^{--}\right) J_n(t_{i-1}) \tag{6.41}$$

$$\breve{Y}_{mn} = \frac{\Delta t}{2}\left[\phi_{mn}^{++}\left(1 - S_{mn}^{++}\right)^2 + \phi_{mn}^{+-}\left(1 - S_{mn}^{+-}\right)^2 \right.$$
$$\left. + \phi_{mn}^{-+}\left(1 - S_{mn}^{-+}\right)^2 + \phi_{mn}^{--}\left(1 - S_{mn}^{--}\right)^2\right] \tag{6.42}$$

$$\breve{B}_m(t_{i-i}) = \frac{\Delta t}{2} \sum_{n=1}^{N}\left(\phi_{mn}^{++}\left(2 - S_{mn}^{++2}\right) + \phi_{mn}^{+-}\left(2 - S_{mn}^{+-2}\right)\right.$$
$$\left. + \phi_{mn}^{-+}\left(2 - S_{mn}^{-+2}\right) + \phi_{mn}^{--}\left(2 - S_{mn}^{--2}\right)\right) J_n(t_{i-1}) \tag{6.43}$$

$$\breve{D}_m(t_{i-2}) = \Delta t \sum_{n=1}^{N} \left[\left(\phi_{mn}^{++} + \phi_{mn}^{+-} + \phi_{mn}^{-+} + \phi_{mn}^{--} \right) \sum_{k=0}^{i-2} J_n(t_k) \right] \qquad (6.44)$$

In Eqs. (6.38)–(6.44), the symbol $\smallsmile$ on top of the elements is used to indicate the elements for which $R_{mn}^{pq} < c\Delta t$ and the symbol $\smallfrown$ is for elements when $R_{mn}^{pq} \geq c\Delta t$. The terms $\hat{A}_m(t_i)$ and $\hat{\phi}_m(t_i)$ in Eqs. (6.38) and (6.39) for which $R_{mn}^{pq} \geq c\Delta t$, are of the same form as in Eqs. (6.22) and (6.23). A matrix equation is obtained by substituting Eqs. (6.38) and (6.39) into Eq. (6.19) as

$$\left[\alpha_{mn}^{E} \right] \left[J_n(t_i) \right] = \left[\beta_m^{E} \right] \qquad (6.45a)$$

where

$$\alpha_{mn}^{E} = \breve{X}_{mn} + \Delta t (1 - v) \breve{Y}_{mn} \qquad (6.45b)$$

$$\begin{aligned} \beta_{mn}^{E} = {} & \Delta t V_m(t_{i-v}) + A_m(t_{i-1}) - \Delta t v \phi_m(t_{i-1}) \\ & - \left(\hat{A}_m(t_i) + \breve{P}_m(t_{i-1}) \right) - \Delta t (1 - v) \left(\hat{\phi}_m(t_i) + \breve{B}_m(t_{i-1}) + \breve{D}_m(t_{i-2}) \right) \end{aligned} \qquad (6.45c)$$

Note that when $v = 1$ for the forward difference case, the terms related to the scalar potential $\breve{Y}_{mn}$ in Eq. (6.45b) and $\hat{\phi}_m(t_i) + \breve{B}_m(t_{i-1}) + \breve{D}_m(t_{i-2})$ in Eq. (6.45c) vanish and do not contribute to α_{mn}^{E} and β_m^{E}, respectively. When $v = 0$ for the backward difference approximation, the scalar potential term in Eq. (6.45c), $\phi_m(t_{i-1})$, does not contribute to β_m^{E}. However, when $v = 1/2$ for the central difference case, the scalar potential contributes to both the matrices $\left[\alpha_{mn}^{E} \right]$ and $\left[\beta_m^{E} \right]$. And the incident field is considered at the midpoint of each time step. Matrix $\left[\alpha_{mn}^{E} \right]$ is not a function of time, and hence, the inverse of the matrix needs to be computed only once at the beginning of the computation.

6.4.2 Solving the TD-MFIE by the MOT Method

In this section, the procedures for solving the TD-MFIE given by Eq. (6.11) using explicit or implicit schemes are presented. The TD-MFIE is valid only for closed scatterers. The surface of the closed structure to be analyzed is modeled by planar triangular patches, and the unknown surface current $\mathbf{J}$ is approximated by Eq. (6.14) in terms of the RWG basis function given by Eq. (6.13).

By applying Galerkin's method, the expansion functions $\mathbf{f}_m$ are also chosen to be the testing functions. Applying the testing procedure to Eq. (6.11) results in

$$C_m(t) = \gamma_m(t) + B_m(t) \tag{6.46a}$$

where

$$C_m(t) = \left\langle \mathbf{f}_m, \frac{\mathbf{J}(\mathbf{r},t)}{2} \right\rangle \tag{6.46b}$$

$$\gamma_m(t) = \left\langle \mathbf{f}_m, \mathbf{n} \times \mathbf{H}^i(\mathbf{r},t) \right\rangle \tag{6.46c}$$

$$B_m(t) = \left\langle \mathbf{f}_m, \mathbf{n} \times \frac{1}{4\pi} \int_{S_0} \nabla \times \frac{\mathbf{J}(\mathbf{r}',\tau)}{R} dS' \right\rangle \tag{6.46d}$$

6.4.2.1 Explicit Solution. For solving Eq. (6.46) using the explicit scheme, the RWG basis functions defined in Eq. (6.17) is substituted into Eq. (6.46b), and then $C_m(t)$ becomes

$$C_m(t) = \sum_{n=1}^{N} C_{mn} J_n(t) \tag{6.47a}$$

where

$$C_{mn} = \sum_{p,q} C_{mn}^{pq} = C_{mn}^{++} + C_{mn}^{+-} + C_{mn}^{-+} + C_{mn}^{--} \tag{6.47b}$$

$$C_{mn}^{pq} = \frac{1}{2} \left\langle \mathbf{f}_m^p, \mathbf{f}_n^q \right\rangle \tag{6.47c}$$

where p and q are $+$ or $-$. Applying Eq. (6.17b), the inner product integral C_{mn}^{pq} defined in Eq. (6.47c) can be further expanded analytically, and the result is given by [22]

$$C_{mn}^{pq} = \frac{l_m l_n}{8 a_n^q} \left\{ \frac{3}{4} |\mathbf{r}_m^{cp}|^2 + \frac{1}{12} \left(|\mathbf{r}_{m1}|^2 + |\mathbf{r}_{m2}|^2 + |\mathbf{r}_{m3}|^2 \right) \right.$$
$$\left. -\mathbf{r}_m^{cp} \cdot (\mathbf{r}_{m1} + \mathbf{r}_{n1}) + \mathbf{r}_{m1} \cdot \mathbf{r}_{n1} \right\} \tag{6.48}$$

where $\mathbf{r}_{m1}$, $\mathbf{r}_{m2}$, and $\mathbf{r}_{m3}$ are the position vectors of the three vertices of the triangle T_m^p, and $\mathbf{r}_m^{cp}$ is the centroid of triangle T_m^p. $\mathbf{r}_{m1}$ and $\mathbf{r}_{n1}$ are the position vectors of the free vertex of the triangle T_m^p and T_n^q, respectively. Note that if the field point does not lie on the triangle, T_n^q (i.e. $\mathbf{r} \notin T_n^q$) then the result is $C_{mn}^{pq} = 0$. This is evident from the definition of the expansion functions in Eq. (6.17b).

Next, performing a similar expansion procedure for the term $\gamma_m(t)$ involving an inner product in Eq. (6.46c) results in

$$\gamma_m(t) = \sum_p \gamma_m^p = \gamma_m^+ + \gamma_m^- \qquad (6.49a)$$

where

$$\gamma_m^p = \frac{l_m}{2\,a_m^p} \int_{T_m^p} \boldsymbol{\rho}_m^p \cdot \mathbf{n} \times \mathbf{H}^i(\mathbf{r},t)\,dS \qquad (6.49b)$$

In Eq. (6.49b), the integral is evaluated by approximating the integrand by the value at the centroid of the testing triangle T_m^p.

Finally, consider the inner product term $B_m(t)$ in Eq. (6.46d). Using a vector identity, the curl term inside the integral is given by [3]

$$\nabla \times \frac{\mathbf{J}(\mathbf{r}',\tau)}{R} = \frac{1}{c} \frac{\partial \mathbf{J}(\mathbf{r}',\tau)}{\partial \tau} \times \frac{\hat{\mathbf{R}}}{R} + \frac{\mathbf{J}(\mathbf{r}',\tau) \times \hat{\mathbf{R}}}{R^2} \qquad (6.50)$$

where $\hat{\mathbf{R}}$ is a unit vector along the direction of $\mathbf{r} - \mathbf{r}'$. Thus,

$$\frac{1}{4\pi} \int_{S_0} \nabla \times \frac{\mathbf{J}(\mathbf{r}',\tau)}{R}\,dS' = \frac{1}{4\pi} \left(\frac{1}{c} \int_{S_0} \frac{\partial \mathbf{J}(\mathbf{r}',\tau)}{\partial \tau} \times \frac{\hat{\mathbf{R}}}{R}\,dS' + \int_{S_0} \frac{\mathbf{J}(\mathbf{r}',\tau) \times \hat{\mathbf{R}}}{R^2}\,dS' \right) \qquad (6.51)$$

Now, by substituting Eq. (6.51) into Eq. (6.46d), one obtains

$$B_m(t) = \frac{1}{4\pi} \int_S \mathbf{f}_m \cdot \mathbf{n} \times \left\{ \frac{1}{c} \int_{S_0} \frac{\partial \mathbf{J}(\mathbf{r}',\tau)}{\partial \tau} \times \frac{\hat{\mathbf{R}}}{R}\,dS' + \int_{S_0} \frac{\mathbf{J}(\mathbf{r}',\tau) \times \hat{\mathbf{R}}}{R^2}\,dS' \right\} dS \qquad (6.52)$$

Substituting the current expansion given by Eq. (6.14) into Eq. (6.52) and approximating the value of the retarded time as

$$\tau = t - \frac{R}{c} \rightarrow \tau_{mn}^{pq} = t - \frac{R_{mn}^{pq}}{c}$$

where $R_{mn}^{pq} = \left| \mathbf{r}_m^{cp} - \mathbf{r}_n^{cq} \right|$ is the distance between the centroids of triangle T_m^p and T_n^q, one obtains

$$B_m(t) = \sum_{n=1}^N \sum_{p,q} B_{mn}^{pq} = \sum_{n=1}^N \left(B_{mn}^{++} + B_{mn}^{+-} + B_{mn}^{-+} + B_{mn}^{--} \right) \qquad (6.53a)$$

where

$$B_{mn}^{pq} = \frac{l_m l_n}{16\pi a_m^p a_n^q} \int_{T_m^p} \boldsymbol{\rho}_m^p \cdot \mathbf{n} \times \left\{ \frac{1}{c} \int_{T_n^q} \frac{\boldsymbol{\rho}_n^q \times \hat{\mathbf{R}}}{R} dS' \frac{\partial J_n(\tau_{mn}^{pq})}{\partial \tau_{mn}^{pq}} \right.$$

$$\left. + \int_{T_n^q} J_n(\tau_{mn}^{pq}) \frac{\boldsymbol{\rho}_n^q \times \hat{\mathbf{R}}}{R^2} dS' \right\} dS \tag{6.53b}$$

It can be shown that $B_{mn}^{pq} = 0$ if $\mathbf{r} \in T_n^q$, because $\mathbf{n} \times \boldsymbol{\rho}_n^q \times \hat{\mathbf{R}} = 0$. So the non-self term of the integral in Eq. (6.53b) may be evaluated by using a numerical integration. The term $\partial J_n / \partial \tau_{mn}^{pq}$ is approximated at $t = t_i$ by a first-order backward difference as

$$\frac{\partial J_n(\tau_{mn}^{pq})}{\partial \tau_{mn}^{pq}} \approx \frac{J_n(t_i - R_{mn}^{pq}/c) - J_n(t_{i-1} - R_{mn}^{pq}/c)}{\Delta t} \tag{6.54}$$

where Δt is the time interval. Substituting Eq. (6.54) into Eq. (6.53b) results in

$$B_{mn}^{pq} \approx \left(\frac{\zeta_1^{pq}}{c\Delta t} + \zeta_2^{pq} \right) J_n(t_i - R_{mn}^{pq}/c) - \frac{\zeta_1^{pq}}{c\Delta t} J_n(t_{i-1} - R_{mn}^{pq}/c) \tag{6.55a}$$

where

$$\zeta_v^{pq} = \frac{l_m l_n}{16\pi a_m^p a_n^q} \int_{T_m^p} \boldsymbol{\rho}_m^p \cdot \mathbf{n} \times \int_{T_n^q} \frac{\boldsymbol{\rho}_n^q \times \hat{\mathbf{R}}}{R^v} dS' dS, \quad \text{for} \quad v = 1 \text{ or } 2 \tag{6.55b}$$

This integral may be computed using the centroid testing for the unprimed coordinates and numerically using the Gaussian quadrature formula for the primed coordinates.

Using Eq. (6.47a), Eq. (6.46) is written in a matrix form at time $t = t_i$ as

$$\left[\alpha_{mn}^H \right] \left[J_n(t_i) \right] = \left[\beta_{mn}^H(t_i) \right] \tag{6.56a}$$

where the matrix $\left[\alpha_{mn}^H \right]$ is a sparse matrix with

$$\alpha_{mn}^H = C_{mn} \tag{6.56b}$$

$$\beta_m^H(t_i) = \gamma_m(t_i) + B_m(t_i) \tag{6.56c}$$

Note that if $R_{mn}^{pq} > c\Delta t$ for the non-self term ($R_{mn}^{pq} \neq 0$), then Eq. (6.56) can be solved for at the time step $t = t_i$. This is the explicit form of the solution of Eq. (6.46).

6.4.2.2 Implicit Solution. Now the implicit scheme for the TD-MFIE is presented for solving Eq. (6.46). Assume that the currents are linearly interpolated. When $R_{mn}^{pq} \geq c\Delta t$ or $t_{j-1} < \tau_{mn}^{pq} \leq t_j$ where $t_j \leq t_{i-1}$, the current coefficient J_n is approximated as

$$J_n(\tau_{mn}^{pq}) = (1-\delta)J_n(t_{j-1}) + \delta J_n(t_j) \tag{6.57a}$$

where

$$\delta = \frac{\tau_{mn}^{pq} - t_{j-1}}{\Delta t} \tag{6.57b}$$

when $R_{mn}^{pq} < c\Delta t$ or $t_{i-1} < \tau_{mn}^{pq} \leq t_i$ (i.e., the retarded time is just before the current time step t_i) then for the non-self term, the current is approximated by

$$J_n(\tau_{mn}^{pq}) = S_{mn}^{pq} J_n(t_{i-1}) + \left(1 - S_{mn}^{pq}\right) J_n(t_i) \tag{6.58a}$$

where

$$S_{mn}^{pq} = \frac{R_{mn}^{pq}}{c\Delta t} \tag{6.58b}$$

With Eqs. (6.57) and (6.58), Eq. (6.53a) may be expressed by a sum of the known and the unknown terms as:

$$B_m(t_i) = \sum_{n=1}^{N} \breve{X}_{mn} J_n(t_i) + \breve{P}_m(t_{i-1}) + \hat{B}_m(t_i) - D_m(t_{i-1}) \tag{6.59}$$

where

$$\breve{X}_{mn} = A_{mn}^{++}\left(1 - S_{mn}^{++}\right) + A_{mn}^{+-}\left(1 - S_{mn}^{+-}\right) + A_{mn}^{-+}\left(1 - S_{mn}^{-+}\right) + A_{mn}^{--}\left(1 - S_{mn}^{--}\right) \tag{6.60}$$

$$\breve{P}_m(t_{i-1}) = \sum_{n=1}^{N} \left(A_{mn}^{++} S_{mn}^{++} + A_{mn}^{+-} S_{mn}^{+-} + A_{mn}^{-+} S_{mn}^{-+} + A_{mn}^{--} S_{mn}^{--}\right) J_n(t_{i-1}) \tag{6.61}$$

$$A_{mn}^{pq} = \frac{\zeta_1^{pq}}{c\Delta t} + \zeta_2^{pq} \tag{6.62}$$

$$\hat{B}_m(t_i) = \sum_{n=1}^{N} \left(\hat{B}_{mn}^{++} + \hat{B}_{mn}^{+-} + \hat{B}_{mn}^{-+} + \hat{B}_{mn}^{--}\right) \tag{6.63}$$

$$\hat{B}_{mn}^{pq} = A_{mn}^{pq} J_n(t_i - R_{mn}^{pq}/c) \tag{6.64}$$

$$D_m\left(t_{i-1}\right) = \sum_{n=1}^{N}\left(D_{mn}^{++} + D_{mn}^{+-} + D_{mn}^{-+} + D_{mn}^{--}\right) \tag{6.65}$$

$$D_{mn}^{pq} = \frac{\zeta_1^{pq}}{c\Delta t}J_n\left(t_{i-1} - R_{mn}^{pq}/c\right). \tag{6.66}$$

In Eqs. (6.59)–(6.64), $\breve{X}_{mn}$ and $\breve{P}_m$, which are marked with the symbol $\smile$ on the top, are obtained by collecting the terms that satisfy the condition $R_{mn}^{pq} < c\Delta t$ and $\widehat{B}_m$, which is marked with the symbol $\frown$ on the top, are obtained by collecting the terms that satisfy the condition $R_{mn}^{pq} \geq c\Delta t$.

Finally, by substituting Eqs. (6.47) and (6.59) into Eq. (6.46), a matrix equation is obtained as

$$\left[\alpha_{mn}^{H}\right]\left[J_n(t_i)\right] = \left[\beta_{mn}^{H}(t_i)\right] \tag{6.67a}$$

where

$$\alpha_{mn}^{H} = C_{mn} - \breve{X}_{mn} \tag{6.67b}$$

$$\beta_m(t_i) = \gamma_m(t_i) + \breve{P}_m(t_{i-1}) + \widehat{B}_m(t_i) + D_m(t_{i-1}) \tag{6.67c}$$

Eq. (6.67) constitutes the implicit solution of the TD-MFIE.

6.4.3   Solving the TD-CFIE by the MOT Method

Because the TD-CFIE is obtained by a linear combination of the EFIE and MFIE, the solution of the CFIE can thus be obtained directly by a combination of the solutions for the EFIE of Eq. (6.45) and the explicit solution of MFIE of Eq. (6.56) or the implicit solution of MFIE of Eq. (6.57). A matrix equation can be obtained directly by combining Eqs. (6.56) and (6.67) as

$$\left[\alpha_{mn}^{C}\right]\left[J_n(t_i)\right] = \left[\beta_{mn}^{C}(t_i)\right] \tag{6.68a}$$

where

$$\alpha_{mn}^{C} = (1-\kappa)\alpha_{mn}^{E} + \kappa\eta\alpha_{mn}^{H} \tag{6.68b}$$

$$\beta_m(t_i) = (1-\kappa)\beta_m^{E}(t_i) + \kappa\eta\beta_m^{H}(t_i) \tag{6.68c}$$

6.5   COMPUTED NUMERICAL SOLUTIONS BY THE MOD

The TDIEs obtained for arbitrarily shaped 3-D conducting structures can alternatively be solved by using the MOD method, which has been outlined in

Chapter 5 for the analysis of a conducting wire. The computational procedure for analysis of 3-D conducting objects is similar to that for a conducting wire. In this section, the procedures for solving the TD-EFIE, TD-MFIE, and TD-CFIE using the MOD method will be described. The first step in the MOD and the MOT methods are similar. The surface of the structure to be analyzed is modeled by planar triangular patches. The MOD procedures are based on the Galerkin's method which involves separate spatial and temporal testing procedures. Piecewise triangle basis functions are used for spatial expansion and testing, whereas the time-domain variation is approximated and tested by a temporal basis function set derived from the associated Laguerre functions. Also, a new Hertz vector $\mathbf{u}(\mathbf{r}, t)$ is introduced in the MOD method for operational convenience; in particular it is good for handling the time integral in the TD-EFIE analytically. The current is related to the Hertz vector by

$$\mathbf{J}(\mathbf{r},t) = \frac{\partial}{\partial t}\mathbf{u}(\mathbf{r},t) \tag{6.69}$$

and the relation between this Hertz vector and the electric charge density $Q(\mathbf{r}, t)$ is given through

$$Q(\mathbf{r}, t) = -\nabla \cdot \mathbf{u}(\mathbf{r}, t) \tag{6.70}$$

Using the RWG basis functions of Eq. (6.13), the Hertz vector is spatially expanded as

$$\mathbf{u}(\mathbf{r},t) = \sum_{n=1}^{N} u_n(t)\mathbf{f}_n(\mathbf{r}) \tag{6.71}$$

Next, the TDIEs are tested with the same RWG function. Then, the temporal variation of the Hertz vector is expanded in terms of an orthonormal associated Laguerre functions set $\phi_j(t)$, which is derived from the Laguerre polynomials through the representation [19]

$$\phi_j(t) = e^{-t/2}L_j(t) \tag{6.72}$$

where $L_j(t)$ is the Laguerre polynomial of degree j. By using this basis function, the transient coefficient for the Hertz vector introduced in Eq. (6.71) is expanded as

$$u_n(t) = \sum_{j=0}^{\infty} u_{n,j}\,\phi_j(st) \tag{6.73}$$

where s is a scaling factor. By controlling this factor s, the support provided by the expansion can be increased or decreased. Applying the derivative property for the associated Laguerre functions given in the appendix of Chapter 5, the first and second derivatives of the transient coefficient $u_n(t)$ are obtained as

$$\frac{d}{dt} u_n(t) = s \sum_{j=0}^{\infty} \left(\frac{1}{2} u_{n,j} + \sum_{k=0}^{j-1} u_{n,k} \right) \phi_j(st) \tag{6.74}$$

$$\frac{d^2}{dt^2} u_n(t) = s^2 \sum_{j=0}^{\infty} \left(\frac{1}{4} u_{n,j} + \sum_{k=0}^{j-1} (j-k) u_{n,k} \right) \phi_j(st) \tag{6.75}$$

The TDIEs are then tested with the temporal basis function $\phi_j(t)$. By substituting Eqs. (6.73), (6.74), and (6.75) in the temporal testing of the TDIEs, the unknown is expressed in terms of a matrix equation that can be solved iteratively as the degree of the polynomial increases from a small to a large value.

6.5.1 Solving the TD-EFIE by the MOD Method

The TD-EFIE of Eq. (6.7) can be solved by applying the Galerkin's method. Using the RWG vector functions $\mathbf{f}_m(\mathbf{r})$ for both spatial expansion and testing, Eq. (6.7) yields

$$\left\langle \mathbf{f}_m(\mathbf{r}), \frac{\partial}{\partial t} \mathbf{A}(\mathbf{r},t) \right\rangle + \left\langle \mathbf{f}_m(\mathbf{r}), \nabla \Phi(\mathbf{r},t) \right\rangle = \left\langle \mathbf{f}_m(\mathbf{r}), \mathbf{E}^i(\mathbf{r},t) \right\rangle \tag{6.76}$$

Substitute Eqs. (6.69), (6.70), and (6.71) into the first term of Eq. (6.76). Also, use Eq. (6.3) for the magnetic vector potential $\mathbf{A}$ to yield

$$\left\langle \mathbf{f}_m(\mathbf{r}), \frac{\partial}{\partial t} \mathbf{A}(\mathbf{r},t) \right\rangle = \sum_{n=1}^{N} \frac{\mu}{4\pi} \int_S \mathbf{f}_m(\mathbf{r}) \cdot \int_S \frac{d^2}{dt^2} u_n(\tau) \frac{\mathbf{f}_n(\mathbf{r}')}{R} \, dS' \, dS \tag{6.77}$$

By assuming that the unknown transient quantity does not change appreciably within a triangle, one can obtain

$$\tau = t - \frac{R}{c} \quad \rightarrow \quad \tau_{mn}^{pq} = t - \frac{R_{mn}^{pq}}{c}, \quad R_{mn}^{pq} = \left| \mathbf{r}_m^{cp} - \mathbf{r}_n^{cq} \right| \tag{6.78}$$

where p and q are $+$ or $-$, and $\mathbf{r}_m^{c\pm}$ is the position vector for the centroid of the triangle $T_n^{\pm}$. Eq. (6.77) can be rewritten as

$$\left\langle \mathbf{f}_m(\mathbf{r}), \frac{\partial}{\partial t}\mathbf{A}(\mathbf{r},t)\right\rangle = \sum_{n=1}^{N}\sum_{p,q}\mu A_{mn}^{pq}\frac{d^2}{dt^2}u_n(\tau_{mn}^{pq}) \qquad (6.79a)$$

where

$$A_{mn}^{pq} = \frac{1}{4\pi}\int_S \mathbf{f}_m^p(\mathbf{r})\cdot\int_S \frac{\mathbf{f}_n^q(\mathbf{r}')}{R}\,dS'\,dS \qquad (6.79b)$$

Next, consider the testing of the gradient of the scalar potential $\Phi(\mathbf{r},t)$ in the second term of Eq. (6.76). Using the vector identity $\nabla\cdot\phi\mathbf{A} = \mathbf{A}\cdot\nabla\phi+\phi\nabla\cdot\mathbf{A}$ and the property of the spatial basis function [20], this term can be written as

$$\langle \mathbf{f}_m(\mathbf{r}), \nabla\Phi(\mathbf{r},t)\rangle = -\int_S \nabla\cdot\mathbf{f}_m(\mathbf{r})\Phi(\mathbf{r},t)\,dS$$
$$= \sum_{n=1}^{N}\frac{1}{4\pi\,\varepsilon}\int_S \nabla\cdot\mathbf{f}_m(\mathbf{r})\int_S u_n(\tau)\frac{\nabla'\cdot\mathbf{f}_n(\mathbf{r}')}{R}\,dS'\,dS \qquad (6.80)$$

Based on the assumption defined in Eq. (6.78), Eq. (6.80) can be written as

$$\langle \mathbf{f}_m(\mathbf{r}), \nabla\Phi(\mathbf{r},t)\rangle = \sum_{n=1}^{N}\sum_{p,q}\frac{B_{mn}^{pq}}{\varepsilon}u_n(\tau_{mn}^{pq}) \qquad (6.81a)$$

where

$$B_{mn}^{pq} = \frac{1}{4\pi}\int_S \nabla\cdot\mathbf{f}_m^p(\mathbf{r})\int_S \frac{\nabla'\cdot\mathbf{f}_n^q(\mathbf{r}')}{R}\,dS'\,dS \qquad (6.81b)$$

Similarly, the term in the right side of Eq. (6.76) can be expressed as

$$\langle \mathbf{f}_m(\mathbf{r}), \mathbf{E}^i(\mathbf{r},t)\rangle = V_m^E(t) \qquad (6.82a)$$

where

$$V_m^E(t) = \int_S \mathbf{f}_m(\mathbf{r})\cdot\mathbf{E}^i(\mathbf{r},t)\,dS \qquad (6.82b)$$

Finally, substituting Eqs. (6.79), (6.81), and (6.82) into Eq. (6.76) one finally obtains the result as

$$\sum_{n=1}^{N}\sum_{p,q}\left[\mu A_{mn}^{pq}\frac{d^2}{dt^2}u_n(\tau_{mn}^{pq})+\frac{B_{mn}^{pq}}{\varepsilon}u_n(\tau_{mn}^{pq})\right] = V_m^E(t) \qquad (6.83a)$$

where

$$A_{mn}^{pq} = \frac{1}{4\pi}\int_S \mathbf{f}_m^p(\mathbf{r})\cdot\int_S \frac{\mathbf{f}_n^q(\mathbf{r}')}{R}\,dS'\,dS \qquad (6.83b)$$

$$B_{mn}^{pq} = \frac{1}{4\pi} \int_S \nabla \cdot \mathbf{f}_m^p(\mathbf{r}) \int_S \frac{\nabla' \cdot \mathbf{f}_n^q(\mathbf{r}')}{R} \, dS' \, dS \qquad (6.83c)$$

$$V_m^E(t) = \int_S \mathbf{f}_m(\mathbf{r}) \cdot \mathbf{E}^i(\mathbf{r},t) \, dS \qquad (6.83d)$$

The detailed evaluations of these integrals are described in [20] and [21]. Then the temporal expansion and testing procedures are now carried out using the associated Laguerre functions. The terms $u_n(\tau_{mn}^{pq})$ and $d^2 u_n(\tau_{mn}^{pq})/dt^2$ in Eq. (6.83a) are substituted by Eqs. (6.73) and (6.75) and the temporal testing with $\phi_i(st)$ is taken, resulting in

$$\sum_{n=1}^{N} \sum_{p,q} \sum_{j=0}^{\infty} \left(\left(\frac{s^2}{4} \mu A_{mn}^{pq} + \frac{B_{mn}^{pq}}{\varepsilon} \right) u_{n,j} + s^2 \mu A_{mn}^{pq} \sum_{k=0}^{j-1} (j-k) u_{n,k} \right) I_{ij}(sR_{mn}^{pq}/c)$$
$$= V_{m,i}^E \qquad (6.84a)$$

where

$$I_{ij}(sR_{mn}^{pq}/c) = \int_{sR_{mn}^{pq}/c}^{\infty} \phi_i(st) \, \phi_j(st - sR_{mn}^{pq}/c) \, d(st) \qquad (6.84b)$$

$$V_{m,i}^E = \int_0^{\infty} \phi_i(st) \, V_m^E(t) \, d(st) \qquad (6.84c)$$

The integral in Eq. (6.84b) can be computed by applying the integral property of the associated Laguerre functions given in the appendix of Chapter 5. Note that $I_{ij} = 0$ when $j > i$. Therefore, the upper limit of the third summation symbol in Eq. (6.84a) can be written as i instead of ∞. Moving those terms including $u_{n,j}$ for $j < i$, which are known, to the right-hand side of Eq. (6.84a) can result in

$$\sum_{n=1}^{N} \sum_{p,q} \left(\frac{s^2}{4} \mu A_{mn}^{pq} + \frac{B_{mn}^{pq}}{\varepsilon} \right) u_{n,i} \, I_{ii}(sR_{mn}^{pq}/c)$$

$$= V_{m,i}^E - \sum_{n=1}^{N} \sum_{p,q} \sum_{j=0}^{i-1} \left(\frac{s^2}{4} \mu A_{mn}^{pq} + \frac{B_{mn}^{pq}}{\varepsilon} \right) u_{n,j} \, I_{ij}(sR_{mn}^{pq}/c)$$

$$- \sum_{n=1}^{N} \sum_{p,q} \sum_{j=0}^{i} s^2 \mu A_{mn}^{pq} \sum_{k=0}^{j-1} (j-k) u_{n,k} \, I_{ij}(sR_{mn}^{pq}/c) \qquad (6.85)$$

When Eq. (6.85), is rewritten, we obtain

$$\sum_{n=1}^{N} \alpha_{mn}^{E} u_{n,i} = V_{m,i}^{E} + P_{m,i}^{E} \qquad (6.86a)$$

where

$$\alpha_{mn}^{E} = \sum_{p,q} \left(\frac{s^2}{4} \mu A_{mn}^{pq} + \frac{B_{mn}^{pq}}{\varepsilon} \right) e^{\left(-sR_{mn}^{pq}/(2c) \right)} \qquad (6.86b)$$

$$P_{m,i}^{E} = -\sum_{n=1}^{N} \sum_{p,q} \left(\left(\frac{s^2}{4} \mu A_{mn}^{pq} + \frac{B_{mn}^{pq}}{\varepsilon} \right) \sum_{j=0}^{i-1} u_{n,j} \, I_{ij}\left(sR_{mn}^{pq}/c\right) \right.$$
$$\left. + s^2 \mu A_{mn}^{pq} \sum_{j=0}^{i} \sum_{k=0}^{j-1} (j-k) u_{n,k} \, I_{ij}\left(sR_{mn}^{pq}/c\right) \right) \qquad (6.86c)$$

In deriving Eq. (6.86), the property of the integral related to the temporal basis function given by Eq. (5.A15) is used. The term $I_{ii}\left(sR_{mn}^{pq}/c\right)$ can be represented by an exponential term because of $I_{ii}(y) = e^{-y/2}$ for $j = i$. Finally, Eq. (6.86) can be written into a matrix form as

$$\left[\alpha_{mn}^{E} \right] \left[u_{n,i} \right] = \left[\gamma_{m,i}^{E} \right] \qquad (6.87)$$

where $\gamma_{m,i}^{E} = V_{m,i}^{E} + P_{m,i}^{E}$ and $i = 0, 1, \ldots, \infty$. It is important to note that $[\alpha_{mn}^{E}]$ is not a function of the degree of the temporal testing function. After the temporal testing procedure, the unknown coefficients $u_n(t)$ can be obtained by solving Eq. (6.87) by increasing the degree of the temporal testing functions.

6.5.2   Solving the TD-MFIE by the MOD Method

For solving the TD-MFIE given by Eq. (6.11), the spatial testing procedure is first applied resulting in

$$\left\langle \mathbf{f}_m(\mathbf{r}), \frac{\mathbf{J}(\mathbf{r},t)}{2} \right\rangle - \left\langle \mathbf{f}_m(\mathbf{r}), \mathbf{n} \times \frac{1}{4\pi} \int_{S_0} \nabla \times \frac{\mathbf{J}(\mathbf{r}',\tau)}{R} dS' \right\rangle = \left\langle \mathbf{f}_m(\mathbf{r}), \mathbf{n} \times \mathbf{H}^i(\mathbf{r},t) \right\rangle \qquad (6.88)$$

Substituting the electric current $\mathbf{J}$ and the Hertz vector $\mathbf{u}(\mathbf{r},t)$ given by Eqs. (6.69) and (6.71) into Eq. (6.88), the first term of Eq. (6.88) becomes

$$\left\langle \mathbf{f}_m(\mathbf{r}), \frac{\mathbf{J}(\mathbf{r},t)}{2} \right\rangle = \sum_{n=1}^{N} C_{mn} \frac{d}{dt} u_n(t) \qquad (6.89a)$$

where

$$C_{mn} = \sum_{p,q} C_{mn}^{pq} = C_{mn}^{++} + C_{mn}^{+-} + C_{mn}^{-+} + C_{mn}^{--} \tag{6.89b}$$

$$C_{mn}^{pq} = \frac{1}{2} \int_S \mathbf{f}_m^p(\mathbf{r}) \cdot \mathbf{f}_n^q(\mathbf{r}) \, dS \tag{6.89c}$$

In the second term of Eq. (6.88), the curl operator inside the integral can be expressed as [24]

$$\nabla \times \frac{\mathbf{J}(\mathbf{r}',\tau)}{R} = \frac{1}{c} \frac{\partial}{\partial t} \mathbf{J}(\mathbf{r}',\tau) \times \frac{\hat{\mathbf{R}}}{R} + \mathbf{J}(\mathbf{r}',\tau) \times \frac{\hat{\mathbf{R}}}{R^2} \tag{6.90}$$

where $\hat{\mathbf{R}}$ is a unit vector along the direction $\mathbf{r} - \mathbf{r}'$. Therefore,

$$\left\langle \mathbf{f}_m(\mathbf{r}), \mathbf{n} \times \frac{1}{4\pi} \int_{S_0} \nabla \times \frac{\mathbf{J}(\mathbf{r}',\tau)}{R} \, dS' \right\rangle$$

$$= \sum_{n=1}^{N} \frac{1}{4\pi} \int_S \mathbf{f}_m(\mathbf{r}) \cdot \mathbf{n} \times \int_{S_0} \left(\frac{1}{c} \frac{d^2}{dt^2} u_n(\tau) \mathbf{f}_n(\mathbf{r}') \times \frac{\hat{\mathbf{R}}}{R} \right.$$

$$\left. + \frac{d}{dt} u_n(\tau) \mathbf{f}_n(\mathbf{r}') \times \frac{\hat{\mathbf{R}}}{R^2} \right) dS' dS \tag{6.91}$$

Assume that the retarded time factor can be written as in Eq. (6.78) to obtain

$$\left\langle \mathbf{f}_m(\mathbf{r}), \mathbf{n} \times \frac{1}{4\pi} \int_{S_0} \nabla \times \frac{\mathbf{J}(\mathbf{r}',\tau)}{R} \, dS' \right\rangle$$

$$= \sum_{n=1}^{N} \sum_{p,q} \left(\frac{\zeta_1^{pq}}{c} \frac{d^2}{dt^2} u_n(\tau_{mn}^{pq}) + \zeta_2^{pq} \frac{d}{dt} u_n(\tau_{mn}^{pq}) \right) \tag{6.92a}$$

where

$$\zeta_v^{pq} = \frac{1}{4\pi} \int_S \mathbf{f}_m^p(\mathbf{r}) \cdot \mathbf{n} \times \int_S \mathbf{f}_n^q(\mathbf{r}') \times \frac{\hat{\mathbf{R}}}{R^v} \, dS' dS, \quad v = 1, 2 \tag{6.92b}$$

Substituting Eqs. (6.89) and (6.92) into Eq. (6.88), one obtains for the spatial testing as

$$\sum_{n=1}^{N} \left[C_{mn} \frac{d}{dt} u_n(t) - \sum_{p,q} \left(\frac{\zeta_1^{pq}}{c} \frac{d^2}{dt^2} u_n(\tau_{mn}^{pq}) + \zeta_2^{pq} \frac{d}{dt} u_n(\tau_{mn}^{pq}) \right) \right] = V_m^H(t) \tag{6.93a}$$

where

$$C_{mn}^{pq} = \frac{1}{2} \int_S \mathbf{f}_m^p(\mathbf{r}) \cdot \mathbf{f}_n^q(\mathbf{r}) \, dS \tag{6.93b}$$

$$\zeta_v^{pq} = \frac{1}{4\pi} \int_S \mathbf{f}_m^p(\mathbf{r}) \cdot \mathbf{n} \times \int_S \mathbf{f}_n^q(\mathbf{r}') \times \frac{\hat{\mathbf{R}}}{R^v} dS' \, dS \,, \quad v = 1, 2 \tag{6.93c}$$

$$V_m^H(t) = \int_S \mathbf{f}_m(\mathbf{r}) \cdot \mathbf{n} \times \mathbf{H}^i(\mathbf{r}, t) \, dS \tag{6.93d}$$

The integral in Eq. (6.93b) can be computed analytically, and the result is given in [22]. The integrals in Eqs. (6.93c) and (6.93d) may be evaluated numerically using the Gaussian quadrature for the unprimed and primed coordinates. Then, the temporal expansion and testing procedures for the TD-MFIE are applied in which the basis function $\phi_j(t)$ defined in Eq. (6.72) and the expansion of the coefficient $u_n(t)$ given by Eq. (6.73) are used. By substituting the first and second derivatives of $u_n(t)$ given by Eqs. (6.74) and (6.75) into Eq. (6.93) and performing a temporal testing with $\phi_i(st)$, one obtains

$$\sum_{n=1}^{N} \left[s \, C_{mn} \sum_{j=0}^{\infty} \left(\frac{1}{2} u_{n,j} + \sum_{k=0}^{j-1} u_{n,k} \right) \delta_{ij} \right.$$

$$-\sum_{p,q} \left(s^2 \frac{\zeta_1^{pq}}{c} \sum_{j=0}^{\infty} \left(\frac{1}{4} u_{n,j} + \sum_{k=0}^{j-1} (j-k) u_{n,k} \right) I_{ij}(s \, R_{mn}^{pq}/c) \right.$$

$$\left. \left. + s \, \zeta_2^{pq} \sum_{j=0}^{\infty} \left(\frac{u_{n,j}}{2} + \sum_{k=0}^{j-1} u_{n,k} \right) I_{ij}(s R_{mn}^{pq}/c) \right) \right] = V_{m,i}^H \tag{6.94}$$

where I_{ij} is given by Eq. (6.84b), and

$$V_{m,i}^H = \int_0^{\infty} \phi_i(st) V_m^H(t) \, d(st) \tag{6.95}$$

Because $I_{ij} = 0$ when $j > i$ from Eq. (5.A15), the upper limit in the summation of Eq. (6.94) can be written as i instead of ∞. In this expression, moving those terms including $u_{n,j}$ for $j < i$, which are known, to the right-hand side of the equation, one obtains

$$\sum_{n=1}^{N} \left[\frac{s}{2} C_{mn} u_{n,i} - \sum_{p,q} \left(\frac{s^2}{4} \frac{\zeta_1^{pq}}{c} + \frac{s}{2} \zeta_2^{pq} \right) u_{n,i} I_{ii} \left(s R_{mn}^{pq}/c \right) \right] = V_{m,i}$$

$$-\sum_{n=1}^{N} \left[s \, C_{mn} \sum_{k=0}^{i-1} u_{n,k} - \sum_{p,q} \left(\sum_{j=0}^{i} \sum_{k=0}^{j-1} \left(s^2 \frac{\zeta_1^{pq}}{c} (j-k) + s \zeta_2^{pq} \right) u_{n,k} \, I_{ij} \left(s R_{mn}^{pq}/c \right) \right) \right]$$

$$+\sum_{j=0}^{i-1}\left[\frac{s^2}{4}\frac{\varsigma_1^{pq}}{c}+\frac{s}{2}\varsigma_2^{pq}\right]u_{n,j}I_{ij}\left(sR_{mn}^{pq}/c\right)\right)\right] \tag{6.96}$$

Rewrite Eq. (6.96) to yield

$$\sum_{n=1}^{N}\alpha_{mn}^{H}u_{n,i}=V_{m,i}^{H}+P_{m,i}^{H} \tag{6.97a}$$

where

$$\alpha_{mn}^{H}=\frac{s}{2}C_{mn}-\sum_{p,q}\left(\frac{s^2}{4}\frac{\varsigma_1^{pq}}{c}+\frac{s}{2}\varsigma_2^{pq}\right)e^{\left(-sR_{mn}^{pq}/(2c)\right)} \tag{6.97b}$$

$$P_{m,i}^{H}=-\sum_{n=1}^{N}\left[sC_{mn}\sum_{k=0}^{i-1}u_{n,k}-\sum_{p,q}\left(\left(\frac{s^2}{4}\frac{\varsigma_1^{pq}}{c}+\frac{s}{2}\varsigma_2^{pq}\right)\sum_{j=0}^{i-1}u_{n,j}I_{ij}\left(sR_{mn}^{pq}/c\right)\right.\right.$$

$$\left.\left.+s^2\frac{\varsigma_1^{pq}}{c}\sum_{j=0}^{i}\sum_{k=0}^{j-1}(j-k)u_{n,k}I_{ij}\left(sR_{mn}^{pq}/c\right)+s\varsigma_2^{pq}\sum_{j=0}^{i}\sum_{k=0}^{j-1}u_{n,k}I_{ij}\left(sR_{mn}^{pq}/c\right)\right)\right] \tag{6.97c}$$

Equation (6.97) can be written in a matrix form as

$$\left[\alpha_{mn}^{H}\right]\left[u_{n,i}\right]=\left[\gamma_{m,i}^{H}\right] \tag{6.98}$$

where $\gamma_{m,i}^{H}=V_{m,i}^{H}+P_{m,i}^{H}$ and $i=0,1,\ldots,\infty$

6.5.3 Solving the TD-CFIE by the MOD Method

A matrix equation for the TD-CFIE based on the MOD method can be obtained directly from the linear combination of Eqs. (6.87) and (6.98) as

$$\left[\alpha_{mn}\right]\left[u_{n,i}\right]=\left[\gamma_{m,i}\right] \tag{6.99a}$$

where

$$\alpha_{mn}=\kappa\alpha_{mn}^{E}+\eta(1-\kappa)\alpha_{mn}^{H} \tag{6.99b}$$

$$\gamma_{m,i}=\kappa\gamma_{m,i}^{E}+\eta(1-\kappa)\gamma_{m,i}^{H} \tag{6.99c}$$

where $i=0,1,\ldots,\infty$. In computing Eq. (6.99), the minimum number of temporal basis functions M needs to be known. This parameter is dependent on the time duration of the transient response and the bandwidth of the excitation signal. The excitation waveform is considered to have a bandwidth B in the frequency domain and is of time duration T_f. When this signal is represented by a Fourier series, the range of the sampling frequency is limited by $-B \leq k\Delta f \leq B$, where k is an integer and $\Delta f = 1/T_f$. The result is $|k| \leq B/T_f$. Hence, the minimum number of temporal basis functions required is $M = 2BT_f + 1$. It is noted that the upper limit of the integral $V_{m,i}^E$ in Eq. (6.84c) and $V_{m,i}^H$ in Eq. (6.95) can be replaced by the time duration T_f instead of infinity.

6.5.4   Expressions for the Current and the Far Field

By solving the matrix equation in Eq. (6.99), using the MOD procedure using M temporal basis functions, the transient coefficient $u_n(t)$ can be solved for. Using Eqs. (6.69) and (6.71), the electric current $\mathbf{J}(\mathbf{r},t)$ can be expressed as

$$\mathbf{J}(\mathbf{r},t) = \sum_{n=1}^{N} \frac{\partial}{\partial t} u_n(t) \, \mathbf{f}_n(\mathbf{r}) \tag{6.100}$$

By substituting the derivative term by Eq. (6.75), Eq. (6.100) can be represented as

$$\mathbf{J}(\mathbf{r},t) = \sum_{n=1}^{N} J_n(t) \, \mathbf{f}_n(\mathbf{r})$$

where

$$J_n(t) = \frac{d}{dt} u_n(t) = s \sum_{j=0}^{M-1} \left(\frac{1}{2} u_{n,j} + \sum_{k=0}^{j-1} u_{n,k} \right) \phi_j(st) \tag{6.101}$$

The unknown current coefficients $J_n(t)$ can be obtained once $u_n(t)$ is solved using Eq. (6.99). The far fields can then be expressed by neglecting the scalar potential term in Eq. (6.2) as

$$\mathbf{E}^s(\mathbf{r},t) \approx -\frac{\partial}{\partial t} \mathbf{A}(\mathbf{r},t) \tag{6.102}$$

Substituting the magnetic vector potential $\mathbf{A}(\mathbf{r},t)$ given by Eq. (6.3), the electric current $\mathbf{J}(\mathbf{r},t)$ given by Eq. (6.101) and the RWG vector functions $\mathbf{f}_n(\mathbf{r})$ characterized by Eq. (6.13a), into Eq. (6.102) yields

$$\mathbf{E}^s(\mathbf{r},t) \approx -\frac{\mu}{4\pi} \sum_{n=1}^{N} \sum_{q} \int_S \frac{d^2}{dt^2} u_n(\tau) \frac{\mathbf{f}_n^q(\mathbf{r}')}{R} \, dS' \tag{6.103}$$

The following approximation is now made in the far field:

$R \approx r - \mathbf{r}' \cdot \hat{\mathbf{r}}$ for the time retardation term $t - R/c$

$R \approx r$ for the amplitude term $1/R$

where $\hat{\mathbf{r}} = \mathbf{r}/r$ is a unit vector along the direction of the radiation. The integral in Eq. (6.103) is evaluated by approximating the integrand by the value at the centroid of the source triangle T_n^q. Substituting Eq. (6.13b) for $\mathbf{f}_n^q(\mathbf{r}')$ into Eq. (6.103) and approximating $\mathbf{r}' \approx \mathbf{r}_n^{cq}$ and $\boldsymbol{\rho}_n^q \approx \boldsymbol{\rho}_n^{cq}$, the far field can be obtained as

$$\mathbf{E}^s(\mathbf{r},t) \approx -\frac{\mu}{8\pi r} \sum_{n=1}^N l_n \sum_q \boldsymbol{\rho}_n^{cq} \frac{d^2}{dt^2} u_n(\tau_n^q) \tag{6.104a}$$

where

$$\frac{d^2}{dt^2} u_n(\tau_n^q) = s^2 \sum_{j=0}^{M-1} \left[\frac{1}{4} u_{n,j} + \sum_{k=0}^{j-1} (j-k) u_{n,k} \right] \phi_j(s\tau_n^q) \tag{6.104b}$$

$$\tau_n^q \approx t - (r - \mathbf{r}_n^{cq} \cdot \hat{\mathbf{r}})/c \tag{6.104c}$$

6.6   ALTERNATIVE FORMULATION FOR THE MOD METHOD

In this section, an alternative approach is presented, which directly deals with the unknown electric current vector $\mathbf{J}(\mathbf{r},t)$, instead of using any additional intermediate vector like the Hertz vector. The current vector is expanded by the RWG basis function as

$$\mathbf{J}(\mathbf{r},t) = \sum_{n=1}^N J_n(t) \mathbf{f}_n(\mathbf{r}) \tag{6.105}$$

Using the temporal basis function $\phi_j(t)$ defined by Eq. (6.72), the transient coefficient $J_n(t)$ in Eq. (6.105) can be written as

$$J_n(t) = \sum_{j=0}^{\infty} J_{n,j} \phi_j(st) \tag{6.106}$$

where s is a scaling factor and is used to increase or decrease the temporal expansion. The first and second derivatives of the transient coefficient $J_n(t)$ are given by

$$\frac{d}{dt} J_n(t) = s \sum_{j=0}^{\infty} \left[\frac{1}{2} J_{n,j} + \sum_{k=0}^{j-1} J_{n,k} \right] \phi_j(st) \tag{6.107}$$

$$\frac{d^2}{dt^2} J_n(t) = s^2 \sum_{j=0}^{\infty} \left(\frac{1}{4} J_{n,j} + \sum_{k=0}^{j-1} (j-k) J_{n,k} \right) \phi_j(st) \tag{6.108}$$

6.6.1 Alternate Solution of the TD-EFIE Using the MOD Method

Differentiating the electric field integral equation given by Eq. (6.7), yields

$$\left(\frac{\partial^2}{\partial t^2} \mathbf{A}(\mathbf{r},t) + \nabla \frac{\partial}{\partial t} \Phi(\mathbf{r},t) \right)_{tan} = \left(\frac{\partial}{\partial t} \mathbf{E}^i(\mathbf{r},t) \right)_{tan} \tag{6.109}$$

where $\mathbf{r} \in S$. Using the same spatial expansion function $\mathbf{f}_m(\mathbf{r})$ defined in Section 6.3, the result of the spatial testing from Eq. (6.109) is given as

$$\left\langle \mathbf{f}_m(\mathbf{r}), \frac{\partial^2}{\partial t^2} \mathbf{A}(\mathbf{r},t) \right\rangle + \left\langle \mathbf{f}_m(\mathbf{r}), \nabla \frac{\partial}{\partial t} \Phi(\mathbf{r},t) \right\rangle = \left\langle \mathbf{f}_m(\mathbf{r}), \frac{\partial}{\partial t} \mathbf{E}^i(\mathbf{r},t) \right\rangle \tag{6.110}$$

Substitution of Eqs. (6.7b) and (6.7c) into Eq. (6.110), results in an equation with an unknown current vector $\mathbf{J}(\mathbf{r},t)$. With the assumption for the retarded time given by Eq. (6.78), one can obtain

$$\sum_{n=1}^{N} \sum_{p,q} \left(\mu A_{mn}^{pq} \frac{d^2}{dt^2} J_n(\tau_{mn}^{pq}) + \frac{B_{mn}^{pq}}{\varepsilon} J_n(\tau_{mn}^{pq}) \right) = V_m^E(t) \tag{6.111a}$$

where

$$A_{mn}^{pq} = \frac{1}{4\pi} \int_S \mathbf{f}_m^p(\mathbf{r}) \cdot \int_S \frac{\mathbf{f}_n^q(\mathbf{r}')}{R} \, dS' \, dS \tag{6.111b}$$

$$B_{mn}^{pq} = \frac{1}{4\pi} \int_S \nabla \cdot \mathbf{f}_m^p(\mathbf{r}) \int_S \frac{\nabla' \cdot \mathbf{f}_n^q(\mathbf{r}')}{R} \, dS' \, dS \tag{6.111c}$$

$$V_m^E(t) = \int_S \mathbf{f}_m(\mathbf{r}) \cdot \frac{\partial \mathbf{E}^i(\mathbf{r},t)}{\partial t} \, dS \tag{6.111d}$$

It is observed that Eq. (6.111) is of the same form as Eq. (6.83), particularly the coefficients A_{mn}^{pq} and B_{mn}^{pq} is exactly the same as those expressed in Eqs. (6.83b) and (6.83c). However, because this EFIE is differentiated at the first step in this alternative method, the term $V_m^E(t)$ given by Eq. (6.111d) is different from that in Eq. (6.83d). The difference is that the presence of the derivative term $\partial \mathbf{E}^i(\mathbf{r},t)/\partial t$ in one instead of $\mathbf{E}^i(\mathbf{r},t)$.

As the spatial testing results of this alternative method is of the same form of as that of the MOD method described in the previous section, the temporal testing carried out using $\phi_j(t)$ can directly be obtained by employing the results given in Eq. (6.86).

$$\sum_{n=1}^{N} \alpha_{mn}^{E} J_{n,i} = V_{m,i}^{E} + P_{m,i}^{E} \tag{6.112a}$$

where

$$\alpha_{mn}^{E} = \sum_{p,q} \left(\frac{s^2}{4} \mu A_{mn}^{pq} + \frac{B_{mn}^{pq}}{\varepsilon} \right) e^{\left(-s R_{mn}^{pq}/(2c)\right)} \tag{6.112b}$$

$$P_{m,i}^{E} = -\sum_{n=1}^{N} \sum_{p,q} \left[\left(\frac{s^2}{4} \mu A_{mn}^{pq} + \frac{B_{mn}^{pq}}{\varepsilon} \right) \sum_{j=0}^{i-1} J_{n,j} I_{ij} \left(s R_{mn}^{pq}/c \right) \right.$$

$$\left. + s^2 \mu A_{mn}^{pq} \sum_{j=0}^{i} \sum_{k=0}^{j-1} (j-k) J_{n,k} I_{ij} \left(s R_{mn}^{pq}/c \right) \right] \tag{6.112c}$$

Finally, Eq. (6.112) can be written in a matrix form as

$$\left[\alpha_{mn}^{E} \right] \left[J_{n,i} \right] = \left[\gamma_{m,i}^{E} \right] \tag{6.113}$$

where $\gamma_{m,i}^{E} = V_{m,i}^{E} + P_{m,i}^{E}$ and $i = 0, 1, \ldots, \infty$

6.6.2   Alternate Solution of the TD-MFIE Using the MOD Method

The TD-MFIE given by Eq. (6.11) can be solved in an alternate fashion by setting up a matrix equation for the current vector directly. First, the spatial testing of TD-MFIE given in Eq. (6.88) is considered as

$$\left\langle \mathbf{f}_m(\mathbf{r}), \frac{\mathbf{J}(\mathbf{r},t)}{2} \right\rangle - \left\langle \mathbf{f}_m(\mathbf{r}), \mathbf{n} \times \frac{1}{4\pi} \int_{S_0} \nabla \times \frac{\mathbf{J}(\mathbf{r}',\tau)}{R} dS' \right\rangle = \left\langle \mathbf{f}_m(\mathbf{r}), \mathbf{n} \times \mathbf{H}^i(\mathbf{r},t) \right\rangle \tag{6.114}$$

When the electric current is used directly, the result of the spatial testing can be obtained directly from Eq. (6.93) by replacing $\partial u_n(t)/\partial t$ with $J_n(t)$ and $\partial^2 u_n(t)/\partial t^2$ with $\partial J_n(t)/\partial t$. Thus, the first term can be expressed as

$$\sum_{n=1}^{N} \left[C_{mn} J_n(t) - \sum_{p,q} \left(\frac{\zeta_1^{pq}}{c} \frac{d}{dt} J_n(\tau_{mn}^{pq}) + \zeta_2^{pq} J_n(\tau_{mn}^{pq}) \right) \right] = V_m^H(t) \tag{6.115a}$$

where

$$C_{mn}^{pq} = \frac{1}{2} \int_S \mathbf{f}_m^p(\mathbf{r}) \cdot \mathbf{f}_n^q(\mathbf{r}) \, dS \tag{6.115b}$$

$$\zeta_v^{pq} = \frac{1}{4\pi} \int_S \mathbf{f}_m^p(\mathbf{r}) \cdot \mathbf{n} \times \int_S \mathbf{f}_n^q(\mathbf{r}') \times \frac{\hat{\mathbf{R}}}{R^v} \, dS' \, dS \,, \quad v = 1, 2 \tag{6.115c}$$

$$V_m^H(t) = \int_S \mathbf{f}_m(\mathbf{r}) \cdot \mathbf{n} \times \mathbf{H}^i(\mathbf{r}, t) \, dS \tag{6.115d}$$

Substituting Eqs. (6.106) and (6.107) into Eq. (6.115) and performing the temporal testing with $\phi_i(st)$, one obtains

$$\sum_{n=1}^N \left[C_{mn} \sum_{j=0}^{\infty} J_{n,j} \, \delta_{ij} - \sum_{p,q} \sum_{j=0}^{\infty} \left(\left[\frac{s}{2} \frac{\zeta_1^{pq}}{c} + \zeta_2^{pq} \right] J_{n,j} \right. \right.$$

$$\left. \left. + s \frac{\zeta_1^{pq}}{c} \sum_{k=0}^{j-1} J_{n,k} \right) I_{ij} \left(sR_{mn}^{pq} / c \right) \right] = V_{m,i}^H \tag{6.116}$$

Because $I_{ij} = 0$ when $j > i$, the upper limit for the summation can be replaced by i instead of ∞ in Eq. (6.116). In this result, moving the terms including $J_{n,j}$, which is known for $j < i$, to the right-hand side, one can obtain

$$\sum_{n=1}^N \left[C_{mn} - \sum_{p,q} \left[\frac{s}{2} \frac{\zeta_1^{pq}}{c} + \zeta_2^{pq} \right] I_{ii} \left(sR_{mn}^{pq} / c \right) \right] J_{n,i} = V_{m,i}^H$$

$$+ \sum_{n=1}^N \sum_{p,q} \left(\frac{s}{2} \frac{\zeta_1^{pq}}{c} + \zeta_2^{pq} \right) \sum_{j=0}^{i-1} J_{n,j} I_{ij} \left(sR_{mn}^{pq} / c \right) + \sum_{n=1}^N \sum_{p,q} s \frac{\zeta_1^{pq}}{c} \sum_{j=0}^{i} \sum_{k=0}^{j-1} J_{n,k} I_{ij} \left(sR_{mn}^{pq} / c \right)$$

$$\tag{6.117}$$

Rewriting Eq. (6.117) as

$$\sum_{n=1}^N \alpha_{mn}^H J_{n,i} = V_{m,i}^H + P_{m,i}^H \tag{6.118a}$$

where

$$\alpha_{mn}^H = C_{mn} - \sum_{p,q} \left[\frac{s}{2} \frac{\zeta_1^{pq}}{c} + \zeta_2^{pq} \right] e^{\left(-sR_{mn}^{pq} / (2c) \right)} \tag{6.118b}$$

$$P_{m,i}^H = \sum_{n=1}^N \sum_{p,q} \left(\left[\frac{s}{2} \frac{\zeta_1^{pq}}{c} + \zeta_2^{pq} \right] \sum_{j=0}^{i-1} J_{n,j} \, I_{ij} \left(sR_{mn}^{pq} / c \right) \right)$$

$$+s\frac{\zeta_i^{pq}}{c}\sum_{j=0}^{i}\sum_{k=0}^{j-1}J_{n,k}I_{ij}\left(sR_{mn}^{pq}/c\right)\Bigg] \qquad (6.118c)$$

Furthermore, Eq. (6.118) can be written into a matrix form as

$$\left[\alpha_{mn}^{H}\right]\left[J_{n,i}\right]=\left[\gamma_{m,i}^{H}\right] \qquad (6.119)$$

where $\gamma_{m,i}^{H}=V_{m,i}^{H}+P_{m,i}^{H}$ and $i=0,1,\ldots,\infty$

6.6.3   Alternate Solution of the TD-CFIE Using the MOD Method

The matrix equation of the alternative TD-CFIE is obtained by combining Eqs. (6.113) and (6.119), which results in

$$\left[\alpha_{mn}\right]\left[J_{n,i}\right]=\left[\gamma_{m,i}\right] \qquad (6.120a)$$

where

$$\alpha_{mn}=\kappa\alpha_{mn}^{E}+\eta(1-\kappa)\alpha_{mn}^{H} \qquad (6.120b)$$

$$\gamma_{m,i}=\kappa\gamma_{m,i}^{E}+\eta(1-\kappa)\gamma_{m,i}^{H} \qquad (6.120c)$$

where the expressions of α_{mn} and $\gamma_{m,i}$ are of the same form as Eqs. (6.99b) and (6.99c).

6.6.4   Expressions for the Current and the Far Field

By solving Eq. (6.120) by a MOD algorithm with M temporal basis functions, the transient current coefficient $J_n(t)$ can be obtained approximately from Eq. (6.106) as

$$J_n(t)=\sum_{j=0}^{M-1}J_{n,j}\,\phi_j(st). \qquad (6.121)$$

The far field expression is obtained using a similar procedure described in Section 6.5. Neglect the scalar potential term in Eq. (6.2) and obtain the far field as

$$\mathbf{E}^s(\mathbf{r},t)\approx-\frac{\partial}{\partial t}\mathbf{A}(\mathbf{r},t) \qquad (6.122)$$

Substituting the magnetic vector potential $\mathbf{A}(\mathbf{r},t)$ given by Eqs. (6.3), the electric current $\mathbf{J}(\mathbf{r},t)$ given by Eq. (6.105) and the RWG vector function $\mathbf{f}_n(\mathbf{r})$ given by Eq. (6.13a) into Eq. (6.122) yields

$$\mathbf{E}^s(\mathbf{r},t) \approx -\frac{\mu}{8\pi r}\sum_{n=1}^{N} l_n \sum_{q} \mathbf{\rho}_n^{cq}\frac{d}{dt}J_n(\tau_n^q) \tag{6.123a}$$

where

$$\frac{d}{dt}J_n(\tau_n^q) = s\sum_{j=0}^{M-1}\left(\frac{1}{2}J_{n,j}+\sum_{k=0}^{j-1}J_{n,k}\right)\phi_j(s\tau_n^q) \tag{6.123b}$$

6.7   NUMERICAL EXAMPLES

6.7.1   Scattering Analysis by Solving the TD-EFIE Using the MOT Method

In this section, we present some numerical results for transient scattering from objects using the TD-EFIE solved by the MOT method. Specifically, the problems chosen are for the solution of the currents induced on a flat plate, cube, and sphere using backward ($v=0$) and central difference ($v=1/2$) schemes. We do not show the result for the forward difference formula ($v=1$) because of the instability in the results from its early time oscillations.

As a first example, consider a flat, 2 m $\times$ 2 m square plate, located in the xy plane as shown in Figure 6.1. Eight subsections are made along the x direction and seven along the y direction, resulting in 112 triangular patches with 153 common edges. The value for R_{min} is 12.65 cm, where R_{min} represents the minimum distance between any two patches. The plate is illuminated by a sinusoidal plane wave given by

$$\mathbf{E}^i(\mathbf{r},t) = \mathbf{E}_0 \sin \omega(t - \frac{\mathbf{r}\cdot\hat{\mathbf{k}}}{c}) \tag{6.124}$$

with $\mathbf{E}_0 = \hat{\mathbf{x}}$, $\omega = 2\pi f$, and $f = 50$ MHz. $\hat{\mathbf{k}}$ is the unit propagation vector. The field is incident from $\phi = 0°$ and $\theta = 0°$. The time step is chosen so that $c\Delta t = 2R_{min}$ to generate an implicit solution. Figure 6.2 plots the transient response for an x-directed current located at the center of the conducting plate of Figure 6.1 illuminated by a sinusoidal wave. The problem is solved using the backward and central difference scheme until 100 lm (light meter). We can observe that the transient current obtained by EFIE using the central difference is very stable. But the current computed by the backward scheme develops spurious oscillations in the solution after only one period. When steady state is reached in the time domain, the amplitude of the current should correspond to the phasor value obtained in a frequency domain solution when the same plate is excited in the frequency domain by a plane wave of amplitude 1 V/m excited with a frequency of 50 MHz. From Figure 6.2 we can know that the amplitude of the fourth positive peak is 9.365 mA. When a plane wave excitation is chosen for the solution of the current on the plate using a frequency domain EFIE solution procedure, one obtains a value of 9.374 mA/m at the same edge. It therefore validates the TD-EFIE formulation using the central difference along with the frequency domain EFIE.

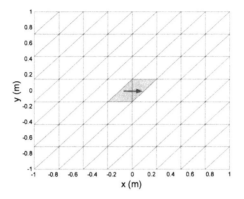

Figure 6.1. Triangle patching of a conducting plate (2 m × 2 m) resulting in 112 patches and 153 common edges.

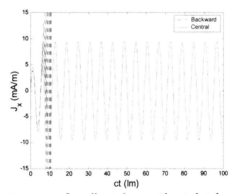

Figure 6.2. Transient response of a x-directed current located at the center of a conducting plate (Figure 6.1) illuminated by a sinusoidal wave with a frequency of 50 MHz.

Next, we present numerical results for the currents induced on conducting objects when illuminated by a Gaussian plane wave of the form

$$\mathbf{E}^i(\mathbf{r},t) = \mathbf{E}_0 \frac{4}{\sqrt{\pi T}} e^{-\gamma^2} \quad \text{and} \quad \gamma = \frac{4}{T}\left(ct - ct_0 - \mathbf{r} \cdot \hat{\mathbf{k}}\right) \tag{6.127}$$

with $\mathbf{E}_0 = \hat{\mathbf{x}}$, $T = 4$ lm, and $ct_0 = 6$ lm. The field is incident from $\phi = 0°$ and $\theta = 0°$. For comparison, we present the results obtained by the inverse discrete Fourier transform (IDFT) of the frequency domain solution. We first consider the same plate of Figure 6.1. Figure 6.3 shows the transient response from an x-directed current induced at the center of the plate as a function of time obtained by using double-and single-precision computations. It is evident from the figures that the result from the TD-EFIE with central difference is accurate and stable. Note the absence of any late time instability in the result even for single precision. However, for the backward difference scheme, use of double precision only delays the start on time for the spurious oscillations.

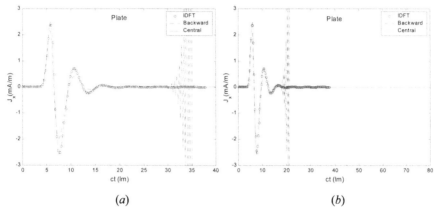

(a)   (b)

Figure 6.3. Transient response for a x-directed current located at the center of a conducting plate (Figure 6.1). The results have been computed using (a) double-precision and (b) single-precision arithmetic.

The second structure chosen is a conducting cube as shown in Figure 6.4. It is 1 m on a side and is centered at the origin. There are four, five, and four subdivisions of the structure along the x, y, and z directions, respectively. This represents a total of 224 patches and 336 common edges. The time step has been chosen as $c\Delta t = 2R_{min}$, where $R_{min} = 10.67$ cm. The x-directed current at the center of the top face and the z-directed current at the center of the side face of the cube are seen Figures 6.5 and 6.6, respectively. Here the agreement with the frequency domain data is very good. Even though we are using single precision, the transient response for the current is accurate.

As a third geometry, Figure 6.7 shows a conducting sphere of radius 0.5 m centered at the origin. There are eight subdivisions along the θ directions and 12 subdivisions along the ϕ direction. This results in a total of 168 patches and 152 common edges. The time step is chosen as $c\Delta t = 4R_{min}$, where $R_{min} = 6.37$ cm. Figure. 6.8 plots the θ and ϕ components of the currents as indicated by the arrows in Figure 6.7. The agreement between the results from EFIE with central difference using single precision computation and IDFT is very good, but the result from EFIE with backward difference has late time oscillations, irrespective of whether single or double precision arithmetic is used in the computation.

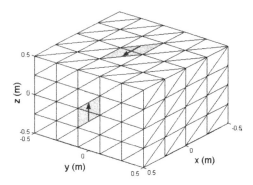

Figure 6.4. Triangle patching scheme for a conducting cube (1 m × 1 m × 1m) with 224 patches having 336 common edges.

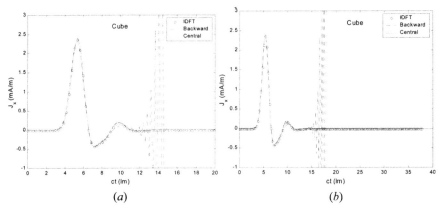

Figure 6.5. Transient response for a x-directed current at the center (0,0,0.5) of the top face of a conducting cube (Figure 6.4). The results have been computed using (a) single precision and (b) double precision.

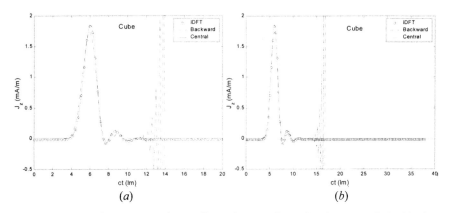

Figure 6.6. Transient response for a z-directed current located at the center (0.5,0,0) of a side face of a conducting cube (Figure 6.4). The results have been computed using (a) single precision and (b) double precision.

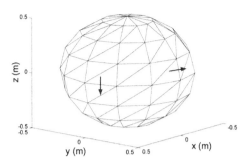

Figure 6.7. Triangle patching of a conducting sphere with a radius 0.5 m resulting in 168 patches and 252 common edges.

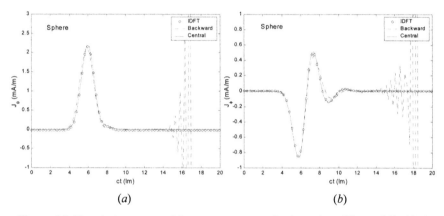

$$(a) \qquad\qquad\qquad (b)$$

Figure 6.8. Transient response of the current on a conducting sphere (Figure 6.7). (a) θ-component of current at $\theta = 90°$, $\phi = 15°$. (b) ϕ-component of current at $\theta = 78.75°$, $\phi = 90°$.

6.7.2 Scattering Analysis by Solving the TD-MFIE Using the MOT Method

In this section, we present the numerical results for three-dimensional closed scatterers, viz. a cube, a sphere, and a cylinder. The scatterers are illuminated by a Gaussian plane wave in which the magnetic field is given by

$$\mathbf{H}^i(\mathbf{r},t) = \frac{1}{\eta}\hat{\mathbf{k}} \times \mathbf{E}_0 \frac{4}{\sqrt{\pi T}} e^{-\gamma^2} \quad \text{and} \quad \gamma = \frac{4}{T}\left(ct - ct_0 - \mathbf{r}\cdot\hat{\mathbf{k}}\right) \qquad (6.128)$$

with $\mathbf{E}_0 = \hat{\mathbf{x}}$, $T = 4$ lm, and $ct_0 = 6$ lm. $\hat{\mathbf{k}}$ is the unit propagation vector. The field is incident from $\phi = 0°$ and $\theta = 0°$. For comparison, we present the results obtained by TD-EFIE and the IDFT of the frequency domain solution calculated in the 0–500 MHz interval using 128 samples. There are several methods to deal with a magnetic vector potential term in the EFIE formulation [1–4]. We use EFIE with a central finite difference scheme, which gives stable results when using an implicit solution. All results in this section have been obtained using single-precision computation.

As a first example, consider a conducting cube, 1 m on a side, centered about the origin as shown in Figure 6.4. The cube has a total of 224 patches and 336 common edges, and $R_{min} = 10.67$ cm, where R_{min} represents the minimum distance between any two patches.

The x-directed current at the center of the top face is indicated by an arrow in Figure 6.4. Figure 6.9(a) plots the transient response using an explicit solution as the time step is chosen such that $c\Delta t = 0.5R_{min}$. Here, the agreement in the result between the TD-MFIE and the IDFT of the frequency domain solution calculated using EFIE and MFIE in the frequency domain are very good. It is important to note that both the explicit solutions of the TD-EFIE and TD-MFIE display instability. However, the TD-MFIE breaks down in later time than for the TD-EFIE. Figure 6.9(b) and Figure 6.10(a) show the transient response obtained by choosing $c\Delta t = R_{min}$ and $c\Delta t = 2R_{min}$, which correspond to an implicit solution. With the increase of the time step, the results of TD-EFIE become more stable, but still display late-time oscillations. We can see that the solutions of TD-MFIE are very stable and agree with IDFT well. The results for TD-MFIE in Figure 6.10 are stable until 80 lm. However, it is important to point out that as we increase ct, the TD-MFIE eventually becomes unstable.

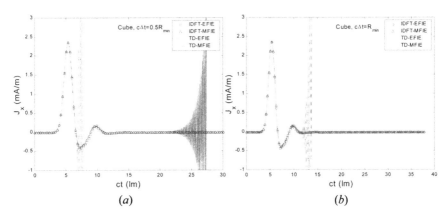

Figure 6.9. Transient response for a x-directed current located at the center of the top face on a conducting cube: (a) $c\Delta t = 0.5R_{min}$. (b) $c\Delta t = R_{min}$.

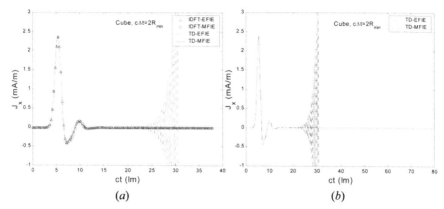

Figure 6.10. Transient response for a x-directed current located at the center of the top face on a conducting cube with $c\Delta t = 2R_{min}$. (a) Simulation upto 40 lm. (b) Simulation upto 80 lm.

As a second example, we consider a conducting sphere of radius 0.5 m centered at the origin as shown in Figure 6.7. There are 10 divisions in θ and ϕ directions with equal intervals. We divide the sphere by using 6, 10, and 14 subdivisions to investigate the response with the refinement of the mesh. Six divisions result in a total of 60 patches, 90 common edges, and $R_{min} = 13.95$ cm, 10 divisions result in a total of 180 patches, 270 common edges, and $R_{min} = 6.055$ cm, and 14 divisions result in a total of 364 patches, 546 common edges, and $R_{min} = 3.218$ cm. The time step was selected as $c\Delta t = 2R_{min}$. Figure 6.11 shows the θ-component of the current at $\theta = 90°$ and $\phi = 30°$ on the sphere with $N = 90$ edges. Figure 6.12 shows the θ-component of the current at $\theta = 90°$ and $\phi = 18°$ on the sphere with $N = 270$ edges. Figure 6.13 shows the θ-component of current at $\theta = 90°$ and $\phi = 13°$ on the sphere with $N = 546$ edges. The agreement with the result from the IDFT solution calculated using both EFIE and MFIE in the frequency domain is also presented. It is shown that refinement of the mesh produces oscillations in early time for the TD-EFIE, whereas the TD-MFIE remains stable. We note that there is a difference between the peak amplitudes of the current between EFIE and MFIE. It is evident that this difference in the solutions is smaller with the increase of the number of patches. But the interesting point is that the time domain solution and the inverse Fourier transform of the frequency domain solution coincide with each other.

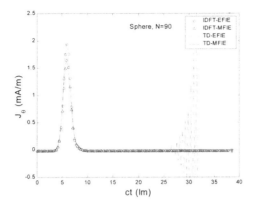

Figure 6.11. Transient response of a θ-directed current located at $\theta = 90°$, $\phi = 30°$ on a conducting sphere. $c\Delta t = 2R_{min}$. $R_{min} = 13.95$ cm., 90 common edges.

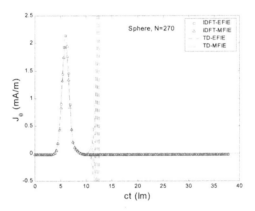

Figure 6.12. Transient response of a θ-directed current located at $\theta = 90°$, $\phi = 18°$ on a conducting sphere. $c\Delta t = 2R_{min}$. $R_{min} = 6.055$ cm., 270 common edges.

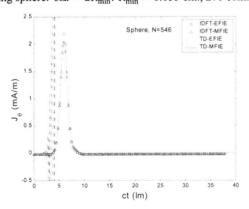

Figure 6.13. Transient response of θ-directed current at $\theta = 90°$, $\phi = 13°$ on a conducting sphere. $c\Delta t = 2R_{min}$. $R_{min} = 3.218$ cm., 546 common edges.

As a final example, we plot the transient response for a conducting cylinder as shown in Figure 6.14 with a radius of 0.5 m and a height of 1 m, centered at the origin. We subdivide the cylinder into 2, 12, and 4 divisions along r, ϕ, and z directions, respectively. This represents a total of 168 patches with 252 common edges, and $R_{min} = 8.33$ cm. Figure 6.15 plots the transient response of the x- and z-directed current at a point indicated by an arrow in Figure 6.14. The time step has been chosen as $c\Delta t = 2R_{min}$. The agreement of the result between the IDFT solution calculated using both the EFIE and MFIE in the frequency domain, with the time domain EFIE, and the MFIE solution is good. The only difference observed is in the peak value of the currents for TD-EFIE and TD-MFIE, and this difference becomes smaller when the patch size decreases.

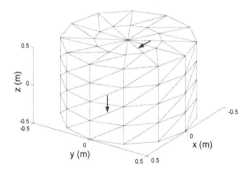

Figure 6.14. Triangular patching of a conducting cylinder (radius = 0.5 m, height = 1 m) with 168 patches and 252 common edges. R_{min} = 8.33 cm.

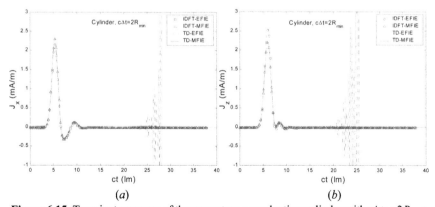

Figure 6.15. Transient response of the current on a conducting cylinder with $c\Delta t = 2R_{min}$: (a) x-directed current at the top face. (b) z-directed current at the side.

6.7.3   Scattering Analysis by Solving the TD-CFIE Using the MOT Method

In this section, we present the numerical results for representative 3-D closed scatterers, viz. a cube and a sphere. The scatterers are illuminated by a Gaussian plane wave in which the electric and the magnetic fields are given by Eqs. (6.127) and (6.128).

For the examples of this section, the incident field is arriving from $\phi = 0°$ and $\theta = 0°$ with $\hat{\mathbf{k}} = -\hat{\mathbf{z}}$ and $\mathbf{E}_0 = \hat{\mathbf{x}}$. We use two different types of pulse excitation, pulses of widths $T = 8$ lm with $ct_0 = 12$ lm and 2 lm with $ct_0 = 3$ lm, respectively. The first one has a frequency spectrum of 125 MHz, and the second one contains a frequency spectrum of 500 MHz. When the explicit solution is used, the stability of the transient response is dependent on the parameter κ. The parameter κ is varied from 0 (EFIE) to 1 (MFIE) at intervals of 0.1. It is important to note that the explicit solutions are most stable for $\kappa = 0.8$. For the implicit case, the acceptable range of κ is relatively wider, and we choose $\kappa = 0.5$. For comparison, we present the results obtained from the IDFT solution calculated from the frequency domain CFIE with $\kappa = 0.5$ in the 0–500 MHz interval with 128 samples. All results have been obtained using single-precision computation.

As a first example, consider a conducting cube, 1 m on a side, centered about the origin as shown in Figure 6.16. The first three internal resonant frequencies of this cube are 212, 260, and 335 MHz. There are 8, 9, and 8 divisions along the x, y, and z directions, respectively. This represents a total of 832 patches and 1248 common edges, and $R_{min} = 5.57$ cm, where R_{min} represents the minimum distance between any two distinct patch centers. The average area of the patches is 72 cm^2. The x- and z-directed currents at the center of the top and side faces of the cube are indicated by arrows in Figure 6.16. Figure 6.17(a) plots the transient response for the x-directed current at the center of the top face when $T = 8$ lm. The time step is chosen as $c\Delta t = 0.5R_{min}$ to generate the explicit solution. Here the agreement between the results from the MFIE and CFIE is very good even for late times. It is important to note that all three explicit solutions show instability. However, the MFIE breaks down later in time than for the EFIE and the CFIE breaks down much later in time than the MFIE. Figure 6.17(b) shows the transient response obtained by choosing $c\Delta t = 2R_{min}$, which corresponds to an implicit solution. The result of EFIE becomes more stable than that of the explicit case, but has still late-time oscillations with the increase of the time step. We can see that solutions of the MFIE and CFIE are very stable and that the agreement between them is good.

Figure 6.18 plots the explicit and the implicit solutions when a Gaussian plane wave with $T = 2$ lm is incident on the cube. This incident pulse contains several resonant frequencies of the cube. The explicit solutions in Figure 6.18(a) are similar to those of Figure 6.17(a) ($T = 8$ lm) except that a very small oscillation is observed in the MFIE solution. A small oscillation exists for the implicit solution of the MFIE in Figure 6.18(b). We can observe this slowly varying oscillation in the MFIE solution clearly in Figure 6.19.

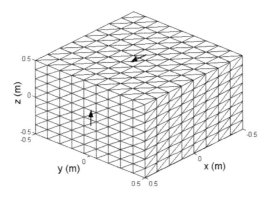

Figure 6.16. Triangular patching of a conducting cube (side 1 m) with 832 patches and 1248 common edges.

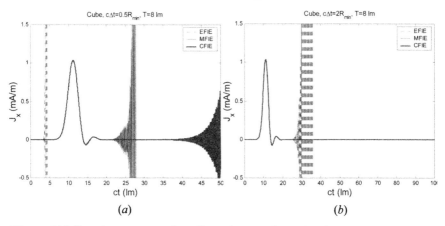

Figure 6.17. Transient response of a x-directed current located at the center of the top face on a conducting cube using (a) explicit method and (b) implicit method.

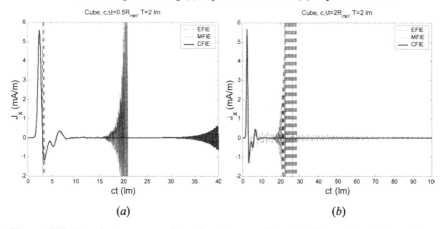

Figure 6.18. Transient response of a x-directed current located at the center of the top face on a conducting cube using (a) explicit method, and (b) implicit method.

Figure 6.20 shows similar results for the z-directed current at the center of the side face on a cube using an implicit method when $T = 2$ lm. Also, we note that the results of EFIE and CFIE agree well. As expected, the results using CFIE are better than those from EFIE and MFIE because the incident wave with $T = 2$ lm includes several resonant frequencies of the object. Figure 6.21(a) compares the explicit and implicit solutions along with the IDFT of the frequency domain CFIE solution for the transient current at the center of the top face. Also, Figure 6.21(b) shows the z-directed current on the side face. All three solutions—explicit, implicit, and IDFT—agree reasonably well as is evident from the figures. It is important to notice that the explicit solution is in better agreement with the IDFT solution.

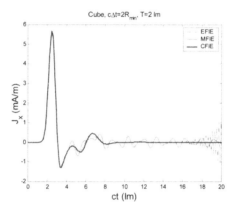

Figure 6.19. Solutions displayed in Figure 6.20(b) are plotted until 20 lm.

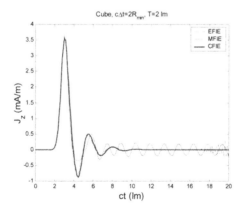

Figure 6.20. Transient response of a z-directed current located at the center of the side face of a conducting cube using an implicit method excited with a 2-lm-wide pulse.

As a second example, we consider a conducting sphere of radius 0.5 m centered at the origin as shown in Figure 6.22. The first three internal resonant frequencies of this sphere are located at 262, 369, and 429 MHz. There are 12 and

24 divisions along the θ and ϕ directions with equal intervals. This results in a total of 528 patches and 792 common edges, and $R_{min} = 2.23$ cm. The average patch area is 59 cm². The θ-directed current at $\theta = 90°$ and $\phi = 7.5°$, and ϕ-directed current at $\theta = 7.5°$ and $\phi = 90°$ on the sphere are indicated by arrows in Figure 6.22. Figure 6.23(a) shows the transient response for the ϕ-directed current when $T = 8$ lm. The time step is chosen such that $c\Delta t = 0.5R_{min}$ to generate the explicit solution. It is important to note that all three explicit solutions show instability. However, the MFIE formulation breaks down later in time than the EFIE and CFIE breaks down still much later in time than MFIE. Figure 6.23(b) shows the transient response for the ϕ-directed current obtained by choosing $c\Delta t = 4R_{min}$, which corresponds to an implicit solution. The result of the EFIE becomes more stable than that of the explicit case, but it still has the late-time oscillations even for the implicit case. We can see that the solutions of the MFIE and CFIE are very stable and that the agreement between them is good.

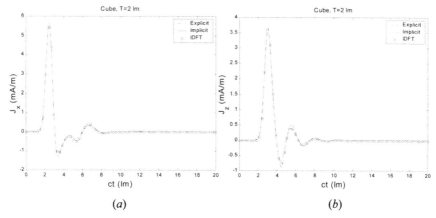

Figure 6.21. Comparison of time-domain and IDFT solution of CFIE for the current on a conducting cube: (a) x-directed at the center of top face. (b) z-directed at the center of side face.

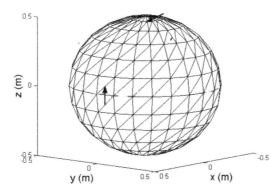

Figure 6.22. Triangle patching of a conducting sphere (radius = 0.5 m) with 528 patches and 792 common edges.

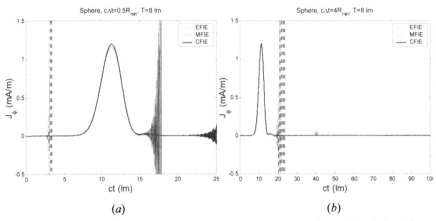

Figure 6.23. Transient response of a ϕ-directed current located at $\theta = 7.5°$ and $\phi = 90°$ on a conducting sphere using (a) explicit method, and (b) implicit method.

Figure 6.24 plots the explicit and the implicit solutions when a Gaussian plane wave with $T = 2$ lm is incident on the sphere. This incident pulse contains several internal resonant frequencies of the sphere. The explicit solutions in Figure 6.24(a) are similar to those in Figure 6.23(a) ($T = 8$ lm). But we can see that there are some small oscillations in the implicit solution of the MFIE in Figure 6.24(b). We can observe this clearly in Figure 6.25, which is plotted until 20 lm along the time axis using the same parameters as in Figure 6.24(b).

Figure 6.26 plots the results for a θ-directed current on the sphere using an implicit method when $T = 2$ lm. Also, the solutions for the EFIE and CFIE agree well. As expected, the results using the CFIE are better than those of the EFIE and MFIE in this case because the incident wave with $T = 2$ lm includes several internal resonant frequencies of the sphere. Figure 6.27(a) compares the explicit and implicit solutions with the IDFT of the frequency-domain CFIE solution for the ϕ-directed current on the sphere. Also, Figure 6.27(b) compares the result for the θ-directed current on the sphere obtained by the different methods. All three solutions—explicit, implicit, and IDFT—agree reasonably well, as is evident from the figures, except for the late-time oscillation obtained for the explicit solutions. It is important to note that the explicit solution is in better agreement with the IDFT solution.

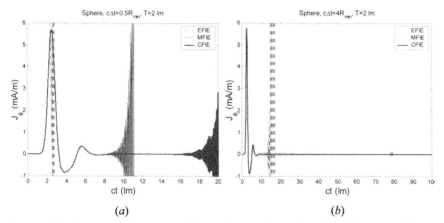

Figure 6.24. Transient response of a ϕ-directed current located at $\theta = 7.5°$ and $\phi = 90°$ on a conducting sphere using (a) explicit method and (b) implicit method.

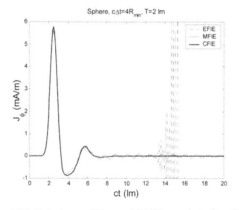

Figure 6.25. Solutions of Figure 6.24(b) are plotted until 20 lm.

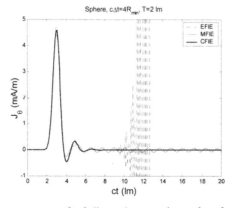

Figure 6.26. Transient response of a θ-directed current located at $\theta = 90°$ and $\phi = 7.5°$ on a conducting sphere using an implicit method.

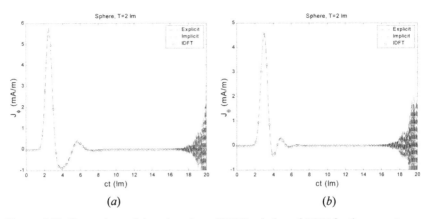

Figure 6.27. Comparison of time-domain and IDFT solution of CFIE for the current on a conducting sphere (*a*) ϕ-directed, at $\theta = 7.5°$ and $\phi = 90°$. (*b*) θ-directed, at $\theta = 90°$, and $\phi = 7.5°$

6.7.4   Scattering Analysis by Solving the TDIEs Using the MOD Method

In this section, numerical results from three representative 3-D closed conducting scatterers, viz. a sphere, a cube, and a cylinder, are presented. The scatterers are illuminated by a Gaussian plane wave in which the electric and magnetic fields are given by Eqs. (6.127) and (6.128). The incident field is arriving from $\phi = 0°$ and $\theta = 0°$ and $\mathbf{E}_0 = \hat{\mathbf{x}}$. First, in the numerical computation, we use a Gaussian pulse of $T = 8$ lm and $ct_0 = 12$ lm. This pulse has a frequency spectrum of 125 MHz, so the internal resonance effects are excluded for the structures. Next, a Gaussian pulse with $T = 2$ lm and $ct_0 = 3$ lm is used. This pulse has a frequency spectrum of 500 MHz, which encompasses several internal resonant frequencies of the structures. The numerical results to be plotted are the transient currents on the surface and θ- (or x-) components of the normalized far fields computed by EFIE, MFIE, and CFIE formulations. These far scattered fields from the structures are obtained along the backward direction and, hence, represent the back-scattered fields. We also present the monostatic radar cross section (RCS) by considering both θ- and ϕ-components of the far fields. All numerical solutions have been computed using the methods presented and are compared with IDFT of the solutions obtained in the frequency-domain as well as the transient currents using the MOT method [18]. In addition, the far field solutions for the sphere are compared with the Mie series solutions. We set $s = 10^9$ and $M = 80$, which is sufficient to get accurate solutions. In all legends of the figures to be shown, the numbers '1' and '2' denote results computed using the MOD formulation of Section 6.5 and the alternative MOD formulation of section 6.6, respectively.

6.7.5   Response in the Non-resonance Region

When the spectrum of the incident field excludes the internal resonant frequencies of the structure, an incident Gaussian pulse of $T = 8$ lm is used. The purpose of this computation is to check the accuracy and validity of the TD-EFIE and TD-MFIE formulations when the resonance effects are not taken into account.

As a first example, we consider the sphere of Figure 6.22. Figure 6.28 plots the transient θ-directed current on the sphere computed by the TD-EFIE and compares it with the IDFT solution of the frequency FD-EFIE. The solutions of the two described TD-EFIE methods are stable and the agreement with the IDFT solution is very good. Figure 6.29 compares the transient field response of the two presented TD-EFIE methods along with the Mie series solution and the IDFT of the FD-EFIE solution for the normalized far scattered field and the monostatic RCS. All four solutions agree well, as is evident from the figures.

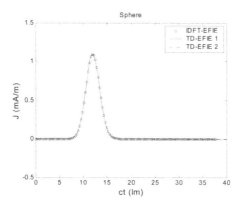

Figure 6.28. Transient current induced on the sphere computed by EFIE.

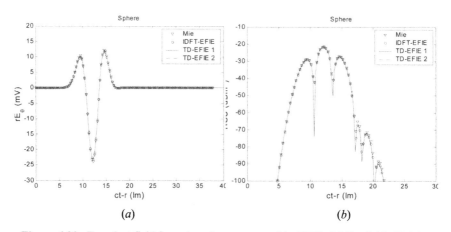

Figure 6.29. Transient field from the sphere computed by EFIE: (*a*) Far field. (*b*) RCS.

Figure 6.30 plots the transient response for the θ-directed current on the sphere computed by TD-MFIE and compared with the IDFT solution of the FD-MFIE. The solutions from the two presented TD-MFIE methods are stable, and the agreement with the IDFT solution is good.

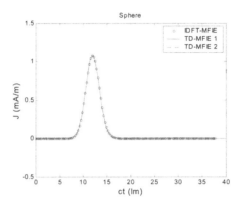

Figure 6.30. Transient current induced on the sphere computed by MFIE.

Figure 6.31 compares the transient response computed by two TD-MFIE methods along with the Mie series solution and the IDFT of the FD-MFIE solution for the normalized far scattered field and the monostatic RCS of the sphere. All four solutions agree well, as is evident from the figures.

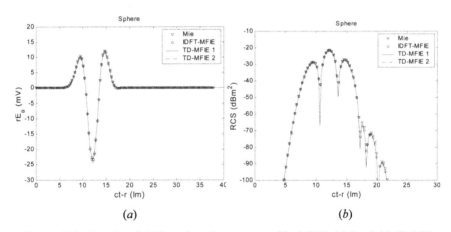

(a)   *(b)*

Figure 6.31. Transient field from the sphere computed by MFIE: *(a)* Far field. *(b)* RCS.

Figure 6.32 plots the transient θ-directed current on the sphere computed by the TD-CFIE and compares it with the IDFT of the FD-CFIE. The solutions obtained from the two TD-CFIE methods are stable and the agreement with the IDFT solution is good.

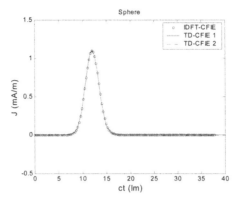

Figure 6.32. Transient current on the sphere computed by CFIE.

Figure 6.33 compares the transient field from the two TD-CFIE methods along with the Mie series solution and the IDFT of the FD-CFIE solution for the normalized far scattered field and the monostatic RCS from the sphere. All four solutions agree well, as is evident from the figure.

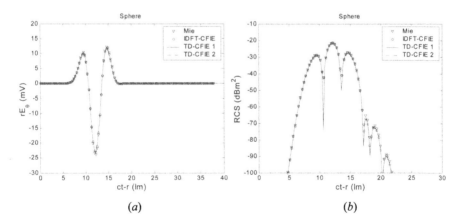

(a) (b)

Figure 6.33. Transient field from the sphere computed by CFIE: (a) Far field. (b) RCS.

As a second example, consider the cube of Figure 6.16. Figure 6.34 plots the z-directed current on the cube computed by TD-EFIE and compares it with the IDFT solution of the FD-EFIE. It is shown that the solutions for the two TD-EFIE methods are stable and are in agreement with the IDFT solution. Figure 6.35 compares the transient field computed by the two TD-EFIE methods along with the IDFT of the FD-EFIE solution for the normalized far scattered field and the monostatic RCS from the cube. All three solutions agree well, as is evident from this figure. Figure 6.36 displays the z-directed current on the cube computed by the TD-MFIE and compares the IDFT solution of the FD-MFIE. The solutions of the two TD-MFIE methods are stable and the agreement with the IDFT solution is good. Figure 6.37 compares the response from the two TD-MFIE methods with the IDFT of the FD-MFIE solution for the normalized far scattered field and the monostatic RCS from the cube. All three solutions agree well.

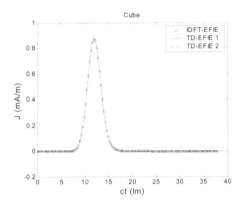

Figure 6.34. Transient current on the cube computed by EFIE.

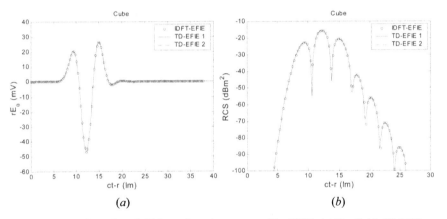

Figure 6.35. Transient field from the cube computed by EFIE: (*a*) Far field. (*b*) RCS.

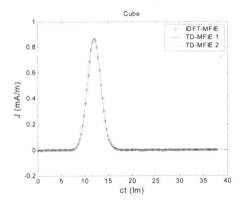

Figure 6.36. Transient current on the cube computed by MFIE.

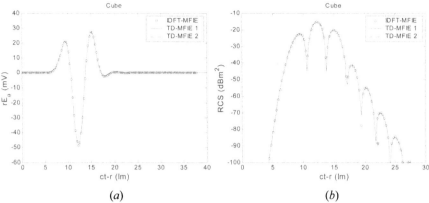

(a)   (b)

Figure 6.37. Transient field from the cube computed by MFIE: (*a*) Far field. (*b*) RCS.

Figure 6.38 displays the *z*-directed current on the cube computed by the TD-CFIE and compares it with the IDFT solution of the FD-CFIE. The solutions obtained from the two TD-CFIE methods are stable and the agreement with the IDFT solution is good. Figure 6.39 compares the transient field computed by the two TD-CFIE methods along with the IDFT of the FD-CFIE solution for the normalized far scattered field and the monostatic RCS from the cube. All three solutions agree well, as is evident from the figure.

As a third example, consider the cylinder of Figure 6.40 with a radius of 0.5 m and height of 1 m, centered at the origin. The first resonant frequency of this cylinder is at 230 MHz. The cylinder is subdivided into 4, 24, and 8 divisions along *r*, ϕ, and *z* directions, respectively. This represents a total of 720 patches with 1080 common edges. Figure 6.41 plots the transient *z*-directed current on the cylinder computed by the TD-EFIE and compares it with the IDFT solution of the FD-EFIE. The solutions of the two TD-EFIE methods are stable and the agreement with the IDFT solution is good. Figure 6.42 compares the transient fields computed by the two TD-EFIE methods with the IDFT of the FD-EFIE solution for the normalized far scattered field and monostatic RCS from the cylinder. All three solutions agree well. Figure 6.43 displays the transient *z*-directed current on the cylinder computed by the TD-MFIE and compares it with the IDFT solution of the FD-MFIE. The solutions of the two TD-MFIE methods are stable and the agreement with the IDFT solution is good. Figure 6.44 compares the transient field computed by the two TD-MFIE methods with the IDFT of the FD-MFIE solution for the normalized far scattered field and the monostatic RCS from the cylinder. All three solutions agree well, as is evident from the figure.

Figure 6.45 plots the transient *z*-directed current on the cylinder computed by the TD-CFIE and compares it with the IDFT of the FD-CFIE. The solutions of the two TD-CFIE methods are stable and the agreement with the IDFT solution is good. Figure 6.46 compares the transient response of the two TD-CFIE methods with the IDFT of the FD-CFIE solution for the normalized far scattered field and the monostatic RCS from the cylinder. All three solutions agree well, as is evident from the figure.

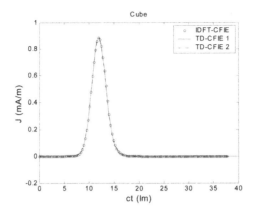

Figure 6.38. Transient current on the cube computed by CFIE.

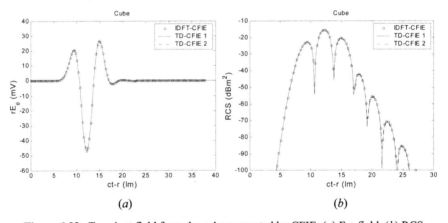

Figure 6.39. Transient field from the cube computed by CFIE: (*a*) Far field. (*b*) RCS.

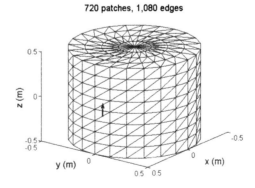

Figure 6.40. Triangular patch modeling of a cylinder.

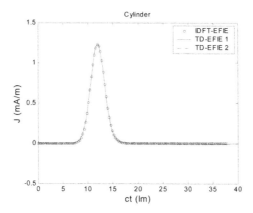

Figure 6.41. Transient current on the cylinder computed by EFIE.

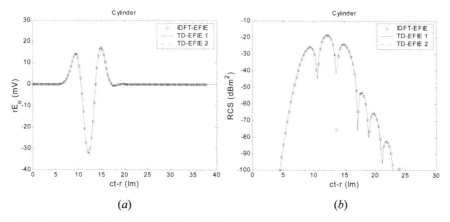

Figure 6.42. Transient field from the cylinder computed by EFIE: (*a*) Far field. (*b*) RCS.

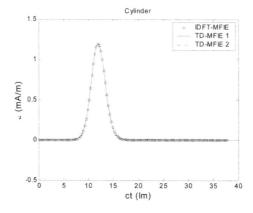

Figure 6.43. Transient current on the cylinder computed by MFIE.

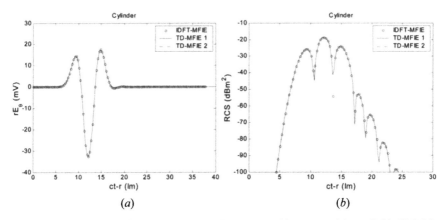

Figure 6.44. Transient field from the cylinder computed by MFIE: (*a*) Far field. (*b*) RCS.

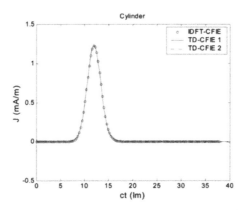

Figure 6.45. Transient current on the cylinder computed by CFIE.

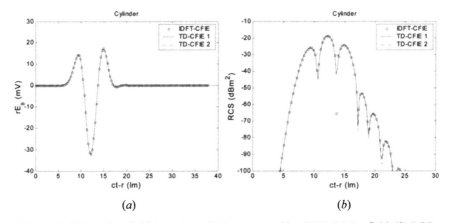

Figure 6.46. Transient field from the cylinder computed by CFIE: (*a*) Far field. (*b*) RCS.

6.7.6 Responses Resulting From a Wide-Band Incident Waveform

When the bandwidth of the incident signal is large, it can excite numerical internal resonances for the structures to be analyzed. In this case, a Gaussian pulse with $T = 2$ lm is used as an excitation, whose spectrum overlaps over several internal resonant frequencies of the structures.

As a first example, consider the sphere of Figure 6.22. The time step in the MOT computation is chosen such that $c\Delta t = 4R_{min}$ to generate an implicit solution where R_{min} represents the minimum distance between any two distinct patch centers and R_{min} is 2.23 cm in the sphere model. Figure 6.47 plots the transient θ-directed current on the sphere computed by the TD-EFIE and compares it with the IDFT solution of the FD-EFIE and the MOT solution of the TD-EFIE. The solutions for TD-EFIE using the two MOD methods are stable and are in agreement with the IDFT and MOT solutions. The agreement is very good, except a late-time oscillation for the MOT solution appears.

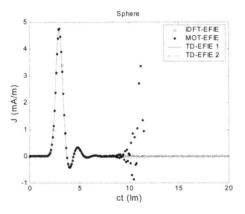

Figure 6.47. Transient current on the sphere computed by EFIE.

Figure 6.48 compares the transient response from the two TD-EFIE methods along with the Mie series solution and the IDFT of the FD-EFIE solution for the normalized far scattered field and the monostatic RCS of the sphere. All four solutions agree well, as is evident from the figure. Figure 6.49 plots the transient θ-directed current on the sphere computed by the TD-MFIE and compares it with the IDFT of the FD-MFIE and the MOT solution of the TD-MFIE. The solutions of the two TD-MFIE methods are stable and the agreement with the IDFT is very good, whereas the MOT solution is not as accurate as it displays slowly varying oscillations. Figure 6.50 compares the transient field from the two TD-MFIE methods along with the Mie series solution and the IDFT of the FD-MFIE solution for the normalized far scattered field and the monostatic RCS from the sphere. All four solutions show a good agreement for the far field in Figure 6.50(a). In Figure 6.50(b), however, it is noted that the IDFT solution of the FD-MFIE is different from the Mie series solution for smaller values of the RCS.

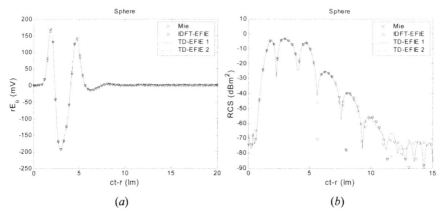

Figure 6.48. Transient field from the sphere computed by EFIE: (a) Far field. (b) RCS.

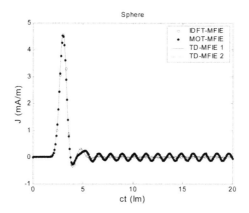

Figure 6.49. Transient current on the sphere computed by MFIE.

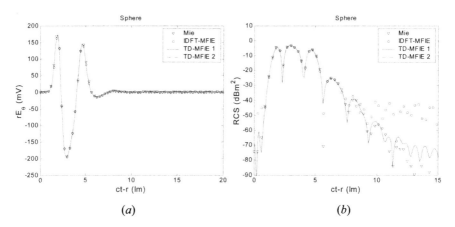

Figure 6.50. Transient field from the sphere computed by MFIE: (*a*) Far field. (*b*) RCS.

Figure 6.51 plots the transient θ-directed current on the sphere computed by the TD-CFIE and compares it with the IDFT solution of the FD-CFIE and the MOT solution of the TD-CFIE. The solutions for the two TD-CFIE methods are stable and the agreement with the IDFT and MOT solutions is good. Figure 6.52 compares the transient field from the two TD-CFIE methods along with the Mie series solution and the IDFT of the FD-CFIE solution for the normalized far scattered field and the monostatic RCS from the sphere. All four solutions agree well as is evident from the figure. It is important to note that the TD-CFIE solutions are more accurate than those of the TD-EFIE and the TD-MFIE for the smaller values for the RCS.

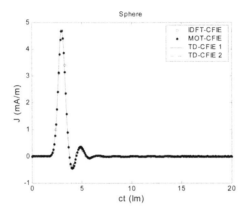

Figure 6.51. Transient current on the sphere computed by CFIE.

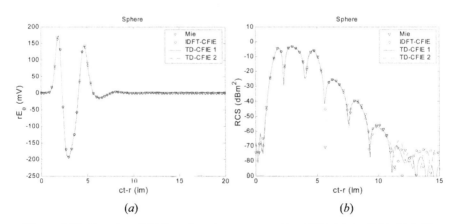

Figure 6.52. Transient field from the sphere computed by CFIE: (*a*) Far field. (*b*) RCS.

As a second example, consider the cube of Figure 6.16. The time step in the MOT computation is chosen such that $c\Delta t = 2R_{min}$ to generate the implicit solution where R_{min} is 5.57 cm in the cube model. Figure 6.53 plots the transient z-directed current on the cube computed by the TD-EFIE and compares it with the

IDFT solution of the FD-EFIE and the MOT solution of the TD-EFIE. The solutions for the two TD-EFIE methods are stable and the agreement with the IDFT and MOT solutions is very good, except for the late-time oscillations displayed by the MOT solution. Figure 6.54 compares the transient response computed by the two TD-EFIE methods along with the IDFT of the FD-EFIE solution for the normalized far scattered field and the monostatic RCS from the cube. All three solutions agree well as is evident from the figure. Figure 6.55 plots the transient z-directed current on the cube computed by the TD-MFIE and compares it with the IDFT solution of the FD-MFIE and the MOT solution of the TD-MFIE. The solutions computed by the two TD-MFIE methods are stable and the agreement with the IDFT is good, whereas the MOT solution is not accurate, as it contains some slowly varying oscillations.

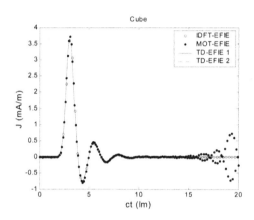

Figure 6.53. Transient current on the cube computed by EFIE.

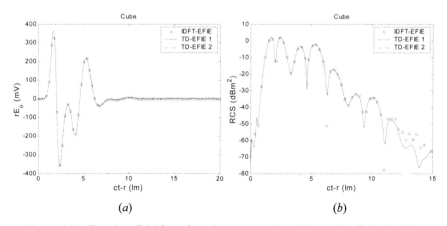

(a)   (b)

Figure 6.54. Transient field from the cube computed by EFIE: (a) Far field. (b) RCS.

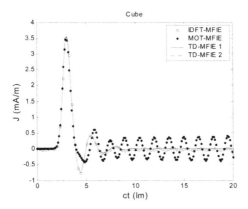

Figure 6.55. Transient current on the cube computed by MFIE.

Figure 6.56 compares the transient response computed by the two TD-MFIE methods along with the IDFT of the FD-MFIE solution for the normalized far scattered field and the monostatic RCS from the cube. All three solutions show a good agreement for the far field, although there is a small fluctuation of the IDFT solution as shown in Figure 6.56(a). However, we note that there is a difference between the TD-MFIE solutions and the IDFT solution of FD-MFIE for the small values of the RCS as shown in Figure 6.56(b).

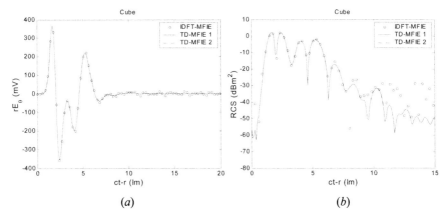

(a) (b)

Figure 6.56. Transient field from the cube computed by MFIE: (a) Far field. (b) RCS.

Figure 6.57 plots the transient z-directed current on the cube computed by the TD-CFIE and compares it with the IDFT solution of the FD-CFIE and the MOT solution of the TD-CFIE. The solutions computed by the two TD-CFIE methods are stable and the agreement with the IDFT solution is very good, whereas the MOT solution is stable but not accurate. The MOT solution has small differences from the other solutions in the second and third peaks as seen in Figure 6.57. This is a result of the instabilities of the TD-MFIE. When using the explicit

MOT scheme along with the TD-CFIE, the solution become more accurate, which has been pointed out in [2]. Figure 6.58 compares the transient field computed by the two TD-CFIE methods along with the IDFT of the FD-CFIE solution for the normalized far scattered field and the monostatic RCS for the cube. All three solutions show a good agreement for the far field as observed in Figure 6.58(a). It is important to note that solutions for the TD-CFIE show better agreement with the IDFT solution than those from the TD-EFIE and TD-MFIE for small values of the RCS.

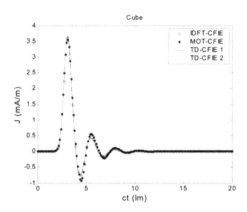

Figure 6.57. Transient current on the cube computed by CFIE.

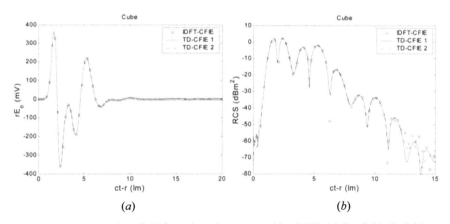

Figure 6.58. Transient field from the cube computed by CFIE: (a) Far field. (b) RCS.

As a third example, consider the cylinder of Figure 6.40. The time step in the MOT computation is chosen such that $c\Delta t = 4R_{min}$ to generate the implicit solution where R_{min} is 2.15 cm for the cylinder model. Figure 6.59 plots the transient response for the z-directed current on the cylinder computed by the TD-

EFIE and compares it with the IDFT solution of the FD-EFIE and the MOT solution of the TD-EFIE. The solutions computed by the two TD-EFIE methods are stable and the agreement with the IDFT and MOT solutions is very good, except for the late-time oscillations displayed by the MOT solution.

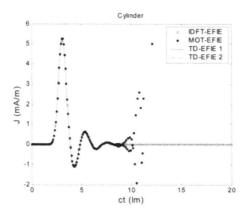

Figure 6.59. Transient current on the cylinder computed by EFIE.

Figure 6.60 compares the transient field computed by the two TD-EFIE methods along with the IDFT of the FD-EFIE solution for the normalized far scattered field and the monostatic RCS of the cylinder. All three solutions agree well as is evident from the figure. Figure 6.61 displays the transient response for the z-directed current on the cylinder computed by the TD-MFIE and compares it with the IDFT solution of the FD-MFIE and the MOT solution of the TD-MFIE. The solutions computed by the two TD-MFIE methods are stable and the agreement with IDFT is very good, whereas the MOT solution is not accurate as it contains some slowly varying oscillations.

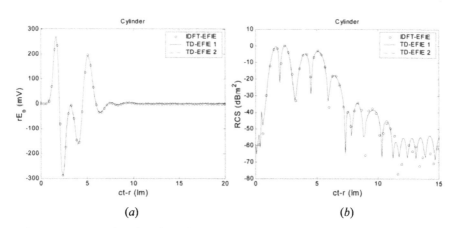

(a)   (b)

Figure 6.60. Transient field from the cylinder computed by EFIE: (a) Far field. (b) RCS.

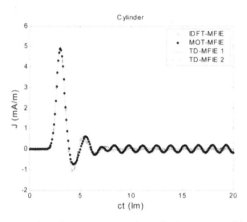

Figure 6.61. Transient current on the cylinder computed by MFIE.

Figure 6.62 compares the transient response computed by the two TD-MFIE methods along with the IDFT of the FD-MFIE solution for the normalized far scattered field and the monostatic RCS from the cylinder. All three solutions show a good agreement for the far field with a small fluctuation of the IDFT solution as seen in Figure 6.62(a). Also, it is noted that there is a difference between the TD-MFIE solutions and the IDFT solution of FD-MFIE for small values of the RCS, as shown in Figure 6.62(b). Figure 6.63 plots the transient response for the z-directed current on the cylinder computed by the TD-CFIE and compares it with the IDFT solution of the FD-CFIE and the MOT solution of the TD-CFIE. The solutions computed by the two TD-CFIE methods are stable and the agreement with the IDFT solution is very good, whereas the MOT solution is stable but has a small difference in the plots related to the second and third peaks as seen in Figure 6.63.

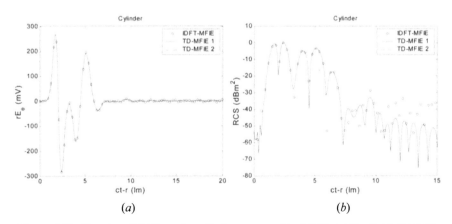

(a)   (b)

Figure 6.62. Transient field from the cylinder computed by MFIE: (a) Far field. (b) RCS.

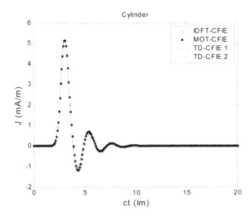

Figure 6.63. Transient current on the cylinder computed by CFIE.

Figure 6.64 compares the transient field computed by the two TD-CFIE methods with the IDFT of the FD-CFIE solution for the normalized far scattered field and the monostatic RCS from the cylinder. All three solutions show a good agreement for the far field as seen in Figure 6.64(a). However, it is noted that the solutions computed by the two TD-CFIE methods show better agreement with the IDFT solution than those computed using the TD-EFIE and TD-MFIE, particularly for the small values of the RCS.

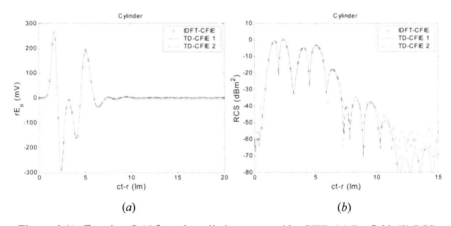

(a) (b)

Figure 6.64. Transient field from the cylinder computed by CFIE: (a) Far field. (b) RCS.

Finally, the far field computed by the TD-EFIE, TD-MFIE, and TD-CFIE for the sphere is compared, as this problem has an analytic solution. The three computed results are then compared with the analytic solution to assess the accuracy of this time domain methodology. Figure 6.65 plots the results for the far field and the RCS when the sphere is illuminated with a Gaussian pulse of width $T = 8$ lm. In this example, because of the exclusion of the internal resonant frequencies in the excitation, the agreement between the solutions is excellent. The results for the RCS also show good agreement with the Mie series solution as shown in Figure 6.65(*b*).

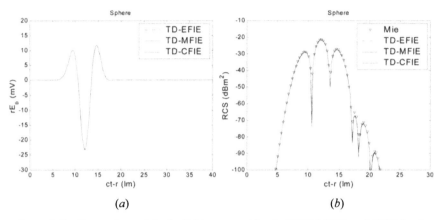

(*a*)   (*b*)

Figure 6.65. Comparison of the far field computed by the EFIE, MFIE, and CFIE for the sphere: (*a*) Far field. (*b*) RCS.

Figure 6.66 plots the results when the width of the incident exciting Gaussian pulse is small (i.e., $T = 2$ lm). So the incident waveform has a wider bandwidth. Even though this excitation covers the internal resonant frequencies of the sphere, solutions of the TD-EFIE and the TD-MFIE agree well with the TD-CFIE in Figure 6.66(*a*). All three solutions agree well with the Mie series solution except for the extremely low values of the RCS as shown in Figure 6.66(*b*).

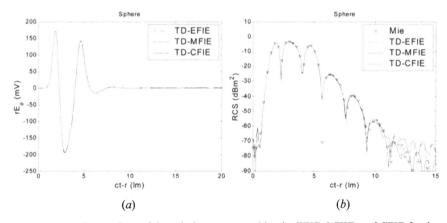

(*a*)   (*b*)

Figure 6.66. Comparison of the solutions computed by the EFIE, MFIE, and CFIE for the sphere when the excitation pulse has a wider bandwidth: (*a*) Far field. (*b*) RCS.

6.8   CONCLUSION

Two different time domain techniques (the marching-on-in-time and marching-on-in-degree) have been presented to solve the time domain integral equations for arbitrarily shaped 3-D conducting structures. In the application of the MoM procedure, the triangular patch functions have been used as both spatial basis and testing functions. A temporal basis function set derived from the Laguerre polynomials has been introduced in the marching-on-in-degree method. The advantage of this new method is to guarantee the late time stability of the computed solutions. With the representation of the derivative of the transient coefficient in an analytic form, the temporal derivatives in the integral equations can be treated analytically when using the associated Laguerre functions as both temporal testing and basis functions. In this way, one can do away with the Courant–Friedrich–Levy criteria relating the spatial discretizations with the temporal ones. Transient electric current and the far field computed by the two methods have been accurate and stable. The agreement between the solutions obtained using the two presented methods with the IDFT of the frequency domain solution are very good.

REFERENCES

[1]   S. M. Rao and D. R. Wilton, "Transient scattering by conducting surfaces of arbitrary shape," *IEEE Trans. Antennas Prop.*, Vol. 39, pp. 56–61, 1991.

[2]   D. A. Vechinski and S. M. Rao, "A stable procedure to calculate the transient scattering by conducting surfaces of arbitrary shape," *IEEE Trans. Antennas Prop.*, Vol. 40, pp. 661–665, 1992.

[3]   D. A. Vechinski, *Direct Time Domain Analysis of Arbitrarily Shaped Conducting or Dielectric Structures Using Patch Modeling Techniques*. PhD Dissertation, Auburn, Ala., 1992.

[4]   S. M. Rao and T. K. Sarkar, "An alternative version of the time-domain electric field integral equation for arbitrarily shaped conductors," *IEEE Trans. Antennas Prop.*, Vol. 41, pp. 831–834, 1993.

[5]   S. M. Rao, *Time Domain Electromagnetics*. Academic Press, New York, 1999.

[6]   A. Sadigh and E. Arvas, "Treating the instabilities in marching-on-in-time method from a different perspective," *IEEE Trans. Antennas Prop.*, Vol. 41, pp. 1695–1702, 1993.

[7]   P. J. Davies, "On the stability of time-marching schemes for the general surface electric-field integral equation," *IEEE Trans. Antennas Prop.*, Vol. 44, pp. 1467–1473, 1996.

[8]   S. M. Rao and T. K. Sarkar, "An efficient method to evaluate the time-domain scattering from arbitrarily shaped conducting bodies," *Microw. Opt. Tech. Lett.*, Vol. 17, pp. 321–325, 1998.

[9]   S. M. Rao, D. A. Vechinski, and T. K. Sarkar, "Transient scattering by conducting cylinders-implicit solution for the transverse electric case," *Microw. Opt. Tech. Lett.*, Vol. 21, pp. 129–134, 1999.

[10]   T. K. Sarkar, W. Lee, and S. M. Rao, "Analysis of transient scattering from composite arbitrarily shaped complex structures," *IEEE Trans. Antennas Prop.*, Vol. 48, pp. 1625–1634, 2000.

[11]   S. M. Rao and T. K. Sarkar, "Time-domain modeling of two-dimensional conducting cylinders utilizing an implicit scheme-TM incidence," *Microw. Opt. Tech. Lett.*, Vol. 15, pp. 342–347, 1997.

[12]   S. M. Rao and T. K. Sarkar, "Transient analysis of electromagnetic scattering from wire structures utilizing an implicit time-domain integral-equation technique," *Microw. Opt. Tech. Lett.*, Vol. 17, pp. 66–69, 1998.

[13]   B. H. Jung and T. K. Sarkar, "Time-domain electric-field integral equation with central finite difference," *Microw. Opt. Tech. Lett.*, Vol. 31, pp. 429–435, 2001.

[14]   B. H. Jung, T. K. Sarkar, Y.-S. Chung, and Z. Ji, "An accurate and stable implicit solution for transient scattering and radiation from wire structures," *Microw. Opt. Tech. Lett.*, Vol. 34, pp. 354–359, 2002.

[15]   B. H. Jung and T. K. Sarkar, "Transient scattering from three-dimensional conducting bodies by using magnetic field integral equation," *J. Electromagn. Waves Appl.*, Vol. 16, pp. 111–128, 2002.

[16]   D. A. Vechinski and S. M. Rao, "Transient scattering by dielectric cylinders: e-field, h-field, and combined field solutions," *Radio Sci.*, Vol. 27, pp. 611–622, 1992.

[17]   B. Shankar, A. A. Ergin, K. Aygun, and E. Michielssen, "Analysis of transient electromagnetic scattering from closed surfaces using a combined field integral equation," *IEEE Trans. Antennas Prop.*, Vol. 48, pp. 1064–1074, 2000.

[18]   B. H. Jung and T. K. Sarkar, "Time-domain CFIE for the analysis of transient scattering from arbitrarily shaped 3D conducting objects," *Microw. Opt. Tech. Lett.*, Vol. 34, pp. 289–296, 2002.

[19]   T. K. Sarkar and J. Koh, "Generation of a wide-band electromagnetic response through a Laguerre expansion using early-time and low-frequency data," *IEEE Trans. Antennas Prop.*, Vol. 50, pp. 1408–1416, 2002.

[20]   S. M. Rao, D. R. Wilton and A. W. Glisson, "Electromagnetic scattering by surfaces of arbitrary shape," *IEEE Trans. Antennas Prop.*, Vol. 30, pp. 409–418, 1982.

[21]   D. R. Wilton, S. M. Rao, A. W. Glisson, D. H. Schaubert, O. M. Al-Bundak, and C. M. Butler, "Potential integrals for uniform and linear source distributions on polygonal and polyhedral domains," *IEEE Trans. Antennas Prop.*, Vol. 32, pp. 276–281, 1984.

[22]   S. M. Rao, *Electromagnetic Scattering and Radiation of Arbitrarily-Shaped Surfaces by Triangular Patch Modeling.* PhD Dissertation, Univ. Mississippi, Aug. 1980.

[23]   A. J. Poggio and E. K. Miller, "Integral equation solutions of three dimensional scattering problems," in *Computer Techniques for Electromagnetics.* Pergamon Press, New York, 1973.

[24]   J. Van Bladel, *Electromagnetic Fields.* Hemisphere Publishing Corporation, New York, 1985.

7

ANALYSIS OF DIELECTRIC STRUCTURES IN THE TIME DOMAIN

7.0 SUMMARY

For the analysis of transient scattering, the marching-on-in-time (MOT) and the marching-on-in-degree (MOD) techniques have been introduced in the previous chapters. From the numerical examples presented in Chapters 5 and 6, it can be observed that the MOD approach provides better late-time stability than the MOT technique. Hence, this chapter will focus on the application of the MOD method for the analysis of transient scattering from arbitrary shaped three-dimensional (3-D) dielectric structures. Formulations for solving the various time domain integral equations (TDIEs) including the time domain electric field integral equation (TD-EFIE), the time domain magnetic field integral equation (TD-MFIE), the time domain combined field integral equation (TD-CFIE) and the time domain Poggio–Miller–Chang–Harrington–Wu integral equation (TD-PMCHW) are presented. In the solution procedures for all integral equations, two sets of point-wise orthogonal triangular patch RWG vector functions (RWG and $\mathbf{n} \times \mathrm{RWG}$) are used for spatial expansions and testings of the unknown electric and magnetic currents. The temporal variables are approximated by using a set of orthonormal basis functions called the associated Laguerre functions that is derived from the Laguerre polynomials. These basis functions also are used for the temporal testing. Alternative solution procedures for the TD-EFIE, the TD-MFIE, and the TD-PMCHW also are described. Numerical results involving the equivalent currents and the far fields computed with the various integral equations are presented and compared with the results obtained from the inverse discrete Fourier transform (IDFT) of the respective frequency domain solutions to demonstrate the accuracy of these methods.

7.1 INTRODUCTION

The MOT method has been used widely for solving TDIEs developed for the analysis of electromagnetic scattering from arbitrarily shaped 3-D dielectric structures [1–7]. The solution computed by this method suffers from late-time instabilities that manifest in the form of high-frequency oscillations. To improve the late-time stability, many schemes have been investigated. Two popular methods are the use of the backward finite difference approximation for the magnetic vector potential term using an implicit technique [8,9] and the central finite difference approximation for the time derivative term operating on the magnetic vector potential in a TD-EFIE [10]. However, none of these schemes acceptably can solve the numerical instability problem, and thus, it leads to a search for better methodologies. Recently, a method called the MOD, which uses the Laguerre polynomials as temporal basis functions [11], has been introduced. In the MOD, the unknown temporal coefficients are obtained by solving the final matrix equation recursively using a finite number of basis functions. Because of its intrinsic causality and convergence to zero for late times, the use of the associated Laguerre functions in the MOD technique presents the potential for a promising solution for stable analysis of transient scattering.

In this chapter, the procedures for analyzing arbitrarily shaped 3-D dielectric structures are presented using the TD-EFIE, the TD-MFIE, the TD-CFIE [12,13], and the TD-PMCHW formulations. Two sets of point-wise orthogonal RWG vector functions (RWG and $\mathbf{n} \times \text{RWG}$) are used for spatial expansions of the electric and the magnetic currents induced on the surface of the dielectric structures. The RWG function is used to perform the spatial testing for the TD-EFIE, the TD-MFIE, and the TD-PMCHW formulations, whereas a combination of RWG and $\mathbf{n} \times \text{RWG}$ functions is used in the TD-CFIE formulation [14,15]. The temporal variable is characterized by a set of orthogonal basis functions that are derived from the Laguerre polynomials. Using the Galerkin's method, a temporal testing procedure also is introduced, which is similar to the spatial testing procedure used in the MoM. The advantage of using the associated Laguerre functions is that the spatial and the temporal variables can be decoupled, and hence, there is no need for the Courant–Friedrich–Levy sampling criteria to be employed in this solution procedure

In section 7.2, the TD-EFIE, the TD-MFIE, the TD-PMCHW, and the TD-CFIE formulations for the analysis of 3-D arbitrarily shaped dielectric bodies are presented. The details on how to obtain the matrix equations from the integral equations will be given in section 7.3, whereas alternative approaches based on different expansion functions or use of the differential form of the integral equation will be presented in section 7.4. Finally, numerical examples obtained from the different integral equation formulations are presented in section 7.5, followed by some conclusions in section 7.6.

7.2   TIME DOMAIN INTEGRAL EQUATIONS (TDIES) FOR DIELECTRIC STRUCTURES

Consider a homogeneous dielectric body with a permittivity ε_2 and a permeability μ_2 placed in an infinite homogeneous medium with permittivity ε_1 and permeability μ_1 (refer to Figure 3.1). Similar to the scattering analysis of the dielectric structure in the frequency domain as mentioned in Chapter 3, two sets of integral equations related to the fields using the equivalent electric current $\mathbf{J}$ and the magnetic current $\mathbf{M}$ on the surface S of the dielectric body can be obtained by invoking the equivalence principle. Applying the equivalence principle to the exterior region of the dielectric body, equivalent electric current $\mathbf{J}$ and magnetic current $\mathbf{M}$ are required on the surface S of the dielectric body to preserve the continuity of the tangential electric and magnetic fields across the boundary. Therefore,

$$\left(-\mathbf{E}_1^s(\mathbf{r},t)\right)_{\text{tan}} = \left(\mathbf{E}^i(\mathbf{r},t)\right)_{\text{tan}} \tag{7.1}$$

$$\left(-\mathbf{H}_1^s(\mathbf{r},t)\right)_{\text{tan}} = \left(\mathbf{H}^i(\mathbf{r},t)\right)_{\text{tan}} \tag{7.2}$$

for $\mathbf{r} \in S$, where $\mathbf{E}^i$ and $\mathbf{H}^i$ are the incident electric and magnetic fields and $\mathbf{E}^s$ and $\mathbf{H}^s$ are the scattered electric and magnetic fields caused by the equivalent currents $\mathbf{J}$ and $\mathbf{M}$. The subscript "1" denotes that the scattered field $(\mathbf{E}_1^s, \mathbf{H}_1^s)$ is for the external region, and "tan" refers to the tangential component. Using similar procedures for the equivalence to the interior region and enforcing the continuity of the tangential electric and magnetic fields at the surface S, yields

$$\left(-\mathbf{E}_2^s(\mathbf{r},t)\right)_{\text{tan}} = 0 \tag{7.3}$$

$$\left(-\mathbf{H}_2^s(\mathbf{r},t)\right)_{\text{tan}} = 0 \tag{7.4}$$

for $\mathbf{r} \in S$, where the subscript "2" indicates that the scattered fields $(\mathbf{E}_2^s, \mathbf{H}_2^s)$ are for the internal regions.

In Eqs. (7.1)–(7.4), the scattered electric field on the surfaces of the internal and external boundaries can be expressed analytically in terms of the electric current $\mathbf{J}$ and the magnetic current $\mathbf{M}$ from

$$\mathbf{E}_v^s(\mathbf{r},t) = -\frac{\partial}{\partial t}\mathbf{A}_v(\mathbf{r},t) - \nabla\Phi_v(\mathbf{r},t) - \frac{1}{\varepsilon_v}\nabla\times\mathbf{F}_v(\mathbf{r},t) \tag{7.5}$$

and the scattered magnetic field is given by

$$\mathbf{H}_v^s(\mathbf{r},t) = -\frac{\partial}{\partial t}\mathbf{F}_v(\mathbf{r},t) - \nabla\Psi_v(\mathbf{r},t) + \frac{1}{\mu_v}\nabla\times\mathbf{A}_v(\mathbf{r},t) \tag{7.6}$$

In Eqs. (7.5) and (7.6), the symbol v is used to represent the region where the currents $\mathbf{J}$ or $\mathbf{M}$ are radiating and can be assigned as either 1 for the exterior region or 2 for the interior region. The parameters $\mathbf{A}_v$, $\mathbf{F}_v$, Φ_v, and Ψ_v are the magnetic vector potential, electric vector potential, electric scalar potential, and magnetic scalar potential, respectively, and are mathematically given by

$$A_v(\mathbf{r}, t) = \frac{\mu_v}{4\pi} \int_S \frac{\mathbf{J}(\mathbf{r}', \tau_v)}{R} dS' \tag{7.7}$$

$$F_v(\mathbf{r}, t) = \frac{\varepsilon_v}{4\pi} \int_S \frac{\mathbf{M}(\mathbf{r}', \tau_v)}{R} dS' \tag{7.8}$$

$$\Phi_v(\mathbf{r}, t) = \frac{1}{4\pi\varepsilon_v} \int_S \frac{Q_e(\mathbf{r}', \tau_v)}{R} dS' \tag{7.9}$$

$$\Psi_v(\mathbf{r}, t) = \frac{1}{4\pi\mu_v} \int_S \frac{Q_m(\mathbf{r}', \tau_v)}{R} dS' \tag{7.10}$$

$$R = |\mathbf{r} - \mathbf{r}'| \tag{7.11}$$

$$\tau_v = t - R/c_v \tag{7.12}$$

where R represents the distance between the observation point $\mathbf{r}$ and the source point $\mathbf{r}'$. The retarded time is denoted by τ_v and the velocity of the electromagnetic wave propagating in the space with medium parameters (ε_v, μ_v) is $c_v = 1/\sqrt{\varepsilon_v \mu_v}$. From the equation of continuity, the electric surface charge density Q_e is related to the electric current density $\mathbf{J}$, whereas the magnetic surface charge density Q_m is related to the magnetic current density $\mathbf{M}$ by

$$\nabla \cdot \mathbf{J}(\mathbf{r}, t) = -\frac{\partial Q_e(\mathbf{r}, t)}{\partial t} \tag{7.13}$$

and

$$\nabla \cdot \mathbf{M}(\mathbf{r}, t) = -\frac{\partial Q_m(\mathbf{r}, t)}{\partial t} \tag{7.14}$$

By taking the various combinations of the four boundary conditions given by Eqs. (7.1)–(7.4) associated with the electric and the magnetic fields for the exterior and the interior problems results, one can obtain the time domain electric field integral equation (TD-EFIE), the time domain magnetic field integral equation (TD-MFIE), the time domain combined field integral equation (TD-CFIE) and the time domain Poggio–Miller–Chang–Harrington–Wu (TD-PMCHW) formulations for a dielectric body.

Next we describe these four different formulations and their associated integral equation that needs to be solved numerically.

7.2.1 TD-EFIE

When combining the two boundary conditions of Eqs. (7.1) and (7.3) that are related only to the electric field, the TD-EFIE is formed. It gives the relation in the time domain between the incident electric field and the scattered electric field as a result of the induced electric and magnetic currents. Further substituting the expression for the scattered electric field given by Eq. (7.5), one obtains the TD-EFIE as

$$
\left(\frac{\partial}{\partial t} \mathbf{A}_\nu(\mathbf{r}, t) + \nabla \Phi_\nu(\mathbf{r}, t) + \frac{1}{\varepsilon_\nu} \nabla \times \mathbf{F}_\nu(\mathbf{r}, t) \right)_{\tan} = \begin{cases} \left(\mathbf{E}^i(\mathbf{r}, t) \right)_{\tan}, & \nu = 1 \\ 0, & \nu = 2 \end{cases} \tag{7.15}
$$

7.2.2 TD-MFIE

Dual to the TD-EFIE formulation, the MFIE formulation is obtained by selecting the two boundary conditions on the magnetic field only (i.e., using Eq. (7.2) and Eq. (7.4)). Substituting the expression for the scattered magnetic field given by Eq. (7.6), one obtains the TD-MFIE as

$$
\left(\frac{\partial}{\partial t} \mathbf{F}_\nu(\mathbf{r}, t) + \nabla \Psi_\nu(\mathbf{r}, t) - \frac{1}{\mu_\nu} \nabla \times \mathbf{A}_\nu(\mathbf{r}, t) \right)_{\tan} = \begin{cases} \left(\mathbf{H}^i(\mathbf{r}, t) \right)_{\tan}, & \nu = 1 \\ 0, & \nu = 2 \end{cases} \tag{7.16}
$$

7.2.3 TD-CFIE

By linearly combining the TD-EFIE of Eq. (7.15) and the TD-MFIE of Eq. (7.16), the combined field integral equation is obtained as follows[15]:

$$
(1-\kappa)\left(-\mathbf{E}_\nu^s(\mathbf{r}, t) \right)_{\tan} + \kappa \eta_\nu \left(-\mathbf{H}_\nu^s(\mathbf{r}, t) \right)_{\tan}
$$
$$
= \begin{cases} (1-\kappa)\left(\mathbf{E}^i(\mathbf{r}, t) \right)_{\tan} + \kappa \eta_\nu \left(\mathbf{H}^i(\mathbf{r}, t) \right)_{\tan}, & \nu = 1 \\ 0, & \nu = 2 \end{cases} \tag{7.17}
$$

where κ is the usual combination parameter that can have any value between 0 and 1, and η_ν is the wave impedance of region ν.

7.2.4 TD-PMCHW

An alternative way to combine the four equations given by Eqs. (7.1)–(7.4) yields the PMCHW formulation. In this formulation, the set of four equations is reduced to two by adding Eq. (7.1) to Eq. (7.3) and Eq. (7.2) to Eq. (7.4) to yield

$$\sum_{\nu=1}^{2}\left(\frac{\partial}{\partial t}\mathbf{A}_\nu(\mathbf{r},t)+\nabla\Phi_\nu(\mathbf{r},t)+\frac{1}{\varepsilon_\nu}\nabla\times\mathbf{F}_\nu(\mathbf{r},t)\right)_{\tan}=\left(\mathbf{E}^i(\mathbf{r},t)\right)_{\tan} \qquad (7.18a)$$

$$\sum_{\nu=1}^{2}\left(\frac{\partial}{\partial t}\mathbf{F}_\nu(\mathbf{r},t)+\nabla\Psi_\nu(\mathbf{r},t)-\frac{1}{\mu_\nu}\nabla\times\mathbf{A}_\nu(\mathbf{r},t)\right)_{\tan}=\left(\mathbf{H}^i(\mathbf{r},t)\right)_{\tan} \qquad (7.18b)$$

7.3   SOLUTION OF THE TDIES BY USING THE MOD METHOD

Next, the solution of the TDIEs described in the previous section for a dielectric structure is obtained by using the MOD method that has been introduced in Chapters 5 and 6. In the MOD method, the surface of the dielectric structure is approximated by planar triangular patches, and the triangular vector functions called the RWG functions are used as the spatial basis and testing functions. The triangular vector function $\mathbf{f}_n$ associated with the nth common edge is defined by [16]

$$\mathbf{f}_n(\mathbf{r})=\mathbf{f}_n^+(\mathbf{r})+\mathbf{f}_n^-(\mathbf{r}) \qquad (7.19a)$$

$$\mathbf{f}_n^\pm(\mathbf{r})=\begin{cases}\dfrac{l_n}{2A_n^\pm}\boldsymbol{\rho}_n^\pm, & \mathbf{r}\in T_n^\pm\\[2mm]0, & \mathbf{r}\notin T_n^\pm\end{cases} \qquad (7.19b)$$

where l_n and $A_n^\pm$ are the length of the edge n and the area of the triangle $T_n^\pm$, respectively. The position vector, $\boldsymbol{\rho}_n^\pm$, is defined with respect to the free vertex of $T_n^\pm$. For a dielectric body, another set of vector basis function $\mathbf{g}_n$ is required, and it is defined by

$$\mathbf{g}_n(\mathbf{r})=\mathbf{n}\times\mathbf{f}_n(\mathbf{r}) \qquad (7.20a)$$

where

$$\mathbf{g}_n(\mathbf{r})=\mathbf{g}_n^+(\mathbf{r})+\mathbf{g}_n^-(\mathbf{r}) \qquad (7.20b)$$

$$\mathbf{g}_n^\pm(\mathbf{r})=\mathbf{n}\times\mathbf{f}_n^\pm(\mathbf{r}) \qquad (7.20c)$$

where $\mathbf{n}$ is the unit normal pointing outward from the surface at $\mathbf{r}$. The functions $\mathbf{f}_n$ and $\mathbf{g}_n$ are pointwise orthogonal within a pair of triangular patches.

7.3.1   Solving the TD-EFIE

7.3.1.1 Spatial Expansion and Testing. To solve the TD-EFIE for dielectric structures given by Eq. (7.15), the electric current $\mathbf{J}$ and the magnetic current $\mathbf{M}$

on the dielectric structure first are approximated by the two spatial basis functions
as

$$\mathbf{J}(\mathbf{r}, t) = \sum_{n=1}^{N} J_n(t)\, \mathbf{f}_n(\mathbf{r}) \tag{7.21a}$$

$$\mathbf{M}(\mathbf{r}, t) = \sum_{n=1}^{N} M_n(t)\, \mathbf{g}_n(\mathbf{r}) \tag{7.21b}$$

where $J_n(t)$ and $M_n(t)$ are the unknown coefficients to be determined and N is
the number of edges on the triangular patches approximating the surface of the
dielectric body. These basis functions also have been used in the frequency
domain formulation [17,18].

  To handle the time derivative of the magnetic vector potential $\mathbf{A}(\mathbf{r}, t)$ in
Eq. (7.15) analytically, the Hertz vector $\mathbf{u}(\mathbf{r}, t)$ is introduced for the electric
current and is defined by

$$\mathbf{J}(\mathbf{r}, t) = \frac{\partial\, \mathbf{u}(\mathbf{r}, t)}{\partial t} \tag{7.22}$$

And the relation between the Hertz vector and the electric charge density is given
by

$$Q_e(\mathbf{r}, t) = -\nabla \bullet \mathbf{u}(\mathbf{r}, t) \tag{7.23}$$

Similarly, another Hertz vector related to the magnetic current is defined by

$$\mathbf{M}(\mathbf{r}, t) = \frac{\partial\, \mathbf{v}(\mathbf{r}, t)}{\partial t} \tag{7.24}$$

Using the basis function defined by Eq. (7.21), the Hertz vectors for the electric
and the magnetic currents are expressed as

$$\mathbf{u}(\mathbf{r}, t) = \sum_{n=1}^{N} u_n(t)\, \mathbf{f}_n(\mathbf{r}) \tag{7.25a}$$

$$\mathbf{v}(\mathbf{r}, t) = \sum_{n=1}^{N} v_n(t)\, \mathbf{g}_n(\mathbf{r}) \tag{7.25b}$$

Therefore, the solution for the unknown current coefficients $J_n(t)$ and $M_n(t)$
implies that one needs to determine $u_n(t)$ and $v_n(t)$. The next step in the
numerical implementation is to develop a testing procedure to transform the
operator equation of Eq. (7.15) into a matrix equation using the method of moment

(MoM). By choosing the expansion function $\mathbf{f}_m$ also as the testing functions results in

$$\left\langle \mathbf{f}_m(\mathbf{r}), \frac{\partial}{\partial t} \mathbf{A}_v(\mathbf{r},t) \right\rangle + \left\langle \mathbf{f}_m(\mathbf{r}), \nabla \Phi_v(\mathbf{r},t) \right\rangle + \left\langle \mathbf{f}_m(\mathbf{r}), \frac{1}{\varepsilon_v} \nabla \times \mathbf{F}_v(\mathbf{r},t) \right\rangle$$
$$= \begin{cases} \left\langle \mathbf{f}_m(\mathbf{r}), \mathbf{E}^i(\mathbf{r},t) \right\rangle, & v = 1 \\ 0, & v = 2 \end{cases} \tag{7.26}$$

where $m = 1, 2, ..., N$. Note that Eq. (7.26) is very similar to Eq. (6.76) of Chapter 6, which is obtained for the testing of the TD-EFIE for a conducting structure, except that in Eq. (7.26), there is a new term for the electric vector potential $\mathbf{F}_v(\mathbf{r},t)$. Assuming that the unknown transient quantities do not change appreciably within a given triangular patch, one can approximate the retardation factor by

$$\tau_v = t - \frac{R}{c_v} \quad \rightarrow \quad \tau_{mn,v}^{pq} = t - \frac{R_{mn}^{pq}}{c_v}, \quad R_{mn}^{pq} = \left| \mathbf{r}_m^{cp} - \mathbf{r}_n^{cq} \right| \tag{7.27}$$

where p and q are $+$ or $-$. The vector $\mathbf{r}_n^{c\pm}$ is the position vector of the center of triangle $T_n^{\pm}$. In Eq. (7.26), the terms for the magnetic vector potential $\mathbf{A}(\mathbf{r},t)$, the electric scalar potential $\nabla \Phi(\mathbf{r},t)$, and the incident electric field $\mathbf{E}^i(\mathbf{r},t)$ can be obtained from Eqs. (6.79), (6.81), and (6.82), respectively, from Chapter 6. For the term related to the electric vector potential $\mathbf{F}_v(\mathbf{r},t)$, it can be rewritten by extracting the Cauchy principal value from the curl term for a planar surface as

$$\frac{1}{\varepsilon_v} \nabla \times \mathbf{F}_v(\mathbf{r},t) = \pm \frac{1}{2} \mathbf{n} \times \mathbf{M}(\mathbf{r},t) + \frac{1}{\varepsilon_v} \nabla \times \tilde{\mathbf{F}}_v(\mathbf{r},t) \tag{7.28}$$

where a positive sign is used when $v = 1$ and a negative sign is used for $v = 2$. Also, the second term on the right-hand side is given by

$$\frac{1}{\varepsilon_v} \nabla \times \tilde{\mathbf{F}}_v(\mathbf{r},t) = \frac{1}{4\pi} \int_{S_0} \nabla \times \frac{\mathbf{M}(\mathbf{r}',\tau_v)}{R} dS' \tag{7.29}$$

where S_0 denotes a surface with the removal of the singularity at $\mathbf{r} = \mathbf{r}'$ or $R = 0$ from S. Through the use of a vector identity, the curl term inside the integral in Eq. (7.29) can be written as

$$\nabla \times \frac{\mathbf{M}(\mathbf{r}',\tau_v)}{R} = \frac{1}{c} \frac{\partial}{\partial t} \mathbf{M}(\mathbf{r}',\tau_v) \times \frac{\hat{\mathbf{R}}}{R} + \mathbf{M}(\mathbf{r}',\tau_v) \times \frac{\hat{\mathbf{R}}}{R^2} \tag{7.30}$$

where $\hat{\mathbf{R}}$ is a unit vector along the direction $\mathbf{r} - \mathbf{r}'$. Thus, using Eq. (7.28), the third term in Eq. (7.26) becomes

$$\left\langle \mathbf{f}_m(\mathbf{r}), \frac{1}{\varepsilon_v} \nabla \times \mathbf{F}_v(\mathbf{r},t) \right\rangle = \left\langle \mathbf{f}_m(\mathbf{r}), \pm \frac{1}{2} \mathbf{n} \times \mathbf{M}(\mathbf{r},t) \right\rangle + \left\langle \mathbf{f}_m(\mathbf{r}), \frac{1}{\varepsilon_v} \nabla \times \tilde{\mathbf{F}}_v(\mathbf{r},t) \right\rangle \quad (7.31)$$

Substituting the magnetic current $\mathbf{M}(\mathbf{r},t)$ given by Eqs. (7.24) and the Hertz vector $\mathbf{v}(\mathbf{r},t)$ given by (7.25b) into Eq. (7.31), the first term of the right-hand side in Eq. (7.31) becomes

$$\left\langle \mathbf{f}_m(\mathbf{r}), \pm \frac{1}{2} \mathbf{n} \times \mathbf{M}(\mathbf{r},t) \right\rangle = \sum_{n=1}^{N} C_{mn,v} \frac{d}{dt} v_n(t) \quad (7.32a)$$

where

$$C_{mn,v} = \begin{cases} +C_{mn}, & v = 1 \\ -C_{mn}, & v = 2 \end{cases} \quad (7.32b)$$

$$C_{mn} = \sum_{p,q} C_{mn}^{pq} = C_{mn}^{++} + C_{mn}^{+-} + C_{mn}^{-+} + C_{mn}^{--} \quad (7.32c)$$

$$C_{mn}^{pq} = \frac{1}{2} \int_S \mathbf{f}_m^p(\mathbf{r}) \cdot \mathbf{n} \times \mathbf{g}_n^q(\mathbf{r}) dS \quad (7.32d)$$

where p and q can be either $+$ or $-$. The integral of Eq. (7.32d) can be computed analytically [19]. Using Eq. (7.29), the second term of the right-hand side in Eq. (7.31) can be written as

$$\left\langle \mathbf{f}_m(\mathbf{r}), \frac{1}{\varepsilon_v} \nabla \times \tilde{\mathbf{F}}_v(\mathbf{r}, t) \right\rangle = \sum_{n=1}^{N} \frac{1}{4\pi} \int_S \mathbf{f}_m(\mathbf{r}) \cdot \int_{S_0} \left(\frac{1}{c_v} \frac{d^2}{dt^2} v_n(\tau_v) \mathbf{g}_n^q(\mathbf{r}') \times \frac{\hat{\mathbf{R}}}{R} \right.$$
$$\left. + \frac{d}{dt} v_n(\tau_v) \mathbf{g}_n^q(\mathbf{r}') \times \frac{\hat{\mathbf{R}}}{R^2} \right) dS' dS \quad (7.33)$$

By defining the retarded time by Eq. (7.27), one obtains

$$\left\langle \mathbf{f}_m(\mathbf{r}), \frac{1}{\varepsilon_v} \nabla \times \tilde{\mathbf{F}}_v(\mathbf{r}, t) \right\rangle \quad (7.34a)$$

$$= \sum_{n=1}^{N} \sum_{p,q} \left(\frac{\zeta_1^{pq}}{c_v} \frac{d^2}{dt^2} v_n(\tau_{mn,v}^{pq}) + \zeta_2^{pq} \frac{d}{dt} v_n(\tau_{mn,v}^{pq}) \right)$$

where

$$\zeta_v^{pq} = \frac{1}{4\pi} \int_S \mathbf{f}_m^p(\mathbf{r}) \cdot \int_S \mathbf{g}_n^q(\mathbf{r}') \times \frac{\hat{\mathbf{R}}}{R^v} dS' dS \quad (7.34b)$$

The integral in Eq. (7.34b) may be evaluated using the Gaussian quadrature for the unprimed and the primed coordinates numerically. Substituting Eqs. (7.32) and (7.34) into Eq. (7.31), one obtains

$$
\left\langle \mathbf{f}_m(\mathbf{r}), \frac{1}{\varepsilon_v} \nabla \times \mathbf{F}_v(\mathbf{r},t) \right\rangle
$$
$$
= \sum_{n=1}^{N} \left(C_{mn,v} \frac{d}{dt} v_n(t) + \sum_{p,q} \left(\frac{\varsigma_1^{pq}}{c_v} \frac{d^2}{dt^2} v_n(\tau_{mn,v}^{pq}) + \varsigma_2^{pq} \frac{d}{dt} v_n(\tau_{mn,v}^{pq}) \right) \right) \tag{7.35}
$$

Lastly, substituting Eq (7.33) into Eq. (7.24) along with Eqs. (6.79), (6.81), and (6.82) of Chapter 6, one obtains

$$
\sum_{n=1}^{N} \sum_{p,q} \left[\mu_v A_{mn}^{pq} \frac{d^2}{dt^2} u_n(\tau_{mn,v}^{pq}) + \frac{B_{mn}^{pq}}{\varepsilon_v} u_n(\tau_{mn,v}^{pq}) \right] + \sum_{n=1}^{N} C_{mn,v} \frac{d}{dt} v_n(t)
$$
$$
+ \sum_{n=1}^{N} \sum_{p,q} \left[\frac{\varsigma_1^{pq}}{c_v} \frac{d^2}{dt^2} v_n(\tau_{mn,v}^{pq}) + \varsigma_2^{pq} \frac{d}{dt} v_n(\tau_{mn,v}^{pq}) \right] = V_m^v(t) \tag{7.36a}
$$

where

$$
A_{mn}^{pq} = \frac{1}{4\pi} \int_S \mathbf{f}_m^p(\mathbf{r}) \cdot \int_S \frac{\mathbf{f}_n^q(\mathbf{r}')}{R} \, dS' dS \tag{7.36b}
$$

$$
B_{mn}^{pq} = \frac{1}{4\pi} \int_S \nabla \cdot \mathbf{f}_m^p(\mathbf{r}) \int_S \frac{\nabla' \cdot \mathbf{f}_n^q(\mathbf{r}')}{R} \, dS' \, dS \tag{7.36c}
$$

$$
C_{mn}^{pq} = \frac{1}{2} \int_S \mathbf{f}_m^p(\mathbf{r}) \cdot \mathbf{n} \times \mathbf{g}_n^q(\mathbf{r}) \, dS \tag{7.36d}
$$

$$
\varsigma_v^{pq} = \frac{1}{4\pi} \int_S \mathbf{f}_m^p(\mathbf{r}) \cdot \int_S \mathbf{g}_n^q(\mathbf{r}') \times \frac{\hat{\mathbf{R}}}{R^v} \, dS' dS \tag{7.36e}
$$

$$
V_m^v(t) = \begin{cases} \int_S \mathbf{f}_m(\mathbf{r}) \cdot \mathbf{E}^i(\mathbf{r}, t) \, dS, & v = 1 \\ 0, & v = 2 \end{cases} \tag{7.36f}
$$

7.3.1.2 Temporal Expansion and Testing.

Now, consider the choice of the temporal basis function and the temporal testing function. An orthonormal basis function set derived from the Laguerre polynomials called the associated Laguerre functions can be written as [11]

$$
\phi_j(t) = e^{-t/2} L_j(t) \tag{7.37}
$$

The coefficients for the temporal expansion functions $u_n(t)$ and $v_n(t)$ introduced in Eqs. (7.25a) and (7.25b), which are assumed to be causal responses (i.e., exist only for $t \geq 0$) can be expanded using Eq. (7.37) as

$$u_n(t) = \sum_{j=0}^{\infty} u_{n,j} \phi_j(st) \tag{7.38a}$$

$$v_n(t) = \sum_{j=0}^{\infty} v_{n,j} \phi_j(st) \tag{7.38b}$$

where $u_{n,j}$ and $v_{n,j}$ are the fixed coefficients to be determined, and s is a scaling factor. Applying the derivative property of the associated Laguerre functions given in the appendix of Chapter 5 [20–22], the first and second derivatives of $u_n(t)$ can be written explicitly in terms of the time domain coefficient $u_{n,j}$ as [20]

$$\frac{d}{dt} u_n(t) = s \sum_{j=0}^{\infty} \left(\frac{1}{2} u_{n,j} + \sum_{k=0}^{j-1} u_{n,k} \right) \phi_j(st) \tag{7.39a}$$

$$\frac{d^2}{dt^2} u_n(t) = s^2 \sum_{j=0}^{\infty} \left(\frac{1}{4} u_{n,j} + \sum_{k=0}^{j-1} (j-k) u_{n,k} \right) \phi_j(st) \tag{7.39b}$$

Similarly, the first and the second derivatives of $v_n(t)$ are given as

$$\frac{d}{dt} v_n(t) = s \sum_{j=0}^{\infty} \left(\frac{1}{2} v_{n,j} + \sum_{k=0}^{j-1} v_{n,k} \right) \phi_j(st) \tag{7.40a}$$

$$\frac{d^2}{dt^2} v_n(t) = s^2 \sum_{j=0}^{\infty} \left(\frac{1}{4} v_{n,j} + \sum_{k=0}^{j-1} (j-k) v_{n,k} \right) \phi_j(st) \tag{7.40b}$$

In deriving these expressions, it has been assumed that both the function and its first derivative are zero at $t = 0$ because of causality. Substituting Eqs. (7.38)–(7.40) into the spatial testing of Eq. (7.36) and performing a temporal testing, which means multiplying by $\phi_i(st)$ and integrating from zero to infinity, one obtains

$$\sum_{n=1}^{N} \sum_{p,q} \sum_{j=0}^{\infty} \left[\frac{s^2}{4} \mu_v A_{mn}^{pq} u_{n,j} + s^2 \mu_v A_{mn}^{pq} \sum_{k=0}^{j-1} (j-k) u_{n,k} \right.$$

$$+ \frac{B_{mn}^{pq}}{\varepsilon_v} u_{n,j} + \frac{s^2}{4} \frac{\zeta_1^{pq}}{c_v} v_{n,j} + s^2 \frac{\zeta_1^{pq}}{c_v} \sum_{k=0}^{j-1} (j-k) v_{n,k}$$

$$\left. + \frac{s}{2} \zeta_2^{pq} v_{n,j} + s \zeta_2^{pq} \sum_{k=0}^{j-1} v_{n,k} \right] I_{ij} (sR_{mn}^{pq}/c_v)$$

$$+ \sum_{n=1}^{N} \sum_{j=0}^{\infty} \left[\frac{s}{2} C_{mn,v} v_{n,j} + s C_{mn,v} \sum_{k=0}^{j-1} v_{n,k} \right] \delta_{ij} = V_{m,i}^{(v)} \qquad (7.41a)$$

where

$$I_{ij} \left(sR_{mn}^{pq}/c_v \right) = \int_{sR_{mn}^{pq}/c}^{\infty} \phi_i(st) \phi_j(st - s R_{mn}^{pq}/c_v) \, d(st) \qquad (7.41b)$$

$$V_{m,i}^v = \int_0^{\infty} \phi_i(st) V_m^v(t) \, d(st) \qquad (7.41c)$$

The integral in Eq. (7.41*b*) can be computed by applying the integral property of the associated Laguerre functions given in the appendix of Chapter 5. Note that $I_{ij} = 0$ when $j > i$, so that, in Eq. (7.41), the upper limit for the summation over j is replaced by i instead of ∞. By moving the terms associated with $u_{n,j}$ and $v_{n,j}$ for $j < i$ to the right-hand side, one can obtain a $2N \times 2N$ system of linear equations leading to

$$\sum_{n=1}^{N} \sum_{p,q} \left(\frac{s^2}{4} \mu_v A_{mn}^{pq} u_{n,i} + \frac{B_{mn}^{pq}}{\varepsilon_v} u_{n,i} + \frac{s^2}{4} \frac{\zeta_1^{pq}}{c_v} v_{n,i} + \frac{s}{2} \zeta_2^{pq} v_{n,i} \right) I_{ii} (sR_{mn}^{pq}/c_v)$$

$$+ \sum_{n=1}^{N} \frac{s}{2} C_{mn,v} v_{n,i} = V_{m,i}^{(v)} - \sum_{n=1}^{N} \sum_{p,q} \sum_{j=0}^{i-1} \left(\frac{s^2}{4} \mu_v A_{mn}^{pq} u_{n,j} + \frac{B_{mn}^{pq}}{\varepsilon_v} u_{n,j} + \frac{s^2}{4} \frac{\zeta_1^{pq}}{c_v} v_{n,j} \right.$$

$$\left. + \frac{s}{2} \zeta_2^{pq} v_{n,j} \right) I_{ij} (sR_{mn}^{pq}/c_v) - \sum_{n=1}^{N} \sum_{p,q} \sum_{j=0}^{i} \sum_{k=0}^{j-1} \left(s^2 \mu_v A_{mn}^{pq} (j-k) u_{n,k} \right.$$

$$+ s^2 \frac{\zeta_1^{pq}}{c_v} (j-k) v_{n,k} + s \zeta_2^{pq} v_{n,k} \bigg) I_{ij} (sR_{mn}^{pq}/c_v) - \sum_{n=1}^{N} \sum_{k=0}^{i-1} s C_{mn,v} v_{n,k} \qquad (7.42)$$

Rewriting Eq. (7.40) as

$$\sum_{n=1}^{N} \alpha_{mn}^{(v)} u_{n,i} + \sum_{n=1}^{N} \beta_{mn}^{(v)} v_{n,i} = \gamma_{m,i}^{(v)} \qquad (7.43a)$$

where

$$\gamma_{m,i}^{(v)} = V_{m,i}^{(v)} + P_{m,i}^{(v)} + Q_{m,i}^{(v)} \tag{7.43b}$$

$$\alpha_{mn}^{(v)} = \sum_{p,q} \left(\frac{s^2}{4} \mu_v A_{mn}^{pq} + \frac{B_{mn}^{pq}}{\varepsilon_v} \right) e^{\left(-s R_{mn}^{pq}/(2c_v) \right)} \tag{7.43c}$$

$$\beta_{mn}^{(v)} = \frac{s}{2} C_{mn,v} + \sum_{p,q} \left(\frac{s^2}{4} \frac{\zeta_1^{pq}}{c_v} + \frac{s}{2} \zeta_2^{pq} \right) e^{\left(-s R_{mn}^{pq}/(2c_v) \right)} \tag{7.43d}$$

$$P_{m,i}^{(v)} = -\sum_{n=1}^{N} \sum_{p,q} \sum_{j=0}^{i-1} \left(\frac{s^2}{4} \mu_v A_{mn}^{pq} u_{n,j} + \frac{B_{mn}^{pq}}{\varepsilon_v} u_{n,j} \right) I_{ij}(s R_{mn}^{pq}/c_v)$$

$$-\sum_{n=1}^{N} \sum_{p,q} \sum_{j=0}^{i} \sum_{k=0}^{j-1} \left(s^2 \mu_v A_{mn}^{pq} (j-k) u_{n,k} \right) I_{ij}(s R_{mn}^{pq}/c_v) \tag{7.43e}$$

$$Q_{m,i}^{(v)} = -\sum_{n=1}^{N} \sum_{p,q} \sum_{j=0}^{i-1} \left(\frac{s^2}{4} \frac{\zeta_1^{pq}}{c_v} v_{n,j} + \frac{s}{2} \zeta_2^{pq} v_{n,j} \right) I_{ij}(s R_{mn}^{pq}/c_v)$$

$$-\sum_{n=1}^{N} \sum_{p,q} \sum_{j=0}^{i} \sum_{k=0}^{j-1} \left(s^2 \frac{\zeta_1^{pq}}{c_v} (j-k) v_{n,k} + s \zeta_2^{pq} v_{n,k} \right) I_{ij}(s R_{mn}^{pq}/c$$

$$-\sum_{n=1}^{N} \sum_{k=0}^{i-1} s C_{mn,v} v_{n,k} \tag{7.43f}$$

Equation (7.43) can be written in a matrix form as

$$\begin{bmatrix} \left[\alpha_{mn}^{(1)} \right] & \left[\beta_{mn}^{(1)} \right] \\ \left[\alpha_{mn}^{(2)} \right] & \left[\beta_{mn}^{(2)} \right] \end{bmatrix} \begin{bmatrix} \left[u_{n,i} \right] \\ \left[v_{n,i} \right] \end{bmatrix} = \begin{bmatrix} \left[\gamma_{m,i}^{(1)} \right] \\ \left[\gamma_{m,i}^{(2)} \right] \end{bmatrix} \tag{7.44}$$

where $i = 0, 1, 2, \ldots, \infty$. It is important to note that the system matrix is independent of the order of the Laguerre polynomials and is only a function of the spatial discretization. The LU decomposition is used to solve this matrix only once for $i = 0$, and one uses the same decomposition for all the other values of i. The number of values of i is bounded by the time-bandwidth product M of the transient response that we are going to approximate. This parameter M is dependent on the time duration of the transient response and the bandwidth of the excitation signal. Consider a signal with a frequency bandwidth $2B$ in the frequency domain, which is of time duration T_f in the time domain. When this signal is represented by a Fourier series, the range of the sampling frequency varies from $-B \le k \Delta f \le B$, where k is an integer and $\Delta f = 1/T_f$. So we get $|k| \le B/T_f$. Hence, the minimum number of temporal basis functions to approximate the incident signal becomes $M = 2BT_f + 1$. Note that the upper limit of the integral $V_{m,i}^V$ in Eq. (7.41c) can be replaced by the time duration sT_f instead of infinity.

An implicit MOT method also will have similar computational operations. However, at stage $i = 0$ for the implicit MOT method, one will be solving a sparse matrix instead of a dense one. In addition to the number of timesteps at which the implicit MOT scheme solves the problem, at each timestep either is related to the rise time of the pulse or related to the spatial discretizations determined by the Courant condition. Typically, the spatial discretization requires at least 10 subsections per wavelength. The timestep is determined by this spatial discretization divided by the velocity of light. The Courant condition instead of the rise time of the incident pulse will dictate the value for the timestep. For the MOD method, however, the solution needs to be computed at a total number of M steps, which is determined by the time bandwidth product of the waveform, which is generally much smaller than the number of timesteps for an implicit MOT scheme.

7.3.1.3 Computation of the Scattered Far Field. By solving Eq. (7.44), using a *LU* decomposition, in a marching-on-in-degree manner with M temporal basis functions, the Hertz vector coefficients $u_{n,j}$ and $v_{n,j}$ are obtained. Then, the electric and magnetic current coefficients $J_n(t)$ and $M_n(t)$ can be expressed using Eqs. (7.22) and (7.24), Eqs. (7.25a) and (7.25b) with Eqs. (7.39a) and (7.40a), respectively, resulting in

$$J_n(t) = \frac{\partial}{\partial t} u_n(t) = s \sum_{j=0}^{M-1} \left(\frac{1}{2} u_{n,j} + \sum_{k=0}^{j-1} u_{n,k} \right) \phi_j(st) \qquad (7.45a)$$

$$M_n(t) = \frac{\partial}{\partial t} v_n(t) = s \sum_{j=0}^{M-1} \left(\frac{1}{2} v_{n,j} + \sum_{k=0}^{j-1} v_{n,k} \right) \phi_j(st). \qquad (7.45b)$$

Once the equivalent currents on the dielectric structure have been determined, the far scattered fields can be computed. These fields may be thought as the superposition of the fields separately produced by only the electric and the magnetic currents. The analytical form of the far fields can be derived directly from the transient coefficients $u_n(t)$ and $v_n(t)$, which can be obtained by solving Eq. (7.44) with Eq. (7.38). The scattered field caused by the electric currents alone at a point $\mathbf{r}$ is given by

$$\mathbf{E}_{\mathbf{J}}^s(\mathbf{r}, t) \approx -\frac{\partial \mathbf{A}(\mathbf{r}, t)}{\partial t} \qquad (7.46)$$

where the subscript **J** refers to the electric current. Substituting Eqs. (7.7), (7.22), and (7.25a) into Eq. (7.46) along with Eq. (7.19a) results in

$$\mathbf{E}_{\mathbf{J}}^{s}(\mathbf{r}, t) \approx -\frac{\mu_1}{4\pi} \sum_{n=1}^{N} \sum_{q} \int_{S} \frac{d^2}{dt^2} u_n(\tau) \frac{\mathbf{f}_n^q(\mathbf{r}')}{R} dS' \qquad (7.47)$$

The following approximations are made in the far field calculations:

$R \approx r - \mathbf{r}' \cdot \hat{\mathbf{r}}$ for the time retardation term $t - R/c_1$

$R \approx r$ for the amplitude term $1/R$

where $\hat{\mathbf{r}} = \mathbf{r}/r$ is a unit vector along the direction of the radiation. The integral in Eq. (7.47) is evaluated by approximating the integrand by the value at the center of the source triangle T_n^q. Substituting Eq. (7.19b) into Eq. (7.47) and approximating $\mathbf{r}' \approx \mathbf{r}_n^{cq}$ and $\rho_n^q \approx \rho_n^{cq}$, one obtains

$$\mathbf{E}_{\mathbf{J}}^{s}(\mathbf{r}, t) \approx -\frac{\mu_1}{8\pi r} \sum_{n=1}^{N} l_n \sum_{q} \rho_n^{cq} \frac{d^2}{dt^2} u_n(\tau_n^q) \qquad (7.48)$$

where $\tau_n^q \approx t - (r - \mathbf{r}_n^{cq} \cdot \hat{\mathbf{r}})/c_1$. The scattered magnetic field is given by

$$\mathbf{H}_{\mathbf{M}}^{s}(\mathbf{r}, t) \approx -\frac{\partial \mathbf{F}(\mathbf{r}, t)}{\partial t} \qquad (7.49)$$

where the subscript $\mathbf{M}$ refers to the magnetic current. By following a similar procedure as performed for the electric currents or applying the duality theorem to Eq. (7.47), the magnetic far field is given by

$$\mathbf{H}_{\mathbf{M}}^{s}(\mathbf{r}, t) \approx -\frac{\varepsilon_1}{4\pi} \sum_{n=1}^{N} \sum_{q} \int_{S} \frac{d^2}{dt^2} v_n(\tau) \frac{\mathbf{g}_n^q(\mathbf{r}')}{R} dS'$$

$$= -\frac{\varepsilon_1}{8\pi r} \sum_{n=1}^{N} l_n \sum_{q} \mathbf{n} \times \rho_n^{cq} \frac{d^2}{dt^2} v_n(\tau_n^q) \qquad (7.50)$$

where $\mathbf{n}$ is the unit normal pointing outward from the triangle T_n^q. Using Eq. (7.50), the electric field is given by

$$\mathbf{E}_{\mathbf{M}}^{s}(\mathbf{r}, t) = \eta_1 \mathbf{H}_{\mathbf{M}}^{s}(\mathbf{r}, t) \times \hat{\mathbf{r}} \qquad (7.51)$$

where η_1 is the wave impedance in the medium surrounding the scatterer. Finally, the total field scattered from the dielectric body may be obtained by adding Eqs. (7.48) and (7.51) as

$$\mathbf{E}^s(\mathbf{r},t) \approx -\frac{1}{8\pi c_1 r} \sum_{n=1}^{N} \sum_{q} \left[l_n \, \eta_1 \, \mathbf{\rho}_n^q \frac{d^2}{dt^2} u_n(\tau_n^q) + l_n \, \mathbf{n} \times \mathbf{\rho}_n^q \frac{d^2}{dt^2} v_n(\tau_n^q) \times \hat{\mathbf{r}} \right] \tag{7.52}$$

The second derivative of the time domain coefficient is obtained from Eq. (7.39b) with $M-1$ as the upper limit for the summation over j instead of ∞.

7.3.2 Solving the TD-MFIE

7.3.2.1 Spatial Expansion and Testing. The analysis of the dielectric structure in the time domain also can be performed using the TD-MFIE given by Eq. (7.16). The surface of the dielectric structure to be analyzed is approximated by planar triangular patches. The pointwise orthogonal vector functions $\mathbf{f}_n$ and $\mathbf{g}_n$ defined by Eqs. (7.19) and (7.20) are used as the spatial basis function for approximating the electric current $\mathbf{J}$ and the magnetic current $\mathbf{M}$ over the patch as

$$\mathbf{J}(\mathbf{r},t) = \sum_{n=1}^{N} J_n(t) \, \mathbf{g}_n(\mathbf{r}) \tag{7.53a}$$

$$\mathbf{M}(\mathbf{r},t) = \sum_{n=1}^{N} M_n(t) \, \mathbf{f}_n(\mathbf{r}) \tag{7.53b}$$

where $J_n(t)$ and $M_n(t)$ are unknown constants to be determined, and N is the number of edges related to the triangulated model approximating the surface of the dielectric body. These expansions are dual to Eq. (7.21) for the spatial expansion of the TD-EFIE.

The two Hertz vectors $\mathbf{u}(\mathbf{r},t)$ and $\mathbf{v}(\mathbf{r},t)$ given by Eqs. (7.22) and (7.24), respectively, also are used to represent the electric current $\mathbf{J}(\mathbf{r},t)$ and the magnetic current $\mathbf{M}(\mathbf{r},t)$ in the TD-MFIE, and the relation between the Hertz vector and the magnetic charge density is given by

$$Q_m(\mathbf{r},t) = -\nabla \cdot \mathbf{v}(\mathbf{r},t) \tag{7.54}$$

These electric and the magnetic Hertz vectors are rewritten as

$$\mathbf{u}(\mathbf{r},t) = \sum_{n=1}^{N} u_n(t) \, \mathbf{g}_n(\mathbf{r}) \tag{7.55a}$$

$$\mathbf{v}(\mathbf{r},t) = \sum_{n=1}^{N} v_n(t) \, \mathbf{f}_n(\mathbf{r}) \tag{7.55b}$$

By choosing the expansion function $\mathbf{f}_m$ also as a testing function, the testing of the TD-MFIE yields

$$\left\langle \mathbf{f}_m(\mathbf{r}), \frac{\partial}{\partial t} \mathbf{F}_v(\mathbf{r}, t) \right\rangle + \left\langle \mathbf{f}_m(\mathbf{r}), \nabla \Psi_v(\mathbf{r}, t) \right\rangle - \left\langle \mathbf{f}_m(\mathbf{r}), \frac{1}{\mu_v} \nabla \times \mathbf{A}_v(\mathbf{r}, t) \right\rangle$$

$$= \begin{cases} \left\langle \mathbf{f}_m(\mathbf{r}), \mathbf{H}^i(\mathbf{r}, t) \right\rangle, & v = 1 \\ 0, & v = 2 \end{cases} \tag{7.56}$$

where $m = 1, 2, \ldots, N$. The next step in the MoM procedure is to substitute the expansion functions of the Hertz vectors defined in Eqs. (7.55) into Eq. (7.56). First consider the testing of the integral related to the electric vector potential $\mathbf{F}(\mathbf{r}, t)$. Using Eqs. (7.8), (7.53b), and (7.55b), results in

$$\left\langle \mathbf{f}_m(\mathbf{r}), \frac{\partial}{\partial t} \mathbf{F}_v(\mathbf{r}, t) \right\rangle = \sum_{n=1}^{N} \frac{\varepsilon_v}{4\pi} \int_S \mathbf{f}_m(\mathbf{r}) \cdot \int_S \frac{d^2}{dt^2} v_n(\tau_v) \frac{\mathbf{f}_n(\mathbf{r}')}{R} dS' dS \tag{7.57}$$

Assuming that the unknown transient quantity does not change appreciably within the triangular patch as in Eq. (7.27), Eq. (7.57) can be written as

$$\left\langle \mathbf{f}_m(\mathbf{r}), \frac{\partial}{\partial t} \mathbf{F}_v(\mathbf{r}, t) \right\rangle = \sum_{n=1}^{N} \sum_{p,q} \varepsilon_v A_{mn}^{pq} \frac{d^2}{dt^2} v_n(\tau_{mn,v}^{pq}) \tag{7.58a}$$

where

$$A_{mn}^{pq} = \frac{1}{4\pi} \int_S \mathbf{f}_m^p(\mathbf{r}) \cdot \int_S \frac{\mathbf{f}_n^q(\mathbf{r}')}{R} dS' dS \tag{7.58b}$$

Next consider the testing of the gradient of the scalar potential $\Psi(\mathbf{r}, t)$ in Eq. (7.57). Using the vector identity $\nabla \cdot \Psi \mathbf{A} = \mathbf{A} \cdot \nabla \Psi + \Psi \nabla \cdot \mathbf{A}$ and the property of the spatial basis function [16], one can write

$$\left\langle \mathbf{f}_m(\mathbf{r}), \nabla \Psi_v(\mathbf{r}, t) \right\rangle = -\int_S \nabla \cdot \mathbf{f}_m(\mathbf{r}) \Psi_v(\mathbf{r}, t) dS$$

$$= \sum_{n=1}^{N} \frac{1}{4\pi \mu_v} \int_S \nabla \cdot \mathbf{f}_m(\mathbf{r}) \int_S v_n(\tau_v) \frac{\nabla' \cdot \mathbf{f}_n(\mathbf{r}')}{R} dS' dS \tag{7.59}$$

With the assumption for the retarded time given by Eq. (7.27), Eq. (7.59) can be written as

$$\left\langle \mathbf{f}_m(\mathbf{r}), \nabla \Psi_v(\mathbf{r}, t) \right\rangle = \sum_{n=1}^{N} \sum_{p,q} \frac{B_{mn}^{pq}}{\mu_v} v_n(\tau_{mn,v}^{pq}) \tag{7.60a}$$

where

$$B_{mn}^{pq} = \frac{1}{4\pi} \int_S \nabla \cdot \mathbf{f}_m^p(\mathbf{r}) \int_S \frac{\nabla' \cdot \mathbf{f}_n^q(\mathbf{r}')}{R} \, dS' \, dS \tag{7.60b}$$

The integrals in Eqs. (7.58b) and (7.60b) may be evaluated by the method described in Ref. [16]. Extracting the Cauchy principal value from the curl term for a planar surface, the third term of Eq. (7.56) for the testing of magnetic vector potential $\mathbf{A}(\mathbf{r}, t)$ may be written as

$$\frac{1}{\mu_\nu} \nabla \times \mathbf{A}_\nu(\mathbf{r}, t) = \pm \frac{1}{2} \mathbf{n} \times \mathbf{J} + \frac{1}{\mu_\nu} \nabla \times \tilde{\mathbf{A}}_\nu(\mathbf{r}, t) \tag{7.61}$$

where the positive sign is used when $\nu = 1$ and the negative sign is used for $\nu = 2$. The second term of the right-hand side in Eq. (7.61) is given by

$$\frac{1}{\mu_\nu} \nabla \times \tilde{\mathbf{A}}_\nu(\mathbf{r}, t) = \frac{1}{4\pi} \int_{S_0} \nabla \times \frac{\mathbf{J}(\mathbf{r}', \tau_\nu)}{R} \, dS' \tag{7.62}$$

where S_0 denotes the surface with the removal of the contribution due to the singularity at $\mathbf{r} = \mathbf{r}'$ or $R = 0$ from the surface S. Using a vector identity [23], the curl term inside the integral in Eq. (7.62) is given by

$$\nabla \times \frac{\mathbf{J}(\mathbf{r}', \tau_\nu)}{R} = \frac{1}{c} \frac{\partial}{\partial t} \mathbf{J}(\mathbf{r}', \tau_\nu) \times \frac{\hat{\mathbf{R}}}{R} + \mathbf{J}(\mathbf{r}', \tau_\nu) \times \frac{\hat{\mathbf{R}}}{R^2} \tag{7.63}$$

where $\hat{\mathbf{R}}$ is a unit vector along the direction $\mathbf{r} - \mathbf{r}'$. Thus, using Eq. (7.61), the third term in Eq. (7.56) becomes

$$\left\langle \mathbf{f}_m(\mathbf{r}), \frac{1}{\mu_\nu} \nabla \times \mathbf{A}_\nu(\mathbf{r}, t) \right\rangle = \left\langle \mathbf{f}_m(\mathbf{r}), \pm \frac{1}{2} \mathbf{n} \times \mathbf{J} \right\rangle + \left\langle \mathbf{f}_m(\mathbf{r}), \frac{1}{\mu_\nu} \nabla \times \tilde{\mathbf{A}}_\nu(\mathbf{r}, t) \right\rangle \tag{7.64}$$

Now, consider the first term in the right-hand side of Eq. (7.64). Substituting the Hertz vector $\mathbf{u}(\mathbf{r}, t)$ in terms of the electric current $\mathbf{J}(\mathbf{r}, t)$ of Eqs. (7.22) and expanding the Hertz vector $\mathbf{u}(\mathbf{r}, t)$ using (7.55a) yields

$$\left\langle \mathbf{f}_m(\mathbf{r}), \pm \frac{1}{2} \mathbf{n} \times \mathbf{J} \right\rangle = \sum_{n=1}^{N} C_{mn,\nu} \frac{d}{dt} u_n(t) \tag{7.65a}$$

where

$$C_{mn,\nu} = \begin{cases} +C_{mn}, & \nu = 1 \\ -C_{mn}, & \nu = 2 \end{cases} \tag{7.65b}$$

$$C_{mn} = \sum_{p,q} C_{mn}^{pq} = C_{mn}^{++} + C_{mn}^{+-} + C_{mn}^{-+} + C_{mn}^{--} \tag{7.65c}$$

$$C_{mn}^{pq} = \frac{1}{2} \int_S \mathbf{f}_m^p(\mathbf{r}) \cdot \mathbf{n} \times \mathbf{g}_n^q(\mathbf{r}) \, dS \tag{7.65d}$$

The integral of Eq. (7.65d) can be computed analytically [19]. Using Eq. (7.63), the second term on the right-hand side in Eq. (7.64) becomes

$$\left\langle \mathbf{f}_m(\mathbf{r}), \frac{1}{\mu_v} \nabla \times \tilde{\mathbf{A}}_v(\mathbf{r},t) \right\rangle = \sum_{n=1}^N \frac{1}{4\pi} \int_S \mathbf{f}_m(\mathbf{r}) \cdot \int_{S_0} \left[\frac{1}{c_v} \frac{d^2}{dt^2} u_n(\tau_v) \mathbf{g}_n^q(\mathbf{r}') \times \frac{\hat{\mathbf{R}}}{R} \right.$$
$$\left. + \frac{d}{dt} u_n(\tau_v) \mathbf{g}_n^q(\mathbf{r}') \times \frac{\hat{\mathbf{R}}}{R^2} \right] dS' dS \tag{7.66}$$

By approximating the retarded time as in Eq. (7.27), one obtains

$$\left\langle \mathbf{f}_m(\mathbf{r}), \frac{1}{\mu_v} \nabla \times \tilde{\mathbf{A}}_v(\mathbf{r},t) \right\rangle = \sum_{n=1}^N \sum_{p,q} \left(\frac{\zeta_1^{pq}}{c_v} \frac{d^2}{dt^2} u_n(\tau_{mn,v}^{pq}) + \zeta_2^{pq} \frac{d}{dt} u_n(\tau_{mn,v}^{pq}) \right) \tag{7.67a}$$

where

$$\zeta_v^{pq} = \frac{1}{4\pi} \int_S \mathbf{f}_m^p(\mathbf{r}) \cdot \int_S \mathbf{g}_n^q(\mathbf{r}') \times \frac{\hat{\mathbf{R}}}{R^v} \, dS' dS \tag{7.67b}$$

The integral in Eq. (7.67b) may be evaluated using the Gaussian quadrature numerically for the unprimed and the primed coordinates. Substituting Eqs. (7.65) and (7.67) into Eq. (7.64), one obtains

$$\left\langle \mathbf{f}_m(\mathbf{r}), \frac{1}{\mu_v} \nabla \times \mathbf{A}_v(\mathbf{r},t) \right\rangle = \sum_{n=1}^N C_{mn,v} \frac{d}{dt} u_n(t)$$
$$+ \sum_{n=1}^N \sum_{p,q} \left(\frac{\zeta_1^{pq}}{c_v} \frac{d^2}{dt^2} u_n(\tau_{mn,v}^{pq}) + \zeta_2^{pq} \frac{d}{dt} u_n(\tau_{mn,v}^{pq}) \right) \tag{7.68}$$

Finally, substituting Eqs. (7.58), (7.60), and (7.68) into Eq. (7.56), one obtains

$$\sum_{n=1}^N \sum_{p,q} \left(\varepsilon_v A_{mn}^{pq} \frac{d^2}{dt^2} v_n(\tau_{mn,v}^{pq}) + \frac{B_{mn}^{pq}}{\mu_v} v_n(\tau_{mn,v}^{pq}) \right) - \sum_{n=1}^N C_{mn,v} \frac{d}{dt} u_n(t)$$
$$- \sum_{n=1}^N \sum_{p,q} \left(\frac{\zeta_1^{pq}}{c_v} \frac{d^2}{dt^2} u_n(\tau_{mn,v}^{pq}) + \zeta_2^{pq} \frac{d}{dt} u_n(\tau_{mn,v}^{pq}) \right) = V_m^v(t) \tag{7.69a}$$

where

$$
V_m^v(t) = \begin{cases} \int_S \mathbf{f}_m(\mathbf{r}) \cdot \mathbf{H}^i(\mathbf{r}, t)\, dS, & v = 1 \\[2mm] 0, & v = 2 \end{cases} \tag{7.69b}
$$

7.3.2.2 Temporal Expansion and Testing. Next, the temporal basis and the testing functions are described. The orthonormal functions defined by Eq. (7.37) are used for temporal expansion and testing functions. The transient coefficients $u_n(t)$ and $v_n(t)$ then are expanded using this function as shown in Eq. (7.38). The first and second derivatives of the coefficients are given by Eqs. (7.39) and (7.40), respectively. Substituting Eqs. (7.38)–(7.40) into the spatial testing result in Eq. (7.69) and then performing a temporal testing with the testing function $\phi_i(st)$, one obtains

$$
\sum_{n=1}^N \sum_{p,q} \sum_{j=0}^\infty \left(\frac{s^2}{4} \varepsilon_v A_{mn}^{pq} v_{n,j} + \frac{B_{mn}^{pq}}{\mu_v} v_{n,j} + s^2 \varepsilon_v A_{mn}^{pq} \sum_{k=0}^{j-1}(j-k)v_{n,k} - \frac{s^2 \zeta_1^{pq}}{4 c_v} u_{n,j} \right.
$$

$$
\left. - \frac{s^2 \zeta_1^{pq}}{c_v} \sum_{k=0}^{j-1}(j-k)u_{n,k} - \frac{s \zeta_2^{pq}}{2} u_{n,j} - s \zeta_2^{pq} \sum_{k=0}^{j-1} u_{n,k} \right) I_{ij}\!\left(s R_{mn}^{pq}/c_v\right)
$$

$$
- \sum_{n=1}^N \sum_{j=0}^\infty s\, C_{mn,v} \left(\frac{u_{n,j}}{2} + \sum_{k=0}^{j-1} u_{n,k} \right) \delta_{ij} = V_{m,i}^{(v)} \tag{7.70a}
$$

where

$$
I_{ij}\!\left(s R_{mn}^{pq}/c_v\right) = \int_{s R_{mn}^{pq}/c}^\infty \phi_i(st)\, \phi_j(st - s R_{mn}^{pq}/c_v)\, d(st) \tag{7.70b}
$$

$$
V_{m,i}^v = \int_0^\infty \phi_i(st) V_m^v(t)\, d(st) \tag{7.70c}
$$

Note that the integrals I_{ij} in Eq. (7.70b) and $V_{m,i}^v$ in Eq. (7.70c) are the same as in Eqs. (7.41b) and (7.41c). I_{ij} can be computed by applying the integral property of the associated Laguerre functions and observe that $I_{ij} = 0$ when $j > i$. Therefore, in Eq. (7.71), the upper limit for the summation over j is replaced by i instead of ∞. By moving the terms associated with $u_{n,j}$ and $v_{n,j}$, which is known for $j < i$, to the right-hand side, one obtains a $2N \times 2N$ system of linear equations as follows:

$$\sum_{n=1}^{N}\sum_{p,q}\left(\frac{s^2}{4}\varepsilon_v A_{mn}^{pq}v_{n,j}+\frac{B_{mn}^{pq}}{\mu_v}v_{n,j}-\frac{s^2\zeta_1^{pq}}{4c_v}u_{n,j}-\frac{s\zeta_2^{pq}}{2}u_{n,j}\right)I_{ii}(sR_{mn}^{pq}/c_v)$$

$$-\sum_{n=1}^{N}sC_{mn,v}\frac{u_{n,j}}{2}=V_{m,i}^{(v)}-\sum_{n=1}^{N}\sum_{p,q}\sum_{j=0}^{i-1}\left(\frac{s^2}{4}\varepsilon_v A_{mn}^{pq}v_{n,j}+\frac{B_{mn}^{pq}}{\mu_v}v_{n,j}-\frac{s^2\zeta_1^{pq}}{4c_v}u_{n,j}\right.$$

$$\left.-\frac{s\zeta_2^{pq}}{2}u_{n,j}\right)I_{ij}(sR_{mn}^{pq}/c_v)-\sum_{n=1}^{N}\sum_{p,q}\sum_{j=0}^{i}\sum_{k=0}^{j-1}\left(s^2\varepsilon_v A_{mn}^{pq}(j-k)v_{n,k}\right.$$

$$\left.-\frac{s^2\zeta_1^{pq}}{c_v}(j-k)u_{n,k}-s\zeta_2^{pq}u_{n,k}\right)I_{ij}(sR_{mn}^{pq}/c_v)+\sum_{n=1}^{N}\sum_{k=0}^{i-1}sC_{mn,v}u_{n,k} \qquad (7.71)$$

Rewriting Eq. (7.71) in a compact form results in

$$\sum_{n=1}^{N}\beta_{mn}^{(v)}u_{n,i}+\sum_{n=1}^{N}\alpha_{mn}^{(v)}v_{n,i}=\gamma_{m,i}^{(v)} \qquad (7.72a)$$

where

$$\gamma_{m,i}^{(v)}=V_{m,i}^{(v)}+P_{m,i}^{(v)}+Q_{m,i}^{(v)} \qquad (7.72b)$$

$$\alpha_{mn}^{(v)}=\sum_{p,q}\left(\frac{s^2}{4}\varepsilon_v A_{mn}^{pq}+\frac{B_{mn}^{pq}}{\mu_v}\right)e^{\left(-sR_{mn}^{pq}/(2c_v)\right)} \qquad (7.72c)$$

$$\beta_{mn}^{(v)}=-\sum_{p,q}\left(\frac{s^2\zeta_1^{pq}}{4c_v}+\frac{s\zeta_2^{pq}}{2}\right)e^{\left(-sR_{mn}^{pq}/(2c_v)\right)}-\frac{s}{2}C_{mn,v} \qquad (7.72d)$$

$$P_{m,i}^{(v)}=-\sum_{n=1}^{N}\sum_{p,q}\sum_{j=0}^{i-1}\left(\frac{s^2}{4}\varepsilon_v A_{mn}^{pq}v_{n,j}+\frac{B_{mn}^{pq}}{\mu_v}v_{n,j}\right)I_{ij}(sR_{mn}^{pq}/c_v)$$

$$-\sum_{n=1}^{N}\sum_{p,q}\sum_{j=0}^{i}\sum_{k=0}^{j-1}\left(s^2\varepsilon_v A_{mn}^{pq}(j-k)v_{n,k}\right)I_{ij}(sR_{mn}^{pq}/c_v) \qquad (7.72e)$$

$$Q_{m,i}^{(v)}=\sum_{n=1}^{N}\sum_{p,q}\sum_{j}^{i-1}\left(\frac{s^2\zeta_1^{pq}}{4c_v}u_{n,j}+\frac{s\zeta_2^{pq}}{2}u_{n,j}\right)I_{ij}(sR_{mn}^{pq}/c_v)$$

$$+\sum_{n=1}^{N}\sum_{p,q}\sum_{j}^{i}\sum_{k=0}^{j-1}\left(\frac{s^2\zeta_1^{pq}}{c_v}(j-k)u_{n,k}+s\zeta_2^{pq}u_{n,k}\right)I_{ij}(sR_{mn}^{pq}/c_v)$$

$$ \qquad (7.72f)$$

$$+\sum_{n=1}^{N}\sum_{k=0}^{i-1}sC_{mn,v}u_{n,k}$$

Eq. (7.72) now can be written in a matrix form as

$$
\begin{bmatrix} \left[\beta_{mn}^{(1)}\right] & \left[\alpha_{mn}^{(1)}\right] \\ \left[\beta_{mn}^{(2)}\right] & \left[\alpha_{mn}^{(2)}\right] \end{bmatrix} \begin{bmatrix} \left[u_{n,i}\right] \\ \left[v_{n,i}\right] \end{bmatrix} = \begin{bmatrix} \left[\gamma_{m,i}^{(1)}\right] \\ \left[\gamma_{m,i}^{(2)}\right] \end{bmatrix}
\tag{7.73}
$$

where $i = 0, 1, 2, \cdots, \infty$. Similar to the TD-EFIE, the minimum number of temporal basis functions, M, to approximate the incident signal can be estimated by $M = 2BT_f + 1$.

7.3.2.3 Computation of the Scattered Far Field.

The scattered far field can be computed from the equivalent currents on the surface of the dielectric scatterer given by the solution of the TD-MFIE. By solving the TD-MFIE, using the MOD methodology with M temporal basis functions, the Hertz vector coefficients $u_{n,j}$ and $v_{n,j}$ are obtained. Then, the electric and the magnetic transient current coefficients $J_n(t)$ and $M_n(t)$ can be evaluated using the relationship between the transient current and the Hertz vectors of Eq. (7.22) and (7.24) and by using Eqs. (7.45a) and (7.45b). The first derivatives of the Hertz vectors then are given by Eqs. (7.39a) and (7.40a). Because the spatial basis functions $\mathbf{f}_n$ and $\mathbf{g}_n$ for expanding the Hertz vectors are dual to that in the TD-EFIE, the field computations can be obtained based on the procedure described in Section 7.3.1.3 and apply the duality to the spatial basis functions to have the electric field caused by electric current as

$$
\mathbf{E}_{\mathbf{J}}^s(\mathbf{r}, t) \approx -\frac{\mu_1}{4\pi} \sum_{n=1}^{N} \sum_q \int_S \frac{d^2}{dt^2} u_n(\tau) \frac{\mathbf{g}_n^q(\mathbf{r}')}{R} dS'
\tag{7.74}
$$

and simplifies to

$$
\mathbf{E}_{\mathbf{J}}^s(\mathbf{r}, t) \approx -\frac{\mu_1}{8\pi r} \sum_{n=1}^{N} l_n \sum_q \mathbf{n} \times \boldsymbol{\rho}_n^{cq} \frac{d^2}{dt^2} u_n(\tau_n^q)
\tag{7.75}
$$

The magnetic field caused by the magnetic current is given by

$$
\mathbf{H}_{\mathbf{M}}^s(\mathbf{r}, t) \approx -\frac{\varepsilon_1}{4\pi} \sum_{n=1}^{N} \sum_q \int_S \frac{d^2}{dt^2} v_n(\tau) \frac{\mathbf{f}_n^q(\mathbf{r}')}{R} dS'
$$
$$
= -\frac{\varepsilon_1}{8\pi r} \sum_{n=1}^{N} l_n \sum_q \boldsymbol{\rho}_n^{cq} \frac{d^2}{dt^2} v_n(\tau_n^q)
\tag{7.76}
$$

and the total scattered electric field is obtained as

$$
\mathbf{E}^s(\mathbf{r}, t) \approx -\frac{1}{8\pi c_1 r} \left(\eta_1 \sum_{n=1}^{N} \mathbf{A}_n + \sum_{n=1}^{N} \mathbf{F}_n \times \hat{\mathbf{r}} \right)
\tag{7.77a}
$$

$$\mathbf{A}_n = l_n \sum_q \mathbf{n} \times \boldsymbol{\rho}_n^q \frac{d^2}{dt^2} u_n(\tau_n^q) \tag{7.77b}$$

$$\mathbf{F}_n = l_n \sum_q \boldsymbol{\rho}_n^q \frac{d^2}{dt^2} v_n(\tau_n^q) \tag{7.77c}$$

7.3.3 Solving the TD-CFIE

7.3.3.1 Spatial Expansion and Testing. To solve TD-CFIE of Eq. (7.15), the electric and the magnetic currents are approximated by the functions $\mathbf{f}_n$ defined in Eq. (7.17), resulting in

$$\mathbf{J}(\mathbf{r}, t) = \sum_{n=1}^{N} J_n(t) \, \mathbf{f}_n(\mathbf{r}) \tag{7.78a}$$

$$\mathbf{M}(\mathbf{r}, t) = \sum_{n=1}^{N} M_n(t) \, \mathbf{f}_n(\mathbf{r}) \tag{7.78b}$$

The Hertz vectors $\mathbf{u}(r,t)$ and $\mathbf{v}(r,t)$ defined in Eqs. (7.22) and (7.24) are used and are given by

$$\mathbf{u}(\mathbf{r}, t) = \sum_{n=1}^{N} u_n(t) \, \mathbf{f}_n(\mathbf{r}) \tag{7.79a}$$

$$\mathbf{v}(\mathbf{r}, t) = \sum_{n=1}^{N} v_n(t) \, \mathbf{f}_n(\mathbf{r}) \tag{7.79b}$$

Similar to the testing of the CFIE in the frequency domain presented in section 3.6 [15], the spatial testing of TD-CFIE formulation uses a combination of $\mathbf{f}_n$ and its orthogonal vector $\mathbf{g}_n$ as the testing functions to convert the TD-CFIE into a matrix equation. A general expression for the spatial testing of TD-CFIE using the four parameters, f_E, g_E, f_H, and g_{II}, in conjunction with the testing functions $\mathbf{f}_n$ and $\mathbf{g}_n$ may be written as

$$(1-\kappa)\left\langle f_E\mathbf{f}_m + g_E\mathbf{g}_m, -\mathbf{E}_v^s(\mathbf{r}, t)\right\rangle + \kappa\eta_v\left\langle f_H\mathbf{f}_m + g_H\mathbf{g}_m, -\mathbf{H}_v^s(\mathbf{r}, t)\right\rangle =$$

$$\begin{cases} (1-\kappa)\left\langle f_E\mathbf{f}_m + g_E\mathbf{g}_m, \mathbf{E}^i(\mathbf{r}, t)\right\rangle + \kappa\eta_1\left\langle f_H\mathbf{f}_m + g_H\mathbf{g}_m, \mathbf{H}^i(\mathbf{r}, t)\right\rangle & \text{for } v=1 \\ 0 & v=2 \end{cases} \tag{7.80}$$

where $m = 1, 2, \cdots, N$ and the testing coefficients f_E, g_E, f_H, and g_H, may be $+1$ or -1. This equation is termed as the TENE-THNH formulation as in Ref. [14]. To simplify the transformation of Eq. (7.80) into a matrix equation, the testing procedure is separated into two steps that are valid only for those terms related to either the electric field or the magnetic field.

First, the portion related to only the electric field is tested with the testing function $f_E \mathbf{f}_m + g_E \mathbf{g}_m$, and assuming that the transient quantities do not change appreciably within a given triangular patch as in Eq. (7.27), result in

$$\sum_{n=1}^{N} \sum_{p,q} \left[\mu_v A_{mn}^{pq} \frac{d^2}{dt^2} u_n(\tau_{mn,v}^{pq}) + \frac{F_{mn}^{pq}}{\varepsilon_v} u_n(\tau_{mn,v}^{pq}) + \frac{G_{mn}^{pq}}{c_v \varepsilon_v} \frac{d}{dt} u_n(\tau_{mn,v}^{pq}) \right.$$

$$\left. + \frac{D_{mn}^{pq}}{c_v} \frac{d^2}{dt^2} v_n(\tau_{mn,v}^{pq}) + E_{mn}^{pq} \frac{d}{dt} v_n(\tau_{mn,v}^{pq}) \right] + \sum_{n=1}^{N} C_{mn,v} \frac{d}{dt} v_n(t) = V_m^{E(v)}(t) \qquad (7.81)$$

Equation (7.81) is called the TENE formulation [14]. The expressions for the coefficients A_{mn}^{pq}, F_{mn}^{pq}, G_{mn}^{pq}, $C_{mn,v}$, D_{mn}^{pq}, and E_{mn}^{pq}, and $V_m^{E(v)}(t)$ are given by Eqs. (7.A1)–(7.A6) and (7.A24) in the appendix at the end of this chapter.

7.3.3.2 Temporal Expansion and Testing.
For the temporal testing of the TENE formulation, the functions $\phi_j(t)$ given by Eq. (7.37) are used. Using Eqs. (7.38)-(7.40) in Eq. (7.81) and performing a temporal testing with the testing function $\phi_i(st)$, one obtains

$$\begin{bmatrix} \left[\alpha_{mn}^{E(1)}\right] & \left[\beta_{mn}^{E(1)}\right] \\ \left[\alpha_{mn}^{E(2)}\right] & \left[\beta_{mn}^{E(2)}\right] \end{bmatrix} \begin{bmatrix} \left[u_{n,i}\right] \\ \left[v_{n,i}\right] \end{bmatrix} = \begin{bmatrix} \left[\gamma_{m,i}^{E(1)}\right] \\ \left[\gamma_{m,i}^{E(2)}\right] \end{bmatrix} \qquad (7.82a)$$

where

$$\gamma_{m,i}^{E(v)} = V_{m,i}^{E(v)} + P_{m,i}^{E(v)} + Q_{m,i}^{E(v)} \qquad (7.82b)$$

$$\alpha_{mn}^{E(v)} = \sum_{p,q} \left(\frac{s^2 \mu_v A_{mn}^{pq}}{4} + \frac{F_{mn}^{pq}}{\varepsilon_v} + \frac{s G_{mn}^{pq}}{2 c_v \varepsilon_v} \right) e^{\left(-s R_{mn}^{pq}/(2c_v)\right)} \qquad (7.82c)$$

$$\beta_{mn}^{E(v)} = \frac{s C_{mn,v}}{2} + \sum_{p,q} \left(\frac{s^2 D_{mn}^{pq}}{4 c_v} + \frac{s E_{mn}^{pq}}{2} \right) e^{\left(-s R_{mn}^{pq}/(2c_v)\right)} \qquad (7.82d)$$

$$V_{m,i}^{E(v)} = \int_0^\infty \phi_i(st) V_m^{E(v)}(t) \, d(st) \qquad (7.82e)$$

$$P_{m,i}^{E(v)} = -\sum_{n=1}^{N}\sum_{p,q}\sum_{j=0}^{i-1}\left(\frac{s^2\mu_v A_{mn}^{pq}}{4} + \frac{F_{mn}^{pq}}{\varepsilon_v} + \frac{s\,G_{mn}^{pq}}{2c_v\varepsilon_v}\right)u_{n,j}\,I_{ij}(sR_{mn}^{pq}/c_v)$$

$$+\sum_{n=1}^{N}\sum_{p,q}\sum_{j=0}^{i}\sum_{k=0}^{j-1}\left(s^2\mu_v A_{mn}^{pq}(j-k) + \frac{s\,G_{mn}^{pq}}{c_v\varepsilon_v}\right)u_{n,k}\,I_{ij}(sR_{mn}^{pq}/c_v) \qquad (7.82f)$$

$$Q_{m,i}^{E(v)} = -\sum_{n=1}^{N}\sum_{p,q}\sum_{j=0}^{i-1}\left(\frac{s^2 D_{mn}^{pq}}{4c_v} + \frac{s\,E_{mn}^{pq}}{2}\right)v_{n,j}\,I_{ij}(sR_{mn}^{pq}/c_v)$$

$$-\sum_{n=1}^{N}\sum_{p,q}\sum_{j=0}^{i}\sum_{k=0}^{j-1}\left(\frac{s^2 D_{mn}^{pq}}{c_v}(j-k)v_{n,k} + s\,E_{mn}^{pq}v_{n,k}\right)I_{ij}(sR_{mn}^{pq}/c_v)$$

$$-\sum_{n=1}^{N}\sum_{k=0}^{i-1}s\,C_{mn,v}\,v_{n,k} \qquad (7.82g)$$

$$I_{ij}(sR_{mn}^{pq}/c_v) = \phi_{i-j}(sR_{mn}^{pq}/c_v) - \phi_{i-j-1}(sR_{mn}^{pq}/c_v), \quad j \le i \qquad (7.82h)$$

Following a similar procedure for the testing of the part only related to the magnetic field in the TD-CFIE, a matrix equation can be written in which $f_H\mathbf{f}_m + g_H\mathbf{g}_m$ is used as the spatial testing function and $\phi_j(t)$ is the temporal basis function. This equation is termed the THNH formulation [14], where

$$\begin{bmatrix}\left[\beta_{mn}^{H(1)}\right] & \left[\alpha_{mn}^{H(1)}\right]\\[4pt] \left[\beta_{mn}^{H(2)}\right] & \left[\alpha_{mn}^{H(2)}\right]\end{bmatrix}\begin{bmatrix}\left[u_{n,i}\right]\\[4pt] \left[v_{n,i}\right]\end{bmatrix} = \begin{bmatrix}\left[\gamma_{m,i}^{H(1)}\right]\\[4pt] \left[\gamma_{m,i}^{H(2)}\right]\end{bmatrix} \qquad (7.83a)$$

$$\gamma_{m,i}^{H(v)} = V_{m,i}^{H(v)} + P_{m,i}^{H(v)} + Q_{m,i}^{H(v)} \qquad (7.83b)$$

$$\alpha_{mn}^{H(v)} = \sum_{p,q}\left(\frac{s^2\varepsilon_v A_{mn}^{pq}}{4} + \frac{F_{mn}^{pq}}{\mu_v} + \frac{s\,G_{mn}^{pq}}{2c_v\mu_v}\right)e^{\left(-sR_{mn}^{pq}/(2c_v)\right)} \qquad (7.83c)$$

$$\beta_{mn}^{H(v)} = -\frac{s\,C_{mn,v}}{2} - \sum_{p,q}\left(\frac{s^2 D_{mn}^{pq}}{4c_v} + \frac{s\,E_{mn}^{pq}}{2}\right)e^{\left(-sR_{mn}^{pq}/(2c_v)\right)} \qquad (7.83d)$$

$$V_{m,i}^{H(v)} = \int_0^{\infty}\phi_i(st)V_m^{H(v)}(t)\,d(st) \qquad (7.83e)$$

$$P_{m,i}^{H(v)} = -\sum_{n=1}^{N}\sum_{p,q}\sum_{j=0}^{i-1}\left(\frac{s^2\varepsilon_v A_{mn}^{pq}}{4} + \frac{F_{mn}^{pq}}{\mu_v} + \frac{s\,G_{mn}^{pq}}{2c_v\mu_v}\right)v_{n,j}\,I_{ij}(sR_{mn}^{pq}/c_v)$$

$$-\sum_{n=1}^{N}\sum_{p,q}\sum_{j=0}^{i}\sum_{k=0}^{j-1}\left(s^2\varepsilon_v A_{mn}^{pq}(j-k) + \frac{s\,G_{mn}^{pq}}{c_v\mu_v}\right)v_{n,k}\,I_{ij}(sR_{mn}^{pq}/c_v) \qquad (7.83f)$$

$$Q_{m,i}^{H(v)} = \sum_{n=1}^{N} \sum_{p,q} \sum_{j=0}^{i-1} \left(\frac{s^2 D_{mn}^{pq}}{4c_v} + \frac{s E_{mn}^{pq}}{2} \right) u_{n,j} I_{ij}(sR_{mn}^{pq}/c_v)$$

$$+ \sum_{n=1}^{N} \sum_{p,q} \sum_{j=0}^{i} \sum_{k=0}^{j-1} \left[s^2 \frac{D_{mn}^{pq}}{c_v}(j-k) + s E_{mn}^{pq} \right] u_{n,k} I_{ij}(sR_{mn}^{pq}/c_v) \qquad (7.83g)$$

$$+ \sum_{n=1}^{N} \sum_{k=0}^{i-1} s C_{mn,v} u_{n,k}$$

In Eq. (7.83), the elements A_{mn}^{pq}, F_{mn}^{pq}, G_{mn}^{pq}, $C_{mn,v}$, D_{mn}^{pq}, E_{mn}^{pq}, and $V_m^{H(v)}(t)$ are given by Eqs. (7.A7)–(7.A12) and (7.A25) as described in the appendix. The matrix equations of Eqs. (7.82) and (7.83) can be written as

$$\left[\alpha_{mn}^E\right]\left[w_{n,i}\right] = \left[\gamma_{m,i}^E\right] \qquad (7.84a)$$

$$\left[\alpha_{mn}^H\right]\left[w_{n,i}\right] = \left[\gamma_{m,i}^H\right] \qquad (7.84b)$$

where $w_{n,i} = u_{n,i}$ for $n = 1, 2, ..., N$ and $w_{n,i} = v_{n,i}$ for $n = N+1, 2, ..., 2N$.

Finally, a general matrix equation for the TD-CFIE can be obtained by combining the TENE and THNH formulations given by Eqs. (7.84a) and (7.84b) as

$$\left[\alpha_{mn}\right]\left[w_{n,i}\right] = \left[\gamma_{m,i}\right], \quad i = 0, 1, 2, \cdots, \infty \qquad (7.85a)$$

where

$$\alpha_{mn} = (1-\kappa)\alpha_{mn}^E + \kappa\eta_v\alpha_{mn}^H \qquad (7.85b)$$

$$\gamma_{m,i} = (1-\kappa)\gamma_{m,i}^E + \kappa\eta_v\gamma_{m,i}^H \qquad (7.85c)$$

By solving the matrix equations Eq. (7.85) using MOD with M temporal basis functions, one obtains first the coefficients $u_{n,i}$ and $v_{n,i}$. Then the electric and the magnetic current coefficients are derived using the relations in Eqs. (7.78)–(7.79) and Eq. (7.38) [13].

7.3.3.3 Computation of the Scattered Far Field.

Once the equivalent currents on the dielectric scatterer have been determined using the TD-CFIE, the far scattered fields can be computed. These fields may be thought of as the superposition of the fields caused by the electric and magnetic currents obtained from the solutions of TD-EFIE and TD-MFIE of Eqs. (7.44) and (7.73). Also, when we consider a signal with time duration T_f in the time domain, we note that the upper limit of the integral in Eqs. (7.82e) and (7.83e) can be replaced by the time duration sT_f instead of infinity.

7.3.4   Solving the TD-PMCHW

7.3.4.1 Spatial Expansion and Testing. To solve the TD-PMCHW given by Eq. (7.18), the electric and the magnetic currents are expanded by the triangular vector function $\mathbf{f}_n$ as shown in Eq. (7.78). The Hertz vectors $\mathbf{u}(r,t)$ and $\mathbf{v}(r,t)$ defined in Eqs. (7.22) and (7.24) are used to substitute the electric current $\mathbf{J}(\mathbf{r}, t)$ and the magnetic currents $\mathbf{M}(\mathbf{r}, t)$, and they are expanded by the spatial basis function $\mathbf{f}_n$ as in Eq. (7.79).

Consider the first equation Eq. (7.18a) of the TD-PMCHW formulation, which relates to the electric field only. Substituting the expression of the electric and magnetic Hertz vectors into Eq. (7.18a) and testing with the spatial functions $\mathbf{f}_m$ results in

$$\sum_{v=1}^{2}\sum_{n=1}^{N}\sum_{p,q}\left(\mu_v A_{mn}^{pq} \frac{d^2}{dt^2} u_n(\tau_{mn,v}^{pq}) + \frac{B_{mn}^{pq}}{\varepsilon_v} u_n(\tau_{mn,v}^{pq}) \right.$$

$$\left. + \frac{\varsigma_1^{pq}}{c_v} \frac{d^2}{dt^2} v_n(\tau_{mn,v}^{pq}) + \varsigma_2^{pq} \frac{d}{dt} v_n(\tau_{mn,v}^{pq}) \right) = V_m^E(t), \; m=1,2,\cdots,N \qquad (7.86a)$$

where

$$A_{mn}^{pq} = \frac{1}{4\pi}\int_S \mathbf{f}_m^p(\mathbf{r})\cdot\int_S \frac{\mathbf{f}_n^q(\mathbf{r}')}{R}dS'dS \qquad (7.86b)$$

$$B_{mn}^{pq} = \frac{1}{4\pi}\int_S \nabla\cdot\mathbf{f}_m^p(\mathbf{r})\int_S \frac{\nabla'\cdot\mathbf{f}_n^q(\mathbf{r}')}{R}dS' dS \qquad (7.86c)$$

$$\varsigma_k^{pq} = \frac{1}{4\pi}\int_S \mathbf{f}_m^p(\mathbf{r})\cdot\int_S \mathbf{f}_n^q(\mathbf{r}')\times\frac{\hat{\mathbf{R}}}{R^k}dS'dS, \; k=1,2 \qquad (7.86d)$$

$$V_m^E(t) = \int_S \mathbf{f}_m(\mathbf{r})\cdot\mathbf{E}^i(\mathbf{r}, t)\,dS \qquad (7.86e)$$

Note that Eq. (7.86) is very similar to the spatial testing result obtained for the TD-EFIE of Eq. (7.36).

7.3.4.2 Temporal Expansion and Testing. Using the temporal basis function $\phi_j(st)$ defined by Eq. (7.37) to approximate the transient coefficients of Eq. (7.86) and performing a temporal testing with the testing function $\phi_i(st)$, one obtains

$$\sum_{n=1}^{N}\left(\alpha_{mn}^E u_{n,i} + \beta_{mn}^E v_{n,i}\right) = \gamma_{m,i}^E \qquad (7.87a)$$

where

$$\gamma_{m,i}^{E} = V_{m,i}^{E} + P_{m,i}^{E} + Q_{m,i}^{E} \tag{7.87b}$$

$$\alpha_{mn}^{E} = \sum_{v=1}^{2}\sum_{p,q}\left(\frac{s^2}{4}\mu_v A_{mn}^{pq} + \frac{B_{mn}^{pq}}{\varepsilon_v}\right)e^{\left(-sR_{mn}^{pq}/(2c_v)\right)} \tag{7.87c}$$

$$\beta_{mn}^{E} = \sum_{v=1}^{2}\sum_{p,q}\left(\frac{s^2}{4}\frac{\zeta_1^{pq}}{c_v} + \frac{s}{2}\zeta_2^{pq}\right)e^{\left(-sR_{mn}^{pq}/(2c_v)\right)} \tag{7.87d}$$

$$P_{m,i}^{E} = -\sum_{v=1}^{2}\sum_{n=1}^{N}\sum_{p,q}\sum_{j=0}^{i-1}\left(\frac{s^2}{4}\mu_v A_{mn}^{pq} + \frac{B_{mn}^{pq}}{\varepsilon_v}\right)u_{n,j}I_{ij}\left(sR_{mn}^{pq}/c_v\right)$$
$$-\sum_{v=1}^{2}\sum_{n=1}^{N}\sum_{p,q}\sum_{j=0}^{i}\sum_{k=0}^{j-1}s^2\mu_v A_{mn}^{pq}(j-k)u_{n,k}I_{ij}\left(sR_{mn}^{pq}/c_v\right) \tag{7.87e}$$

$$Q_{m,i}^{E} = -\sum_{v=1}^{2}\sum_{n=1}^{N}\sum_{p,q}\sum_{j=0}^{i-1}\left(\frac{s^2}{4}\frac{\zeta_1^{pq}}{c_v} + \frac{s}{2}\zeta_2^{pq}\right)v_{n,j}I_{ij}\left(sR_{mn}^{pq}/c_v\right)$$
$$-\sum_{v=1}^{2}\sum_{n=1}^{N}\sum_{p,q}\sum_{j=0}^{i}\sum_{k=0}^{j-1}\left(s^2\frac{\zeta_1^{pq}}{c_v}(j-k) + s\zeta_2^{pq}\right)v_{n,k}I_{ij}\left(sR_{mn}^{pq}/c_v\right) \tag{7.87f}$$

$$V_{m,i}^{E} = \int_{0}^{\infty}\phi_i(st)V_m^{E}(t)\,d(st) \tag{7.87g}$$

$$V_m^{E}(t) = \int_{S}\mathbf{f}_m(\mathbf{r})\cdot\mathbf{E}^i(\mathbf{r},t)\,dS \tag{7.87h}$$

Applying a similar procedure to the magnetic field results in

$$\sum_{n=1}^{N}\left(\alpha_{mn}^{H}v_{n,i} + \beta_{mn}^{H}u_{n,i}\right) = \gamma_{m,i}^{H} \tag{7.88a}$$

where

$$\gamma_{m,i}^{H} = V_{m,i}^{H} + P_{m,i}^{H} + Q_{m,i}^{H} \tag{7.88b}$$

$$\alpha_{mn}^{H} = \sum_{v=1}^{2}\sum_{p,q}\left(\frac{s^2}{4}\varepsilon_v A_{mn}^{pq} + \frac{B_{mn}^{pq}}{\mu_v}\right)e^{\left(-sR_{mn}^{pq}/(2c_v)\right)} \tag{7.88c}$$

$$\beta_{mn}^{H} = -\sum_{v=1}^{2}\sum_{p,q}\left(\frac{s^2}{4}\frac{\zeta_1^{pq}}{c_v} + \frac{s}{2}\zeta_2^{pq}\right)e^{\left(-sR_{mn}^{pq}/(2c_v)\right)} \tag{7.88d}$$

$$P_{m,i}^H = -\sum_{v=1}^{2}\sum_{n=1}^{N}\sum_{p,q}\sum_{j=0}^{i-1}\left(\frac{s^2}{4}\varepsilon_v A_{mn}^{pq} + \frac{B_{mn}^{pq}}{\mu_v}\right)v_{n,j}I_{ij}\left(sR_{mn}^{pq}/c_v\right)$$

$$+\sum_{v=1}^{2}\sum_{n=1}^{N}\sum_{p,q}\sum_{j=0}^{i}\sum_{k=0}^{j-1}s^2\varepsilon_v A_{mn}^{pq}(j-k)v_{n,k}I_{ij}\left(sR_{mn}^{pq}/c_v\right) \qquad (7.88e)$$

$$Q_{m,i}^H = \sum_{v=1}^{2}\sum_{n=1}^{N}\sum_{p,q}\sum_{j=0}^{i-1}\left(\frac{s^2}{4}\frac{\zeta_1^{pq}}{c_v} + \frac{s}{2}\zeta_2^{pq}\right)u_{n,j}I_{ij}\left(sR_{mn}^{pq}/c_v\right)$$

$$+\sum_{v=1}^{2}\sum_{n=1}^{N}\sum_{p,q}\sum_{j=0}^{i}\sum_{k=0}^{j-1}\left(s^2\frac{\zeta_1^{pq}}{c_v}(j-k)+s\zeta_2^{pq}\right)u_{n,k}I_{ij}(sR_{mn}^{pq}/c_v) \qquad (7.88f)$$

$$V_{m,i}^H = \int_0^\infty \phi_i(st)V_m^H(t)\,d(st) \qquad (7.88g)$$

$$V_m^H(t) = \int_S \mathbf{f}_m(\mathbf{r})\cdot\mathbf{H}^i(\mathbf{r},t)\,dS \qquad (7.88h)$$

Eq. (7.88) is similar to Eq. (7.72) for the TD-MFIE. Combining Eqs. (7.87) and (7.88), the following matrix equation for the TD-PMCHW is obtained as

$$\begin{bmatrix} \left[\alpha_{mn}^E\right] & \left[\beta_{mn}^E\right] \\ \left[\beta_{mn}^H\right] & \left[\alpha_{mn}^H\right] \end{bmatrix}\begin{bmatrix} \left[u_{n,i}\right] \\ \left[v_{n,i}\right] \end{bmatrix} = \begin{bmatrix} \left[\gamma_{m,i}^E\right] \\ \left[\gamma_{m,i}^H\right] \end{bmatrix}, \; i=0,1,2,\cdots,\infty \qquad (7.89)$$

where the minimum degree or number of temporal basis functions, M, is given by $M = 2BT_f + 1$.

7.3.4.3 Computation of the Scattered Far Field. The far field computation for the TD-PMCHW can be obtained directly from sections 7.3.1.3 and 7.3.2.3, which are related to the far field computation for TD-EFIE and TD-MFIE. The scattered electric field caused by the electric current $\mathbf{E}_J^s(\mathbf{r},t)$ is the same as that from the TD-EFIE given by Eq. (7.48). The scattered magnetic and electric fields due to the magnetic current $\mathbf{H}_M^s(\mathbf{r},t)$ and $\mathbf{E}_M^s(\mathbf{r},t)$ are the same as those from TD-MFIE given by Eqs. (7.76) and (7.77). Therefore, the total scattered field from the dielectric body can be obtained by adding Eqs. (7.48) and (7.77) as

$$\mathbf{E}^s(\mathbf{r},t) \approx -\frac{1}{8\pi c_1 r}\left(\eta_1\sum_{n=1}^{N}\mathbf{A}_n + \sum_{n=1}^{N}\mathbf{F}_n\times\hat{\mathbf{r}}\right) \qquad (7.90a)$$

where

$$\mathbf{A}_n = l_n\sum_q \boldsymbol{\rho}_n^q\frac{d^2}{dt^2}u_n(\tau_n^q) \qquad (7.90b)$$

$$\mathbf{F}_n = l_n \sum_q \rho_n^q \frac{d^2}{dt^2} v_n(\tau_n^q) \tag{7.90c}$$

7.4 AN ALTERNATIVE MOD APPROACH

7.4.1 Alternate Solution to the TD-EFIE

In this section, an alternate MOD method is presented for solution of the TD-EFIE without using the Hertz vector to represent the magnetic current. With this modification, all terms related to the magnetic current in the MOD formulation for TD-EFIE needs to be revised. In the spatial testing, the third term in Eq. (7.26), which is related to the electric vector potential $\mathbf{F}(r,t)$ is rewritten. By substituting the magnetic current expansion described in Eq. (7.21b) into this term, the following result can be obtained based on Eq. (7.35) as:

$$\left\langle \mathbf{f}_m(\mathbf{r}), \frac{1}{\varepsilon_v} \nabla \times \mathbf{F}_v(\mathbf{r}, t) \right\rangle$$

$$= \sum_{n=1}^{N} \left[C_{mn,v} M_n(t) + \sum_{p,q} \left(\frac{\zeta_1^{pq}}{c_v} \frac{d}{dt} M_n(\tau_{mn,v}^{pq}) + \zeta_2^{pq} M_n(\tau_{mn,v}^{pq}) \right) \right] \tag{7.91}$$

In the temporal testing, similar to Eq. (7.38b) and (7.40a), the time domain magnetic current coefficient and its derivative can be expanded as

$$M_n(t) = \sum_{j=0}^{\infty} M_{n,j} \phi_j(st) \tag{7.92}$$

$$\frac{d}{dt} M_n(t) = s \sum_{j=0}^{\infty} \left(\frac{1}{2} M_{n,j} + \sum_{k=0}^{j-1} M_{n,k} \right) \phi_j(st) \tag{7.93}$$

Following a similar procedure as in section 7.3.1, and replacing Eqs. (7.35), (7.38b), and (7.40a) with Eqs. (7.91)–(7.93) results in a matrix equation

$$\begin{bmatrix} \left[\alpha_{mn}^{(1)}\right] & \left[\beta_{mn}^{(1)}\right] \\ \left[\alpha_{mn}^{(2)}\right] & \left[\beta_{mn}^{(2)}\right] \end{bmatrix} \begin{bmatrix} \left[u_{n,i}\right] \\ \left[M_{n,i}\right] \end{bmatrix} = \begin{bmatrix} \left[\gamma_{m,i}^{(1)}\right] \\ \left[\gamma_{m,i}^{(2)}\right] \end{bmatrix} \tag{7.94}$$

where the elements related to the magnetic current are recomputed as

$$\beta_{mn}^{(v)} = C_{mn,v} + \sum_{p,q}\left(\frac{s}{2}\frac{\zeta_1^{pq}}{c_v} + \zeta_2^{pq}\right)e^{\left(-sR_{mn}^{pq}/(2c_v)\right)} \tag{7.95}$$

$$Q_{m,i}^{(v)} = -\sum_{n=1}^{N}\sum_{p,q}\left(\left(\frac{s}{2}\frac{\zeta_1^{pq}}{c_v} + \zeta_2^{pq}\right)\sum_{j=0}^{i-1}M_{n,j}I_{ij}(sR_{mn}^{pq}/c_v)\right.$$

$$\left. +s\frac{\zeta_1^{pq}}{c_v}\sum_{j=0}^{i}\sum_{k=0}^{j-1}M_{n,k}\,I_{ij}(sR_{mn}^{pq}/c_v)\right) \tag{7.96}$$

and other elements are the same as those used in Eq. (7.44). By solving Eq. (7.94) with a MOD algorithm with M temporal basis functions, the magnetic current coefficient can be obtained directly as

$$M_n(t) = \sum_{j=0}^{M-1}M_{n,j}\phi_j(st) \tag{7.97}$$

and therefore in the far field computation, the electric vector potential $\mathbf{F}_n$ in Eq. (7.52) can be written as

$$\mathbf{F}_n = l_n\sum_q\mathbf{n}\times\rho_n^q\frac{d}{dt}M_n(\tau_n^q) \tag{7.98}$$

where the first derivative of the magnetic current coefficient is given in Eq. (7.93) with $M-1$ as the upper limit of the summation over j instead of ∞.

7.4.2 Alternate Solution to the TD-MFIE

Similarly, an alternate MOD method is presented for the TD-MFIE, which applies directly to the electric current without using the Hertz vector to represent the electric current. Because of this modification, all terms related to the electric current in the MOD formulation for the TD-MFIE needs to be revised. In the spatial testing, the third term in Eq. (7.46), which is related to the electric vector potential $\mathbf{A}(r,t)$ needs to be rewritten. By substituting the electric current expansion of Eq. (7.53a), the following result can be obtained based on Eq. (7.58):

$$\left\langle \mathbf{f}_m(\mathbf{r}), \frac{1}{\mu_v}\nabla\times\mathbf{A}_v(\mathbf{r},t)\right\rangle$$

$$= \sum_{n=1}^{N}\left(C_{mn,v}J_n(t) + \sum_{p,q}\left(\frac{\zeta_1^{pq}}{c_v}\frac{d}{dt}J_n(\tau_{mn,v}^{pq}) + \zeta_2^{pq}J_n(\tau_{mn,v}^{pq})\right)\right) \tag{7.99}$$

For the temporal testing, the time domain electric current coefficient and its derivative are expanded as

$$J_n(t) = \sum_{j=0}^{\infty} J_{n,j} \, \phi_j(st) \tag{7.100}$$

$$\frac{d}{dt} J_n(t) = s \sum_{j=0}^{\infty} \left(\frac{1}{2} J_{n,j} + \sum_{k=0}^{j-1} J_{n,k} \right) \phi_j(st) \tag{7.101}$$

Following a similar procedure as in section 7.3.2 to arrive at the matrix formulation and replacing Eqs. (7.58), (7.38a), and (7.39a) with Eqs. (7.99)–(7.101) results in a matrix equation

$$\begin{bmatrix} \left[\beta_{mn}^{(1)}\right] & \left[\alpha_{mn}^{(1)}\right] \\ \left[\beta_{mn}^{(2)}\right] & \left[\alpha_{mn}^{(2)}\right] \end{bmatrix} \begin{bmatrix} \left[J_{n,i}\right] \\ \left[v_{n,i}\right] \end{bmatrix} = \begin{bmatrix} \left[\gamma_{m,i}^{(1)}\right] \\ \left[\gamma_{m,i}^{(2)}\right] \end{bmatrix} \tag{7.102}$$

where the elements related to the electric current are modified as

$$\beta_{mn}^{(v)} = -C_{mn,v} - \sum_{p,q} \left(\frac{s}{2} \frac{\zeta_1^{pq}}{c_v} + \zeta_2^{pq} \right) e^{\left(-sR_{mn}^{pq}/(2c_v)\right)} \tag{7.103}$$

$$Q_{m,i}^{(v)} = \sum_{n=1}^{N} \sum_{p,q} \left(\left(\frac{s}{2} \frac{\zeta_1^{pq}}{c_v} + \zeta_2^{pq} \right) \sum_{j=0}^{i-1} M_{n,j} \, I_{ij}(sR_{mn}^{pq}/c_v) \right.$$

$$\left. + s \frac{\zeta_1^{pq}}{c_v} \sum_{j=0}^{i} \sum_{k=0}^{j-1} M_{n,k} \, I_{ij}(sR_{mn}^{pq}/c_v) \right) \tag{7.104}$$

and other elements are same to those used in Eq. (7.63). Solving Eq. (7.102) with a MOD procedure with M temporal basis functions, one can obtain the electric current coefficient directly as

$$J_n(t) = \sum_{j=0}^{M-1} J_{n,j} \, \phi_j(st) \tag{7.105}$$

The far field computed by this alternative MOD method can use Eq. (7.77), described in section 7.3.2.3, except Eq. (7.77c) needs to be modified as

$$\mathbf{A}_n = l_n \sum_q \mathbf{n} \times \boldsymbol{\rho}_n^q \frac{d}{dt} J_n(\tau_n^q) \tag{7.106}$$

where the first derivative of the electric current coefficient is given in Eq. (7.101) with $M-1$ as the upper limit of the summation over j instead of ∞.

7.4.3   Alternate Solution to the TD-PMCHW

In this section, an alternative MOD method for TD-PMCHW is presented that uses the expansion for the equivalent currents $\mathbf{J}(r,t)$ and $\mathbf{M}(r,t)$ directly instead of using the two Hertz vectors $\mathbf{u}(r,t)$ and $\mathbf{v}(r,t)$. Taking a derivative with respect to time, the following integral equations are derived from Eqs. (7.18):

$$\sum_{v=1}^{2}\left[\frac{\partial^2}{\partial t^2}\mathbf{A}_v(\mathbf{r},t)+\nabla\frac{\partial}{\partial t}\Phi_v(\mathbf{r},t)+\frac{1}{\varepsilon_v}\nabla\times\frac{\partial}{\partial t}\mathbf{F}_v(\mathbf{r},t)\right]_{\text{tan}}=\left(\frac{\partial}{\partial t}\mathbf{E}^i(\mathbf{r},t)\right)_{\text{tan}} \quad (7.107a)$$

$$\sum_{v=1}^{2}\left[\frac{\partial^2}{\partial t^2}\mathbf{F}_v(\mathbf{r},t)+\nabla\frac{\partial}{\partial t}\Psi_v(\mathbf{r},t)-\frac{1}{\mu_v}\nabla\times\frac{\partial}{\partial t}\mathbf{A}_v(\mathbf{r},t)\right]_{\text{tan}}=\left(\frac{\partial}{\partial t}\mathbf{H}^i(\mathbf{r},t)\right)_{\text{tan}} \quad (7.107b)$$

The time domain electric and magnetic current coefficients $J_n(t)$ and $M_n(t)$ defined in Eqs. (7.21) are written as

$$J_n(t)=\sum_{j=0}^{\infty}J_{n,j}\,\phi_j(st) \quad (7.108)$$

$$M_n(t)=\sum_{j=0}^{\infty}M_{n,j}\,\phi_j(st) \quad (7.109)$$

where $J_{n,j}$ and $M_{n,j}$ are the coefficients to be determined. Following a similar procedure as outlined in section 7.3.4 and using the new boundary conditions given by Eq. (7.107) and the new current coefficient expansions given by Eqs. (7.108)–(7.109), a new matrix equation for the TD-PMCHW is obtained

$$\begin{bmatrix}\left[\alpha_{mn}^E\right] & \left[\beta_{mn}^E\right] \\ \left[\beta_{mn}^H\right] & \left[\alpha_{mn}^H\right]\end{bmatrix}\begin{bmatrix}\left[J_{n,i}\right] \\ \left[M_{n,i}\right]\end{bmatrix}=\begin{bmatrix}\left[\gamma_{m,i}^E\right] \\ \left[\gamma_{m,i}^H\right]\end{bmatrix},\ i=0,1,2,...,\infty \quad (7.110)$$

where the matrix elements on the left-hand side are the same as those used in Eq. (7.89) of section 7.3.4 and the elements related to the right-hand side are obtained by replacing $J_{n,j}$ and $M_{n,j}$ instead of $u_{n,j}$ and $v_{n,j}$, respectively. Also

$$V_m^E(t)=\int_S \mathbf{f}_m(\mathbf{r})\cdot\frac{\partial\mathbf{E}^i(\mathbf{r},t)}{\partial t}\,dS \quad (7.111)$$

$$V_m^H(t) = \int_S \mathbf{f}_m(\mathbf{r}) \cdot \frac{\partial \mathbf{H}^i(\mathbf{r}, t)}{\partial t} dS \tag{7.112}$$

By solving Eq. (7.110) with a MOD algorithm with M temporal basis functions, the electric and magnetic current coefficients can be obtained directly from Eqs. (7.108) and (7.109) by using an upper limit of $M-1$ instead of ∞. The far field is obtained by using the same expression as Eq. (7.90), where

$$\mathbf{A}_n = l_n \sum_q \mathbf{\rho}_n^q \frac{d}{dt} J_n(\tau_n^q) \tag{7.113}$$

$$\mathbf{F}_n = l_n \sum_q \mathbf{\rho}_n^q \frac{d}{dt} M_n(\tau_n^q) \tag{7.114}$$

The first derivative of the equivalent current coefficients has the same form as Eq. (7.39a) or (7.40a) with $M-1$ as the upper limit of the summation over j instead of ∞.

7.5   NUMERICAL EXAMPLES

7.5.1   Using the TD-EFIE Formulations

In this section, numerical results are presented for three representative dielectric scatterers, viz: a sphere, a cube, and a cylinder with a relative permittivity $\varepsilon_r = 2$, placed in free space, as shown in Figure 7.1. Figure 7.1(a) shows a dielectric sphere having a radius of 0.5 m centered at the origin. It has a total of 528 patches and 792 common edges. The dielectric cube with a 1 m length on each side centered at the origin is shown in Figure 7.1(b). It has a total of 768 patches and 1152 common edges. Figure 7.1(c) shows a dielectric cylinder with a radius of 0.5 m and a height of 1 m centered at the origin. It has a total of 720 patches with 1080 common edges. The shaded patches in Figure 7.1 reflect where the current is to be observed on that structure. Each of the three scatterers are illuminated by a Gaussian plane wave in which the electric field is given by

$$\mathbf{E}^i(\mathbf{r}, t) = \mathbf{E}_0 \frac{4}{\sqrt{\pi}T} e^{-\gamma^2} \quad \text{and} \quad \gamma = \frac{4}{T}(ct - ct_0 - \mathbf{r} \cdot \hat{\mathbf{k}}) \tag{7.115}$$

where $\hat{\mathbf{k}}$ is the unit vector along the direction of the wave propagation. T is the pulse width of the Gaussian impulse, and t_0 is a time delay that represents the time at which the pulse peaks at the origin. The field is incident from $\phi = 0°$ and $\theta = 0°$ with $\hat{\mathbf{k}} = -\hat{\mathbf{z}}$ and $\mathbf{E}_0 = \hat{\mathbf{x}}$. In the numerical computation, a Gaussian pulse with $T = 8$ lm and $ct_0 = 12$ lm is used to eliminate the internal resonances of the structure. This pulse has a bandwidth of 125 MHz, which excludes any internal resonant frequencies of the structures. The transient electric and magnetic currents on the dielectric surface and θ- (or x-) components of the normalized far fields along the backward direction is plotted. All solutions computed by the present method are compared with the inverse discrete Fourier transform (IDFT) of the frequency domain solutions of the FD-EFIE [15]. In addition, the far field for the sphere is compared with the Mie series solution. Setting $s = 10^9$ and $M = 80$ is sufficient to get accurate solutions. In all figures, a label with "(1)" and "(2)" mean results computed by the formulation in section 7.3 and the alternative formulation of section 7.4, respectively.

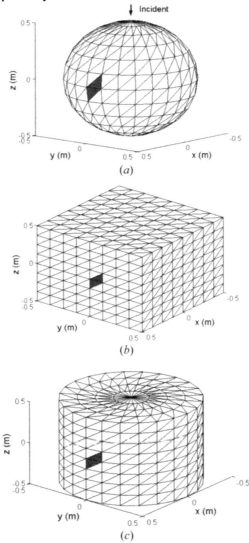

Figure 7.1. Triangle patching of three dielectric objects: (a) Sphere, (b) Cube, and (c) Cylinder.

7.5.1.1 Response from a Sphere. The sphere in Figure 7.1(a) is analyzed using the TD-EFIE formulations given by Eqs. (7.44) and (7.94). Figure 7.2 plots the transient responses for the θ-directed electric and the ϕ-directed magnetic currents on the surface of the sphere computed by these two TD-EFIE formulations, and their results are compared with the IDFT solution. The solutions for the two methods are stable and are in agreement with the IDFT of the solution obtained by the frequency domain electric field integral equation (FD-EFIE). Figure 7.3 plots the transient response for the far field computed by the two TD-EFIE techniques and is plotted against the Mie series solution and the IDFT solution. All four solutions agree well.

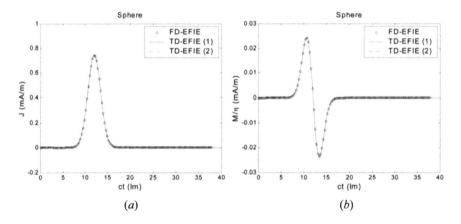

Figure 7.2. Transient currents on the dielectric sphere: (a) θ-directed electric current. (b) ϕ-directed magnetic current.

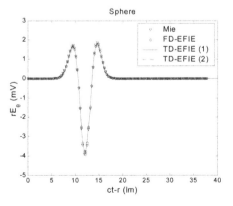

Figure 7.3. Scattered far field from the dielectric sphere along the backward direction.

7.5.1.2 Response from a Cube. The dielectric cube of Figure 7.1(*b*) is analyzed using the two TD-EFIE formulations. The transient response for the *z*-directed electric and the *y*-directed magnetic currents on the cube are computed and compared with the IDFT solution in Figure 7.4. The solutions for the two TD-EFIE formulations are stable and are in good agreement with the IDFT solution. Figure 7.5 compares the transient response of the two TD-EFIE solutions along with the IDFT solution for the normalized far scattered field from the cube. All three solutions show a good agreement.

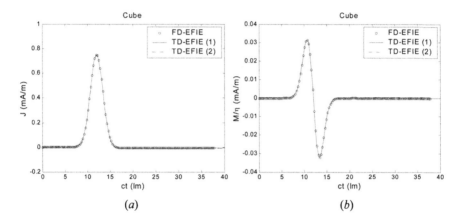

Figure 7.4. Transient currents on the dielectric cube: (*a*) *z*-directed electric current. (*b*) *y*-directed magnetic current.

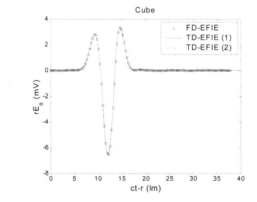

Figure 7.5. Scattered far field from the dielectric cube along the backward direction.

7.5.1.3 Response from a Cylinder. The transient response for the z-directed electric and the ϕ-directed magnetic currents are computed for the cylinder shown in Figure 7.3(c). The currents computed by the two TD-EFIE methods are shown in Figure 7.6 and compare them with the IDFT solution. The solutions for the two methods are stable and are in good agreement with the IDFT solution. Figure 7.7 plots the transient response of the two TD-EFIE formulations against the IDFT solution for the normalized scattered far field from the cylinder. All three solutions show a good agreement.

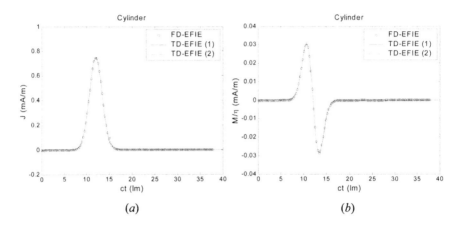

(a) $\qquad\qquad\qquad\qquad\qquad\qquad\qquad\qquad$ (b)

Figure 7.6. Transient currents on the dielectric cylinder: (a) z-directed electric current. (b) ϕ-directed magnetic current.

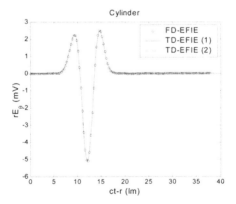

Figure 7.7. Scattered far field from the dielectric cylinder along the backward direction.

7.5.2   Using the TD-MFIE Formulations

In this section, the numerical results for three 3-D dielectric scatterers are presented. They include a sphere and a cube as shown in Figure 7.1(a) and (b) and a cone–cylinder shown in Figure 7.8. The three dielectric scatterers have a relative permittivity $\varepsilon_r = 2$ and are placed in the free space. The results are computed using the two TD-MFIE formulations of Eqs. (7.73) and (7.102). The surface of the dielectric structures is modeled using planar triangular patches. The discretizations for the dielectric sphere and the cube are the same as stated in section 7.5.1. The dielectric cone–cylinder has a radius of 0.5 m, a height of 0.5 m for both the cone and the cylinder. It has a total of 576 patches with 864 edges.

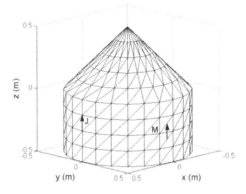

Figure 7.8. Triangle patching of a cone–cylinder.

All scatterers are illuminated by a Gaussian plane wave in which the magnetic field is given by

$$\mathbf{H}^i(\mathbf{r}, t) = \frac{1}{\eta}\hat{\mathbf{k}} \times \mathbf{E}_0 \frac{4}{\sqrt{\pi T}} e^{-\gamma^2} \quad \text{and} \quad \gamma = \frac{4}{T}(ct - ct_0 - \mathbf{r} \cdot \hat{\mathbf{k}}) \tag{7.116}$$

where $\hat{\mathbf{k}}$ is the unit vector along the direction of the wave propagation. T is the pulse width of the Gaussian impulse, and t_0 is a time delay that represents the time at which the pulse peaks at the origin. The field is incident from $\phi = 0°$ and $\theta = 0°$ with $\hat{\mathbf{k}} = -\hat{\mathbf{z}}$, so that $\mathbf{E}_0 = \hat{\mathbf{x}}$. In the numerical computation, $T = 8$ lm and $ct_0 - 12$ lm is used. This bandwidth of the pulse is then 125 MHz, so it does not include any internal resonant frequencies of the structures. The numerical results to be shown are the transient electric and magnetic currents on the dielectric surface, θ- (or x-) components of the normalized far field in the backward direction, and the radar cross section (RCS). All solutions computed by the TD-MFIE formulations are compared with the IDFT of the frequency domain solutions given by the frequency domain magnetic field integral equation (FD-MFIE) [15]. In addition, the far field solution for the sphere is compared with the Mie series solution. Setting $s = 10^9$ and $M = 80$ is sufficient to obtain accurate solutions. In addition "(1)" and "(2)" imply results computed by the formulation in section 7.3 and the alternative formulation of section 7.4, respectively, in all legends.

7.5.2.1 Response from a Sphere. Figure 7.9 plots the transient responses for the θ-directed electric and magnetic currents on the surface of the dielectric sphere of Figure 7.1(a) computed by the two TD-MFIE formulations of Eqs. (7.73) and (7.102) and compares them with the IDFT result from the frequency domain solution. The transient solutions for both the currents are stable and are in good agreement with the IDFT of the FD-MFIE solution. Figure 7.10 plots the far field and the RCS and compares them with the Mie series and the IDFT solution. All solutions agree well with each other.

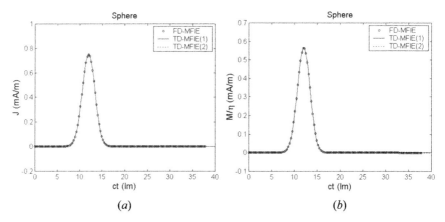

Figure 7.9. Transient currents on the dielectric sphere: (a) θ-directed electric current. (b) θ-directed magnetic current.

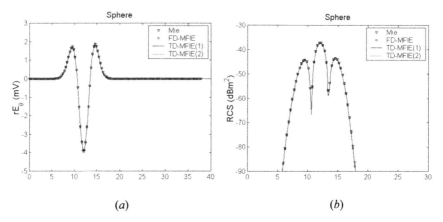

Figure 7.10. (a) Scattered far field from the dielectric sphere along the backward direction. (b) Monostatic radar cross section from the dielectric sphere.

7.5.2.2 *Response from a Cube.* Figure 7.11 plots the transient response for the *z*-directed electric and magnetic currents on the cube of Figure 7.1(*b*) and compares them with the IDFT solution obtained from the FD-MFIE. Time domain solutions for both currents are stable and in good agreement with the IDFT solution. Figure 7.12 plots the transient scattered far field and the monostatic RCS of the dielectric cube and compares them with the IDFT solution obtained from the FD-MFIE. All three solutions agree well with each other.

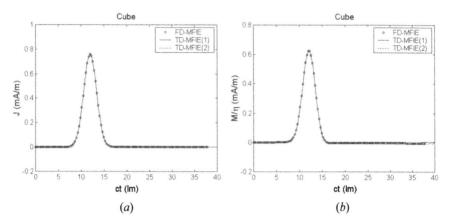

Figure 7.11. Transient currents on the dielectric cube: (*a*) *z*-directed electric current. (*b*) *z*-directed magnetic current.

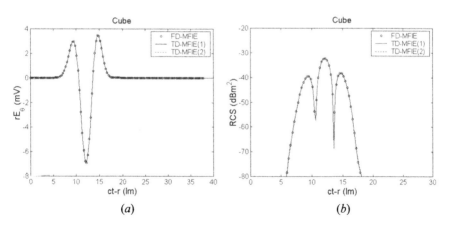

Figure 7.12. (*a*) Scattered far field from the dielectric cube along the backward direction. (*b*) Monostatic radar cross section from the dielectric cube.

7.5.2.3 Response from a Cone–Cylinder. Figure 7.13 plots the transient responses for the z-directed electric and magnetic current on the dielectric cone–cylinder shown in Figure 7.8. The time domain solutions for both the currents are compared with the IDFT solution from the FD-MFIE. The time domain solutions for both the currents are stable and agree well with the IDFT solution. Figure 7.14 plots the transient scattered field and the monostatic RCS for the cone–cylinder computed using the two TD-MFIE formulations and compares them with the IDFT solutions. All the solutions agree well except a little discrepancy in the RCS is observed for the two TD-MFIE formulations at about $ct = 1$ lm.

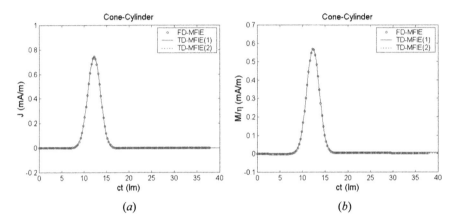

Figure 7.13. Transient currents on the dielectric cone–cylinder. (a) z-directed electric current. (b) z-directed magnetic current.

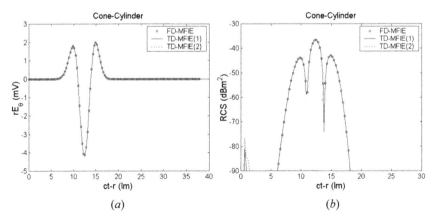

Figure 7.14. (a) Scattered far field from the dielectric cone–cylinder along the backward direction. (b) Monostatic radar cross section from the dielectric cone–cylinder.

7.5.3 Using the TD-PMCHW Formulations

In this section, the numerical results for various 3-D dielectric scatterers are presented using the two TD-PMCHW formulations given by Eqs. (7.89) and (7.110). All scatterers have a relative permittivity $\varepsilon_r = 2$ and are placed in free space. They are illuminated by a Gaussian plane wave defined by Eqs. (7.115) and (7.116). The field is incident from $\phi = 0°$ and $\theta = 0°$ with $\hat{\mathbf{k}} = -\hat{\mathbf{z}}$, and $\mathbf{E}_0 = \hat{\mathbf{x}}$. A pulse width of $T = 2$ lm and a delay of $ct_0 = 4$ lm are chosen for the numerical computation. This pulse has a bandwidth of 500 MHz, which include several internal resonant frequencies of the structures to be analyzed. Setting $s = 2 \times 10^9$ and $M = 80$ is sufficient to get accurate solutions. All solutions are computed using the TD-PMCHW, and the results are compared with the IDFT of the frequency domain solutions obtained by the PMCHW formulation [15]. The frequency domain solution is generated in the range of 0^+–500 MHz with 128 samples. The far field is θ- (or x-) component of the electric field and are computed along the backward direction ($+z$ axis) of the scatterers. The figure legends, "(I)" and "(II)," imply that the results are computed using the PMCHW formulation described in section 7.3 and the alternative formulation of section 7.4, respectively.

First, consider a dielectric sphere as shown in Figure 7.1(a) using the same number of patches and edges used for generating the results of sections 7.5.1 and 7.5.2. Figure 7.15 plots the transient response for the θ-directed electric and magnetic currents on the sphere at $\theta = 90°$ and are computed using the two TD-PMCHW formulations and compared with the IDFT results using the FD-PMCHW solutions. The locations and the directions of the currents on the structure are depicted in the insets of the figures. It can be found that the solutions from the TD-PMCHW formulations are stable and are in good agreement with the IDFT solutions.

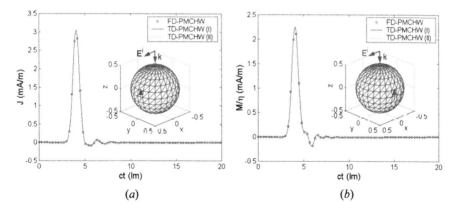

Figure 7.15. Transient currents on the dielectric sphere at $\theta = 90°$: (a) Electric current at $\phi = 7.5°$. (b) Magnetic current at $\phi = 172.5°$.

The transient response for the back-scattered far field from the sphere is presented in Figure 7.16(a). They are compared with the Mie series solution and the IDFT of the frequency domain solution and match with very well each other.

Next, the transient far field response from various dielectric structures are plotted in Figure 7.16 for further verification of the accuracy and stability of the MOD methods for the TD-PMCHW formulation. The various dielectric structures to be analyzed include the following:

- A hemisphere with a radius of 0.5 m, modeled by 432 triangular patches and 648 edges. (shown in the inset of Figure 7.16b).
- A cone with a base radius of 0.5 m and has a height of 1 m along the z-direction is modeled using 624 triangular patches and 936 edges. (as shown in the inset of Figure 7.16c).
- A double cone with a height of 1 m along z-direction and a radius of 0.5 m along the $z = 0$ plane, modeled by 528 triangular patches and 792 edges (as shown in the inset of Figure 7.16d).
- A hemisphere–cone with a radius of 0.5 m and a height along the z-direction of 0.5 m, modeled by 528 triangular patches and 792 edges (as shown in the inset of Figure 7.16e).
- A cylinder with a radius of 0.5 m and a height along the z-direction of 1 m, modeled by 720 triangular patches and 1080 edges (as shown in the inset of Figure 7.16f).
- A hemisphere–cylinder with a radius of 0.5 m and the height of the cylinder along the z-direction is 0.5 m, modeled by 624 triangular patches and 936 edges (as shown in the inset of Figure 7.16g).
- A cone–cylinder with a radius of 0.5 m and the heights of the cylinder and the cone along the z-direction is 0.5 m, modeled by 576 triangular patches and 864 edges (as shown in the inset of Figure 7.16h).

The transient backward scattered fields for all dielectric scatterers presented are computed using the two TD-PMCHW formulations and plotted in the Figures (7.16b)–(7.16h). All computed results are stable and they all agree very well with the FD-PMCHW solutions. In fact, no obvious discrepancy between the three solutions can be observed for all the examples.

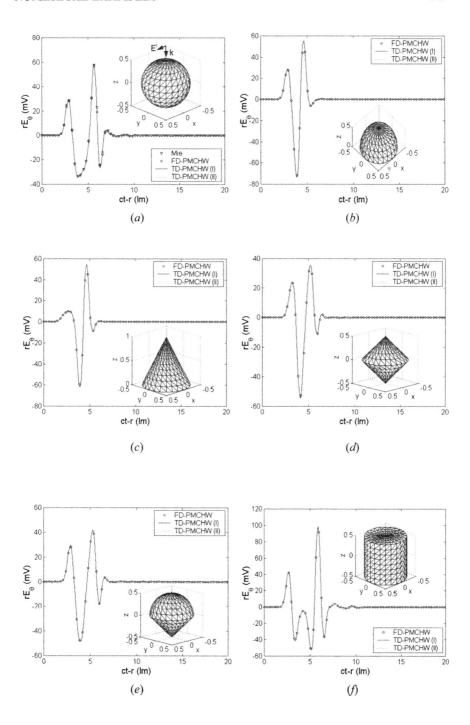

(a)

(b)

(c)

(d)

(e)

(f)

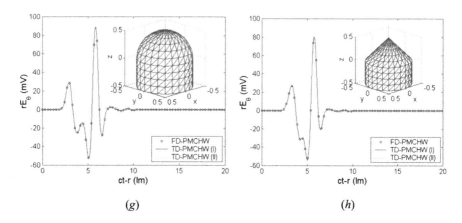

(g)   (h)

Figure 7.16. Backward scattered field along the $+z$ direction from the dielectric structures: (a) Sphere. (b) Hemisphere. (c) Cone. (d) Double cones. (e) Hemisphere–cone (f) Cylinder. (g) Hemisphere–cylinder. (h) Cone–cylinder.

7.5.4 Using the TD-CFIE Formulation

The numerical results for 3-D dielectric scatterers computed using the TD-CFIE formulation given by Eq. (7.85) now are presented. The scatterers are illuminated by a Gaussian plane wave defined by Eq. (7.115) and (7.116). The field is incident from $\phi = 0°$ and $\theta = 0°$ with $\hat{\mathbf{k}} = -\hat{\mathbf{z}}$, and $\mathbf{E}_0 = \hat{\mathbf{x}}$. A pulse width of $T = 4$ lm and a time delay $ct_0 = 6$ lm is chosen for the numerical computation. Setting $s = 0.9 \times 10^9$ and $M = 40$ is sufficient to get accurate solutions. All solutions computed by the TD-CFIE formulation are compared with the IDFT of the FD-CFIE solutions [15] in the bandwidth of 0^+–500 MHz with 128 samples. All dielectric structures have a relative permittivity $\varepsilon_r = 2$ and are placed in free space. The shaded patches on the structure shown in the inset of the figures indicate the location of the current to be observed associated with the common edge. The far field solutions are θ- (or x-) components of the electric field computed along the backward direction ($+z$ axis) from the scatterers. The following two combinations of the spatial testing functions for the TD-CFIE are considered [15]:

$$\langle \mathbf{f}_m + \mathbf{g}_m, \text{EFIE} \rangle + \langle -\mathbf{f}_m + \mathbf{g}_m, \text{MFIE} \rangle \text{ or}$$

$$\langle \mathbf{f}_m - \mathbf{g}_m, \text{EFIE} \rangle + \langle \mathbf{f}_m + \mathbf{g}_m, \text{MFIE} \rangle$$

A general form of the testing can be represented by involving the coefficients f_E, g_E, f_H, and f_H, which can be set as either $+1$ or -1.

$$f_E \langle \mathbf{f}_m, \text{EFIE} \rangle + g_E \langle \mathbf{g}_m, \text{EFIE} \rangle + f_H \langle \mathbf{f}_m, \text{MFIE} \rangle + g_H \langle \mathbf{g}_m, \text{MFIE} \rangle$$

Similar to the FD-CFIE formulation introduced in Chapter 3 [15], it is possible to drop one of the testing terms to generate 16 different formulations (i.e., setting one of the coefficients to zero with the coefficients that can be either $+1$ or -1 [14]). These formulations include TENE-TH for $g_H = 0$, TENE-NH for $f_H = 0$, TE-THNH for $g_E = 0$, and NE-THNH for $f_E = 0$. However, for TD-CFIE, none of the 16 formulations provides valid transient solutions. In this section, numerical results using only the testing coefficients $f_E = 1$, $g_E = 1$, $f_H = -1$, and $g_H = 1$ and the combination parameter of $\kappa = 0.5$ are presented.

7.5.4.1 Response from a Sphere. Consider the dielectric sphere shown in Figure 7.1(a). Figure 7.17 plots the transient response of the θ-directed electric current density at the shaded patch on the sphere ($\theta = 90°$ and $\phi = 7.5°$) computed by the TD-CFIE formulation and it is compared with the IDFT of the frequency domain solutions. In Figure 7.17, it is found that the solutions computed using the MOD method using the TD-CFIE are stable and agree well with the IDFT solution. Figure 7.18 depicts the transient response for the backward scattered far field obtained by the TD-CFIE formulation along with the Mie series solution and the IDFT solution. All three solutions show very good agreement.

7.5.4.2 Response from a Hemisphere–Cone and a Double Cone. The next example deals with a hemisphere terminated by a cone. The radius of the hemisphere is 0.5 m and the height of the cone along the z-direction is 0.5 m. This structure is divided into 480 triangular patches with a total number of 720 edges. Figure 7.19 plots the transient response for the θ-directed electric and magnetic currents at the shaded patch on the sphere ($\theta = 75°$ and $\phi = 7.5°$) computed by the TD-CFIE formulation and compares with the IDFT solution. In Figure 7.19, it is shown that the solutions obtained by the MOD method along with the TD-CFIE are stable and agree well with the IDFT solution. Figure 7.20 presents the transient response for the backward scattered far field from the dielectric hemisphere–cone obtained by the TD-CFIE along with the IDFT of the FD-CFIE solution. Two solutions agree well as is evident from the figure.

     Finally, a double cone dielectric structure is considered, which is shown in Figure 7.21. The height of the cones along the z-direction is 1 m and the radius at $z = 0$ is 0.5 m. This structure is divided into 432 triangular patches with a total of 648 edges. In this figure, the backward scattered field computed using the TD-CFIE formulation is compared with the IDFT of the frequency domain result. The agreement of the solution computed using the TD-CFIE and the IDFT of the FD-CFIE solution is excellent.

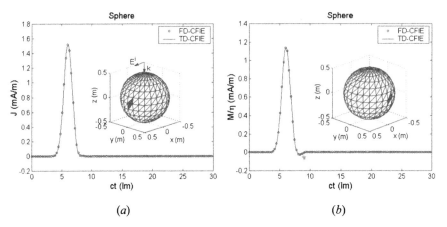

(a)   (b)

Figure 7.17. Transient currents on the dielectric sphere:
(a) Electric current. (b) Magnetic current.

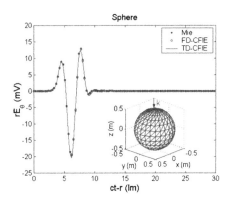

Figure 7.18. Backward scattered fields from the dielectric sphere along the $+z$ direction.

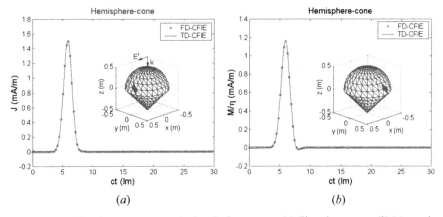

(a)   (b)

Figure 7.19. Transient currents on the hemisphere–cone: (a) Electric current. (b) Magnetic
current.

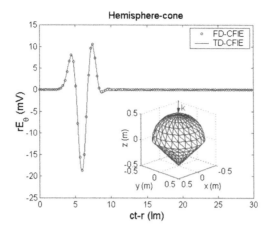

Figure 7.20. Backward scattered fields from the dielectric hemisphere–cone along the $+z$ direction.

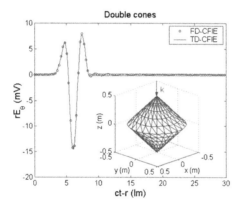

Figure 7.21. Backward scattered field from the double cone along the $+z$ direction.

7.6   CONCLUSION

The MOD approach has been presented to solve the time domain electric field, magnetic field, PMCHW, and the combined field integral equations to analyze the scattering from three-dimensional arbitrarily shaped dielectric structures. To apply a MoM procedure, triangular patch functions have been used as spatial basis and testing functions. A temporal basis function set derived from the Laguerre polynomials has been introduced. The advantage of the proposed method is to guarantee late-time stability. With the representation of the derivative of the transient coefficient in an analytic form, the temporal derivative in the integral equation can be treated analytically. In addition, the time variable can be integrated out completely, and one now solves the problem only in space. This way, one can do away with the Courant–Friedrich–Levy sampling criteria relating the spatial discretizations to the temporal one. Transient equivalent currents and far fields obtained by the two MOD methods presented in this chapter are accurate and stable. However, many formulations that neglect one of the testing terms do not yield a valid solution for the TD-CFIE. The agreement between the solutions obtained using the two proposed methods and the IDFT of the frequency domain solution is excellent.

7.7   APPENDIX

Elements in Eqs. (7.81) and (7.82) for the TENE formulation are given as

$$A_{mn}^{pq} = f_E a_{mn,f}^{pq} + g_E a_{mn,g}^{pq} \tag{7.A1}$$

$$F_{mn}^{pq} = f_E b_{mn,f}^{pq} + g_E g_{mn,g}^{pq} \tag{7.A2}$$

$$G_{mn}^{pq} = g_E f_{mn,g}^{pq} \tag{7.A3}$$

$$C_{mn,v} = \begin{cases} +\sum_{p,q}\left(f_E c_{mn,f}^{pq} + g_E c_{mn,g}^{pq} \right), & v=1 \\ -\sum_{p,q}\left(f_E c_{mn,f}^{pq} + g_E c_{mn,g}^{pq} \right), & v=2 \end{cases} \tag{7.A4}$$

$$D_{mn}^{pq} = f_E d_{mn,f}^{pq} + g_E d_{mn,g}^{pq} \tag{7.A5}$$

$$E_{mn}^{pq} = f_E e_{mn,f}^{pq} + g_E e_{mn,g}^{pq} \tag{7.A6}$$

Elements in Eq. (7.83) for the THNH formulation are given as

$$A_{mn}^{pq} = f_H a_{mn,f}^{pq} + g_H a_{mn,g}^{pq} \tag{7.A7}$$

$$F_{mn}^{pq} = f_H b_{mn,f}^{pq} + g_H g_{mn,g}^{pq} \tag{7.A8}$$

$$G_{mn}^{pq} = g_H f_{mn,g}^{pq} \tag{7.A9}$$

$$C_{mn,v} = \begin{cases} +\sum_{p,q}\left(f_H c_{mn,f}^{pq} + g_H c_{mn,g}^{pq} \right), & v=1 \\ -\sum_{p,q}\left(f_H c_{mn,f}^{pq} + g_H c_{mn,g}^{pq} \right), & v=2 \end{cases} \tag{7.A10}$$

$$D_{mn}^{pq} = f_H d_{mn,f}^{pq} + g_H d_{mn,g}^{pq} \tag{7.A11}$$

$$E_{mn}^{pq} = f_H e_{mn,f}^{pq} + g_H e_{mn,g}^{pq} \tag{7.A12}$$

And the integrals in Eq. (7.A1)–(7.A12) are given by

$$a_{mn,f}^{pq} = \frac{1}{4\pi} \int_S \mathbf{f}_m^p(\mathbf{r}) \cdot \int_S \frac{\mathbf{f}_n^q(\mathbf{r}')}{R} dS' dS \tag{7.A13}$$

$$a_{mn,g}^{pq} = \frac{1}{4\pi} \int_S \mathbf{g}_m^p(\mathbf{r}) \cdot \int_S \frac{\mathbf{f}_n^q(\mathbf{r}')}{R} dS' dS \tag{7.A14}$$

$$b_{mn,f}^{pq} = \frac{1}{4\pi} \int_S \nabla \cdot \mathbf{f}_m^p(\mathbf{r}) \int_S \frac{\nabla' \cdot \mathbf{f}_n^q(\mathbf{r}')}{R} dS' dS \tag{7.A15}$$

$$f_{mn,g}^{pq} = \frac{1}{4\pi} \int_S \mathbf{g}_m^p(\mathbf{r}) \cdot \int_S \nabla' \cdot \mathbf{f}_n^q(\mathbf{r}') \frac{\hat{\mathbf{R}}}{R} dS' dS \tag{7.A16}$$

$$g_{mn,g}^{pq} = \frac{1}{4\pi} \int_S \mathbf{g}_m^p(\mathbf{r}) \cdot \int_S \nabla' \cdot \mathbf{f}_n^q(\mathbf{r}') \frac{\hat{\mathbf{R}}}{R^2} dS' dS \tag{7.A17}$$

$$c_{mn,f}^{pq} = \frac{1}{2} \int_S \mathbf{f}_m^p(\mathbf{r}) \cdot \mathbf{n} \times \mathbf{f}_n^q(\mathbf{r}) dS \tag{7.A18}$$

$$c_{mn,g}^{pq} = \frac{1}{2} \int_S \mathbf{g}_m^p(\mathbf{r}) \cdot \mathbf{n} \times \mathbf{f}_n^q(\mathbf{r}) dS \tag{7.A19}$$

$$d_{mn,f}^{pq} = \frac{1}{4\pi} \int_S \mathbf{f}_m^p(\mathbf{r}) \cdot \int_S \mathbf{f}_n^q(\mathbf{r}') \times \frac{\hat{\mathbf{R}}}{R} dS' dS \tag{7.A20}$$

$$d_{mn,g}^{pq} = \frac{1}{4\pi} \int_S \mathbf{g}_m^p(\mathbf{r}) \cdot \int_S \mathbf{f}_n^q(\mathbf{r}') \times \frac{\hat{\mathbf{R}}}{R} dS' dS \tag{7.A21}$$

$$e_{mn,f}^{pq} = \frac{1}{4\pi} \int_S \mathbf{f}_m^p(\mathbf{r}) \cdot \int_S \mathbf{f}_n^q(\mathbf{r}') \times \frac{\hat{\mathbf{R}}}{R^2} dS' dS \tag{7.A22}$$

$$e_{mn,g}^{pq} = \frac{1}{4\pi} \int_S \mathbf{g}_m^p(\mathbf{r}) \cdot \int_S \mathbf{f}_n^q(\mathbf{r}') \times \frac{\hat{\mathbf{R}}}{R^2} dS' dS \tag{7.A23}$$

where $\hat{\mathbf{R}}$ is a unit vector along the direction $\mathbf{r} - \mathbf{r}'$. The evaluations of Eqs. (7.A13)–(7.A23) have been presented in [24–30]. The integral associated with the incident fields are expressed as

$$V_m^{E(v)}(t) = \begin{cases} \int_S f_E \mathbf{f}_m(\mathbf{r}) + g_E \mathbf{g}_m(\mathbf{r}) \cdot \mathbf{E}^i(\mathbf{r}, t) dS, & v = 1 \\ \\ 0, & v = 2 \end{cases} \tag{7.A24}$$

$$V_m^{H(v)}(t) = \begin{cases} \int_S f_H \mathbf{f}_m(\mathbf{r}) + g_H \mathbf{g}_m(\mathbf{r}) \cdot \mathbf{H}^i(\mathbf{r}, t) dS, & v = 1 \\ \\ 0, & v = 2 \end{cases} \tag{7.A25}$$

REFERENCES

[1]   S. M. Rao, *Time Domain Electromagnetics*, Academic Press, New York, 1999.

[2]   H. Mieras and C. L. Bennet, "Space-time integral equation approach to dielectric targets," *IEEE Trans. Antennas Prop.*, Vol. 30, pp. 2–9, 1982.

[3]   D. A. Vechinski, S. M. Rao, and T. K. Sarkar, "Transient scattering from three-dimensional arbitrary shaped dielectric bodies," *J. Opt. Soc. Amer.*, Vol. 11, pp. 1458–1470, 1994.

[4]   B. P. Rynne, "Time domain scattering from dielectric bodies," *Electromagnetics*, Vol. 14, pp. 181–193, 1994.

[5]   E. Marx, "Integral equation for scattering by a dieleotric," *IEEE Trans. Antennas Prop.*, Vol. 32, pp. 166–172, 1984.

[6]   M. D. Pocock, M. J. Bluck and S. P. Walker, "Electromagnetic scattering from 3-D curved dielectric bodies using time domain integral equations," *IEEE Trans. Antennas Prop.*, Vol. 46, pp. 1212–1219, 1998.

[7]   N. T. Gres, A. A. Ergin, E. Michielssen and B. Shanker, "Volume-integral-equation based analysis of transient electromagnetic scattering from three-dimensional inhomogeneous dielectric objects," *Radio Sci.*, Vol. 36, pp. 379–386, 2001.

[8]   T. K. Sarkar, W. Lee, and S. M. Rao, "Analysis of transient scattering from composite arbitrarily shaped complex structures," *IEEE Trans. Antennas Prop.*, Vol. 48, pp. 1625–1634, 2000.

[9]   S. M. Rao and T. K. Sarkar, "Implicit Solution of time-domain integral equations for arbitrarily shaped dielectric bodies," *Microw. Opt. Tech. Lett.*, Vol. 21, pp. 201–205, 1999.

[10]   B. H. Jung and T. K. Sarkar, "Time-domain electric-field integral equation with central finite difference," *Microw. Opt. Tech. Lett.*, Vol. 31, pp. 429–435, 2001.

[11]   T. K. Sarkar and J. Koh, "Generation of a wide-band electromagnetic response through a Laguerre expansion using early-time and low-frequency data," *IEEE Trans. Microwave Theory Tech.*, Vol. 50, pp. 1408–1416, 2002.

[12]   B. H. Jung, T. K. Sarkar, Y.-S. Chung, Z. Ji, and M. Salazar-Palma, "Analysis of transient electromagnetic scattering from dielectric objects using a combined-field integral equation," *Microw. Opt. Tech. Lett.*, Vol. 40, pp. 476–481, 2004.

[13]   B. H. Jung, Y.-S. Chung, and T. K. Sarkar, "Time-domain EFIE, MFIE, and CFIE formulations using Laguerre polynomials as temporal basis functions for the analysis of transient scattering from arbitrary shaped conducting structures," *J. Electromagn. Waves Appl.*, Vol. 17. pp. 737–739, 2003.

[14]   X. Q. Sheng, J. M. Jin, J. M. Song, W. C. Chew, and C. C. Lu, "Solution of combined-field integral equation using multilevel fast multipole algorithm for scattering by homogeneous bodies," *IEEE Trans. Antennas Prop.*, Vol. 46, pp. 1718–1726, 1998.

[15]   B. H. Jung, T. K. Sarkar, and Y.-S. Chung, "A survey of various frequency domain integral equations for the analysis of scattering from three-dimensional dielectric objects," *J. Electromagn. Waves Appl.*, Vol. 16. pp. 1419–1421, 2002.

[16]   S. M. Rao, D. R. Wilton, and A. W. Glisson, "Electromagnetic scattering by surfaces of arbitrary shape," *IEEE Trans. Antennas Prop.*, Vol. 30, pp. 409–418, 1982.

[17]   T. K. Sarkar, S. M. Rao, and A. R. Djordjevic, "Electromagnetic Scattering and Radiation from Finite Microstrip Structures," *IEEE Transactions on Microwave Theory and Techniques*, Vol. 38, No. 11, pp. 1568–1575, Nov. 1990.

[18]   S. M. Rao and D. R. Wilton, "*E*-field, *H*-field, and combined field solution for arbitrarily shaped three-dimensional dielectric bodies," *Electromagnetics*, Vol. 10, pp. 407–421, 1990.

[19]   S. M. Rao, *Electromagnetic Scattering and Radiation of Arbitrarily-Shaped Surfaces by Triangular Patch Modeling,* PhD Dissertation, University of Mississippi, Miss. 1980.

[20]   L. Debnath, *Integral Transforms and Their Applications*, CRC Press, Boca Raton, FL., 1995.

[21]   A. D. Poularikas, *The Transforms and Applications Handbook*, IEEE Press, New York, 1996.

[22]   I. S. Gradshteyn and I. M. Ryzhik, *Table of Integrals, Series, and Products*, Academic Press, New York, 1980.

[23]   J. Van Bladel, *Electromagnetic Fields*, Hemisphere Publishing Corporation, New York, 1985.

[24]   D. R. Wilton, S. M. Rao, A. W. Glisson, D. H. Schaubert, O. M. Al-Bundak, and C. M. Butler, "Potential integrals for uniform and linear source distributions on polygonal and polyhedral domains," *IEEE Trans. Antennas Prop.*, Vol. 32, pp. 276–281, 1984.

[25]   R. D. Graglia, "Static and dynamic potential integrals for linearly varying source distributions in two- and three-dimensional problems," *IEEE Trans. Antennas Prop.*, Vol. 35, pp. 662–669, 1987.

[26]   S. Caorsi, D. Moreno, and F. Sidoti, "Theoretical and numerical treatment of surface integrals involving the free-space Green's function," *IEEE Trans. Antennas Prop.*, Vol. 41, pp. 1296–1301, 1993.

[27]   R. D. Graglia, "On the numerical integration of the linear shape functions times the 3-D Green's function or its gradient on a plane triangle," *IEEE Trans. Antennas Prop.*, Vol. 41, pp. 1448–1455, 1993.

[28]   T. F. Eibert and V. Hansen, "On the Calculation of potential integrals for linear source distributions on triangular domains," *IEEE Trans. Antennas Prop.*, Vol. 43, pp. 1499–1502, 1995.

[29]   R. E. Hodges and Y. Rahmat-Samii, "The evaluation of MFIE integrals with the use of vector triangle basis functions," *Microw. Opt. Tech. Lett.*, Vol. 14, pp. 9–14, 1997.

[30]   P. Yla-Oijala and M. Taskinen, "Calculation of CFIE impedance matrix elements with RWG and **n**×RWG functions," *IEEE Trans. Antennas Prop.*, Vol. 51, pp. 1837–1846, 2003.

8

AN IMPROVED MARCHING-ON-IN-DEGREE (MOD) METHODOLOGY

8.0   SUMMARY

In Chapters 5–7, the marching-on-in-degree (MOD) algorithm has been illustrated for the solution of the time domain integral equation in lieu of the marching-on-in-time (MOT) method to improve the accuracy and the numerical stability of the solution procedures. In the previous implementations of the MOD algorithms, two shortcomings have been observed. First an approximation has been used in the evaluation of the retarded time at the centroid of the testing functions that may sometimes cause numerical inaccuracies of the time domain solution. The second problem is that the formulation presented for the MOD has some summations over all orders of the basis functions that need to be carried out at each computational instances, which is very time consuming.

In this chapter, an alternative formulation for the MOD methodology is presented that can address these two shortcomings outlined in the previous section resulting in a superior stable accurate time domain integral equation code for the solution of scattering from arbitrary shaped wire, conducting, and dielectric structures. The first shortcoming is removed by performing a method-of-moment (MoM) based time testing first followed by spatial testing. In this way, the time variable along with its derivatives can be completely eliminated from the final computations. This also helps us separate the coupled space–time terms into separate space and time variables, making the final computations to be carried out only over the space variables. Second, by rearranging the summation of the temporal associated Laguerre basis functions in a different computational form, the time for the evaluation of the Green's function can be reduced by an order of magnitude. Illustrative examples are provided to illustrate these two points. Challenging numerical problems are analyzed in the next Chapter 9 to illustrate the potential of this new methodology.

8.1   INTRODUCTION

The MOD method introduced in Chapters 5–7 has demonstrated the fact that the MOD is one of the most accurate and stable computational methodologies for solution of time domain integral equations. The objective of this chapter is to provide the theoretical foundation of a different mathematical form of the MOD solution procedure, which is not only computationally accurate and stable but also takes less time for computation than the previous version. The improvements occur in the following areas:

- *Testing Functions:*
  In the conventional MOD, the spatial testing is usually carried out before the temporal testing. This testing sequence makes the retarded time term difficult to handle. So a central approximation for the spatial integral over the patches is required to evaluate the terms that can result in numerical instabilities. In this chapter, a new testing sequence that performs first the temporal testing and followed by the spatial testing is introduced. This modification helps to handle the retarded time analytically; thus, no more central approximation is needed, and more stable results can be obtained.

- *New Combinational Form of the Temporal Basis Functions:*
  The use of associated Laguerre functions helps one to do away with the Courant–Friedrich–Levy sampling criteria, which relate the spatial to the temporal sampling by the velocity of propagation. This is possible as a result of the decoupling of the spatial and temporal variables. However, the *center processing unit* (CPU) time required to compute the summation of the temporal basis functions is still significant. In this chapter, a new temporal basis function that is formed by a linear combination of three associated Laguerre functions with successive degrees is used to expand the temporal variables. Using this new basis function, the first and second derivatives of the temporal variables can be represented by simpler expressions, as will be outlined in this chapter. This simplification makes the overall computation to the solution of the integral equation more efficient.

- *Green's Function*
  Another improvement in the computational efficiency of the new MOD method is the application of a new form of the Green's function. The time-domain Green's function couples many terms of the matrix equation obtained by the application of the Galerkin's method. In the Green's function, the associated Laguerre functions are not orthogonal for different degrees because of the involvement of the retarded time, so that the computation cannot be simplified using their orthogonality criteria. This leads to a much longer time to complete the computation of the Green's function. By developing a new form of the Green's function that groups the retarded-time terms, the repetitive computations for the Laguerre polynomials of different orders can be minimized to significantly shorten the total computation time.

In this chapter, the new MOD formulation is developed for solving the time domain electric field integral equation (TD-EFIE) for conducting objects including wires and the time domain Poggio–Miller–Chang–Harrington–Wu integral equation (TD-PMCHW) for dielectric objects. In Section 8.1, the new MOD procedure for analyzing conducting objects using the TD-EFIE is described. This new MOD formulation is then extended to deal with conducting objects containing concentrated loads in section 8.2. Finally, the application of the new MOD technique for the analysis of scattering from dielectric objects using the TD-PMCHW formulation is presented in section 8.3 followed by some conclusions in section 8.4. The new MOD approach can also be extended to the solution of other forms of the time domain integral equations by following a similar procedure. However, they are not described in this chapter. To demonstrate the effectiveness of these improved schemes, numerical examples are presented in Chapter 9 where a comparison between the computation time of using the new MOD, the conventional MOD, and the MOT (introduced in Chapter 6) methods for the solution of TDIE are presented. It is shown that the new Green's function introduced in this chapter can significantly improve the computational efficiency by a factor of more than *ten* times over the conventional formulations.

8.2   THE NEW MOD SCHEME FOR CONDUCTING OBJECTS INCLUDING WIRE STRUCTURES

8.2.1   The TD-EFIE for Conducting Structures

Enforcing the boundary condition, that the total tangential electric field is zero on the surface for all time, the time domain electric field integral equation (TD-EFIE) is obtained as follows

$$\left(\frac{\partial}{\partial t} \mathbf{A}(\mathbf{r},t) + \nabla \varPhi(\mathbf{r},t) \right)_{\text{tan}} = \left(\mathbf{E}^{i}(\mathbf{r},t) \right)_{\text{tan}} \tag{8.1a}$$

where

$$\mathbf{A}(\mathbf{r},t) = \frac{\mu}{4\pi} \int_{S} \frac{\mathbf{J}(\mathbf{r}',\tau)}{R} dS' \tag{8.1b}$$

$$\varPhi(\mathbf{r},t) = \frac{1}{4\pi\varepsilon} \int_{S} \frac{Q(\mathbf{r}',\tau)}{R} dS' \tag{8.1c}$$

$$R = \left| \mathbf{r} - \mathbf{r}' \right| \tag{8.1d}$$

$$\tau = t - R / c \tag{8.1e}$$

where $\mathbf{r} \in S$. The surface S of a conducting body is illuminated by a transient electromagnetic wave in which the incident electric field is given by $\mathbf{E}^{i}(\mathbf{r},t)$. The incident field induces an electric current $\mathbf{J}(\mathbf{r},t)$ on the surface of the conducting structure. The electric charge density is denoted by $Q(\mathbf{r},t)$. $\mathbf{A}(\mathbf{r},t)$ and $\varPhi(\mathbf{r},t)$

represent the magnetic vector and electric scalar potentials, respectively. The symbol R represents the distance between the observation point $\mathbf{r}$ and the source point $\mathbf{r'}$. τ is the retarded time resulting from the delay for the transmission of the field from the source point to the field point. μ and ε are the permeability and the permittivity of the medium, and c is the velocity of the propagation of the electromagnetic wave in that space. The subscript "tan" denotes the tangential components. Note that the integrals over the surface S in Eqs. (8.1b) and (8.1c) are transformed into line integrals over the wire length L when the scatterer is a conducting wire.

To simplify the time integral in the TD-EFIE analytically, the Hertz vector $\mathbf{u}(\mathbf{r}, t)$ is introduced in the MOD method and it is defined as [1]

$$\mathbf{J}(\mathbf{r},t) = \frac{\partial}{\partial t}\mathbf{u}(\mathbf{r},t) \tag{8.2}$$

and the relation between the Hertz vector and the electric charge density $Q(\mathbf{r}, t)$ is given by

$$Q(\mathbf{r}, t) = -\nabla \cdot \mathbf{u}(\mathbf{r}, t) \tag{8.3}$$

Substituting Eqs. (8.2) and (8.3) into Eq. (8.1), the TD-EFIE can be expressed in terms of the Hertz vector as

$$\left(\frac{\mu}{4\pi}\frac{\partial^2}{\partial t^2} \int_S \frac{\mathbf{u}(\mathbf{r'},\tau)}{R} dS' - \frac{\nabla}{4\pi\varepsilon} \int_S \frac{\nabla' \cdot \mathbf{u}(\mathbf{r'},\tau)}{R} dS' \right)_{\text{tan}} = \left(\mathbf{E}^i(\mathbf{r},t) \right)_{\text{tan}} \tag{8.4}$$

8.2.2   Expansion Functions in Space and Time

The Hertz vector in Eq. (8.4) is now expanded using the RWG vector functions, which have been introduced in Eq. (6.13) of Section 6.3, as the spatial basis functions. This result in

$$\mathbf{u}(\mathbf{r},t) = \sum_{n=1}^{N} u_n(t)\mathbf{f}_n(\mathbf{r}) \tag{8.5}$$

Note that in Eq. (8.5), the spatial and temporal variables are separated. It is important to note that the spatial basis functions do not need to be limited by the choice of the RWG vector functions [2,3]. In fact, it is possible to choose instead the higher-order basis functions [4] or of some other type.

In the conventional MOD method introduced in Chapters 5–7, the temporal basis functions chosen are the associated Laguerre functions $\phi_j(st)$, which have been described by Eq. (6.72) in section 6.5. Mathematically, they are written as

$$\phi_j(st) = \begin{cases} e^{-st/2} L_j(st) & t \geq 0 \\ 0 & t < 0 \end{cases} \tag{8.6}$$

where $L_j(t)$ is the Laguerre polynomial of degree j and s is a scaling factor. The associated Laguerre function is causal (i.e., defined over the domain $[0,+\infty)$) and convergent to zero as time goes to infinity [5,6]. To improve the computational efficiency, the temporal basis function is represented by a combination of three associated Laguerre functions with successive degrees multiplied by some unknown constants for the transient coefficient $u_n(t)$ as [7]

$$u_n(t) = \sum_{j=0}^{\infty} u_{n,j} \left(\phi_j(st) - 2\phi_{j+1}(st) + \phi_{j+2}(st) \right) \tag{8.7}$$

Referring to the property of the Laguerre polynomials presented in the appendix of Chapter 5, the derivative of the Laguerre polynomials can be expressed as a sum of its lower-order components as

$$\frac{d}{dt} L_j(st) = -L_{j-1}^{(1)}(st) = -\sum_{p=0}^{j-1} L_p(st) \tag{8.8}$$

From Eq. (8.8), the derivative of the associated Laguerre function is written as

$$\frac{d}{dt} \phi_j(st) = -\frac{s}{2} \phi_j(st) - s \sum_{p=0}^{j-1} \phi_p(st) \tag{8.9}$$

Using the property given by Eq. (8.9), the first and second derivatives of the transient coefficient $u_n(t)$ of Eq. (8.7) can be derived and represented as a combination of associated Laguerre functions of different degrees by the following:

$$\frac{d}{dt} u_n(t) = \sum_{j=0}^{\infty} \frac{s}{2} u_{n,j} \left(\phi_j(st) - \phi_{j+2}(st) \right) \tag{8.10a}$$

$$\frac{d^2}{dt^2} u_n(t) = \sum_{j=0}^{\infty} u_{n,j} \frac{s^2}{4} \left(\phi_j(st) + 2\phi_{j+1}(st) + \phi_{j+2}(st) \right) \tag{8.10b}$$

In contrast to the conventional MOD method [2, 3] in which the transient variable is expanded by a single associated Laguerre function set $\phi_j(st)$ as depicted by Eq. (6.73), the new temporal basis function presented here is formed by a combination of three associated Laguerre functions, $\phi_j(st)$, $\phi_{j+1}(st)$, and $\phi_{j+2}(st)$, for each degree j. By using this combinational temporal basis function set, the first and second derivatives of the transient variable $u_n(t)$ expressed in Eqs. (8.10a) and (8.10b) contains less terms in the summation than those using the conventional MOD. It is because of the cancellation of the lower-degree components in the derivatives of the three associated Laguerre functions, $\phi_j(st)$, $\phi_{j+1}(st)$, and $\phi_{j+2}(st)$. This cancellation leads to fewer terms to be computed and results in a shorter time for the computation.

Substituting Eqs. (8.5), (8.7), and (8.10) into Eq. (8.4), the TD-EFIE are expressed by the spatial vector function $\mathbf{f}_n(\mathbf{r})$ and the temporal function $\phi_j(st) - 2\phi_{j+1}(st) + \phi_{j+2}(st)$ as

$$
\left(\frac{\mu}{4\pi} \int_S \sum_{n=1}^{N} \sum_{j=0}^{\infty} u_{n,j} \frac{1}{R} \frac{s^2}{4} \left(\phi_j(s\tau) + 2\phi_{j+1}(s\tau) + \phi_{j+2}(s\tau) \right) \mathbf{f}_n(\mathbf{r}') \, dS' \right.
$$

$$
\left. - \frac{\nabla}{4\pi \varepsilon} \int_S \sum_{n=1}^{N} \sum_{j=0}^{\infty} u_{n,j} \frac{1}{R} \left(\phi_j(s\tau) - 2\phi_{j+1}(s\tau) + \phi_{j+2}(s\tau) \right) \nabla' \cdot \mathbf{f}_n(\mathbf{r}') \, dS' \right)_{\text{tan}}
$$

$$
= \left(\mathbf{E}^i(\mathbf{r},t) \right)_{\text{tan}} \qquad (8.11)
$$

Expanding the terms inside the summation and recombining the terms with the same degree of associated Laguerre functions, Eq. (8.11) can be rewritten by introducing two temporary parameters $e_{n,j}$ and $d_{n,j}$ as

$$
\left(\frac{\mu}{4\pi} \int_S \sum_{n=1}^{N} \sum_{j=0}^{\infty} e_{n,j} \frac{1}{R} \frac{s^2}{4} \phi_j(s\tau) \mathbf{f}_n(\mathbf{r}') \, dS' \right.
$$

$$
\left. - \frac{\nabla}{4\pi \varepsilon} \int_S \sum_{n=1}^{N} \sum_{j=0}^{\infty} d_{n,j} \frac{1}{R} \phi_j(s\tau) \nabla' \cdot \mathbf{f}_n(\mathbf{r}') \, dS' \right)_{\text{tan}} = \left(\mathbf{E}^i(\mathbf{r},t) \right)_{\text{tan}}
$$

$$
(8.12a)
$$

where

$$
e_{n,j} = u_{n,j-2} + 2u_{n,j-1} + u_{n,j} \qquad (8.12b)
$$

$$
d_{n,j} = u_{n,j-2} - 2u_{n,j-1} + u_{n,j} \qquad (8.12c)
$$

where $u_{n,j} = 0$ is assumed for $j < 0$

8.2.3   Testing Procedures

In the conventional MOD procedure, the time domain integral equations are tested spatially first followed by a temporal testing. This order of testing leads to inaccuracies when dealing with the retarded time term so that a central approximation of this quantity is needed. As a result, a relatively larger approximation error is induced. To overcome this problem, the testing sequence is revised to have the temporal testing before the spatial one. Also, through this procedure, the time variable can be completely eliminated from the final computations. Applying the temporal testing with a testing function $\phi_i(st)$ yields

$$
\left(\frac{\mu}{4\pi} \int_S \sum_{n=1}^{N} \sum_{j=0}^{\infty} e_{n,j} \frac{1}{R} \frac{s^2}{4} \int_{sR/c}^{\infty} \phi_i(st)\phi_j(s\tau)d(st) \, \mathbf{f}_n(\mathbf{r}') \, dS' \right.
$$

$$
\left. - \frac{\nabla}{4\pi\varepsilon} \int_S \sum_{n=1}^{N} \sum_{j=0}^{\infty} d_{n,j} \frac{1}{R} \int_{sR/c}^{\infty} \phi_i(st)\phi_j(s\tau)d(st) \, \nabla'\cdot\mathbf{f}_n(\mathbf{r}') \, dS' \right)_{\text{tan}}
$$

$$
= \left(\int_0^{\infty} \phi_i(st) \mathbf{E}^i(\mathbf{r},t)d(st) \right)_{\text{tan}} \qquad (8.13)
$$

It is important to note that the basis functions contain a combination of three associated Laguerre functions, whereas only one is used for testing.

Based on the integral property of the associated Laguerre functions given in the appendix of Chapter 5, the integrals in $I_{ij}(sR/c)$ can be evaluated as

$$
I_{ij}(sR/c) = \int_{sR/c}^{\infty} \phi_i(st)\phi_j(st - sR/c)d(st)
$$

$$
= \begin{cases} e^{(-sR/(2c))}\left(L_{i-j}(sR/c) - L_{i-j-1}(sR/c) \right) & j < i \\ e^{(-sR/(2c))} & j = i \\ 0 & j > i \end{cases} \qquad (8.14)
$$

Substituting Eq. (8.14) into Eq. (8.13) yields

$$
\left(\frac{\mu}{4\pi} \int_S \sum_{n=1}^{N} \sum_{j=0}^{i} e_{n,j} \frac{1}{R} \frac{s^2}{4} I_{ij}(sR/c) \, \mathbf{f}_n(\mathbf{r}') \, dS' \right.
$$

$$
\left. - \frac{\nabla}{4\pi\varepsilon} \int_S \sum_{n-1}^{N} \sum_{j=0}^{i} d_{n,j} \frac{1}{R} I_{ij}(sR/c) \, \nabla'\cdot\mathbf{f}_n(\mathbf{r}') \, dS' \right)_{\text{tan}} = \left(\mathbf{V}^i(\mathbf{r}) \right)_{\text{tan}} \qquad (8.15a)
$$

where

$$
\mathbf{V}^i(\mathbf{r}) = \int_0^{\infty} \phi_i(st) \mathbf{E}^i(\mathbf{r},t)d(st) \qquad (8.15b)
$$

In Eq. (8.15a), the upper limit of the summation over j is replaced by i instead of ∞ because the integral $I_{ij}(sR/c)$ equals to 0 when $j > i$. Furthermore, applying spatial testing to Eq. (8.15) with a testing function $\mathbf{f}_m(\mathbf{r})$ yields

$$\frac{\mu}{4\pi}\int_S\int_S\sum_{n=1}^{N}\sum_{j=0}^{i}e_{n,j}\frac{1}{R}\frac{s^2}{4}I_{ij}(sR/c)\,\mathbf{f}_m(\mathbf{r})\cdot\mathbf{f}_n(\mathbf{r}')\,dS'dS$$

$$-\frac{1}{4\pi\varepsilon}\int_S\mathbf{f}_m(\mathbf{r})\cdot\nabla\int_S\sum_{n=1}^{N}\sum_{j=0}^{i}d_{n,j}\frac{1}{R}I_{ij}(sR/c)\,\nabla'\cdot\mathbf{f}_n(\mathbf{r}')dS'dS$$

$$=\int_S\mathbf{f}_m(\mathbf{r})\cdot\mathbf{V}^i(\mathbf{r})\,dS \qquad (8.16)$$

Equivalently, Eq. (8.16) can be written as

$$\frac{s^2\mu}{4}\sum_{n=1}^{N}\sum_{j=0}^{i}e_{n,j}\,A_{mnij}+\frac{1}{\varepsilon}\sum_{n=1}^{N}\sum_{j=0}^{i}d_{n,j}\,B_{mnij}=\Omega_{m,i} \qquad (8.17a)$$

where

$$A_{mnij}=\int_S\mathbf{f}_m(\mathbf{r})\cdot\int_S\frac{1}{4\pi R}I_{ij}(sR/c)\,\mathbf{f}_n(\mathbf{r}')\,dS'\,dS \qquad (8.17b)$$

$$B_{mnij}=\int_S\nabla\cdot\mathbf{f}_m(\mathbf{r})\int_S\frac{1}{4\pi R}I_{ij}(sR/c)\,\nabla'\cdot\mathbf{f}_n(\mathbf{r}')dS'\,dS \qquad (8.17c)$$

$$\Omega_{m,i}=\int_S\mathbf{f}_m(\mathbf{r})\cdot\mathbf{V}^i(\mathbf{r})\,dS \qquad (8.17d)$$

Using Eq. (8.14), Eqs. (8.17b) and (8.17c) for $j=i$ can be written as

$$\alpha_{mn}=A_{mnij}\Big|_{j=i}=\int_S\mathbf{f}_m(\mathbf{r})\cdot\int_S\frac{1}{4\pi R}e^{(-sR/(2c))}\mathbf{f}_n(\mathbf{r}')\,dS'\,dS \qquad (8.18a)$$

$$\beta_{mn}=B_{mnij}\Big|_{j=i}=\int_S\nabla\cdot\mathbf{f}_m(\mathbf{r})\int_S\frac{1}{4\pi R}e^{(-sR/(2c))}\nabla'\cdot\mathbf{f}_n(\mathbf{r}')\,dS'\,dS \qquad (8.18b)$$

Finally, transforming the variables $e_{n,j}$ and $d_{n,j}$ back into $u_{n,j}$ according to Eqs. (8.12b) and (8.12c) and moving the terms associated with $u_{n,j}$ of degrees smaller than i to the right-hand side yields

$$\sum_{n=1}^{N}u_{n,i}\left(\frac{s^2\mu}{4}\alpha_{mn}+\frac{1}{\varepsilon}\beta_{mn}\right)=\Omega_{m,i}-\frac{s^2\mu}{4}\sum_{n=1}^{N}\left(u_{n,i-2}+2u_{n,i-1}\right)\alpha_{mn}$$

$$-\frac{1}{\varepsilon}\sum_{n=1}^{N}\left(u_{n,i-2}-2u_{n,i-1}\right)\beta_{mn}-\frac{s^2\mu}{4}\sum_{n=1}^{N}\sum_{j=0}^{i-1}\left(u_{n,j-2}+2u_{n,j-1}+u_{n,j}\right)A_{mnij}$$

$$-\frac{1}{\varepsilon}\sum_{n=1}^{N}\sum_{j=0}^{i-1}\left(u_{n,j-2}-2u_{n,j-1}+u_{n,j}\right)B_{mnij} \qquad (8.19)$$

Eq. (8.19) indicates that the unknown coefficient of a temporal degree i can be characterized by its lower degrees, which leads to a marching-on-in-degree algorithm that can now be applied for solving the coefficients of any order.

8.2.4 Green's Functions

Note that the term $I_{ij}(sR/c)/R$, which is the Green's function after testing, exist in all calculations of A_{mnij} and B_{mnij}. In Eq. (8.14), the associated Laguerre functions are not orthogonal because of the involvement of the retarded time factor, so that the computations cannot be simplified using the orthogonality between the functions. This leads to a much longer CPU time to complete the computation of the Green's function. As such, a new Green's function is designed that groups the retarded-time terms to minimize the repetitive computations for the Laguerre polynomials of different degrees and the floating-point number divisions for R. Defining the two parameters k_{mni} and h_{mni} associated with the new Green's functions as

$$k_{mni} = \left(u_{n,i-2} + 2u_{n,i-1}\right)\alpha_{mn} + \sum_{j=0}^{i-1}\left(u_{n,j-2} + 2u_{n,j-1} + u_{n,j}\right)A_{mnij}$$

$$= \int_S \mathbf{f}_m(\mathbf{r}) \cdot \int_S \mathbf{f}_n(\mathbf{r}')\frac{1}{4\pi R}\left[\left(u_{n,i-2} + 2u_{n,i-1}\right)I_{ii}(sR/c)\right.$$

$$\left. + \sum_{j=0}^{i-1}\left(u_{n,j-2} + 2u_{n,j-1} + u_{n,j}\right)I_{ij}(sR/c)\right]dS'\,dS \qquad (8.20)$$

and

$$h_{mni} = \left(u_{n,i-2} - 2u_{n,i-1}\right)\beta_{mn} + \sum_{j=0}^{i-1}\left(u_{n,j-2} - 2u_{n,j-1} + u_{n,j}\right)B_{mnij}$$

$$= \int_S \nabla\cdot\mathbf{f}_m(\mathbf{r})\int_S \nabla'\cdot\mathbf{f}_n(\mathbf{r}')\frac{1}{4\pi R}\left[\left(u_{n,i-2} - 2u_{n,i-1}\right)I_{ii}(sR/c)\right.$$

$$\left. + \sum_{j=0}^{i-1}\left(u_{n,j-2} - 2u_{n,j-1} + u_{n,j}\right)I_{ij}(sR/c)\right]dS'\,dS \qquad (8.21)$$

Eq. (8.19) can be expressed as

$$\sum_{n=1}^{N}u_{n,i}\left(\frac{s^2\mu}{4}\alpha_{mn} + \frac{1}{\varepsilon}\beta_{mn}\right) = \Omega_{m,i} - \frac{s^2\mu}{4}\sum_{n=1}^{N}k_{mni} - \frac{1}{\varepsilon}\sum_{n=1}^{N}h_{mni} \qquad (8.22)$$

where $m = 1, 2, ..., N$ and $i = 0, 1, 2, ..., I$. This can be represented further in a matrix form as

$$[Z_{mn}]_{N\times N}\,[u_{n,i}]_{N\times 1} = [\gamma_{m,i}]_{N\times 1} \qquad (8.23a)$$

where

$$Z_{mn} = \frac{s^2 \mu}{4} \alpha_{mn} + \frac{1}{\varepsilon} \beta_{mn} \qquad (8.23b)$$

$$U_{n,i} = u_{n,i} \qquad (8.23c)$$

$$\gamma_{m,i} = \Omega_{m,i} - \frac{s^2 \mu}{4} \sum_{n=1}^{N} k_{mni} - \frac{1}{\varepsilon} \sum_{n=1}^{N} h_{mni} \qquad (8.23d)$$

For any temporal degree i_0, the coefficients u_{n,i_0} can be computed by a MOD procedure using Eq. (8.23), providing that all lower-degree coefficients for $i = 0, 1, 2,..., i_0 - 1$ has already been computed. Using the computational scheme outlined, the computation time can be significantly reduced by a factor of more than ten times for most applications when contrasted against the use of the conventional MOD.

8.2.5   Improvement of the Computation Time

To quantize the improvement of the computation time for solving the TD-EFIE with the new MOD scheme outlined in sections 8.1.2–8.1.4, the computational efficiency can be approximated in terms of the order of magnitude $\mathcal{O}$ (called big O notation) of the total number of operations, which is mainly dependent on the maximum number of spatial unknowns N and the maximum temporal degree I. From the TD-EFIE formulations given by Eq. (6.86) and Eq. (8.22), the number of operations needed for the floating-point divisions and the calculations of the Laguerre polynomials using the conventional MOD and the new MOD are counted and listed in Table 8.1.

Table 8.1 The number of operations for the conventional and the new MOD methods

	The conventional MOD	The new MOD
Floating point divisions	$\mathcal{O}(N^2 I^2)$	$\mathcal{O}(N^2 I)$
Laguerre polynomial computations	$\mathcal{O}(N^2 I^3)$	$\mathcal{O}(N^2 I^2)$

These two computation tasks are the most time-consuming in the MOD solution procedures as the number of floating-point divisions take a lot of instruction cycles and the calculation of the Laguerre polynomials require a recursive algorithm. From Table 8.1, it can be observed that the number of operations needed for these two computation tasks in the new MOD is relatively much less than those using the conventional MOD. A significant reduction in time for solving the TD-EFIE is achieved. To better illustrate the improvement in the efficiency of the new MOD, numerical examples are shown in Chapter 9 to compare the computation time for these two MOD methods.

8.3   THE NEW MOD METHOD FOR CONDUCTING OBJECTS WITH LOADS

8.3.1   TD-EFIE for Conducting Structures with Loads

In this section, the new MOD method is applied for the analysis of conducting scatterers with loads. The loads to be considered include both distributed and concentrated loads. The distributed load is characterized by an additional impedance distributing throughout the conducting surface; for example, skin effect is one kind of distributed load, whereas the concentrated load refers to the impedances, such as lumped elements connected at a point of the structure. In fact, the concentrated load can be considered as a special case of the distributed load in which the load manifests as a Dirac delta function in space. In the frequency domain, the load is described by a function $Z(\omega)$, and it can be characterized by the transient response $Z(t)$. For conducting objects with a load, the TD-EFIE can be obtained from the boundary conditions of the tangential electric fields as outlined in Section 8.1.1 and generalized as

$$\left(\mathbf{E}^i(\mathbf{r},t)\right)_{\tan} - \left(\frac{\partial}{\partial t}\mathbf{A}(\mathbf{r},t)+\nabla\Phi(\mathbf{r},t)\right)_{\tan} = Z(\mathbf{r},t)*\mathbf{J}(\mathbf{r},t) \qquad (8.24)$$

where $*$ denotes the convolution. The Hertz vector $\mathbf{u}(\mathbf{r},t)$ is substituted into the TD-EFIE in Eq. (8.24) and yields

$$\left(\frac{\mu}{4\pi}\frac{\partial^2}{\partial t^2}\int_S \frac{\mathbf{u}(\mathbf{r}',\tau)}{R}dS' - \frac{\nabla}{4\pi\varepsilon}\int_S \frac{\nabla'\cdot\mathbf{u}(\mathbf{r}',\tau)}{R}dS'\right)_{\tan}$$
$$+ Z(\mathbf{r},t) * \frac{\partial}{\partial t}\mathbf{u}(\mathbf{r},t) = \left(\mathbf{E}^i(\mathbf{r},t)\right)_{\tan} \qquad (8.25)$$

The load $Z(\mathbf{r},t)$ in the time-domain is often described by a set of discrete values on a finite sampling time, which can be approximately represented by a rectangular piecewise basis function set $\Pi(t)$ as

$$Z(\mathbf{r},t)\approx\sum_{k=0}^{K}Z_k(\mathbf{r})\Pi(t-k\Delta t)$$

where

$$\Pi(t) = \begin{cases} 1 & 0 < t < \Delta t \\ 0 & \text{others} \end{cases} \tag{8.26}$$

If Δt is chosen small enough, then the convolution of the load $Z(\mathbf{r},t)$ with the current $\mathbf{J}(\mathbf{r},t)$ is given by

$$Z(\mathbf{r},t) * \frac{\partial}{\partial t}\mathbf{u}(\mathbf{r},t) = \sum_{k=0}^{K} Z_k(\mathbf{r})\frac{\partial}{\partial t}\mathbf{u}(\mathbf{r}, t - k\Delta t) \tag{8.27}$$

8.3.2   Expansion Procedures

Following the similar procedures described in Section 8.1.2, the TD-EFIE in Eq. (8.25) can be expanded by the spatial basis function set $\mathbf{f}_n(\mathbf{r})$ and the temporal basis function set $\phi_j(st) - 2\phi_{j+1}(st) + \phi_{j+2}(st)$, respectively, and yields

$$\left[\frac{\mu}{4\pi}\int_S \sum_{n=1}^{N}\sum_{j=0}^{\infty} u_{n,j}\frac{1}{R}\frac{s^2}{4}\left(\phi_j(s\tau) + 2\phi_{j+1}(s\tau) + \phi_{j+2}(s\tau)\right)\mathbf{f}_n(\mathbf{r}')\,dS' \right.$$

$$\left. - \frac{\nabla}{4\pi\varepsilon}\int_S \sum_{n=1}^{N}\sum_{j=0}^{\infty} u_{n,j}\frac{1}{R}\left(\phi_j(s\tau) - 2\phi_{j+1}(s\tau) + \phi_{j+2}(s\tau)\right)\nabla'\bullet\mathbf{f}_n(\mathbf{r}')\,dS' \right]_{\text{tan}}$$

$$+ \sum_{k=0}^{K} Z_k(\mathbf{r})\sum_{n=1}^{N}\sum_{j=0}^{\infty}\frac{s}{2}u_{n,j}\left(\phi_j(st - sk\Delta t) - \phi_{j+2}(st - sk\Delta t)\right)\mathbf{f}_n(\mathbf{r})$$

$$= \left(\mathbf{E}^i(\mathbf{r},t)\right)_{\text{tan}} \tag{8.28}$$

Using the parameters $e_{n,j}$ and $d_{n,j}$ defined in Eqs. (8.12b) and (8.12c) and assuming that $u_{n,j} = 0$ for $j < 0$, Eq. (8.28) is written as

$$\left[\frac{\mu}{4\pi}\int_S \sum_{n=1}^{N}\sum_{j=0}^{\infty} e_{n,j}\frac{1}{R}\frac{s^2}{4}\phi_j(s\tau)\,\mathbf{f}_n(\mathbf{r}')\,dS' \right.$$

$$\left. - \frac{\nabla}{4\pi\,\varepsilon}\int_S \sum_{n=1}^{N}\sum_{j=0}^{\infty}\frac{1}{R}d_{n,j}\,\phi_j(s\tau)\,\nabla'\bullet\mathbf{f}_n(\mathbf{r}')\,dS' \right]_{\text{tan}}$$

$$+ \sum_{k=0}^{K} Z_k(\mathbf{r})\sum_{n=1}^{N}\sum_{j=0}^{\infty}\frac{s}{2}\left(u_{n,j} - u_{n,j-2}\right)\phi_j(st - sk\Delta t)\,\mathbf{f}_n(\mathbf{r}) = \left(\mathbf{E}^i(\mathbf{r},t)\right)_{\text{tan}} \tag{8.29}$$

8.3.3   Testing Procedures

Applying the new scheme in the testing sequence, which performs the temporal testing with $\phi_i(st)$ before the spatial testing, results in

$$\left(\frac{\mu}{4\pi}\int_S\sum_{n=1}^{N}\sum_{j=0}^{i}e_{n,j}\frac{1}{R}\frac{s^2}{4}I_{ij}(sR/c)\,\mathbf{f}_n(\mathbf{r}')\,dS'\right.$$

$$-\frac{\nabla}{4\pi\varepsilon}\int_S\sum_{n=1}^{N}\sum_{j=0}^{i}d_{n,j}\frac{1}{R}I_{ij}(sR/c)\,\nabla'\bullet\,\mathbf{f}_n(\mathbf{r}')\,dS'\bigg|_{\text{tan}}$$

$$+\frac{s}{2}\sum_{k=0}^{K}\sum_{n=1}^{N}\sum_{j=0}^{i}Z_k(\mathbf{r})\left(u_{n,j}-u_{n,j-2}\right)I_{ij}(sk\Delta t)\,\mathbf{f}_n(\mathbf{r})=\left(\mathbf{V}^i(\mathbf{r})\right)_{\text{tan}}\qquad (8.30)$$

where $\mathbf{V}^i(\mathbf{r})$ is given by Eq. (8.15b). Next, applying the spatial testing with $\mathbf{f}_m(\mathbf{r})$ gives

$$\frac{s^2\mu}{4}\sum_{n=1}^{N}\sum_{j=0}^{i}e_{n,j}A_{mnij}+\frac{1}{\varepsilon}\sum_{n=1}^{N}\sum_{j=0}^{i}d_{n,j}B_{mnij}$$

$$+\frac{s}{2}\sum_{k=0}^{K}\sum_{n=1}^{N}\sum_{j=0}^{i}\left(u_{n,j}-u_{n,j-2}\right)I_{ij}(sk\Delta t)\,Z_{mnk}=\Omega_{m,i}\qquad (8.31a)$$

where

$$A_{mnij}=\int_S\mathbf{f}_m(\mathbf{r})\bullet\int_S\frac{1}{4\pi R}I_{ij}(sR/c)\,\mathbf{f}_n(\mathbf{r}')\,dS'dS\qquad (8.31b)$$

$$B_{mnij}=\int_S\nabla\bullet\mathbf{f}_m(\mathbf{r})\int_S\frac{1}{4\pi R}I_{ij}(sR/c)\,\nabla'\bullet\mathbf{f}_n(\mathbf{r}')dS'dS\qquad (8.31c)$$

$$\Omega_{m,i}=\int_S\mathbf{f}_m(\mathbf{r})\bullet\mathbf{V}^i(\mathbf{r})\,dS\qquad (8.31d)$$

$$Z_{mnk}=\int_S Z_k(\mathbf{r})\,\mathbf{f}_m(\mathbf{r})\bullet\mathbf{f}_n(\mathbf{r})\,dS\qquad (8.31e)$$

In Eq. (8.31), the parameters A_{mnij}, B_{mnij}, and $\Omega_{m,i}$ are the same as Eqs. (8.17b), (8.17c), and (8.17d). Transforming the coefficients $e_{n,j}$ and $d_{n,j}$ back into $u_{n,j}$ according to Eqs. (8.12b) and (8.12c) and moving the terms associated with $u_{n,j}$ of degrees smaller than i to the right-hand side yields

$$\sum_{n=1}^{N}u_{n,i}\left(\frac{s^2\mu}{4}\alpha_{mn}+\frac{1}{\varepsilon}\beta_{mn}+\frac{s}{2}\sum_{k=0}^{K}e^{-\frac{sk\Delta t}{2}}Z_{mnk}\right)=\Omega_{m,i}$$

$$-\frac{s^2\mu}{4}\sum_{n=1}^{N}\left(u_{n,i-2}+2u_{n,i-1}\right)\alpha_{mn}-\frac{1}{\varepsilon}\sum_{n=1}^{N}\left(u_{n,i-2}-2u_{n,i-1}\right)\beta_{mn}$$

$$-\frac{s^2\mu}{4}\sum_{n=1}^{N}\sum_{j=0}^{i-1}\left(u_{n,j-2}+2u_{n,j-1}+u_{n,j}\right)A_{mnij}$$

$$-\frac{1}{\varepsilon}\sum_{n=1}^{N}\sum_{j=0}^{i-1}\left(u_{n,j-2}-2u_{n,j-1}+u_{n,j}\right)B_{mnij}$$

$$-\frac{s}{2}\sum_{k=0}^{K}\sum_{n=1}^{N}\sum_{j=0}^{i-1}I_{ij}(sk\Delta t)\left(u_{n,j}-u_{n,j-2}\right)Z_{mnk}+\sum_{n=1}^{N}\frac{s}{2}\sum_{k=0}^{K}e^{-\frac{sk\Delta t}{2}}u_{n,i-2}Z_{mnk} \qquad (8.32)$$

Eq. (8.32) can be solved with a marching-on-in-degree procedure as the unknown coefficient of a temporal degree i is expressed in terms of only its lower degrees. In addition, the new Green's function given by Eqs. (8.20) and (8.21) are also used to handle the retarded time to improve the computational efficiency as

$$\sum_{n=1}^{N}u_{n,i}\left(\frac{s^2\mu}{4}\alpha_{mn}+\frac{1}{\varepsilon}\beta_{mn}+\frac{s}{2}\sum_{k=0}^{K}e^{-\frac{sk\Delta t}{2}}Z_{mnk}\right)=\Omega_{m,i}-\frac{s^2\mu}{4}\sum_{n=1}^{N}k_{mni}$$

$$-\frac{1}{\varepsilon}\sum_{n=1}^{N}h_{mni}-\frac{s}{2}\sum_{k=0}^{K}\sum_{n=1}^{N}\sum_{j=0}^{i-1}I_{ij}(sk\Delta t)\left(u_{n,j}-u_{n,j-2}\right)Z_{mnk}$$

$$+\sum_{n=1}^{N}\frac{s}{2}\sum_{k=0}^{K}e^{-\frac{sk\Delta t}{2}}u_{n,i-2}Z_{mnk} \qquad (8.33)$$

When there is no loading associated with the conducting structure (i.e., $Z_{mnk}=0$), Eq. (8.33) is exactly the same as Eq. (8.22).

8.4   THE NEW MOD SCHEME FOR DIELECTRIC STRUCTURES

8.4.1   The TD-PMCHW Integral Equations

The TD-PMCHW integral equations are widely used for analyzing dielectric structures and have been introduced in Section 7.2. Some steps in that formulation are repeated here for completeness in the description of the new MOD scheme for dielectric structures. Consider a homogeneous dielectric body with a permittivity ε_2 and a permeability μ_2 placed in an infinite homogeneous medium with permittivity ε_1 and permeability μ_1 (refer to Figure 3.1); the TD-PMCHW integral equations are given by

$$\sum_{v=1}^{2}\left(\frac{\partial}{\partial t}\mathbf{A}_v(\mathbf{r},t)+\nabla\Phi_v(\mathbf{r},t)+\frac{1}{\varepsilon_v}\nabla\times\mathbf{F}_v(\mathbf{r},t)\right)_{tan}=\left(\mathbf{E}^i(\mathbf{r},t)\right)_{tan} \qquad (8.34a)$$

$$\sum_{v=1}^{2} \left[\frac{\partial}{\partial t} \mathbf{F}_v(\mathbf{r},t) + \nabla \Psi_v(\mathbf{r},t) - \frac{1}{\mu_v} \nabla \times \mathbf{A}_v(\mathbf{r},t) \right]_{\tan} = \left(\mathbf{H}^i(\mathbf{r},t) \right)_{\tan} \qquad (8.34b)$$

In Eqs. (8.34a) and (8.34b), the symbol v is used to represent the region where the currents $\mathbf{J}$ or $\mathbf{M}$ are radiating and can be assigned as either 1 for the exterior region or 2 for the interior region. The parameters $\mathbf{A}_v$, $\mathbf{F}_v$, Φ_v, and Ψ_v are the magnetic vector potential, electric vector potentials, electric scalar potential, and magnetic scalar potential, respectively, which are further given by

$$\mathbf{A}_v(\mathbf{r},t) = \frac{\mu_v}{4\pi} \int_S \frac{\mathbf{J}(\mathbf{r}',\tau_v)}{R} dS' \qquad (8.35)$$

$$\mathbf{F}_v(\mathbf{r},t) = \frac{\varepsilon_v}{4\pi} \int_S \frac{\mathbf{M}(\mathbf{r}',\tau_v)}{R} dS' \qquad (8.36)$$

$$\Phi_v(\mathbf{r},t) = \frac{1}{4\pi\varepsilon_v} \int_S \frac{Q_e(\mathbf{r}',\tau_v)}{R} dS' \qquad (8.37)$$

$$\Psi_v(\mathbf{r},t) = \frac{1}{4\pi\mu_v} \int_S \frac{Q_m(\mathbf{r}',\tau_v)}{R} dS' \qquad (8.38)$$

$$\tau_v = t - R/c_v \qquad (8.39)$$

where R represents the distance between the observation point $\mathbf{r}$ and the source point $\mathbf{r}'$, the retarded time is denoted by τ_v and the velocity of the electromagnetic wave propagated in the space with medium parameters (ε_v, μ_v) is $c_v = 1/\sqrt{\varepsilon_v \mu_v}$.

To handle the time derivative of the magnetic vector potential $\mathbf{A}(\mathbf{r},t)$ in Eq. (8.34) analytically, Hertz vectors $\mathbf{u}(\mathbf{r},t)$ and $\mathbf{v}(\mathbf{r},t)$ are introduced for the electric and magnetic currents and defined by

$$\mathbf{J}(\mathbf{r},t) = \frac{\partial}{\partial t} \mathbf{u}(\mathbf{r},t) \qquad (8.40)$$

$$\mathbf{M}(\mathbf{r},t) = \frac{\partial}{\partial t} \mathbf{v}(\mathbf{r},t) \qquad (8.41)$$

and the relation between the Hertz vectors and the electric and magnetic charge densities are given by

$$Q_e(\mathbf{r},t) = -\nabla \cdot \mathbf{u}(\mathbf{r},t) \qquad (8.42)$$

$$Q_m(\mathbf{r},t) = -\nabla \cdot \mathbf{v}(\mathbf{r},t) \qquad (8.43)$$

Substituting Eq. (8.35)–(8.43) into Eq. (8.34), the TD-PMCHW is rewritten as

$$\sum_{v=1}^{2}\left(\frac{\mu_v}{4\pi}\frac{\partial^2}{\partial t^2}\int_S \frac{\mathbf{u}(\mathbf{r}',t)}{R}dS' - \frac{\nabla}{4\pi\varepsilon_v}\int_S \frac{\nabla'\cdot\mathbf{u}(\mathbf{r}',t)}{R}dS'\right.$$

$$\left. +\frac{1}{4\pi}\frac{\partial}{\partial t}\nabla\times\int_S \frac{\mathbf{v}(\mathbf{r}',t)}{R}dS'\right)_{\text{tan}} = \left(\mathbf{E}^i(\mathbf{r},t)\right)_{\text{tan}} \tag{8.44}$$

$$\sum_{v=1}^{2}\left(\frac{\varepsilon_v}{4\pi}\frac{\partial^2}{\partial t^2}\int_S \frac{\mathbf{v}(\mathbf{r}',t)}{R}dS' - \frac{\nabla}{4\pi\mu_v}\int_S \frac{\nabla'\cdot\mathbf{v}(\mathbf{r}',t)}{R}dS'\right.$$

$$\left. -\frac{1}{4\pi}\frac{\partial}{\partial t}\nabla\times\int_S \frac{\mathbf{u}(\mathbf{r}',t)}{R}dS'\right)_{\text{tan}} = \left(\mathbf{H}^i(\mathbf{r},t)\right)_{\text{tan}} \tag{8.45}$$

8.4.2   Expansion Functions

These unknown Hertz vectors $\mathbf{u}(\mathbf{r},t)$ and $\mathbf{v}(\mathbf{r},t)$ are spatially expanded by the RWG vector function set $\mathbf{f}_n(\mathbf{r})$ as

$$\mathbf{u}(\mathbf{r},t)=\sum_{n=1}^{N} u_n(t)\,\mathbf{f}_n(\mathbf{r}) \tag{8.46}$$

$$\mathbf{v}(\mathbf{r},t)=\sum_{n=1}^{N} v_n(t)\,\mathbf{f}_n(\mathbf{r}) \tag{8.47}$$

and the transient coefficients in Eqs. (8.46) and (8.47) are further expanded by using the new temporal basis function $\phi_j(st)-2\phi_{j+1}(st)+\phi_{j+2}(st)$. One obtains

$$u_n(t)=\sum_{j=0}^{\infty} u_{n,j}\left(\phi_j(st)-2\phi_{j+1}(st)+\phi_{j+2}(st)\right) \tag{8.48}$$

$$v_n(t)=\sum_{j=0}^{\infty} v_{n,j}\left(\phi_j(st)-2\phi_{j+1}(st)+\phi_{j+2}(st)\right) \tag{8.49}$$

Substituting Eqs. (8.46)–(8.48) into Eqs. (8.44) and (8.45) results in

$$\sum_{v=1}^{2}\left(\frac{\mu_v}{4\pi}\sum_{n=1}^{N}\int_S \frac{1}{R}\sum_{j=0}^{\infty} u_{n,j}\frac{s^2}{4}\left(\phi_j(st)+2\phi_{j+1}(st)+\phi_{j+2}(st)\right)\mathbf{f}_n(\mathbf{r}')\,dS'\right.$$

$$-\frac{1}{4\pi\varepsilon_v}\nabla\sum_{n=1}^{N}\int_S \frac{1}{R}\sum_{j=0}^{\infty} u_{n,j}\left(\phi_j(st)-2\phi_{j+1}(st)+\phi_{j+2}(st)\right)\nabla'\cdot\mathbf{f}_n(\mathbf{r}')\,dS'$$

$$+\frac{1}{4\pi}\nabla\times\sum_{n=1}^{N}\int_{S}\frac{1}{R}\sum_{j=0}^{\infty}\frac{s}{2}v_{n,j}\left(\phi_{j}(st)-\phi_{j+2}(st)\right)\mathbf{f}_{n}(\mathbf{r}')\,dS'\Bigg|_{\mathrm{tan}}$$

$$= \left(\mathbf{E}^{i}(\mathbf{r},t)\right)_{\mathrm{tan}} \qquad (8.50)$$

and

$$\sum_{v=1}^{2}\left(\frac{\varepsilon_{v}}{4\pi}\sum_{n=1}^{N}\int_{S}\frac{1}{R}\sum_{j=0}^{\infty}v_{n,j}\frac{s^{2}}{4}\left(\phi_{j}(st)+2\phi_{j+1}(st)+\phi_{j+2}(st)\right)\mathbf{f}_{n}(\mathbf{r}')\,dS'\right.$$

$$-\frac{1}{4\pi\mu_{v}}\nabla\sum_{n=1}^{N}\int_{S}\frac{1}{R}\sum_{j=0}^{\infty}v_{n,j}\left(\phi_{j}(st)-2\phi_{j+1}(st)+\phi_{j+2}(st)\right)\nabla'\cdot\mathbf{f}_{n}(\mathbf{r}')\,dS'$$

$$-\frac{1}{4\pi}\nabla\times\sum_{n=1}^{N}\int_{S}\frac{1}{R}\sum_{j=0}^{\infty}\frac{s}{2}u_{n,j}\left(\phi_{j}(st)-\phi_{j+2}(st)\right)\mathbf{f}_{n}(\mathbf{r}')\,dS'\Bigg|_{\mathrm{tan}}$$

$$= \left(\mathbf{H}^{i}(\mathbf{r},t)\right)_{\mathrm{tan}} \qquad (8.51)$$

8.4.3   Testing Functions

Following the testing procedures described in Section 8.1.3, Eq. (8.51) is tested in time with $\phi_{i}(st)$ first and then, defining $t_{d}=sR/c$, yields

$$\sum_{v=1}^{2}\left(\frac{\mu_{v}}{4\pi}\sum_{n=1}^{N}\int_{S}\frac{1}{R}\sum_{j=0}^{\infty}u_{n,j}\frac{s^{2}}{4}\left(I_{ij}(t_{d})+2I_{i,j+1}(t_{d})+I_{i,j+2}(t_{d})\right)\mathbf{f}_{n}(\mathbf{r}')\,dS'\right.$$

$$-\frac{1}{4\pi\varepsilon_{v}}\nabla\sum_{n=1}^{N}\int_{S}\frac{1}{R}\sum_{j=0}^{\infty}u_{n,j}\left(I_{ij}(t_{d})-2I_{i,j+1}(t_{d})+I_{i,j+2}(t_{d})\right)\nabla'\cdot\mathbf{f}_{n}(\mathbf{r}')\,dS'$$

$$+\frac{1}{4\pi}\nabla\times\sum_{n=1}^{N}\int_{S}\frac{1}{R}\sum_{j=0}^{\infty}\frac{s}{2}v_{n,j}\left(I_{ij}(t_{d})-I_{i,j+2}(t_{d})\right)\mathbf{f}_{n}(\mathbf{r}')\,dS'\Bigg|_{\mathrm{tan}}$$

$$= \left(\mathbf{V}_{i}^{E}(\mathbf{r})\right)_{\mathrm{tan}} \qquad (8.52a)$$

where

$$\mathbf{V}_{i}^{E}(\mathbf{r})=\int_{0}^{\infty}\phi_{i}(st)\,\mathbf{E}^{i}(\mathbf{r},t)\,d(st) \qquad (8.52b)$$

and

$$\sum_{v=1}^{2}\left(\frac{\varepsilon_{v}}{4\pi}\sum_{n=1}^{N}\int_{S}\frac{1}{R}\sum_{j=0}^{\infty}v_{n,j}\frac{s^{2}}{4}\left(I_{ij}(t_{d})+2I_{i,j+1}(t_{d})+I_{i,j+2}(t_{d})\right)\mathbf{f}_{n}(\mathbf{r}')\,dS'\right.$$

$$-\frac{1}{4\pi\mu_{v}}\nabla\sum_{n=1}^{N}\int_{S}\frac{1}{R}\sum_{j=0}^{\infty}v_{n,j}\left(I_{ij}(t_{d})-2I_{i,j+1}(t_{d})+I_{i,j+2}(t_{d})\right)\nabla'\cdot\mathbf{f}_{n}(\mathbf{r}')\,dS'$$

$$-\frac{1}{4\pi}\nabla\times\sum_{n=1}^{N}\int_{S}\frac{1}{R}\sum_{j=0}^{\infty}\frac{s}{2}u_{n,j}\left(I_{ij}(t_d)-I_{i,j+2}(t_d)\right)\mathbf{f}_n(\mathbf{r}')\,dS'\Bigg|_{\text{tan}}$$

$$=\left(\mathbf{V}_i^H(\mathbf{r})\right)_{\text{tan}}\qquad(8.53a)$$

where

$$\mathbf{V}_i^H(\mathbf{r})=\int_0^{\infty}\phi_i(st)\mathbf{H}^i(\mathbf{r},t)\,d(st)\qquad(8.53b)$$

Then, performing spatial testing with the testing functions $\mathbf{f}_m(\mathbf{r})$ yields

$$\sum_{v=1}^{2}\Bigg(\frac{\mu_v s^2}{4}\sum_{n=1}^{N}\sum_{j=0}^{i}u_{n,j}\left(A_{mnij}^v+2A_{mni,j+1}^v+A_{mni,j+2}^v\right)$$

$$+\frac{1}{\varepsilon_v}\sum_{n=1}^{N}\sum_{j=0}^{i}u_{n,j}\left(B_{mnij}^v-2B_{mni,j+1}^v+B_{mni,j+2}^v\right)$$

$$-\frac{s}{2}\sum_{n=1}^{N}\sum_{j=0}^{i}v_{n,j}\left(C_{mnij}^v-C_{mni,j+2}^v\right)\Bigg)=\Omega_{m,i}^E\qquad(8.54)$$

and

$$\sum_{v=1}^{2}\Bigg(\frac{\varepsilon_v s^2}{4}\sum_{n=1}^{N}\sum_{j=0}^{i}v_{n,j}\left(A_{mnij}^v+2A_{mni,j+1}^v+A_{mni,j+2}^v\right)$$

$$+\frac{1}{\mu_v}\sum_{n=1}^{N}\sum_{j=0}^{i}v_{n,j}\left(B_{mnij}^v-2B_{mni,j+1}^v+B_{mni,j+2}^v\right)$$

$$-\frac{s}{2}\sum_{n=1}^{N}\sum_{j=0}^{i}u_{n,j}\left(C_{mnij}^v-C_{mni,j+2}^v\right)\Bigg)=\Omega_{m,i}^H\qquad(8.55)$$

The coefficients in Eqs. (8.54) and (8.55) are given by

$$A_{mnij}^v=\int_S\mathbf{f}_m(\mathbf{r})\cdot\int_S\frac{1}{4\pi R}I_{ij}(t_{d,v})\,\mathbf{f}_n(\mathbf{r}')\,dS'\,dS\qquad(8.56)$$

$$B_{mnij}^v=\int_S\nabla\cdot\mathbf{f}_m(\mathbf{r})\int_S\frac{1}{4\pi R}I_{ij}(t_{d,v})\nabla'\cdot\mathbf{f}_n(\mathbf{r}')\,dS'\,dS\qquad(8.57)$$

$$C_{mnij}^v=\int_S\mathbf{f}_m(\mathbf{r})\cdot\int_S\nabla\times\frac{1}{4\pi R}I_{ij}(t_{d,v})\,\mathbf{f}_n(\mathbf{r}')\,dS'\,dS\qquad(8.58)$$

$$\alpha_{mn}^v=A_{mnij}^v\Big|_{j=i}=\int_S\mathbf{f}_m(\mathbf{r})\cdot\int_S\frac{1}{4\pi R}e^{(-t_{d,v}/2)}\mathbf{f}_n(\mathbf{r}')\,dS'\,dS\qquad(8.59)$$

$$\beta_{mn}^{v} = B_{mnij}^{v}\Big|_{j=i} = \int_{S}\nabla\boldsymbol{\cdot}\mathbf{f}_{m}(\mathbf{r})\int_{S}\frac{1}{4\pi R}e^{\left(-t_{d,v}/2\right)}\nabla'\boldsymbol{\cdot}\mathbf{f}_{n}(\mathbf{r}')dS'dS \tag{8.60}$$

$$\gamma_{mn}^{v} = C_{mnij}^{v}\Big|_{j=i} = \int_{S}\mathbf{f}_{m}(\mathbf{r})\boldsymbol{\cdot}\int_{S}\nabla\times\frac{1}{4\pi R}e^{\left(-t_{d,v}/2\right)}\mathbf{f}_{n}(\mathbf{r}')\,dS'dS \tag{8.61}$$

$$\Omega_{m,i}^{E} = \int_{S}\mathbf{f}_{m}(\mathbf{r})\boldsymbol{\cdot}\mathbf{V}_{i}^{E}(\mathbf{r})\,dS \tag{8.62}$$

$$\Omega_{m,i}^{H} = \int_{S}\mathbf{f}_{m}(\mathbf{r})\boldsymbol{\cdot}\mathbf{V}_{i}^{H}(\mathbf{r})\,dS \tag{8.63}$$

$$t_{d,v} = sR/c_{v} \tag{8.64}$$

Transforming Eqs. (8.54) and (8.55) into a matrix form, the TD-PMCHW formulations are expressed into a matrix form as

$$\begin{bmatrix}\begin{bmatrix}Z_{mn}^{E1}\end{bmatrix} & \begin{bmatrix}Z_{mn}^{H1}\end{bmatrix}\\\begin{bmatrix}Z_{mn}^{E2}\end{bmatrix} & \begin{bmatrix}Z_{mn}^{H2}\end{bmatrix}\end{bmatrix}\begin{bmatrix}\begin{bmatrix}u_{n,i}\end{bmatrix}\\\begin{bmatrix}v_{n,i}\end{bmatrix}\end{bmatrix} = \begin{bmatrix}\begin{bmatrix}V_{m,i}^{E}\end{bmatrix}\\\begin{bmatrix}V_{m,i}^{H}\end{bmatrix}\end{bmatrix} \tag{8.65}$$

where

$$Z_{mn}^{E1} = \sum_{v=1}^{2}\left[\frac{\mu_{v}s^{2}}{4}\alpha_{mn}^{v} + \frac{1}{\varepsilon_{v}}\beta_{mn}^{v}\right] \tag{8.66}$$

$$Z_{mn}^{H1} = \sum_{v=1}^{2}\left[-\frac{s}{2}\gamma_{mn}^{v}\right] \tag{8.67}$$

$$Z_{mn}^{E2} = \sum_{v=1}^{2}\left[\frac{s}{2}\gamma_{mn}^{v}\right] \tag{8.68}$$

$$Z_{mn}^{H2} = \sum_{v=1}^{2}\left[\frac{\varepsilon_{v}s^{2}}{4}\alpha_{mn}^{v} + \frac{1}{\mu_{v}}\beta_{mn}^{v}\right] \tag{8.69}$$

$$\begin{aligned}V_{m}^{E} = \Omega_{m,i}^{E} &- \sum_{v=1}^{2}\left(\frac{\mu_{v}s^{2}}{4}\sum_{n=1}^{N}\sum_{j=0}^{i}u_{n,j}\left(A_{mnij}^{v} + 2A_{mni,j+1}^{v} + A_{mni,j+2}^{v}\right)\right.\\&+\frac{1}{\varepsilon_{v}}\sum_{n=1}^{N}\sum_{j=0}^{i}u_{n,j}\left(B_{mnij}^{v} - 2B_{mni,j+1}^{v} + B_{mni,j+2}^{v}\right)\\&\left.-\frac{s}{2}\sum_{n=1}^{N}\sum_{j=0}^{i}v_{n,j}\left(C_{mnij}^{v} - C_{mni,j+2}^{v}\right)\right)\end{aligned} \tag{8.70}$$

$$V_m^H = \Omega_{m,i}^H - \sum_{v=1}^{2} \left[\frac{\varepsilon_v s^2}{4} \sum_{n=1}^{N} \sum_{j=0}^{i} v_{n,j} \left(A_{mnij}^v + 2A_{mni,j+1}^v + A_{mni,j+2}^v \right) \right.$$

$$+ \frac{1}{\mu_v} \sum_{n=1}^{N} \sum_{j=0}^{i} v_{n,j} \left(B_{mnij}^v - 2B_{mni,j+1}^v + B_{mni,j+2}^v \right)$$

$$\left. - \frac{s}{2} \sum_{n=1}^{N} \sum_{j=0}^{i} u_{n,j} \left(C_{mnij}^v - C_{mni,j+2}^v \right) \right] \tag{8.71}$$

Finally, the computational efficiency can be improved further by using the new formulations. Eqs. (8.70) and (8.71) are written as

$$V_m^E = \Omega_{m,i}^E - \sum_{v=1}^{2} \sum_{n=1}^{N} K_{mni}^v \tag{8.72}$$

$$V_m^H = \Omega_{m,i}^H - \sum_{v=1}^{2} \sum_{n=1}^{N} L_{mni}^v \tag{8.73}$$

where

$$K_{mni}^v = \sum_{j=0}^{i} \left[\frac{\mu_v s^2}{4} u_{n,j} \left(A_{mnij}^v + 2A_{mni,j+1}^v + A_{mni,j+2}^v \right) \right.$$

$$\left. + \frac{1}{\varepsilon_v} u_{n,j} \left(B_{mnij}^v - 2B_{mni,j+1}^v + B_{mni,j+2}^v \right) - \frac{s}{2} v_{n,j} \left(C_{mnij}^v - C_{mni,j+2}^v \right) \right] \tag{8.74}$$

$$H_{mni}^v = \sum_{j=0}^{i} \left[\frac{\varepsilon_v s^2}{4} v_{n,j} \left(A_{mnij}^v + 2A_{mni,j+1}^v + A_{mni,j+2}^v \right) \right.$$

$$\left. + \frac{1}{\mu_v} v_{n,j} \left(B_{mnij}^v - 2B_{mni,j+1}^v + B_{mni,j+2}^v \right) - \frac{s}{2} u_{n,j} \left(C_{mnij}^v - C_{mni,j+2}^v \right) \right] \tag{8.75}$$

8.5   CONCLUSION

A new MOD method is described in this chapter for improving the accuracy and efficiency of the conventional MOD technique. By interchanging the sequence of spatial and temporal testing, in the new MOD method, the time variable can be eliminated analytically to improve the stability, and there is no need to introduce the central approximation, which can cause problems. In addition, a new temporal basis function that is formed by three associated Laguerre functions of successive degrees as well as a new time domain Green's function that groups the retarded-time terms are proposed to shorten the computational time, and thus lead to a better efficiency. Marching-on-in-degree formulations based on this new MOD method for analyzing conducting wires or objects with or without loads by TD-

EFIE and dielectric objects by TD-PMCHW are presented. Numerical examples will be given in Chapter 9 to demonstrate the effectiveness of these improvements.

REFERENCES

[1]   E. A. Essex, "Hertz vector potentials of electromagnetic theory," *Am. J. Phys.*, Vol. 45, pp. 1099–1101, 1977.

[2]   Z. Ji, T. K. Sarkar, B. H. Jung, Y.-S. Chung, M. Salazar-Palma, and M. Yuan, "A stable solution of time domain electric field integral equation for thin-wire antennas using the Laguerre polynomials," *IEEE Trans. Antennas Prop.*, Vol. 52, pp. 2641–2649, 2004.

[3]   Z. Ji, T. K. Sarkar, B. H. Jung, M. Yuan, and M. Salazar-Palma, "Solving time domain electric field integral equation without the time variable," *IEEE Trans. Antennas Prop.*, Vol. 54, pp. 258–262, 2006.

[4]   Z. Mei, Y. Zhang, T. K. Sarkar, "Solving time domain EFIE using higher order basis functions and marching-on-in-degree method," *Anten. Prop. Society Int. Sym., 2009*, 2009.

[5]   D. Poularikas, *The Transforms and Applications Handbook*, IEEE Press, Piscataway, N.J., 1996.

[6]   N. N. Lebedev, *Special Functions and Their Applications*, Prentice-Hall, Englewood Cliffs, N.J., 1963.

[7]   B. G. Mikhailenko, "Spectral Laguerre method for the approximate solution of time dependent problems," *Appl. Math. Lett.*, Vol. 12, pp. 105-110, 1999.

9

NUMERICAL EXAMPLES FOR THE NEW AND IMPROVED MARCHING-ON-IN-DEGREE (MOD) METHOD

9.0   SUMMARY

A new and improved marching-on-in-degree (MOD) method was introduced in Chapter 8. Using these new formulations, various conducting and dielectric structures will be analyzed in this chapter. The computed solutions are compared with the results obtained from the conventional MOD, the marching-on-in-time (MOT) method, as well as the inverse discrete Fourier transform (IDFT) of the frequency domain method of moment (MoM) solutions. It will be evident from the results that this new MOD algorithm analyzes the time domain scatterings from arbitrary-shaped conducting and dielectric structures in an accurate, stable, and efficient way. Specifically, the new improved MOD scheme takes at least an order of magnitude less CPU time as the conventional MOD scheme to solve the same problem. It also will be evident that this method takes about the same amount of time as the MOT algorithm even though the latter does not necessarily provide a stable solution. Also, this method can handle discontinuous excitation waveforms and can deal with structures that have both concentrated and distributed impedance loading.

9.1   INTRODUCTION

The MOD algorithm has been introduced in Chapters 5–7 to tackle the instability problems of the MOT method in solving the time domain integral equation (TDIE). As described in Chapter 8, the conventional MOD algorithm is modified to eliminate the intrinsic shortcoming of insufficient accuracy and low computational efficiency. This is achieved by introducing three new schemes that significantly improved its performance. The schemes include the order of implementation of the spatial and temporal testing procedure, a new temporal basis function set, which is a combination of three associated Laguerre functions with successive degrees, and a new form of the Green's function, which is computationally very fast.

In this chapter, the new, improved MOD methodology is employed to analyze scattering from various types of conducting and dielectric structures and is compared with the solutions obtained from the conventional MOD method. The examples related to the conducting structures are analyzed using the time domain electric field integral equation (TD-EFIE), whereas the dielectric structures are analyzed using the time domain Poggio-Miller-Chang-Harrington-Wu (TD-PMCHW) formulations. The results obtained in the time domain then are compared with the IDFT of the frequency domain solutions to validate the accuracy of the results obtained using this new approach. In addition, the results also are compared with the MOT method to assess the performance of these numerical methods in terms of accuracy, stability, and computational efficiency. To make the comparisons realistic, the same integral equation along with its appropriate surface discretization used in the MOD approach will be used for the frequency domain and the MOT methods for the analysis of the conducting and dielectric structures, respectively.

In section 9.2, a Gaussian pulse and the T-pulse will be discussed that have been used to illuminate the various scatterers described in this chapter. Sections 9.3 and 9.4 demonstrate the advantages of the improved MOD method over the conventional one in terms of accuracy and computational efficiency. Analyses for various kinds of conducting structures are described in sections 9.5 and 9.6, which include not only simple-shaped conducting objects, such as a sphere, a cube, and a cylinder but also analyses of large complicated structures, such as an aircraft and a tank, as well as scattering from conducting objects loaded with impedances. Section 9.7 illustrates that in this new MOD method various types of excitation pulses can be used, including discontinuous waveforms, like pulses. The analysis of dielectric structures is given in section 9.8 where results using different values of permittivity and permeability for a dielectric object also are presented. Finally, some conclusions are drawn in section 9.9.

9.2   USE OF VARIOUS EXCITATION SOURCES

Two types of transient plane waves used to illuminate the scatterers are presented in this chapter. They are the Gaussian and T-pulse. The temporal electric field shaped by a Gaussian pulse is given by

$$\mathbf{E}^i(\mathbf{r},t) = \mathbf{E}_0 \frac{4}{\sqrt{\pi T}} e^{-\gamma^2} \tag{9.1}$$

where

$$\gamma = \frac{4}{T}\left(ct - ct_0 - \mathbf{r}\cdot\hat{\mathbf{k}}\right) \tag{9.2}$$

The parameter $\hat{\mathbf{k}}$ is the unit vector along the direction of wave propagation, T is the pulse width of the Gaussian pulse, and t_0 is the time delay what represents the time from the origin at which the pulse reaches its peak.

The T-pulse is a discrete time signal with most of its prescribed energy focused in a given bandwidth [1]. It is a strictly time-limited pulse with the added stipulation that 99% of its signal energy be concentrated in a narrow band. Hence, effectively, the pulse is also approximately band limited. In addition, the pulse can be designed to be orthogonal with its shifted version as well and can have a zero DC bias if required. By interpolating this discrete pulse, a continuous pulse can be generated while its bandwidth remains approximately the same. The mathematical generation of a T-pulse is described in [2]. Figure 9.1 shows the transient and frequency responses of a T-pulse, which is 6 light-meters (lm) in duration. Most of its energy is concentrated in the band from 0 to 200 MHz, and only less than 0.008% of the energy is out of this band so this time-limited pulse is also practically band limited and, thus, is very convenient in system applications.

Next, responses of the system to these pulses are discussed.

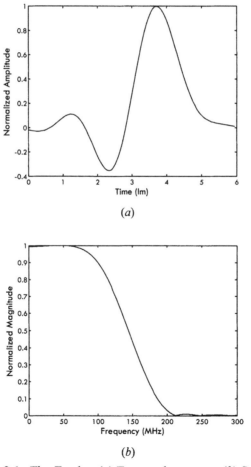

(a)

(b)

Figure 9.1. The T-pulse. (a) Temporal response. (b) Spectrum.

9.3   IMPROVED PERFORMANCE OF THE NEW MOD METHOD IN TERMS OF ACCURACY AND STABILITY

Example 1:   A Pair of Conducting Plates

The first example deals with a pair of perfectly conducting (PEC) plates forming a narrow angle (5°) as shown in Figure 9.2. The objective is to demonstrate that the shortcomings of the conventional MOD in accuracy and stability are alleviated in this new MOD formulation. The surfaces of the two plates are discretized into triangular patches as shown in Figure 9.2. There are 274 patches and 435 edges. Some patches along the connected edge of the two metal plates are in very close proximity to each other, and this result in strong coupling between them. Hence, a highly accurate computation for the spatial integrals over the patches is required.

The incident wave is a θ-polarized Gaussian wave arriving from $\phi = 0°$ and $\theta = 0°$ with a pulse width of 8 lm and delayed by 12 lm. Here, 1 lm is 3.33 nanoseconds. It has an approximate bandwidth of 120 MHz. The induced current on this conducting structure are calculated by the frequency domain MoM, the new MOD, and the conventional MOD method. The current across the edge connecting the nodes (0, 0, 0) and (0.0375, 0, 0) are plotted in Figure 9.3 and the three plots are marked as IDFT, improved MOD, and conventional MOD, respectively. Obviously, in Figure 9.3, the conventional MOD is completely unstable, whereas the new, improved MOD method is stable and its results agree with the IDFT of the frequency domain solution.

When the conventional MOD method is used to analyze this structure, the spatial integrals over the patch are tested at the center of the patch only. This leads to an error that may result in an unstable solution. By employing first the temporal testing followed by the spatial testing, the retarded time terms can be handled analytically in the new MOD scheme as opposed to the conventional one. This example illustrates that because of the approximations used in the conventional MOD, the method may fail for some types of scatterers. However, this shortcoming can be eliminated by using the new MOD methodology.

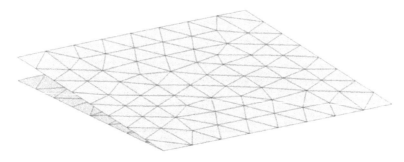

Figure 9.2. Triangular patch model for a pair of 0.6 m × 0.6 m conducting plates connected at an angle of 5°.

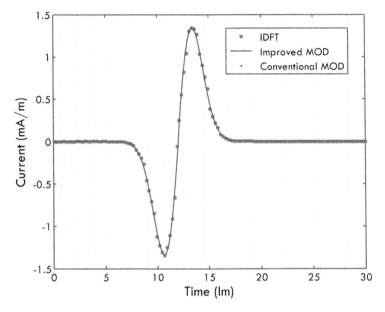

Figure 9.3. Current on the conducting plates computed by the improved MOD method plotted against the results from the conventional MOD and the IDFT of the frequency domain MoM solution.

9.4 IMPROVEMENT IN COMPUTATIONAL EFFICIENCY OF THE NEW MOD METHOD

The utilization of a combination of associated Laguerre functions as temporal basis functions as well as the introduction of a new mathematical form of the Green's functions, leads to a simplified formulation in the new MOD method. This also significantly reduces the computation time by at least a factor of 10, which makes this new method as competitive as the MOT algorithm. We illustrate these claims through some examples.

Example 2: A Dipole

Consider a 1-m long dipole placed along the x-axis that is illuminated by a T-pulse of width 6 lm and starting at 8 lm. The pulse has an approximate bandwidth of 200 MHz. It is incident from the z-axis and is polarized along the x-axis. Figure 9.4 plots the current at the central point of the dipole calculated using the conventional MOD, the new MOD described in Chapter 8, and the IDFT of the frequency domain solution. The highest temporal order of the Laguerre polynomial selected is 50 in the two MOD calculations. The computation time needed for all time domain methods are listed in Table. 9.1. From Figure 9.4, it can be observed that the results of the conventional MOD and the new MOD completely overlap, and they agree well with the IDFT of the frequency domain solution. However, the computation time for the new MOD method over the conventional one is less by a factor of 18 (46.3 s as opposed to 2.45 s). Also, as shown in Figure 9.4, the conventional MOT method becomes unstable for late-time (after 15 lm), although it takes the least amount of computation time.

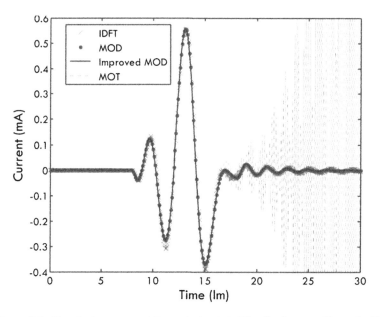

Figure 9.4. Transient response at the central point of the dipole caused by an incident *T*-pulse.

Table 9.1. Comparison of the computation time for the analysis of the dipole using different time domain methods

Methods	Total calculation time (s)
The conventional MOD	46.30
The new MOD	2.45
The MOT	2.04

Example 3:   A Helix

Next, consider the analysis of a helix with a height of 8.8 meters and a radius of 2 m, which contains 100 spatial unknowns. Its axis is oriented along the *z*-axis. A plane wave consisting of a θ-polarized *T*-pulse is incident from $\phi = 0°$ and $\theta = 0°$. The *T*-pulse has a pulse width of 6 lm and starts at 8 lm. It has a bandwidth of 200 MHz. The transient current near the end of the structure is computed using the improved MOD method. It is plotted in Figure 9.5 along with the results from the IDFT of a frequency domain MoM solution, the conventional MOD, and the MOT techniques. The computational time required by all methods is listed in Table 9.2 for comparison. Figure 9.5 shows that a very good agreement can be observed among the solutions obtained by the conventional MOD, the new MOD, and the IDFT of the frequency domain solution. The computational time required by the conventional MOD is 22 times more than that of the new MOD (326.2 s and 14.23 s). Although the MOT can solve the problem much faster, its result starts oscillating at an early time as shown in Figure 9.5.

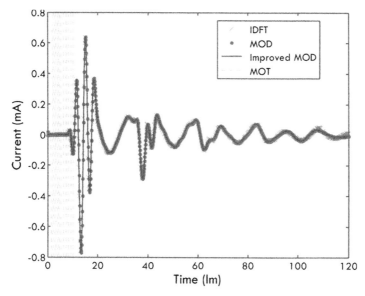

Figure 9.5. Transient response at a point near the end of the helix caused by an incident T-pulse.

Table 9.2. Comparison of computation time for the analysis of the helix using different time domain methods

Methods	Total calculation time (min)
The conventional MOD	326.20
The new MOD	14.23
The MOT	6.66

Example 4:   A Loop Antenna

A loop antenna with a diameter of 0.5 m is placed in the xoy plane. A y-polarized T-pulse of duration 10 lm and delayed by 8 lm is incident from the x-axis. It has a bandwidth of 120 MHz. The highest temporal order of the Laguerre polynomials used is 100. In this example, the solution obtained by the conventional and the improved MOD are transformed into the frequency domain, and compared with the frequency-domain MoM solution. Figure 9.6 plots the currents on the loop at a point (0.25, 0, 0) m in the frequency domain. The results from the two MOD methods agree very well along with the frequency domain MoM solution in both magnitude and phase. The computation time to obtain the time domain solutions using the two MOD and the MOT methods are listed in Table 9.3. The improved MOD method requires a much shorter computation time than the conventional MOD method by a factor of 25. For this structure, the new MOD technique is faster than the MOT method, whose results are not shown because it is highly oscillatory.

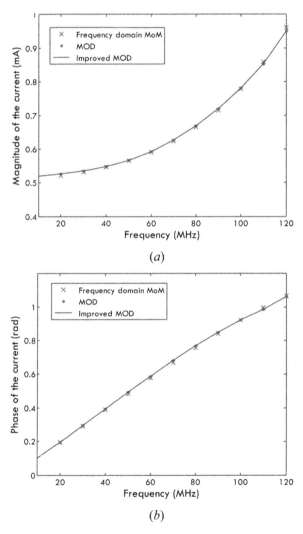

Figure 9.6. Current response in the frequency domain at a point on the loop antenna illuminated by a T-Pulse: (a) Magnitude. (b) Phase.

Table 9.3. Comparison of computation time for the analysis of the loop antenna using different time domain methods

Methods	Total calculation time (s)
The conventional MOD	155.56
The new MOD	6.20
The MOT	6.90

Example 5:   Far Field of a *V*-Shaped Antenna

In this example, the far field of a *V*-shaped wire antenna with two arms of 0.5 m is computed. The angle between the two arms is 60°. The feed point is located at the center of the antenna. A *T*-pulse with a width of 10 lm is fed at this point at $t = 8$ lm. The pulse has a bandwidth of 120 MHz. The antenna is placed horizontally, and the far field along $\theta = 0°$ and $\phi = 0°$ is calculated. In this example, the highest temporal order of the Laguerre polynomials chosen is 100. The IDFT result and the solutions from the conventional and the improved MOD methods are plotted in Figure 9.7 against the solution obtained using the MOT. It is observed that the MOT method fails to get a stable result, whereas both the MOD methods show very good agreement with the IDFT results of the frequency domain MoM solution. The computational time taken by the various methods to analyze this structure is shown in Table 9.4. In this example, the improved MOD requires only 2.43 s in contrast to 33.23 s needed by the conventional MOD, demonstrating a 13-time improvement in the computational efficiency.

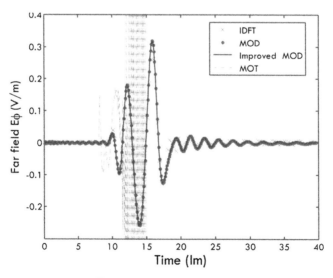

Figure 9.7. The far field E_ϕ of a *V*-shaped antenna along the direction of $\theta = 0°$ and $\phi = 0°$.

Table 9.4. Comparison of computation time for the analysis of the *V*-shaped antenna using different time domain methods

Methods	Total calculation time (s)
The conventional MOD	33.23
The new MOD	2.43
The MOT	2.40

In summary, for the analysis of the four wire structures discussed, the computed results using the new MOD method is not only accurate and stable, but the computation time has been reduced at least by an order of magnitude over the conventional MOD method.

9.5   ANALYSIS OF CONDUCTING STRUCTURES

In this section, results of scattering from various shapes and sizes of conducting structures are presented to demonstrate the improved performance of this new time domain MOD algorithm.

Example 6:   A Conducting Cube

The scattering from a conducting cube with each side of 1.0 m, is shown in Figure 9.8. A total of 852 patches and 1374 edges are used for the triangular patch model of this structure. The transient current is computed at the edge connecting the nodes (0.391, 0.187, –0.5) and (0.375, 0.366, –0.5) when the cube is illuminated by a θ-polarized Gaussian wave incident from $\phi = 0°$ and $\theta = 90°$. The transient current computed by the improved MOD method is plotted in Figure 9.9 against the IDFT of the frequency domain MoM solution. Figure 9.9(*a*) is for a Gaussian incident wave with a pulse width $T = 2$ lm and a delay time $ct_0 = 3$ lm and Figure 9.9(*b*) is for $T = 8$ lm and $ct_0 = 12$ lm. These two Gaussian pulses correspond to a bandwidth of 480 MHz and 120 MHz, respectively. In both situations, the solutions given by the MOD solution match well with the IDFT results.

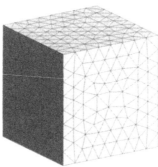

Figure 9.8. Triangular patch model for a PEC cube with a length of 1.0 m.

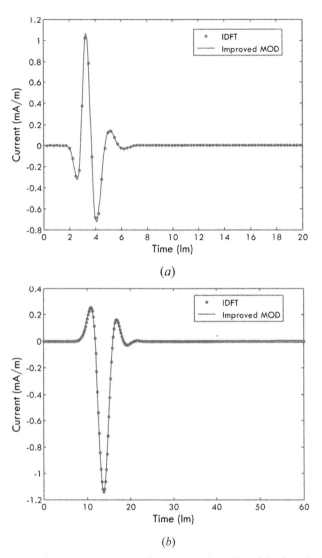

Figure 9.9. Transient current response for a conducting cube with a length of 1.0 m. The plot is due to Gaussian pulse excitations with parameters (a) $T = 2$ lm and $ct_0 = 3$ lm; (b) $T = 8$ lm and $ct_0 - 12$ lm.

Example 7: A Conducting Sphere

A conducting sphere with a radius of 0.5 m is shown in Figure 9.10. The triangular patch model consists of 528 patches and 792 edges. It is illuminated by a θ-polarized Gaussian pulse incident from $\phi = 0°$ and $\theta = 90°$. The transient current is computed at the edge connecting the nodes (0.0, –0.25, –0.433) and (0.0, –0.191, –0.461). For two different θ-polarized incident Gaussian pulses, the current given by the improved MOD method is plotted in Figures 9.11(a) and 9.11(b) along with the IDFT results of the frequency domain MoM solution. In Figure 9.11(a), the current is induced by a pulse of width $T = 2$ lm and a delay time of $ct_0 = 3$ lm. In Figure 9.11(b), the plotted induced current is due to an incident pulse of width $T = 8$ lm and with a delay of $ct_0 = 12$ lm. The two Gaussian pulses correspond to bandwidths of 480 MHz and 120 MHz, respectively. For both these cases, the solutions given by the improved MOD formulation are very accurate and have a good agreement with the IDFT of the frequency domain solution.

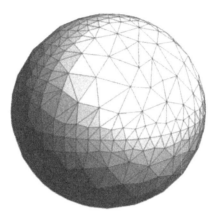

Figure 9.10. Triangular patch model for a PEC sphere with a radius of 0.5 m.

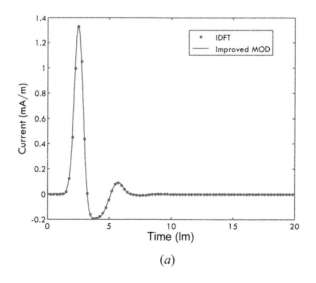

(a)

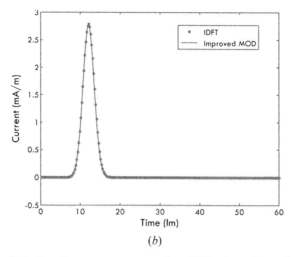

Figure 9.11. Transient current response for a PEC sphere with radius of 0.5 m: (a) $T = 2$ lm and $ct_0 = 3$ lm. (b) $T = 8$ lm and $ct_0 = 12$ lm.

Example 8: A Conducting Cylinder

The next example deals with a conducting cylinder of radius 0.5 m with a height of 1.0 m, as shown in Figure 9.12. There are 982 patches and 1473 edges in this triangular patch model. The transient current is computed at the edge connecting the nodes (–0.498, –0.004, 0.818) and (–0.499, 0.003, 0.926) when the cylinder is illuminated by a θ-polarized Gaussian pulse incident from $\phi = 0°$ and $\theta = 90°$. The transient current computed by the new MOD method is plotted against the IDFT results of the frequency domain MoM solution for two different types of pulses with a pulse width $T = 2$ lm and a delay time $ct_0 = 3$ lm and for $T = 8$ lm and $ct_0 = 12$ lm as shown in Figures 9.13(a) and 9.13(b), respectively. The bandwidth of the first Gaussian pulse is 480 MHz and the second one is 120 MHz. Good agreements are obtained between the MOD solutions and the IDFT of the frequency domain MoM solution, validating the efficiency and accuracy of the new MOD method.

Figure 9.12. Triangular patch model for a PEC cylinder with radius of 0.5 m and a height of 1.0 m.

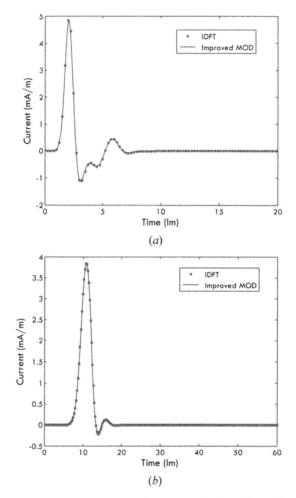

Figure 9.13. Transient current response for a PEC cylinder with a radius of 0.5 m and a height of 1.0 m: (*a*) $T = 2$ lm, $ct_0 = 3$ lm. (*b*) $T = 8$ lm, $ct_0 = 12$ lm.

Example 9:   A Tank

The improved MOD algorithm now is used to analyze complicated conducting structures. Figure 9.14 shows a tank with a dimension of 9.5 m × 3.4 m × 2.3 m. The surface is discretized using a triangular patch model that contains 2,737 patches and 4306 edges, resulting in 3905 unknowns to be solved for. The incident field is a θ-polarized triangular plane wave coming from the front ($\phi = 0°$ and $\theta = 90°$). It has a pulse width of 60 lm and reaches its peak at 90 lm. It has a bandwidth of 50 MHz. The current induced at the edge between the nodes (2.17, −1.7, 1.48) and (2.12, −1.49, 1.48) on the surface of the tank is computed using the new MOD and is plotted in Figure 9.15 along with the IDFT of the frequency domain MoM solution. As shown in Figures 9.15, the two computed currents agree well.

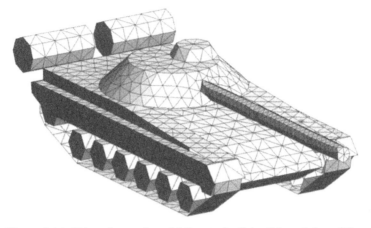

Figure 9.14. Triangular patch model for a tank of size 9.5 m× 3.4 m× 2.3 m.

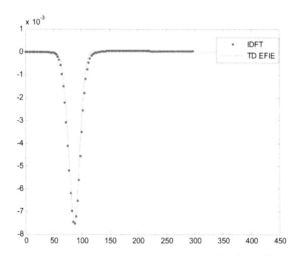

Figure 9.15. Transient current response for the tank.

Example 10: A Fokker Aircraft

In this example, the current distribution on a Fokker aircraft with a dimension of 21.0 m × 15.6 m × 4.7 m as shown in Figure 9.16, is computed using the new MOD technique. The surface is discretized using 2184 triangular patches that have 3424 edges. A θ-polarized Gaussian pulse is incident head on from $\phi = 0°$ and $\theta = 90°$. The excitation pulse has a width $T = 20$ lm, a delay time $ct_0 = 30$ lm, and a bandwidth of 48 MHz. Figure 9.17 plots the current distribution on the surface of the aircraft from $t = 20$ lm to $t = 72$ lm. To display the color contrast, the values of the current for all time instants have been scaled. The darkest (blue) color is referred to as 0 mA/m whereas the lightest (red) color is for all currents larger than 7.5 mA/m. For all the Figures 9.17(a) – 9.17(d), the current increases from the front to the rear as the time varies from $t = 20$ lm to $t = 32$ lm. Then, the current increases from the rear to the wings as time progresses from $t = 32$ lm to $t = 72$ lm, as shown in Figures 9.17(d) to 9.17(f).

Example 11: An AS-322 Helicopter

Next, a helicopter with a size of 19.1 m × 15.3 m × 4.7 m is analyzed. The surface is discretized using 1958 triangular patches with 2937 edges as shown in Figure 9.18. The structure is excited by a θ-polarized triangular pulse with a width $T = 80$ lm and a time delay of $ct_0 = 120$ lm. The pulse has a bandwidth of 37.5 MHz and is incident from $\phi = 0°$ and $\theta = 0°$. The transient current distribution on the structure from $t = 80$ lm to $t = 160$ lm is computed using the new MOD method and is plotted in Figure 9.19. To display the color contrast in the plots, the values of the current for all time instants have been scaled. The darkest (blue) color is defined to be 0 mA/m, whereas the lightest (red) color is used for all currents larger than 1.0 mA/m.

Example 12: A Mirage Aircraft

The next example to be analyzed is a Mirage with a dimension of 11.6 m × 7.0 m × 2.9 m. The triangular patch model of the structure is shown in Figure 9.19. It has 1782 patches and 2673 edges. A θ-polarized triangular pulse with a width $T = 80$ lm, a time delay of $ct_0 = 120$ lm, and a bandwidth of 37.5 MHz; it is incident from $\phi = 0°$ and $\theta = 90°$. The transient current distributions on the structure is plotted in Figure 9.20 from $t = 100$ lm to $t = 160$ lm using the new MOD method. The values of the current for all time instants are scaled (Figure 9.21). The darkest (blue) color is referred to as 0 mA/m, whereas the lightest (red) color is used for all currents larger than 1.5 mA/m.

Example 13: A Boeing-737 Aircraft

In this section, a Boeing-737 aircraft with a size of 26 m × 26 m × 11 m is analyzed. The surface is discretized using 4721 triangular patches with 7327 edges as shown in Figure 9.22. The structure is excited by a T-pulse coming from the head of the plane ($\theta = 90°$ and $\phi = 0°$). The pulse has a pulse-width of 25 lm, a time delay of $ct_0 = 17.5$ lm, and a bandwidth of 50 MHz. The transient current distribution on the structure from $t = 30$ lm to $t = 100$ lm is computed using the new MOD method and is plotted in Figure 9.23. The values of the current for all time instants have been scaled. The darkest (blue) color is defined to be 0 mA/m, whereas the lightest (red) color is used for all currents larger than 0.3 mA/m.

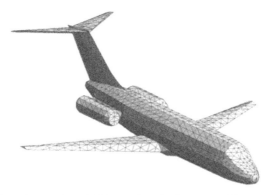

Figure 9.16. Triangular patch model for an aircraft with a size of 21.0 m × 15.6 m × 4.7 m.

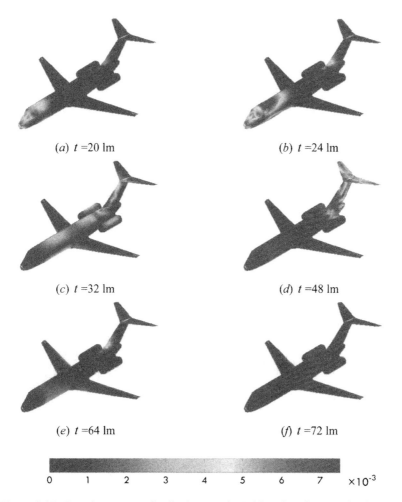

(a) t =20 lm

(b) t =24 lm

(c) t =32 lm

(d) t =48 lm

(e) t =64 lm

(f) t =72 lm

0   1   2   3   4   5   6   7   ×10⁻³

Figure 9.17. Transient current distributions on the Fokker aircraft (See color insert).

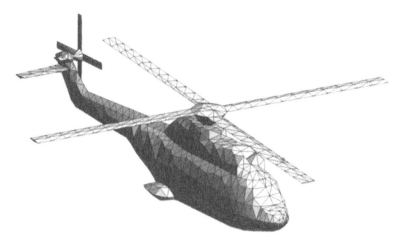

Figure 9.18. Triangular patch model for a AS 322 helicopter of size 19.1m × 15.3m × 4.7m.

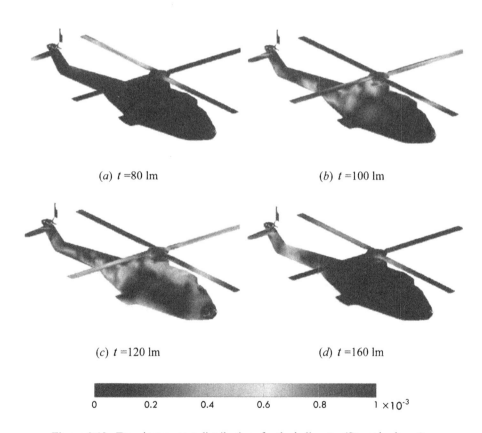

(a) t =80 lm   (b) t =100 lm

(c) t =120 lm   (d) t =160 lm

0   0.2   0.4   0.6   0.8   1 ×10⁻³

Figure 9.19. Transient current distributions for the helicopter (See color insert).

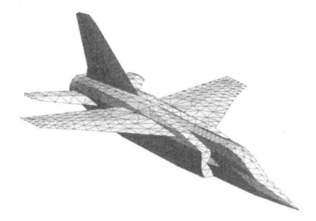

Figure 9.20. Triangular patch model for a Mirage with a size of 11.6m × 7.0m × 2.9m.

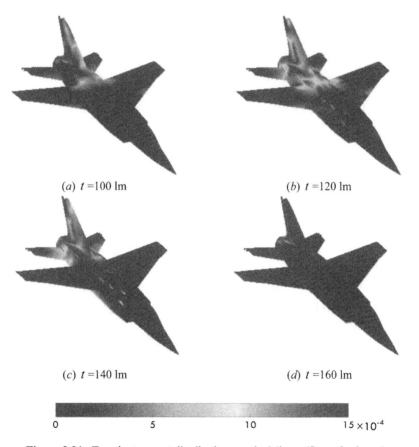

(a) t =100 lm

(b) t =120 lm

(c) t =140 lm

(d) t =160 lm

0   5   10   15 ×10⁻⁴

Figure 9.21. Transient current distributions on the Mirage (See color insert).

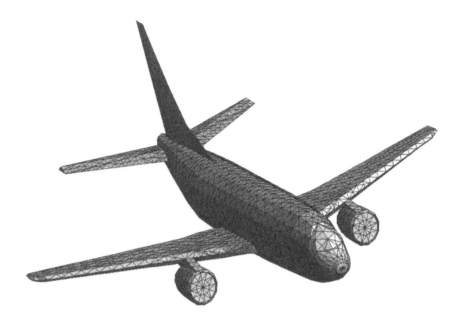

Figure 9.22. Triangular patch model for a Boeing-737 Aircraft with a size of
26 m × 26 m ×11 m.

9.6   ANALYSIS OF CONDUCTING STRUCTURES WITH LOADS

In Chapter 8, the TD-EFIE formulation for analysis of conducting structures with
loads was presented. The loads may include distributed loads like the skin effect
losses or can have a concentrated load. Some examples are presented to illustrate
this methodology.

Example 14: Uniformly Distributed Load

First, consider a dipole of length 1 m that is not PEC but is loaded with a constant
surface resistance distributed uniformly. The surface resistance has a value of 100
Ω/m. The structure is illuminated by a x-polarized T-pulse incident form the y-
axis and starting at 8 lm. The T-pulse has a pulse width of 10 lm and a bandwidth
of 120 MHz. Figure 9.24 plots the current computed by the new MOD method
along with the IDFT of the frequency domain MoM solution. These two results
have good agreement.

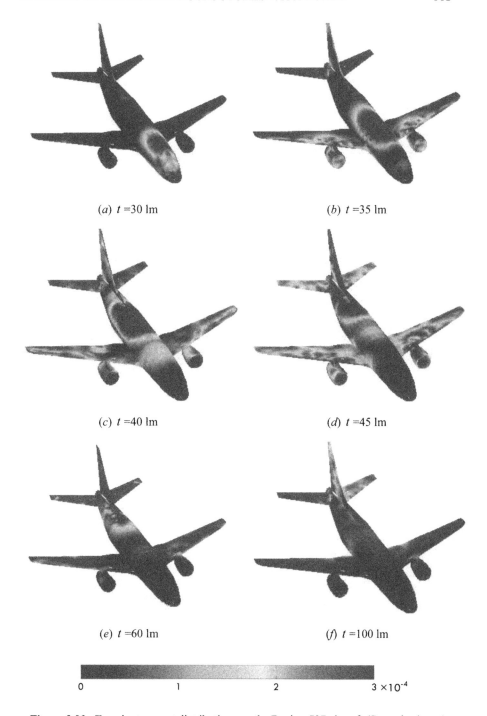

(a) t =30 lm   (b) t =35 lm

(c) t =40 lm   (d) t =45 lm

(e) t =60 lm   (f) t =100 lm

Figure 9.23. Transient current distributions on the Boeing-737 aircraft (See color insert).

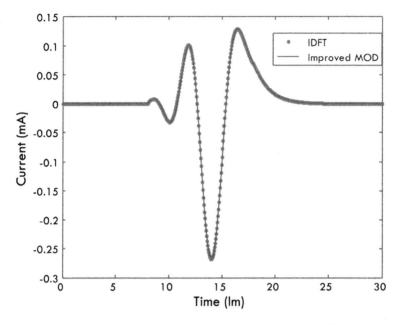

Figure 9.24. Transient current response on a dipole with a uniformly distributed resistance.

Example 15: Skin Effect

Skin effect is a physical phenomenon that is a result of the distribution of an alternating electric current (AC) inside a conductor. Therefore, the cross-sectional resistance is no longer the same like that for the direct current (DC). Particularly, at high frequencies, most of the current is located only within a small depth h on the skin of a conducting structure, called skin depth, which is given by

$$h = \sqrt{\frac{2}{\omega\mu\sigma}} \tag{9.3}$$

where μ is the permeability and σ is the conductivity. So the equivalent resistance can be obtained by

$$R = \frac{1}{\sigma\pi Dh} = \frac{1}{\pi D}\sqrt{\frac{\mu}{2\sigma}}\,\omega \tag{9.4}$$

where D is the diameter of the wire.

For example, consider a 1-m dipole with a conductivity of $\sigma = 1000$ S/m. The excitation is the same T-pulse described in Example 14. It is x-polarized, and is incident from the y-axis and starting at 8 lm. Its width is 10 lm, which corresponds to a bandwidth of 120 MHz. The current at the center of the dipole is computed by using the new, improved MOD method. It is plotted in Figure 9.25 along with the IDFT of the frequency domain solution. The two solutions agree well.

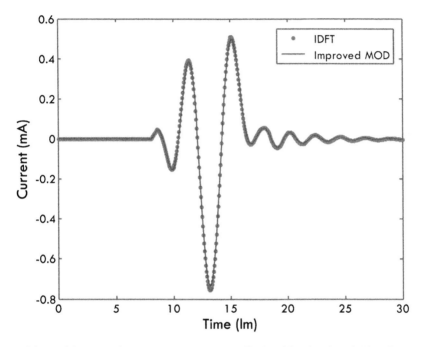

Figure 9.25. Transient current response on a dipole with a developed skin effect.

Example 16: Concentrated Load

A concentrated load represents an impedance connected at a point or at a very small area of a conductor. It can be viewed as the special case of a distributed load with a Dirac delta distribution. This example analyzes a 1-m long dipole with a concentrated load located at its central point. The load is assumed to be resistive and is expressed as $Z(\omega) = 500e^{-(\omega/10^8)}$ Ω. The dipole is excited by a 120 MHz bandwidth T-pulse with a pulse width of 10 lm starting at 8 lm. Figure 9.26 plots the current at the center of the dipole, and it matches very well with the IDFT of the frequency domain MoM solution.

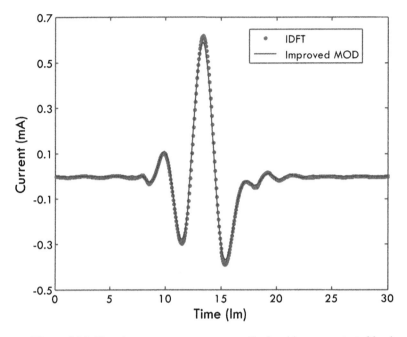

Figure 9.26. Transient current response on a dipole with a concentrated load.

9.7   APPLICATIONS OF DIFFERENT TYPES OF EXCITATION PULSES

The new MOD algorithm also can be used with different types of incident pulse shapes. One novel features of this technique is that the incident pulse could be discontinuous and does not need to be differentiable. In this section, a sphere is illuminated by six types of pulses, including a Gaussian pulse, T-pulse, trapezoidal pulse, triangular pulse, sinusoidal pulse, and a raised sinusoidal pulse.

Example 17: Various Excitation Pulses

A PEC sphere with a diameter of 0.7 m is analyzed using different types of excitation pulses. The sphere is modeled by triangular patches, and the structure contains 192 edges and 128 patches. The transient current is computed at the edge connecting the nodes (0.0, –0.108, –0.333) and (0.0, –0.206, –0.283). Figure 9.27 plots the results computed by the new, improved MOD method and all the computed results agree well with the frequency domain solutions. The excitations used are a Gaussian pulse, T-pulse, trapezoidal pulse, triangular pulse, sinusoidal pulse, and raised sinusoidal pulse. The incident pulses are all θ-polarized and come from $\phi = 0°$ and $\theta = 0°$. The pulse widths of the various pulses are 8 lm, 6 lm, 20 lm, 20 lm, 20 lm, and 20 lm, whereas the center of the pulse is located at 12 lm, 8 lm, 30 lm, 30 lm, 30 lm, and 30 lm. The bandwidths of the pulses are 120 MHz, 200 MHz, 170 MHz, 150 MHz, 140 MHz, and 90 MHz, respectively.

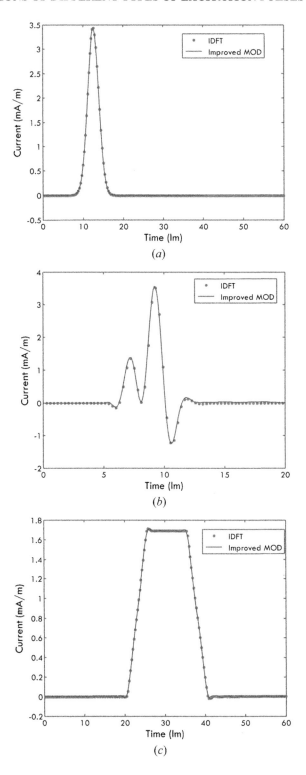

(a)

(b)

(c)

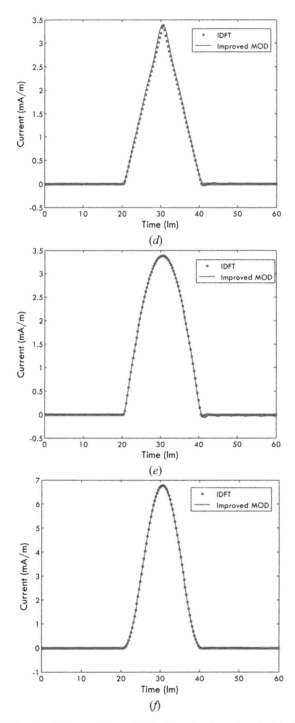

Figure 9.27. Transient current on a PEC sphere due to various incident pulses: (*a*)
Gaussian. (*b*) *T*-pulse. (*c*) Trapezoidal. (*d*) Triangular. (*e*) Sinusoidal. (*f*) Raised sinusoidal.

9.8   ANALYSIS OF MATERIAL STRUCTURES

The TD-PMCHW formulation for analyzing dielectric structures has been presented in Chapter 8. This formulation now is used to analyze a dielectric sphere and a cylinder.

Example 18: A Material Sphere

A dielectric sphere with a radius of 0.1 m, relative permittivity $\varepsilon_r = 1$ and relative permeability $\mu_r = 0.5$ is modeled by triangular patches containing 84 edges and 56 patches. The incident wave exciting the structure is a ϕ-polarized T-pulse with a duration of 10 lm, starting at 4 lm and is arriving from the direction of $\theta = 0°$ and $\phi = 0°$. It has a bandwidth of 120 MHz. Figure 9.28 plots the scattered far field along the direction of $\theta = 180°$ and $\phi = 0°$ computed by the new MOD method. The computed result agrees well with the frequency domain solution.

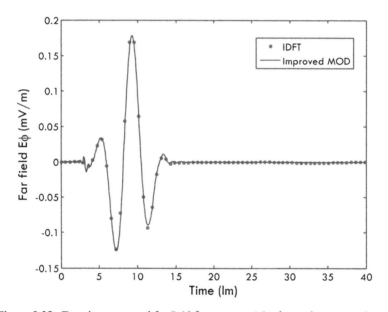

Figure 9.28. Transient scattered far field from a material sphere of $\varepsilon_r = 1$ and $\mu_r = 0.5$.

Example 19: A Dielectric Cylinder

A dielectric cylinder with a radius of 0.1 m and a height of 0.2 m is radiated with a ϕ-polarized T-pulse with a duration of 10 lm, starting at 4 lm, and coming from the direction of $\theta = 0°$ and $\phi = 0°$. It has a relative permittivity $\varepsilon_r = 2$ and permeability $\mu_r = 1$. It has a bandwidth of 120 MHz. The structure is discretized by triangular patches containing 246 edges and 164 patches. The scattered far field along the direction $\theta = 180°$ and $\phi = 0°$ is computed using the new MOD method. The results are plotted in Figure 9.29 along with the IDFT of the frequency domain solution. The two results agree well with each other.

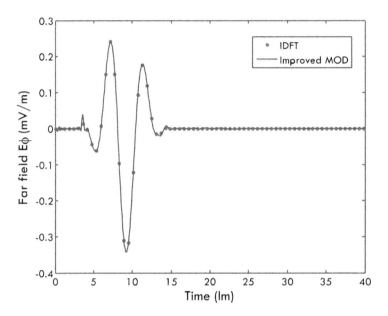

Figure 9.29. Transient scattered far field from a dielectric cylinder of $\varepsilon_r = 2$ and $\mu_r = 1$.

Example 20: Structures with Different Relative Permittivity

The dielectric sphere of Example 18 is studied further using different materials with a relative permeability of $\mu_r = 1$ and relative permittivity of $\varepsilon_r = 0.5$ or 10. The structure is excited with a ϕ-polarized T-pulse starting at 4 lm and is incident from $\theta = 0°$ and $\phi = 0°$. Two versions of the T-pulse are considered with pulse durations of 10 lm and 20 lm. These correspond to a bandwidth of 120 MHz and 60 MHz, respectively. Figure 9.30 plots the scattered far field along the direction $\theta = 180°$ and $\phi = 0°$ using the new MOD method. Both results for $\varepsilon_r = 0.5$ and $\varepsilon_r = 10$ agree well with the frequency domain solutions.

Example 21: Structures with Different Relative Permeability

The same dielectric sphere of the previous example is considered with different material parameters. The sphere has a relative permittivity $\varepsilon_r = 2$ but can have a different relative permeability of $\mu_r = 1$ or 2. The structure is excited with a T-

pulse with the same parameters as in Example 17. Figure 9.31 plots the scattered far field along the direction of $\theta = 180°$ and $\phi = 0°$ computed by the improved MOD method. Both results for $\mu_r = 1$ and $\mu_r = 2$ agree well with the frequency domain solutions.

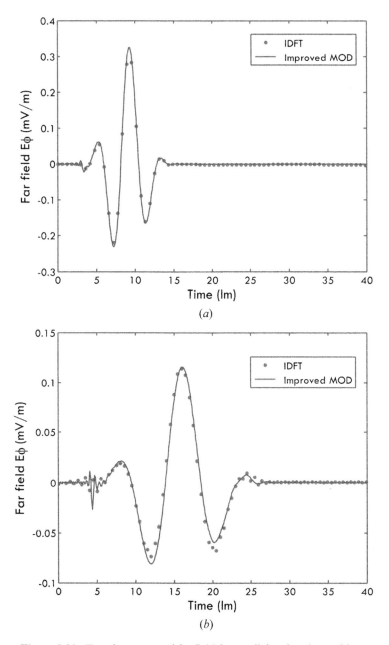

Figure 9.30. Transient scattered far field from a dielectric sphere with $\mu_r = 1$:
(a) $\varepsilon_r = 0.5$. (b) $\varepsilon_r = 10$.

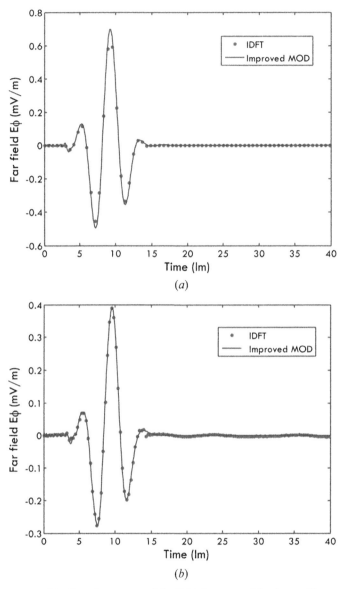

Figure 9.31. Transient scattered far field for a material sphere with $\varepsilon_r = 2$:
(a) $\mu_r = 1$. (b) $\mu_r = 2$.

Example 22: Structures with Negative Relative Permittivity and Permeability

The same dielectric sphere of the previous example is considered with a negative relative permittivity and permeability material. The sphere has a relative permittivity $\varepsilon_r = -4$ and $\mu_r = -4$. The structure is excited with a T-pulse with the same parameters as in Example 17. Figure 9.32 plots the scattered far field along the direction of $\theta = 180°$ and $\phi = 0°$ computed by the improved MOD method. The result for agrees well with the frequency domain solutions.

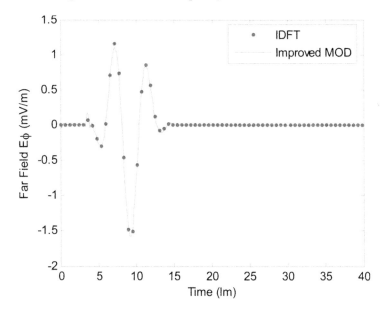

Figure 9.32. Transient scattered far field for a material sphere with $\varepsilon_r = -4, \mu_r = -4$.

9.9   CONCLUSION

In this chapter, the effectiveness of the new improved MOD formulations for conducting and dielectric objects has been verified using various examples. It is evident that the new, improved MOD method provides better stability and efficiency than the conventional time domain MOD method. It also demonstrates that the method can analyze a broad variety of complex conducting and dielectric structures, both with or without loads, and material bodies with different permittivity and permeability. The structures also can be excited using different types of incident pulses, which do not need to be continuous.

REFERENCES

[1]   Y. Hua and T. K. Sarkar, "Design of optimum discrete finite duration orthogonal Nyquist signals," *IEEE Trans. on Acoust, Speech Signal Process.*, Vol. 36, pp. 606–608, 1988.

[2]   T. K. Sarkar, M. Salazar-Palma, and M. C. Wicks, *Wavelet Applications in Engineering Electromagnetics*, Artech House, Norwood, Mass. 2002.

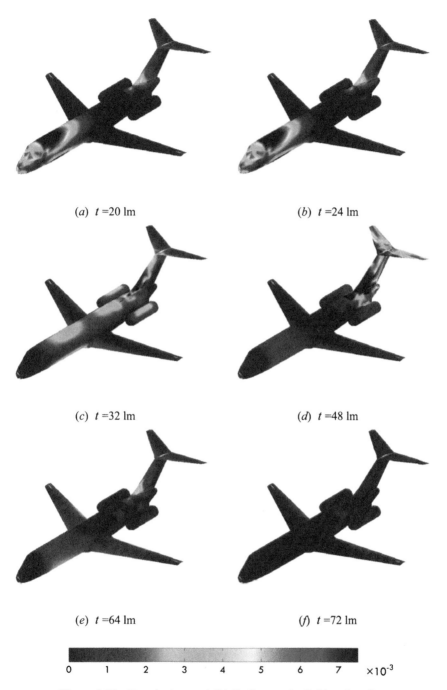

(a) $t = 20$ lm

(b) $t = 24$ lm

(c) $t = 32$ lm

(d) $t = 48$ lm

(e) $t = 64$ lm

(f) $t = 72$ lm

0   1   2   3   4   5   6   7 $\times 10^{-3}$

Figure 9.17.   Transient current distributions on the Fokker aircraft.

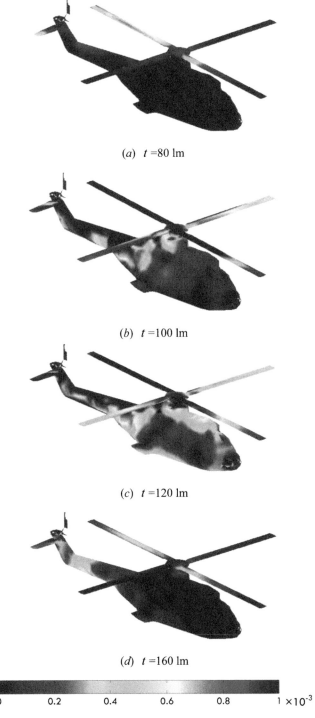

(a) $t = 80$ lm

(b) $t = 100$ lm

(c) $t = 120$ lm

(d) $t = 160$ lm

| 0 | 0.2 | 0.4 | 0.6 | 0.8 | 1 $\times 10^{-3}$ |

Figure 9.19. Transient current distributions for the helicopter.

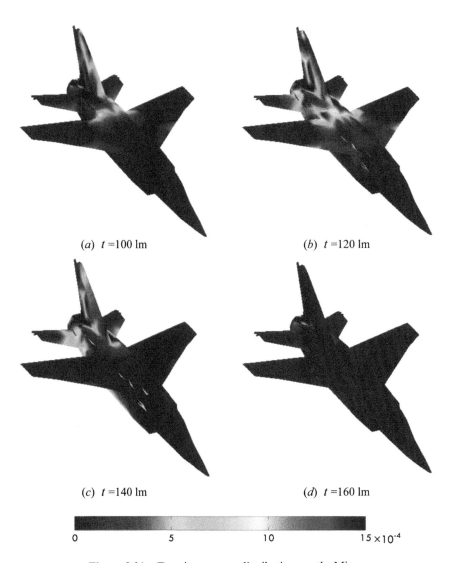

(a) $t = 100$ lm

(b) $t = 120$ lm

(c) $t = 140$ lm

(d) $t = 160$ lm

0   5   10 15×10^{-4}

Figure 9.21. Transient current distributions on the Mirage.

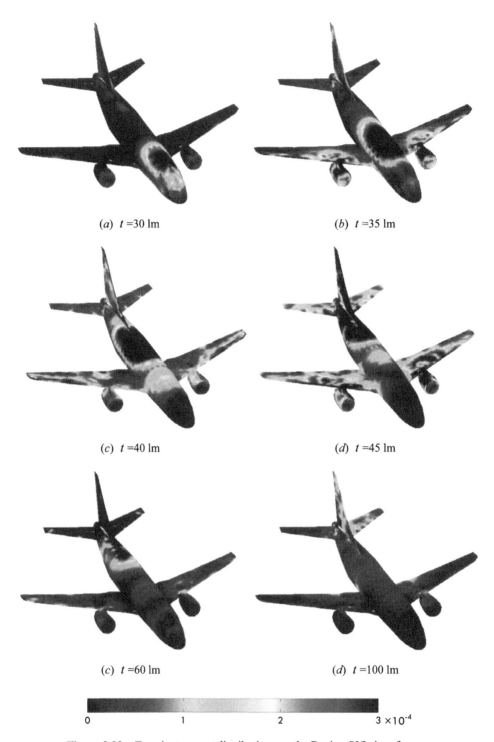

(a) t =30 lm

(b) t =35 lm

(c) t =40 lm

(d) t =45 lm

(c) t =60 lm

(d) t =100 lm

0   1   2   3 ×10^{-4}

Figure 9.23. Transient current distributions on the Boeing-737 aircraft.

10

A HYBRID METHOD USING EARLY-TIME AND LOW-FREQUENCY INFORMATION TO GENERATE A WIDEBAND RESPONSE

10.0 SUMMARY

The objective of this chapter is to present a hybrid methodology for the analysis of electrically large electromagnetic systems irradiated by an ultra-short, ultra-wideband electromagnetic pulse. This is accomplished through the use of a hybrid method that involves simultaneous generation of early-time and low-frequency information. These two data sets, namely early-time and low-frequency, in the original and transform domains are not only easy to generate using an appropriate numerical electromagnetics code but also contain all necessary, mutually, complementary, information to characterize the system completely electromagnetically. The same philosophy also can be applied in measurements by using data from a vector network analyzer and a time domain reflectometer to generate the low frequency and early time data, respectively. This assumes that a sufficient record length is available in both domains. A criterion is provided to assess whether the record lengths are sufficiently long. Two different algorithms are proposed to carry out the required extrapolation using data in the complementary domains, because the given low-frequency data provides the missing late-time information, whereas the early-time data provides the missing high-frequency information of the system. Hence, this procedure is not really data extrapolation but intelligent processing, exploiting all characteristics of the given data sets. Two different algorithms are presented to illustrate how this can be achieved and the respective error bounds for each of the methodologies.

The first hybrid algorithm introduced in this chapter is parametric. An orthogonal series along with its Fourier transform using the same unknown coefficients linearly combining the various functions is used to fit the data both in the time and frequency domains and then solve for the unknown coefficients. These coefficients are then used to extrapolate the responses both for late-time and high frequency. The orthogonal functions generated from the three polynomials (Hermite polynomial, Bessel–Chebyshev polynomial, and Laguerre polynomials) in the continuous domain are presented to illustrate how the extrapolation is carried out. This method is referred to as parametric because the performance of the extrapolation is dependent on the choice of the parameters of the polynomials, which include the scaling factor and the order. By analyzing the properties of the polynomials and their transforms, the converging range of parameters can be determined analytically a priori, and some error bounds can be defined. In addition, an iterative method to generate the solution also is described.

The second hybrid algorithm described is nonparametric. It will be shown that a particular realization of this iterative hybrid method is equivalent to a Neumann-series solution of an integral equation of the second kind. It is shown further that this iterative Neumann expansion is an error-reducing method. Hence, this integral equation can be solved efficiently by employing a conjugate gradient iterative scheme. The convergence of this scheme also is demonstrated. The matrix equations that generate these methodologies are developed, and the equivalence to the integral equations is shown. Also, the method of singular value decomposition (SVD), used for solving the matrix equation, provides accurate and stable results. Error bounds also are provided to assess the accuracy of this methodology.

The computational load for the electromagnetic analysis for this method is modest compared with solving the electromagnetic analysis problem exclusively in either time or frequency domain. Numerical examples are presented to illustrate the principles of each of these methodologies.

10.1   INTRODUCTION

Current trends in computational electromagnetics primarily are confined to the analysis of systems in only one domain. Either one analyzes the electromagnetic system in the frequency domain by using the method of moments (MoM), finite element, or finite difference techniques or solves the Maxwell equations in the time domain using a similar set of methodologies. The strategy then of solving for the electromagnetic response from an electrically large scatterer is to use a bigger and faster supercomputer. Therefore, the response that one can solve for essentially is limited by the hardware of the supercomputer. However, the desired solution may not necessarily be ultra-wideband, but still, one may need to compute the response from an electrically large structure. Hence, we need to look at alternative methodologies, as the traditional tools are inadequate to address this problem because they take a long time to generate a solution.

In electromagnetic analysis, field quantities usually are assumed to be time harmonic. This suggests that the solution lies in the frequency domain. The MoM, which uses an integral equation formulation, can be used to perform the frequency domain analysis as outlined in the first four chapters. However, for broadband analysis, this approach can get very computationally intensive, as the MoM program [1] needs to be executed for each frequency of interest, and for high frequencies, the size of the matrix that needs to be solved for can be very large. So for a given computer system, how large a problem one can analyze is dictated by the memory (both real and virtual) of the computer and assumes that the computer system does not fail during the numerical solution of the problem of interest. To speed up the computations in the frequency domain, Cauchy's method [2,3] was introduced so that one could interpolate/extrapolate the response without the need to solve the problem at each single frequency of interest. The Hilbert transform [4–6] concept also was used to attain the same goal. As opposed to Cauchy's method, which is a model-based parametric methodology and therefore only works adequately for conducting structures, the non-parametric Hilbert transform-based methods are more general in nature, but their applicability is limited as to how far one can extrapolate the response. The time domain [7,8] approach is generally the preferred technique for broadband analysis. Other advantages of a time domain formulation include easier modeling of nonlinear and time-varying media and use of time gating to eliminate unwanted reflections. For a time domain integral equation formulation, the marching-on-in-time (MOT) usually is employed to solve it. A serious drawback of this algorithm is the occurrence of late time instabilities in the form of high-frequency oscillations. To remove this difficulty, generally an implicit technique is proposed as outlined in the Chapters 5 and 6. One way to reduce the numerical instability in the MOT scheme is to use the marching-on-in-order (MOD) methodology as described in Chapters 5–9. An ultra-wideband analysis of an electromagnetic system is limited by the execution times and not so much by the memory as for a direct MOT/MOD solution, a matrix inversion is required only once and not for each instances when new results need to be generated. The time domain computational techniques have been speeded up by using the matrix pencil method [9,10]. The matrix pencil method approximates the late-time domain response (in an interval that occurs after the incident pulse has passed the object) by a sum of complex exponentials and then extrapolates the response for very late times reducing the execution time. The problem with this methodology is to determine what exactly is *late time*. In finding the response from a deep cavity, the incident pulse simply bounces back and forth inside the structure and it is not clear when the resonances are set up, thereby defining the late-time condition. In addition, this model of extrapolating the electromagnetic responses by fitting the late-time domain data by a sum of complex exponentials may not be a good approach if the data are from lossy dielectric bodies. In that case, the extrapolation does not work as the electromagnetic response may have a different functional model other than from a pole singularity!

In this chapter, we present a hybrid methodology that overcomes the limitations of solving the electromagnetic problem exclusively in one domain. The term hybrid technique is used because we do not solve the problem exclusively only in one domain. We use a frequency domain code and generate the low/resonant frequency response, which requires modest computational resources. Next, we take a time domain code and generate the early-time response from the system, which is also easy to carry out. The claim here is that the early-time and the low-frequency information completely determines the entire response in either domain, as these two pieces of information are complementary. The low/resonant frequency data provides the very late-time response, which is missing and the early-time data provides the high-frequency response, which is not in the given data. Therefore, this new hybrid methodology is not an extrapolation scheme as we are not creating any new information but processing the given sets of data in different domains, which contain all the necessary information, in an intelligent fashion. Using early-time and low-frequency data, we obtain stable late-time and broadband information. The overall analysis is thus computationally very efficient.

The time and frequency domain responses of three-dimensional (3D) material bodies are considered in this chapter. It is assumed that a band-limited ultra-wideband pulse excites the material body. The band-limited assumption is not only necessary but also mandatory. The reason for this is that for numerical analysis we need to discretize the structure spatially. To obtain accurate results, the triangular patches modeling the surface of the material body should be at least a tenth of a wavelength (the wavelength is defined in terms of the path length inside the material body) at the highest frequency of interest or smaller. Otherwise, there would be aliasing errors in the response. This spatial discretization of the material body puts a hard limit on the highest frequency that can excite the structure. Therefore, the frequency domain response is strictly band limited and let this be $2B$, as the response in frequency stretches from $-B$ to $+B$. However, that bandwidth B could be very large. On the other hand, the time domain response is not strictly limited in time, but for practical reasons, in a numerical experiment, we want to extrapolate the response up to some time T_0, such that both the time and frequency domain responses are of finite support for all practical purposes. Two methods are described in this chapter to carry out this extrapolation procedure. They are termed parametric and nonparametric.

The first hybrid algorithm introduced in this chapter in section 10.2 is parametric [11–17]. An orthogonal series along with its Fourier transform using the same unknown coefficients linearly combining the various functions is used to fit the data both in the time and frequency domain. One then solves for these unknown coefficients by matching the various polynomials and its transforms using early-time and low-frequency data. These coefficients then are used to extrapolate the response both for late-time and high-frequency. Orthogonal functions based on the three polynomials (namely the Hermite, Bessel–Chebyshev, and Laguerre) in the continuous domain have been presented to achieve this goal of extrapolation [15,16]. The hybrid method in the discrete domain using a discrete Laguerre polynomial has been introduced in Ref. [17]. Fourier Transforms of the presented polynomials are analytic functions. This allows work to be performed simultaneously with time and frequency domain data. This method is referred to as parametric because the performance of the extrapolation is dependent on the choice of the scaling factor and the order of the polynomials. By analyzing the properties of the polynomials and their transforms the region of convergence of the parameters can be determined analytically a priori [15] and some error bounds can be defined. In addition an iterative method to generate the solution also is described.

The second hybrid algorithm described in section 10.3 is nonparametric [18]. This implies that no parameters are optimized to get the appropriate solution. The total solution, however is calculated based on the physics of the problem. It will be shown that a particular realization of this iterative hybrid method is equivalent to a Neumann-series solution of an integral equation of the second kind. It also is shown that this iterative Neumann expansion is an error-reducing method. This integral equation can be solved efficiently by employing a conjugate gradient iterative scheme. The convergence of this scheme also is demonstrated. The matrix equations that generate these methodologies are developed and the equivalence to the integral equations is shown. Also, the method of singular value decomposition (SVD) [19], used for solving the matrix equation, provides accurate and stable results. Error bounds also are provided to assess the accuracy of this methodology.

Computational load for the electromagnetic analysis for this hybrid method using either the parametric or the nonparametric methodology is modest compared with solving the electromagnetic analysis problem exclusively in either time or frequency domain. Numerical examples are presented to illustrate the principles of each of these methodologies. Section 10.4 provides some conclusions followed by some selected references where additional materials can be obtained.

10.2   A PARAMETRIC HYBRID SOLUTION METHODOLOGY

The objective is to generate a broadband electromagnetic (EM) response at various spatial locations on the structures of interest in either time or in the frequency domain. This broadband response is generated using a time-frequency hybrid methodology. This hybrid methodology is a combination of the data generated using early-time and low-frequency information. In summary, we are fitting a given data set along with its transform using a set of orthogonal polynomials along with their transforms containing the same coefficients. Because we fit the data in these two complementary domains by using a set of orthogonal polynomials along with their respective transforms, it is termed a parametric hybrid solution methodology. The philosophy here is that the given data in both the domains time and frequency, will be parameterized by some known quantities with some unknown constants. The various orthogonal polynomials and how they can be used effectively along with the error estimates involved in the approximation are discussed next.

The early-time response generated by a time domain code provides M_1 samples with the sampling step of Δt. This time domain data set is designated by $\mathbf{x}_1$. The low-frequency response at the same spatial location is evaluated at M_2 points with the sampling step of Δf. This frequency domain data set is designated by $\mathbf{X}_1$. The missing late-time and high-frequency data sets are similarly designated by $\mathbf{x}_2$ and $\mathbf{X}_2$, respectively. The complete time and frequency domain responses are given by the functions $x(t)$ and $X(f)$, respectively. The total number of samples that characterize $x(t)$ is M_t, and the total number of samples that characterize $X(f)$ is M_f. These relationships are given by

$$
\begin{cases}
\mathbf{x}_1 = \left\{x(t_0), x(t_1), \ldots, x(t_{M_1-1})\right\}^T \\[4pt]
\mathbf{x}_2 = \left\{x(t_{M_1}), x(t_{M_1+1}), \ldots, x(t_{M_t-1})\right\}^T \\[4pt]
\mathbf{X}_1 = \left\{X(f_0), X(f_1), \ldots, X(f_{M_2-1})\right\}^T \\[4pt]
\mathbf{X}_2 = \left\{X(f_{M_2}), X(f_{M_2+1}), \ldots, X(f_{M_f-1})\right\}^T
\end{cases}
\tag{10.1}
$$

In Eq. (10.1), $t_i = (i-1)\Delta t$ and $f_j = (j-1)\,\Delta f$. The complete data set then is characterized by concatenating the given and the missing data sets as

$$
\mathbf{x} = \begin{Bmatrix} \mathbf{x}_1 \\ \mathbf{x}_2 \end{Bmatrix}
\qquad\qquad
\mathbf{X} = \begin{Bmatrix} \mathbf{X}_1 \\ \mathbf{X}_2 \end{Bmatrix}
\tag{10.2}
$$

The basic philosophy of the parametric hybrid method is to fit $x(t)$ and $X(f)$ by a series of orthogonal polynomials and their transforms, respectively, using the same parameters. Suppose the time domain response is fitted by the expression

$$
x(t) = \sum_{n=0}^{\infty} a_n \phi_n(t/l_1) \triangleq \sum_{n=0}^{\infty} a_n \phi_n(t, l_1)
\tag{10.3}
$$

where $\phi_n(t)$ is the orthogonal basis function of degree n and l_1 is a scaling factor. The objective of the scaling factor is to scale the axis of the data, be it in picoseconds, microseconds, and so on. Therefore, the frequency domain response automatically is fitted with the transform of Eq. (10.3) as

$$
X(f) = \sum_{n=0}^{\infty} a_n \Phi_n(f/l_2) \triangleq \sum_{n=0}^{\infty} a_n \Phi_n(f, l_2)
\tag{10.4}
$$

where $\phi_n(t)$ and $\Phi_n(f)$ are related by the Fourier transform and the two scaling factors l_1 and l_2 are related by

$$
l_2 = 1/(2\pi l_1)
\tag{10.5}
$$

To solve for the unknown coefficients a_n, only the early-time and the low-frequency information is used. For a numerical solution of the coefficients a_n, the infinite upper limit in Eqs. (10.3) and (10.4) are replaced by a finite number N resulting in the following matrix equation:

$$
\begin{bmatrix}
\begin{bmatrix}
\phi_0(t_0,l_1) & \phi_1(t_0,l_1) & \cdots & \phi_{N-1}(t_0,l_1) \\
\phi_0(t_1,l_1) & \phi_1(t_1,l_1) & \cdots & \phi_{N-1}(t_1,l_1) \\
\vdots & \vdots & \vdots & \vdots \\
\phi_0(t_{M_1-1},l_1) & \phi_1(t_{M_1-1},l_1) & \cdots & \phi_{N-1}(t_{M_1-1},l_1)
\end{bmatrix} \\
\mathrm{Re}\begin{pmatrix}
\Phi_0(f_0,l_2) & \Phi_1(f_0,l_2) & \cdots & \Phi_{N-1}(f_0,l_2) \\
\Phi_0(f_1,l_2) & \Phi_1(f_1,l_2) & \cdots & \Phi_{N-1}(f_1,l_2) \\
\vdots & \vdots & \vdots & \vdots \\
\Phi_0(f_{M_2-1},l_2) & \Phi_1(f_{M_2-1},l_2) & \cdots & \Phi_{N-1}(f_{M_2-1},l_2)
\end{pmatrix} \\
\mathrm{Im}\begin{pmatrix}
\Phi_0(f_0,l_2) & \Phi_1(f_0,l_2) & \cdots & \Phi_{N-1}(f_0,l_2) \\
\Phi_0(f_1,l_2) & \Phi_1(f_1,l_2) & \cdots & \Phi_{N-1}(f_1,l_2) \\
\vdots & \vdots & \vdots & \vdots \\
\Phi_0(f_{M_2-1},l_2) & \Phi_1(f_{M_2-1},l_2) & \cdots & \Phi_{N-1}(f_{M_2-1},l_2)
\end{pmatrix}
\end{bmatrix}
\begin{bmatrix} a_0 \\ a_1 \\ \vdots \\ a_{N-1} \end{bmatrix}
=
\begin{bmatrix}
\begin{bmatrix} x(t_0) \\ x(t_1) \\ \vdots \\ x(t_{M_1-1}) \end{bmatrix} \\
\mathrm{Re}\begin{pmatrix} X(f_0) \\ X(f_1) \\ \vdots \\ X(f_{M_2-1}) \end{pmatrix} \\
\mathrm{Im}\begin{pmatrix} X(f_0) \\ X(f_1) \\ \vdots \\ X(f_{M_2-1}) \end{pmatrix}
\end{bmatrix}
$$

$$(10.6)$$

where $\mathrm{Re}(\bullet)$ and $\mathrm{Im}(\bullet)$ stands for the real and imaginary parts of the transfer function. Then, once the N coefficients $\{a_n\}$, have been solved for, they can be used in Eq. (10.3) to extrapolate the late-time response and in Eq. (10.4) to extrapolate the missing high-frequency response. The values $\hat{\mathbf{x}}_2$ and $\hat{\mathbf{X}}_2$ are designated as the corresponding extrapolated data, (i.e., the estimation for $\mathbf{x}_2$ and $\mathbf{X}_2$). Eq. (10.6) shows that $N \leq M_1 + 2M_2$. The question now is how many functions N are required in Eqs. (10.3) and (10.4) instead of ∞.

It is assumed that the response that we are interested in is strictly band-limited up to a frequency B. This is justified, as explained before, because one needs to discretize the structure spatially for computational reasons. Typically, one uses about 10 expansion functions per wavelength in the material body (of course this depends on the numerical method; a value of 10 is chosen for illustrative purposes), and therefore, the spatial discretization of the structure limits the high-frequency response (and let us say it is up to B). If we now are interested in representing the same waveform in the time domain, then it is of infinite duration. However, for a numerical method, we cannot go to infinity and the record is assumed to be limited to a time of duration T_0. Then the question is how many parameters are necessary to represent mathematically that waveform, which is also band limited by B (the frequency goes from $-B$ to $+B$). We now demonstrate that $(2BT_0 + 1)$ pieces of information are sufficient to describe that waveform from a strictly mathematical point of view. To prove this, we represent the real-time domain function $x(t)$ by a Fourier series. This is permissible because the waveform is strictly band limited, so that [20,21] $x(t) = \sum_p C_p \exp(i\,p\,\omega_0\,t)$ where $\omega_0 = \dfrac{2\pi}{T_0}$.

Because $x(t)$ is real, $C_p^* = C_{-p}$ where * represents the complex conjugate. If $x(t)$ is band limited to B Hz, then the range of the summation index p is fixed by $-B \leq p f_0$

$\leq B$, and because $f_0 = 1/T_0$, we obtain $x(t) = \displaystyle\sum_{p=-BT_0}^{BT_0} C_p \exp(i\, p\, \omega_0\, t)$. Hence, there

are exactly $2BT_0 + 1$ terms in this expansion of $x(t)$. Therefore, the dimensionality N in Eq. (10.6) reflects $N = 2BT_0 + 1$. From a strictly mathematical standpoint, one needs a total of $2BT_0+1$ samples to characterize the ultra-wideband waveform of bandwidth $2B$ and of time duration T_0. These $2BT_0 + 1$ samples can be entirely in the time domain or in the frequency domain or a combination, being partially in of both the domains. These samples are then sufficient from a strict mathematical perspective to represent any waveform. However, from a numerical perspective, if these samples are too close to each other, then the solution of Eq. (10.6) would be very ill conditioned and the extrapolation may fail. One way to know whether the samples are spaced adequately and whether their numbers are adequate from a strictly numerical perspective (even though it may be sufficient from a strictly mathematical point of view) is to carry out a total least-squares solution [2,19] of Eq. (10.6) instead of using an iterative method such as the conjugate gradient to solve for the coefficients. The distribution of the singular values in the total least-squares solution [2,19] will provide an automatic methodology to know whether the data lengths are adequate. This point will be illustrated later on. We find that taking samples simultaneously in both domains provides a much more stable solution than just taking them in one domain exclusively. Therefore, we must have the total number of samples M_1 in the time and M_2 in frequency equal to $M_1 + M_2 \geq N = 2BT_0 + 1$.

Finally, what is the optimum sampling rate in time and frequency is still an open question.

To evaluate the performance of this hybrid method, the following normalized mean-squared errors (MSE) are used in the time and frequency domain as

$$E_t = \frac{\left\| \hat{\mathbf{x}}_2 - \mathbf{x}_2 \right\|_2}{\left\| \mathbf{x}_2 \right\|_2} \qquad E_f = \frac{\left\| \hat{\mathbf{X}}_2 - \mathbf{X}_2 \right\|_2}{\left\| \mathbf{X}_2 \right\|_2} \qquad (10.7)$$

where $\left\| \bullet \right\|_2$ is the $\mathcal{L}_2$ norm of a vector. If $\hat{\mathbf{x}}_2$ and $\hat{\mathbf{X}}_2$ are equal to 0 (trivial results), then E_t and E_f will be equal to 1. Therefore, the hybrid method does not converge when $E_t \geq 1$ or $E_f \geq 1$. The goal of this section is to demonstrate through numerical examples that it is possible to carry out such a hybrid solution with good accuracy. In this way, the computational time can be reduced significantly.

The success of this procedure depends on the choices of the parameters Δt, Δf, M_1, M_2, l_1, l_2, and N. The choice of Δt and Δf is restricted by the Nyquist's theory of sampling, which states that $\Delta t \geq 1/(2W)$ and $\Delta f \geq 1/T$. The variables T and W discussed in this chapter are the time support and the bandwidth for the electromagnetic responses $x(t)$ and $X(f)$, respectively. For the choice of M_1 and M_2, it is clear that $(M_1 + 2M_2)$ should be greater than N, whatever that value is.

Notice that the matrix of Eq. (10.6) need not be a square matrix, and it may be ill-conditioned if the time and frequency samples of the response are close to each other. That is why we advocate the use of the singular value decomposition (SVD) method [19] to solve Eq. (10.6). There are two reasons of doing that. First, it can tell us whether we have sufficient data to begin with in both the domains to carry out the extrapolation in the two domains, and second, even if the data sets given are sampled closely, it does not influence the final result, as the SVD still can provide a solution for $\{a_n\}$ in a least-square sense. Hence, the solution, when it exists, has a minimum norm for $\hat{x}_2$ and $\hat{X}_2$. This feature is in accordance with the assumption that the early-time data complements the high-frequency information and the low-frequency data complements the late-time information. Another advantage of using SVD is that the distribution of the singular values can be analyzed. SVD will "provide an automatic methodology to know whether the data lengths are adequate" [10]. A tolerance (*tol*) of the singular values can be chosen to be of the same level as the numerical error and the noise of the original data. If the smallest singular value of the matrix is larger than the tol, it means M_1 or M_2 is not large enough, and the time domain or the frequency domain codes has to be executed to obtain more data and the extrapolation cannot proceed. However, when the value of the smallest singular value is less than the value of *tol*, which implies that that there is sufficient data and a solution can be obtained. In computing that solution using the singular vectors, the information related to the singular values smaller than *tol* is discarded. Therefore, in this procedure, what is early-time and low-frequency is not defined clearly. However, using M_1 and M_2 samples of the data, along with the SVD, one can determine a priori whether the data sets given are sufficient or not. If the data sets are not sufficient by observing the magnitude of the various singular values, then additional data must be generated in either domain.

Next, we choose the value of N, which is required to approximate any time-domain function of duration T and bandwidth $2W$ and is determined by the time-bandwidth product of the signal or by the dimensionality theorem, also called the $2NW$ theorem [20,21]. It is also a different statement for the sampling theorem. Consider a Fourier transform pair

$$h(t) \Leftrightarrow H(f) \tag{10.8}$$

and $H(f)$ is approximately band-limited and so $|f| \leq W$. The energy outside W is small and has an upper bound of ε.

$$\int\limits_{|f|>W} |H(f)|^2 \, df \leq \varepsilon \tag{10.9}$$

Consider G to be the set of finite energy sequences. In addition, h and M is the dimension of the smallest linear space of sequences that approximate G on the index set $(0, N-1)$, with energy less than ε', and N is the number of time samples or the dimension of G. Then

$$\sum_{n=0}^{N-1}\left[g_n - \sum_{j=1}^{M} a_j g_n^{(j)}\right]^2 \leq \varepsilon' \tag{10.10}$$

where g_n is a sequence such that $\{g\} \in G$ and a_j are the coefficients of approximation. Then, the $2NW$ theorem can be stated as follows: For $1/2 \geq W > 0$, the sampling time needs to be less than the Nyquist sampling interval, and when $\varepsilon' > \varepsilon > 0$, we obtain

$$\lim_{N \to \infty} \frac{M}{N} = 2W \tag{10.11}$$

The fact can be summarized succinctly as follows: For a large value of N, the set of sequences of bandwidth W that are confined to an index set of length about M has dimensions approximately equal to the time-bandwidth product of the waveform, which is equal to $2NW$. Here, the sampling interval in time domain is assumed to be unity. If ΔT is the sampling interval in time domain, then the order or the degrees of freedom of the expansion is approximately $2NW\Delta T$, as stated by the sampling theorem.

In summary, as long as the total number of the orthonormal basis functions exceeds the time-bandwidth product of the EM response, they are sufficient to approximate any function in the $\mathscr{L}_2$-norm. The choice of the values of the remaining parameters, l_1 and l_2, is a little involved, as it depends on the type of orthogonal polynomial chosen to approximate the functions. Also, note that it is necessary to choose only one of the scaling factors, as the scaling factor in one domain is related to the other through Eq. (10.5). It is important to note that the expansion given by Eqs. (10.3) and (10.4) only converge for certain values of the parameters and not for all values. Therefore, it is important to reflect on how to arrive at the optimum values of these parameters. Ref. [22] seeks the optimal value for l_1 analytically in synthesis of systems using Laguerre polynomials. The optimal scaling factor should be one of the roots of the following equations:

$$c_L(b) = 0 \tag{10.12}$$

$$c_{L+1}(b) = 0 \tag{10.13}$$

where b is a value relative to the scaling factor and $c_L(b)$ is a polynomial given by

$$c_L(b) = \sqrt{1-b^2} \sum_{i=0}^{N} \sum_{s=0}^{\min(i,L)} (-1)^{L-s} \binom{i}{s}\binom{L-s+i}{i} a_i b^{(i+L-2s)} \tag{10.14}$$

where L is the highest order of the Laguerre polynomial and N is the order of the system. The value a_i is the sequence related to the response of the system. The optimal value b can be solved easily under the following assumptions. First, the order of the polynomial is not very high. Also Eq. (10.14) is an $L + N$ order equation. Typically, $L + N$ may be of the order of hundreds. It is very difficult, if not impossible, to solve for the roots of an equation of so high an order even with double or quad precision. Even if it can be solved, it is possible to obtain hundreds of values for the root. Each root has to be put through the solution process, and the errors have to be computed and plotted. The computational load will be very large and unnecessary. The bottom line is that the whole range of the system response needs to be known in this case, which means that the entire time domain response for the problem must be known. If the data for the whole time-span is known, then it is not required to carry out any extrapolation at all. Hence, this method is useful in system synthesis but not suitable for our methods.

The optimal value l_1 is determined by all the data in the whole range of the time interval. If only the early-time and low-frequency data is known, it may be difficult or impossible to find the optimal value for l_1 analytically. In addition, a numerical search for the optimal values of N, l_1, and l_2 analytically in this application of EM computation is a nonlinear problem, and in most practical cases, it is not feasible. However, there are useful properties of the EM response as well as the polynomials themselves, which can be used to determine this value. We illustrate next how that can be accomplished.

In most situations, the bandwidth W and the time support T of the EM response is known. In addition, the a priori choice of the polynomials used for the extrapolation tells us what their properties are. In section 10.2.1, an a priori choice is proposed for the upper and lower bounds of the parameters which is easy to obtain. The details of the choice of three selected polynomials also will be introduced in this section along with their properties.

As the low-frequency data contains the late-time information as well as the early-time data, a numerical method is needed to obtain the optimum value for the parameters. By knowing the theoretical bound of the parameters of interest to obtain convergent results, the optimal parameters can be obtained iteratively by comparing the error for the approximation of the early-time and low-frequency data. The details of this numerical method will be introduced in section 10.2.2.

10.2.1 Error Bound of the Parameters

In this section, a priori error bound using the properties of both the given data and the set of polynomials is introduced. The given data is assumed to be generated by exciting the EM system with a Gaussian pulse and is discussed in section 10.2.1.1. The three types of polynomials, Hermite, Bessel–Chebyshev, and Laguerre, are introduced in section 10.2.1.2. and their properties are discussed. Error bounds of the parameters are derived and numerical examples are used in section 10.2.1.3 to validate the error bounds.

10.2.1.1 Generation of the Data Resulting From a Gaussian Input. Typically, in solving time domain problems the electromagnetic structure is usually excited by a Gaussian pulse, which is described in the time domain as

$$g(t) = \frac{1}{\sigma\sqrt{\pi}} U_0 e^{-\left(\frac{t-t_0}{\sigma}\right)^2} \tag{10.15}$$

where σ is proportional to the width of the pulse. t_0 is the delay to make $g(t) \approx 0$ for $t < 0$, and U_0 is the scalar amplitude. In frequency domain, the pulse has a form

$$G(f) = U_0 e^{-\left[\frac{(2\pi f \sigma)^2}{4} + j2\pi f t_0\right]} \tag{10.16}$$

Note that $G(f)$ is approximately band limited by $1/\sigma$. Hence, a Gaussian pulse is practically both time and band-limited. However, note that a waveform cannot be simultaneously strictly limited in time and frequency. A waveform is defined to be time limited to T and frequency limited to W with the understanding that

$$\frac{|x(t)|}{\max\{|x(t)|\}} < \varepsilon, \qquad \text{for } t > T,$$

$$\frac{|X(f)|}{\max\{|X(f)|\}} < \varepsilon, \qquad \text{for } |f| > W \tag{10.17}$$

where ε is a small positive number. For our discussions, we assume that ε is approximately 10^{-3}. The EM response observed in the real world has finite energy. It is therefore reasonable to consider wave shapes that are practically time and band limited. Because EM systems are linear systems, W is restricted by the bandwidth of the input $(1/\sigma)$, and W can be known a priori as $W \approx 1/\sigma$. For a pulse input, the EM response will die down after some time. This duration of T can be estimated by using the inverse Fourier transform directly on the known low-frequency data. This estimation is based on the fact that the late-time damping rate should be determined by the low-frequency data. The result of the inverse Fourier transform is equivalent to applying a rectangular window to the original data in the frequency domain and convolving the impulse response of the system with a sinc function. The value T can be estimated by comparing the damping rate of the sinc function and from the result of the inverse Fourier transform. The value W is determined from the spatial discretization of the structure because typically for subsectional basis, one needs to discretize the structure to at least 10 subsections at the highest frequency of interest.

The response $x(t)$ of a system as a result of the Gaussian pulse will be the induced surface currents. The radiation pattern, near field, or admittance parameters related to the structure are derived from the currents.

10.2.1.2 Three Different Choices of Basis Functions and Their Numerical Stability and Error Bounds.
Three different basis functions now are used for the approximation. They are the orthogonal functions generated from the Hermite, Bessel–Chebyshev, and Laguerre polynomials. The three have different properties, which will be discussed separately. The expansions using these basis functions must satisfy the following two criteria:

1. If the functions are of compact support, then the basis functions must approximate the temporal function of interest up to time T_N, which must be larger than T, the duration of the waveform of interest. Its Fourier transform, which approximates the frequency domain waveform, must provide support up to bandwidth W_N, which must be larger than the desired bandwidth W of the original function. If the basis function does not have compact time support, then its variations (first and second derivatives) must be larger than the variations of the original waveform at T. Accordingly in frequency domain, the derivative of the phase of the basis functions at $f = W$ must be larger than that of the original waveform, that is,

$$\left| \text{Arg}\{\Phi_N(f)\}' \Big|_{f=W} \right| > \left| \text{Arg}\{X(f)\}' \Big|_{f=W} \right| \qquad (10.18a)$$

where $\text{Arg}\{X(f)\}$ implies that it is the angle of the complex $X(f)$.

2. The second criterion is straightforward when the waveform is a shifted Gaussian pulse. A time shift will cause a phase change of $2\pi f t_0$ in the frequency domain, and the derivative of this phase change is a constant. This criterion then has the following form

$$\left| \text{Arg}\{X(f)\}' \Big|_{f=W} \right| > C_W = 2\pi t_0 \qquad (10.18b)$$

By combining Eqs. (10.18a) and (10.18b), one obtains

$$\left| \text{Arg}\{\Phi_N(f)\}' \Big|_{f=W} \right| > \left| \text{Arg}\{X(f)\}' \Big|_{f=W} \right| > C_W = 2\pi t_0 \qquad (10.18c)$$

Next the various bounds of interest for the three different choices of the basis functions are outlined.

10.2.1.2.1. Hermite Basis Functions. First, choose the orthogonal basis function as $\varphi_n(t,l_1) = h_n(t,l_1)$ where $h_n(t,l_1)$ are the associated Hermite polynomials. They can be expressed as

$$h_n(t,l_1) = \frac{H_n\left(t/l_1\right) e^{\frac{-t^2}{2l_1^2}}}{\sqrt{2^n n!}\, l_1\sqrt{\sqrt{\pi}}}, \quad n = 0, 1, 2, \dots \tag{10.19}$$

where $H_n(t)$ is the Hermite polynomial. The Hermite polynomial can be computed recursively through [23]

$$H_0(t) = 1, \quad H_1(t) = 2t \tag{10.20}$$
$$H_n(t) = 2tH_{n-1}(t) - 2(n-1)H_{n-2}(t); \quad n = 2,3,4,\dots$$

Thus the associated Hermite polynomials are calculated recursively by

$$h_n(t) = \frac{1}{\sqrt{n}}\left(\sqrt{2}t\, h_{n-1}(t) - \sqrt{n-1}\, h_{n-2}(t)\right); \quad n = 2,3,4,\dots \tag{10.21}$$

Because the associated Hermite polynomials are the eigenfunctions of the Fourier transform operator, their Fourier transforms are given by

$$h_n(t, l_1) \triangleq h_n(t/l_1) \Leftrightarrow (-i)^n h_n(f/l_2) \triangleq (-i)^n h_n(f, l_2) \tag{10.22}$$

So, if the function $x(t)$ can be expanded by the orthogonal basis $h_n(t, l_1)$, its Fourier transform $X(f)$ can be obtained by adding up the terms $(-i)^n h_n(f, l_2)$ with the same coefficients. It is observed that the associated Hermite polynomials are not necessarily causal, and in addition, its two properties in the time domain are:

1. The time support of the basis function T_n increases as the order n increases, assuming the scaling factor l_1 is fixed. This property is because of the following relation:

$$\lim_{t \to \infty} \frac{h_n(t)}{h_{n+1}(t)} = 0; \quad n = 0, 1, \dots \tag{10.23}$$

2. More oscillations occur for the functions as the order n increases. Because of Eqs. (10.22) and (10.23), it can be observed that the basis function with a higher order has more high frequency components.

The time support of these basis functions, T_n as defined in Eq. (10.17) (when the functions have decayed to a level $\varepsilon = 10^{-3}$) for order n (for $l_1 = 1$) is plotted in Figure 10.1 (dashed). From this figure, it is shown that the length of the time support can be approximated by the following empirical formula:

$$T_n \approx \sqrt{\pi n / 1.7} + 1.8 \tag{10.24}$$

which is also shown by the solid line in Figure 10.1.

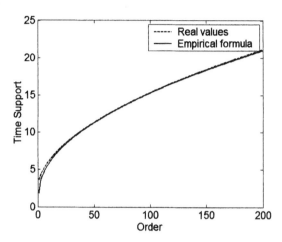

Figure 10.1. The time support of the Hermite basis functions with respect to the order n.

Because the functional forms of these functions are identical in the time and frequency domains, it can be shown that the support of these functions in the time and frequency domains can be given by

$$T_n \approx 2l_1 \left(\sqrt{\pi n / 1.7} + 1.8 \right) \tag{10.25a}$$

$$W_n \approx l_2 \left(\sqrt{\pi n / 1.7} + 1.8 \right) = \frac{\sqrt{\pi n / 1.7} + 1.8}{2\pi l_1} \tag{10.25b}$$

Because $T_N > T$ and $W_N > W$, one obtains

$$\frac{T}{2\left(\sqrt{\pi N / 1.7} + 1.8\right)} < l_1 < \frac{\sqrt{\pi N / 1.7} + 1.8}{2\pi W} \tag{10.26}$$

Eq. (10.26) gives the upper and lower bound of the scaling factor with respect to N for the approximation of $x(t)$ (with time support T and bandwidth W) by the associated Hermite polynomials. The minimum value of N required for the extrapolation is relative to WT. Because only two conditions are considered, $T_N > T$ and $W_N > W$, the implicit lower bound of N in Eq. (10.26) is more relaxed than the value given by the "$2WT$ theorem" [20,21]. The $2WT$ theorem states that the "approximate dimension of the signal" is $N = 2WT$, resulting in a lower bound on N as

$$N > 2WT \tag{10.27}$$

Eq. (10.27) is used as the lower bound for N when employing the Hermite basis functions. Although N has a lower bound only, it cannot be arbitrarily large because the magnitude of the coefficients a_n will have oscillations for large values of N because of computational error.

10.2.1.2.2. Bessel–Chebyshev Functions. Another choice for the orthogonal basis function is $\varphi_n(t,l_1) = (t/l_1)^{-1} J_n(t/l_1)$ where $J_n(t/l_1)$ is a Bessel function of the first kind of degree n. A waveform with a compact time support can be expanded as

$$x(t) \approx \sum_{n=0}^{N} a_n (t/l_1)^{-1} J_n(t/l_1) \triangleq \sum_{n=0}^{N} a_n (t/l_1)^{-1} J_n(t,l_1) \tag{10.28}$$

The Fourier transform of Eq. (10.28) can be evaluated as [15]

$$X(f) = \begin{cases} \displaystyle\sum_{n=0}^{N} a_n \frac{i}{n} (-i)^n \left[1 - (f/l_2)^2\right]^{\frac{1}{2}} U_{n-1}(f/l_2) \ ; \ |f| < l_2 \\ 0 \qquad\qquad\qquad\qquad\qquad\qquad\qquad ; \ |f| > l_2 \end{cases} \tag{10.29}$$

where $l_2 = 1/(2\pi l_1)$ and $U_n(f)$ is the Chebyshev polynomial of the second kind defined by [23]

$$U_n(f) = \frac{\sin[(n+1)\cos^{-1} f]}{(1-f^2)^{\frac{1}{2}}} \tag{10.30}$$

In Eqs. (10.28) and (10.29), the causality in time is not forced, whereas the waveforms being used are causal. The relationship between the first and second kind of Chebyshev polynomials is given by the Hilbert transform that is,

$$\int_{-1}^{1} \frac{\sqrt{1-y^2} U_{n-1}(y)}{y-x} dy = -\pi T_n^C(x) \tag{10.31}$$

where $T_n^c(x)$ is the Chebyshev polynomial of the first kind defined by

$$T_n^C(f) = \cos\left[n\cos^{-1}(f)\right] \tag{10.32}$$

Therefore, a causal time domain waveform and its Fourier transform can be written as

$$x(t) = \sum_{n=0}^{N} a_n (t/l_1)^{-1} J_n(t/l_1); t \geq 0 \tag{10.33}$$

$$X(f) = \begin{cases} \displaystyle\sum_{n=0}^{N} a_n \frac{1}{n}(-i)^n \left\{ i\left(1-\left(\frac{f}{l_2}\right)^2\right)^{1/2} U_{n-1}\left(\frac{f}{l_2}\right) + T_n^C\left(\frac{f}{l_2}\right) \right\}; & |f| < l_2 \\ 0; & ; |f| > l_2 \end{cases} \tag{10.34}$$

Unlike the associated Hermite functions, which have the same form in time and frequency domains with the decay rate of $e^{-t^2/2}$ or $e^{-f^2/2}$ for large t or f, the Bessel–Chebyshev basis functions are strictly band limited, (i.e., $|f| < l_2$). The fact that they are strictly band limited results in infinite time support, or $T_n \to \infty$. The time decay rate $1/t$ of the functions as in Eq. (10.33) is not sufficient to make the basis functions limited in time.

The upper bound of l_1 is determined from the bandwidth of the functions obtained from the condition $W_N > W$. Because $W_n = l_2$ for all n,

$$l_1 < 1/(2\pi W) \tag{10.35}$$

Because, $T_n \to \infty$, the lower bound of l_1 cannot be decided simply by the condition $T_N > T$. The second criterion then is used as stated in the beginning of this section. Bessel functions have the following asymptotic forms:

$$J_n(t) \approx \frac{1}{\Gamma(n+1)}(t/2)^n, \quad t \ll 1 \tag{10.36}$$

$$J_n(t) \approx \sqrt{2/(\pi t)} \cos(t - n\pi/2 - \pi/4), \quad t \gg 1 \tag{10.37}$$

For large values of t ($t \gg 1$), $J_n(t)$ approximates a sinusoidal function with an attenuation $1/\sqrt{t}$. $J_n(t)$ displays a delayed peak with an oscillating tail with the same frequency $\omega = 1$ for each n. This oscillation is strictly band limited by $\omega = 1$ ($l_2 = 1/2\pi$ if $l_1 = 1$) in the frequency domain. The tails of the basis cannot be used to approximate an arbitrary function because all tails have only a single frequency component. Mathematically speaking, the dimension of the set $\{J_n(t) | n = 0, 1, 2, \ldots, N\}$ is limited and cannot form a complete basis set for $t \gg 1$, if N is not large enough. We calculate the effective time support T_n' as the second root of each basis function and impose $T_N' > T$ to obtain the lower bound of l_1. The second root for the Bessel functions can be obtained numerically and can be approximated by the empirical formula

$$T_n' \approx 1.08n + 9 \tag{10.38}$$

as it is illustrated by the solid line in Figure 10.2.

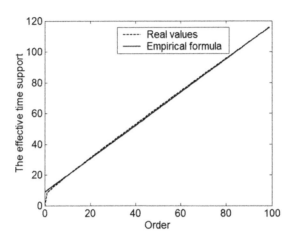

Figure 10.2. The effective time support T_n' of the Bessel functions with respect to order n.

By adding the scaling factor and imposing $T_N' > T$, one obtains

$$l_1 > \frac{T}{1.08N + 9} \tag{10.39}$$

Thus, the upper and lower bounds of l_1 is given by

$$\frac{T}{1.08N + 9} < l_1 < \frac{1}{2\pi W} \tag{10.40}$$

and from Eq. (10.40), one obtains a bound for N as

$$N > \frac{2\pi WT - 9}{1.08} \tag{10.41}$$

Eq. (10.41) shows that when both T and W are large, the number of Bessel functions required for the approximation of the data $x(t)$ is larger than the dimension of $2WT$ as dictated by the "$2WT$ theorem". The value of $2WT$, the lower bound of N only can be reached in the limit $T \to \infty$ or $W \to \infty$. The range of l_1 for the Bessel-Chebyshev basis functions is smaller than that for the Hermite basis functions as defined in Eq. (10.26).

10.2.1.2.3. Laguerre Basis Functions. Next, the Laguerre polynomials of order n are defined by

$$L_n(x) = \frac{1}{n!}e^x \frac{d^n(x^n e^{-x})}{dx^n}, \quad n = 0, 1, 2, \ldots.; \text{ for } x \geq 0 \tag{10.42}$$

These functions are causal, and they can be computed recursively using

$$L_0(t) = 1, \qquad L_1(t) = 1 - t$$

$$L_n(t) = \frac{2n-1-t}{n}L_{n-1}(t) - \frac{n-1}{n}L_{n-2}(t); \quad n = 2, 3, 4, \ldots \tag{10.43}$$

The Laguerre polynomials are orthogonal with respect to a weighting function as

$$\int_0^\infty e^{-t}L_n(t)L_m(t)dt = \delta_{mn} = \begin{cases} 1; & m = n \\ 0; & \text{otherwise} \end{cases} \tag{10.44}$$

An orthonormal basis function called the associated Laguerre functions now can be derived from the Laguerre polynomials through the representation

$$\phi_n(t, l_1) = \left(1/\sqrt{l_1}\right)e^{-t/2l_1}L_n(t/l_1) \tag{10.45}$$

where l_1 is a scaling factor. A causal electromagnetic response $x(t)$ at a particular location in space for $t \geq 0$ can be expanded by a Laguerre series as

$$x(t) = \sum_{n=0}^\infty a_n\left(1/\sqrt{l_1}\right)e^{-t/2l_1}L_n(t/l_1) \tag{10.46}$$

Therefore, the greatest advantage of this representation is that they naturally enforce causality. Also, their modality (number of local maxima and minima) increases with the increase of the order. So a waveform with compact time support can be approximated by

$$x(t) \approx \sum_{n=0}^N a_n\left(1/\sqrt{l_1}\right)e^{-t/2l_1}L_n(t/l_1) \tag{10.47}$$

The Fourier transform of Eq. (10.47) can be evaluated as [22]

$$X(f) \approx \sum_{n=0}^{N} \frac{a_n \left(-0.5 + j f / l_2\right)^n}{\sqrt{2\pi l_2} \left(0.5 + j f / l_2\right)^{n+1}}$$

(10.48)

where $l_2 = 1/(2\pi l_1)$. The time support for these functions can be approximated by T_n (when the functions have decayed to a level $\varepsilon = 10^{-3}$) as

$$T_n = 4.1n + 25$$

(10.49)

This is a good approximation, as shown by the dotted line in Figure 10.3.

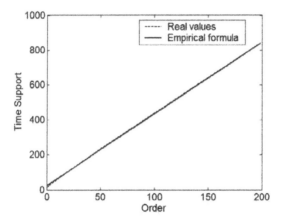

Figure 10.3. The time support of the associated Laguerre functions with respect to order n.

When $T_N > T$, the lower bound for l_1 is given by

$$l_1 > \frac{T}{4.1N + 25}$$

(10.50)

The Fourier transform of the Laguerre functions has a rational form in the frequency domain as

$$\Phi_n(f, l_2) = A(f) e^{j(\alpha(f) - \pi/2)}, \quad \text{where}$$

(10.51)

$$\alpha(f) = (2n+1) \tan^{-1}\left(l_2 / (2f)\right)$$

(10.52)

$$A(f) = \frac{1}{\sqrt{2\pi l_2} \sqrt{\left(f / l_2\right)^2 + 1/4}}$$

(10.53)

The amplitude of $\Phi_n(f,l_2)$ is independent of n. Now we use the second criterion to obtain the upper bound of l_1, which is given by

$$\left| \frac{d\,\alpha(f)}{d\,f} \right|_{f=W} > C_W \tag{10.54}$$

This results in

$$\frac{(2N+1)2l_2}{4W^2 + l_2^2} \approx \frac{4Nl_2}{4W^2 + l_2^2} > C_W \tag{10.55}$$

By solving the inequality in Eq. (10.55), one obtains

$$l_1 < \frac{C_W}{4\pi\left(N - \sqrt{N^2 - W^2 C_W^2}\right)} \tag{10.56}$$

Together with Eq. (10.50), the upper and lower bounds of l_1 is given by

$$\frac{T}{4.1N+25} < l_1 < \frac{C_W}{4\pi\left(N - \sqrt{N^2 - W^2 C_W^2}\right)} \tag{10.57}$$

and the lower bound of the number of basis functions can be found easily from Eq. (10.57) as

$$N > 1.2\sqrt{W^2 T C_W^2 + 6} - 3 \tag{10.58}$$

10.2.1.2.4. Summary. Properties of all three basis functions are summarized in Table 10.1. The upper and lower bounds for l_1 along with the lower bound of N also are given in Table 10.1. Given an EM response with a time support T and one-sided bandwidth W, and for a fixed number N of these basis functions, the Laguerre basis functions have the widest range of l_1 for an accurate approximation of the desired response. The Bessel–Chebyshev basis functions have the narrowest range of l_1. From the bounds on l_1, the lower bounds on N (except for Hermite functions) can be obtained, which also are shown in Table 10.1. Because the lower bound of l_1 is determined by the properties of the basis functions in the time domain, and the upper bound of l_1 is determined from the properties in the frequency domain, the optimal scaling factor is expected to be somewhere in the middle of the two bounds.

An approximate value can be obtained from the point where the lower and upper bounds are equal. In this case, it is easier to obtain the optimal scaling factor for the Bessel–Chebyshev functions, simply as $1/2\pi W$, which is its upper bounds.

The bounds shown in Table 10.1 now are validated through numerical examples presented in the next section.

Table 10.1 The basis functions and the bounds for l_1

Hermite	Basis functions	$\phi_n(t,l_1) = \left(1/\sqrt{l_1}\right) h_n(t,l_1)$
		$\Phi_n(f,l_2) = (-i)^n \left(1/\sqrt{l_2}\right) h_n(f/l_2)$
	Bounds for l_1	$\dfrac{T}{2\left(\sqrt{\pi N/1.7}+1.8\right)} < l_1 < \dfrac{\sqrt{\pi N/1.7}+1.8}{2\pi W}$
	Bounds for N	$N > 2WT$
Bessel–Chebyshev	Basis functions	$\phi_n(t,l_1) = (t/l_1)^{-1} J_n(t,l_1)$
		$\Phi_n(f,l_2) = \dfrac{(-i)^n}{n}\left\{ i\left[1-(f/l_2)^2\right]^{1/2} U_{n-1}(f/l_2) + T_n^C(f/l_2) \right\}$
	Bounds for l_1	$\dfrac{T}{1.08N+9} < l_1 < \dfrac{1}{2\pi W}$
	Bounds for N	$N > \dfrac{2\pi WT - 9}{1.08}$
Laguerre	Basis functions	$\phi_n(t,l_1) = \left(1/\sqrt{l_1}\right) e^{-t/2l_1} L_n(t/l_1)$
		$\Phi_n(f,l_2) = \dfrac{\left[-\frac{1}{2}+jf/l_2\right]^n}{\sqrt{2\pi l_2}\left(\frac{1}{2}+jf/l_2\right)^{n+1}}$
	Bounds for l_1	$\dfrac{T}{4.1N+25} < l_1 < \dfrac{C_W}{4\pi\left(N-\sqrt{N^2-W^2C_W^2}\right)}$
	Bounds for N	$N > 1.2\sqrt{W^2 TC_W^2 + 6} - 3$

10.2.1.3 Numerical Examples.

Example 1. In this example, a microstrip band-pass filter is chosen as the structure of interest. It is shown in Figure 10.4. We want to obtain its broadband response by using early time and low frequency data for the S-parameter. The microstrip band-pass filter is designed at a center frequency of 4.0 GHz with an 8% bandwidth and is fabricated on a substrate with relative dielectric constant of 3.38. The height of the substrate is 1.5 mm. The other relevant dimensions of the filter are $L_1 = 14.5$ mm, $L_2/L_1 = 1.3$, $L_3/L_1 = 1.16$, $W_3 = 20$ mm, $W_2/W_1 = 1.15$, and $Gap_1/Gap_2 = 3.2$.

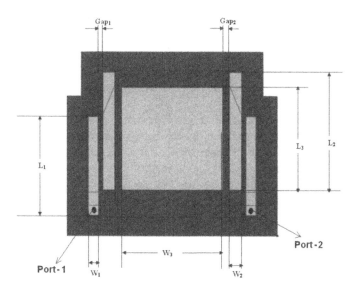

Figure 10.4. A microstrip band-pass filter.

The S parameters for the ports are extrapolated using early-time and low-frequency data and compared with the results obtained from using the MoM [1] solution in the frequency domain and fast Fourier transform (FFT) of the marching-on-in-time solution in the time domain. The extrapolated time domain response is compared with the results obtained using the MOT method [23] and IDFT of the complete frequency domain data. In all the figures displaying the S parameters, the solid lines represent the actual data, and the dots represent the extrapolated data. A vertical line partition the data. The left side of the vertical line shows the original data and the extrapolated data on the right side. The tolerance level is set to $tol = 10^{-3}$ when Eq. (10.6) is solved by using the SVD. When l_1 is larger than the upper bound, more errors will occur in the frequency domain, and when l_1 is smaller than the lower bound, more errors will occur in the time domain. The average of the time and frequency domain errors are defined as

$$E = \frac{E_f + E_t}{2}$$ (10.59)

to express the ultimate performance of the extrapolation.

Each port of the filter is excited by a voltage with the shape of a Gaussian pulse. The parameter of the Gaussian pulse used in this example is $\sigma = 0.4$ ns and $t_0 = 0.6$ ns. The parameter S_{11} is shown by the solid line in Figure 10.5 in the time domain, and the real and imaginary parts of the frequency domain data are shown in Figure 10.6.

From the figures, it is shown that the time support and the bandwidth of the original data are $T \approx 3.5$ ns, and $W \approx 9$ GHz. The sampling time is $\Delta t = 0.024$ ns and the frequency step is $\Delta f = 40$ MHz. The first 60 time domain points are taken as the early-time data, and the first 110 frequency domain points as the low-frequency data. They are extrapolated up to 200 points (5 ns) in time domain and 250 points (10 GHz) in the frequency domain. The performance of the extrapolation with respect to l_1 and N is shown in Figures 10.7–10.9 in terms of log of E defined by Eq. (10.59).

To delineate the range of convergence more clearly, all errors larger than 1 are restricted to 1 and are indicated by the lightly shaded areas. The dashed lines on top of the shaded area in the plots are for the upper bounds, and the solid lines correspond to the lower bounds of the scale factors. The optimal values of l_1 and N are used. They are obtained through the plots of Figures 10.7–10.9. These optimal values are listed in Table 10.2.

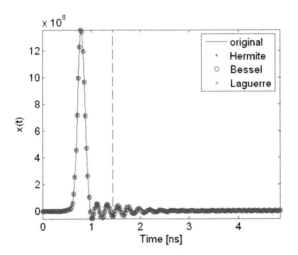

Figure 10.5. S_{11} of the filter in the time domain.

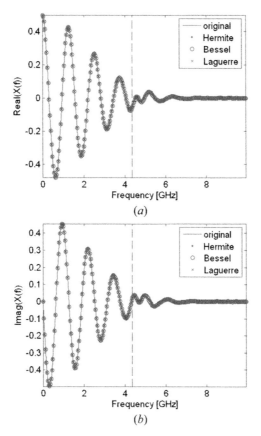

Figure 10.6. S_{11} of the filter in the frequency domain, for (a) real and (b) imaginary parts.

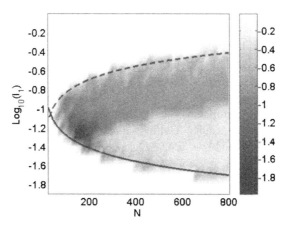

Figure 10.7. The range of l_1 when employing the associated Hermite polynomials. This range can be estimated by the upper (dashed) and the lower (solid) lines.

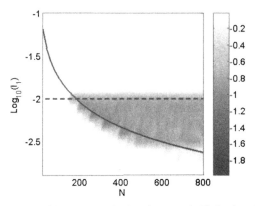

Figure 10.8. The range of l_1 when employing the Bessel–Chebyshev functions. This range can be estimated by the upper (dashed) and the lower (solid) lines.

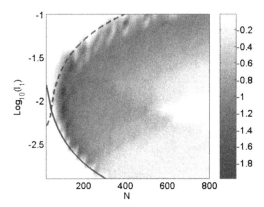

Figure 10.9. The range of l_1 when employing the associated Laguerre functions. This range can be estimated by the upper (dashed) and the lower (solid) lines.

From the three figures the following conclusions can be drawn:

1. The Laguerre basis function has the widest range of convergence, whereas the Bessel–Chebyshev has the narrowest one.
2. The lower and the upper bounds calculated from Table 10.1 are accurate in estimating the range of convergence.
3. The optimal scaling factor is close to the point of intersection of these two bounds.

The estimated optimum values of the parameters for the three basis functions are shown in Table 10.2.

Figure 10.10 plots the coefficients for the three basis functions. It shows that the coefficients display an oscillatory behavior for the lower orders of the Hermite and Laguerre basis functions and approach zero for higher orders. The coefficients for the Bessel–Chebyshev basis functions shown in Figure 10.10 have similar shape as the time domain data of Figure 10.5. To obtain the best performance, we use a large value of N for the Bessel–Chebyshev basis functions, whereas the Laguerre basis functions require smaller values of N. The extrapolated data obtained by the three choices of basis functions also are shown in Figure 10.5 in the time domain and in Figure 10.6 for frequency domain, using the optimal parameters listed in Table 10.2. These plots show that the extrapolated data match the original data well. The simulation of S_{12} provides similar results, which are not shown.

Table 10.2 Optimal values for l_1 and N used in the extrapolation

	Hermite	Bessel	Laguerre
l_1	3×10^7	6×10^6	4.7×10^6
N	150	200	120

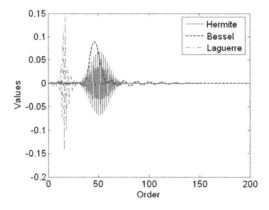

Figure 10.10. Value of the coefficients for the three basis functions using the optimal scaling factors.

Example 2. In this example, we consider the S_{11} parameter of a planar dipole antenna, which is shown in Figure 10.11.

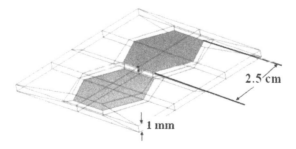

Figure 10.11 A planar dipole antenna with a frequency range of operation between 20 MHz and 10 GHz.

The antenna is made up of two hexagonal elements of sides 1.43 cm, which are connected by the feed wire of length 1 mm and diameter of 0.01 mm at the base edge. The antenna is placed on either side of the dielectric of 3.36 and has a thickness of 1 mm. The parameter of the Gaussian pulse used in this example is $\sigma = 0.4$ ns and $t_0 = 1.2$ ns. The S_{11} parameter is shown by the solid line in Figure 10.12 in the time domain, and the real and imaginary parts in the frequency domain are shown in Figure 10.13. From the figures it can be observed that the time support and the bandwidth of the original data are $T \approx 4$ ns and $W \approx 10$ GHz. The sampling time is $\Delta t = 0.024$ ns, and the frequency step is $\Delta f = 20$ MHz. The first 65 time domain points are taken as the early-time data, and the first 100 frequency domain points are taken as the low-frequency data. They are extrapolated up to 200 points (5 ns) in the time domain and 500 points (10 GHz) in the frequency domain. The performance of the extrapolation with respect to l_1 and N is shown in Figures 10.14–10.16 in terms of log of E defined in Eq. (10.59). One can draw similar conclusions for Example 2 as in Example 1.

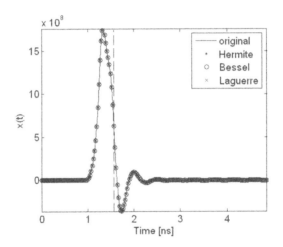

Figure 10.12. S_{11} of the planar dipole in the time domain.

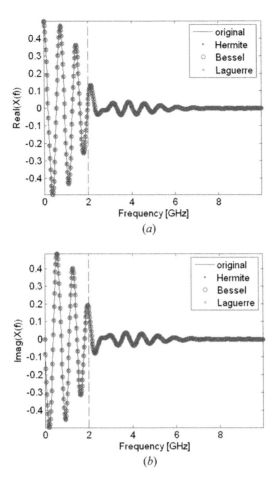

Figure 10.13. S_{11} of the planar dipole in the frequency domain, for the (a) real and (b) imaginary parts.

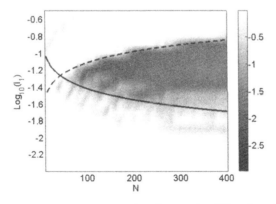

Figure 10.14. The range of l_1 when employing the associated Hermite polynomials. This range can be estimated by the upper (dashed) and the lower (solid) lines.

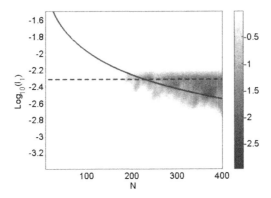

Figure 10.15. The range of l_1 when employing the Bessel–Chebyshev functions. This range can be estimated by the upper (dashed) and the lower (solid) lines.

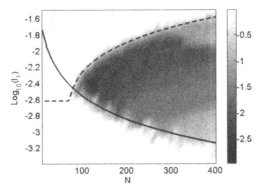

Figure 10.16. The range of l_1 when employing the associated Laguerre functions. This range can be estimated by the upper (dashed) and the lower (solid) lines.

From this example, one can discuss the memory usage and the CPU time gain that can be achieved by using the hybrid method across a purely time domain or a frequency domain computational methodology. The dominant computational load in the hybrid method is in the computation of the SVD of the matrix in Eq. (10.6). It is a real matrix with dimension $(M_1 + 2M_2) \times N$. Because M_1, M_2, and N are not large (usually around several hundreds), the CPU time for the extrapolation is several seconds, which is negligible compared with the solution time required for only the time domain or the frequency domain methods. For MoM in the frequency domain, the number of unknowns, U_n, is proportional to the frequency of operation. The memory usage and the CPU time are proportional to U_n^2 and U_n^3 respectively. In this example, $U_n = 515$ when generating the data using the commercial EM software HOBBIES [1] upto 2 GHz, which is considered to be the low-frequency data. The impedance matrix will occupy 4.2 Mb memory, and the problem will be finished in 3 seconds for one frequency. To simulate the structure at the highest frequency (10 GHz), set $U_n = 2500$, which will occupy 100 Mb memory. The problem will execute for 60 seconds per frequency, which is 20 times slower than analyzing the same problem at a low frequency. If a large array of this antenna is simulated, the CPU time gain can be up to 300 times. The memory usage gain is not obvious in the time domain because a fine mesh is used at the highest frequency to obtain the early-time data. However, this matrix can be saved to the hard disk, and the out-of-core solver can be used [1]. In addition, the explicit MOT can be used to achieve the CPU time gain because it does not require matrix inversion. The most important issue in time domain is the instability in late time, and the hybrid method sometimes can address this problem successfully. However, to obtain an accurate solution under all circumstances, the MOD method is preferred over the MOT technique [25].

10.2.2   An Iterative Method to Find the Optimal Parameters

The objective of the hybrid method is to obtain the data for late-time and high-frequency from the known early-time and low-frequency information. This objective can be achieved as the data in early-time and low-frequency contain all the necessary mutually complimentary information to characterize the system electromagnetically, assuming a sufficient record length is available in both domains. In that case, the error of the extrapolated data is computed only in early-time and low-frequency, as the error of the extrapolation across the whole ranges in both domains, and finds the minimum error to get the best performance. The error is computed as follows. Take the Fourier transform of the data in the time domain (both given and extrapolated) to obtain another set of data in the frequency domain. Now compute the difference between the computed low-frequency data from the time domain response and the given low-frequency data as follows:

$$E_f'(l_1,N)=\left\|\Im\left[x_{ex}(t,l_1,N)_{\text{whole span}}\right]_{\text{low frequency}} - X_{\text{given}}(f)\Big|_{\text{low frequency}}\right\|^2 \quad (10.60)$$

where $x_{ex}(t, l_1, N)$ is the given plus extrapolated data in the time domain and $X_{\text{given}}(f)$ is the given data in the frequency domain. $\Im$ is the Fourier transform operator. This error is computed in terms of the $\mathscr{L}_2$ norm from only the low-frequency information. The error associated with the frequency domain data to predict the early-time response is computed in a similar fashion. This is obtained by extrapolating the frequency domain data and inverse Fourier transforming both the given low-frequency and the extrapolated high-frequency information to the time domain and computing the difference between this response and the given early-time response. The error in the time domain is expressed as follows:

$$E_t'(l_1,N)=\left\|\Im^{-1}\left[X_{ex}(f,l_2,N)_{\text{whole span}}\right]_{\text{early time}} - x_{\text{given}}(t)\Big|_{\text{early time}}\right\|^2 \quad (10.61)$$

where $X_{ex}(t, l_1, N)$ is the extrapolated data in the frequency domain plus the given low-frequency data, and $x_{\text{given}}(t)$ is the given early-time data. Either Eqs.

(10.60) or (10.61) can be used to find the optimal value for l_1 and N when the lowest error is achieved. This is accomplished by carrying out the computations for all possible set is of parameters and then computing the following sum:

$$E'(l_1, N) = E'_f(l_1, N) + E'_t(l_1, N) \tag{10.62}$$

as the two errors in the time and frequency domains are complementary. It is important to note, however, that the error given by Eq. (10.62) also can be computed using either the $\mathscr{L}_1$ or the $\mathscr{L}_\infty$ norm. From a computational standpoint, the matrix containing the unknown coefficients in Eq. (10.6) can be computed at the highest value of N once, for every l_1, to avoid repeated computations. Each iteration will undertake only one matrix inversion. The whole search procedure takes only a fraction of the time as compared with the solution of the original electromagnetics problem caused by MoM or MOT/MOD, especially for evaluating ultra-wide band responses of very complicated structures. Using Eq. (10.62), one can estimate l_1 and N, which can yield values very close to the optimal ones. Simulation results presented later in section 10.5 will validate the concept of this method.

In summary, the optimal choice of the scaling parameter l_1 and N can be obtained using the following recipe:

1. Determine the minimum degree $N_{\min}$ as specified in Table 10.1. There is no need to know the exact value of T and W of the data. A range of the values of N from $N_{\min}$ can be chosen for the numerical search.
2. Use the error bound as given in Table 10.1 to determine the range of l_1 in which the hybrid method will converge.
3. Search for the optimal values of l_1 and N by minimizing the error in Eq. (10.62), by using the range of values of l_1 obtained at step 2 (and the range of N obtained in step 1).

Each of these three steps can be programmed easily and automatically, and the computational cost is a minimum, as a small-dimensioned matrix equation of Eq. (10.6) is solved. Here, the input required is the data specified in the early-time and low-frequency, the sampling interval in time and frequency, ΔT and ΔF, respectively, and the approximate time span T and the bandwidth W of the data.

Finally, these principles are illustrated using some numerical results. Scattering from a dielectric hemisphere of 0.2 m diameter with $\varepsilon_r = 10$ is considered. The configuration of interest along with the discretization is shown in Figure 10.17. The excitation arrives from the direction $\theta = 0$, $\phi = 0$ (i.e., along the negative z direction). The incident electric field is a Gaussian pulse, and it is polarized along the x-axis. A frequency domain MoM EFIE formulation [1] is used to compute the response in the low-frequency region. Also used is a MOT time domain EFIE code [25] to obtain the early-time response. The time step used in the MOT program is 266.7 ps. In this example, $\sigma = 2$ ns and $t_0 = 20$ ns. E_0 is chosen to be 377 V/m. The time step (Δt) is dictated by the discretization used in modeling of the geometry. The conjugate gradient method (CGM) has been used to solve the matrix in Eq. (10.6) until the residue goes below 10^{-6}. The extrapolated time domain response is compared with the output of the time domain response obtained from the MOT program [25], and the frequency domain response is compared with that of the MoM code [1] and the discrete Fourier transform of the time domain results. In all plots, the extrapolated signal refers to the extrapolated response using all polynomial expansions, whereas the original signal refers to the data obtained from the MOT or MoM codes. Also, the vertical dotted lines in all figures indicate beyond which extrapolations have been performed. The normalized error for the time and frequency domain data is defined by Eq. (10.7).

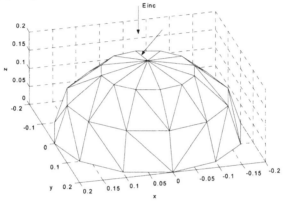

Fig. 10.17 Discretization of a dielectric hemisphere.

Using the MOT algorithm, the time domain data for the electric and the magnetic currents is obtained from $t = 0$ to $t = 21.33$ ns (80 data points). And the frequency domain data is obtained from DC to $f = 900.0$ MHz (143 data points). The results for only the electric current are plotted, as the result for the magnetic current is similar. Assume that only the first 32 time-data points (up to $t = 8.533$ ns) and first 42 frequency-data points (up to $f = 264.3$ MHz) are available. Solving the matrix equation, Eq. (10.6), using the available data, the time domain response is extrapolated to 80 points (21.33 ns) and the frequency domain response is extrapolated to 143 points (900.0 MHz).

Using the dimensionality theorem, choose N for this data set to be 42. To examine how the error scales as a function of N, change N from 5 to 200 with an increment of 5. Hence, for a fixed scale factor, l_1, simulations using 40 different values of N are carried out. The values of the scale factor are set as follows. The range of l_1 for the Laguerre and Bessel polynomials are varied from $1.5\Delta T \times 10^{\{0.1-1\}} \sim 1.5\Delta T \times 10^{\{20 \times 0.1-1\}}$. The range of l_1 for Hermite polynomials are varied from $1.5\Delta T \times 10^{0.1} \sim 1.5\Delta T \times 10^{\{20 \times 0.1\}}$, as for the Hermite polynomials it is larger than that of the other two polynomials. This implies that the scale factor l_1 chosen at each step is incremented by the factor $10^{0.1}$.

Figures 10.18, 10.19, and 10.20 plot the log of the total error as a function of the scale parameter l_1 and the order of the expansion N for the three different types of polynomials. The corresponding coefficients associated with the various polynomials, given by the sequence $\{a_n\}$ is plotted in Figure 10.21 for the Hermite, Figure 10.22 for the Laguerre, and Figure 10.23 for the Bessel. Using these coefficients, the extrapolation is made for late times and high frequencies for the electric current at the location shown by the arrow in Figure 10.17.

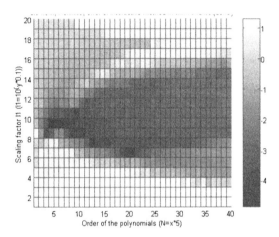

Figure 10.18. Log of the total error in the combined time and frequency domains (Hermite).

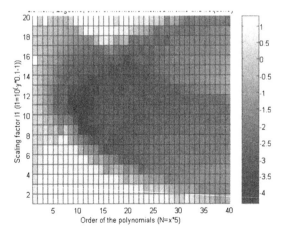

Figure 10.19. Log of the total error in the combined time and frequency domains (Laguerre).

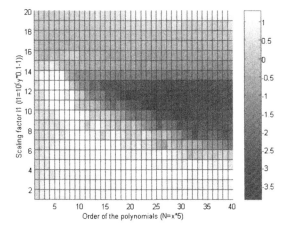

Figure 10.20. Log of the total error in the combined time and frequency domains (Bessel).

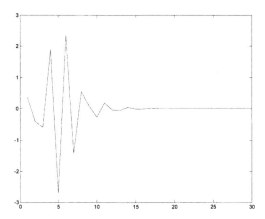

Figure 10.21. Value of the coefficients generated using the estimated l_1 and N (Hermite).

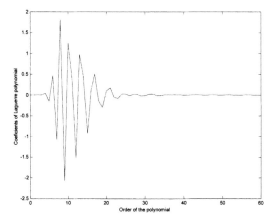

Figure 10.22. Value of the coefficients generated using the estimated l_1 and N (Laguerre).

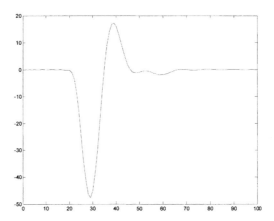

Figure 10.23. Value of the coefficients generated using the estimated l_1 and N (Bessel).

Figure 10.24 in the time domain and Figures 10.25(a) and 10.25(b) for the real and imaginary parts of the current in the frequency domain display the result of the extrapolation using the various polynomials. Because the results of the extrapolation for all three methods are virtually indistinguishable to the naked eye, only one has been plotted for illustration purposes. As is shown the extrapolated response shown by dots is virtually indistinguishable from the actual response in either domain.

Figures 10.26(a) and 10.26(b) plot the error associated with the extrapolation as a function of the choice of the order of the number of expansion functions N at the optimum scaling l_1 in both time and frequency domains, respectively, for all three polynomials. From a numerical point of view, it seems that the optimum Laguerre set shows the best performance compared with the other two.

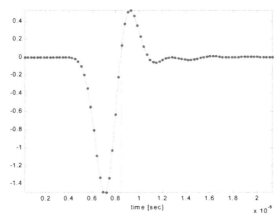

Figure 10.24. Both given and extrapolated response in the time domain using the three orthogonal functions generated from the polynomials (computed using a numerical electromagnetic code: solid line, extrapolated response: dots).

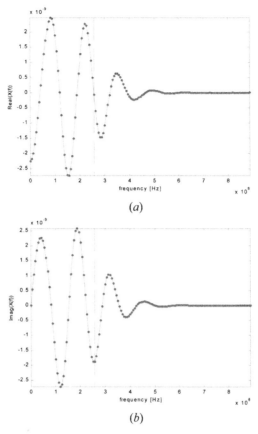

Figure 10.25. Both given and extrapolated response for (a) real and (b) imaginary part in the frequency domain using the three orthogonal functions generated from the polynomials (computed using a numerical electromagnetic code: solid line, extrapolated response: dots).

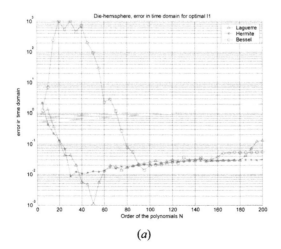

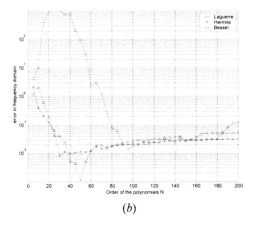

(b)

Figure 10.26. Error in the (a) time and (b) frequency domain plotted against the order of the polynomial N for the optimum scale factor l_1.

10.3   A NON-PARAMETRIC HYBRID METHODOLOGY

In this section, we look at nonparametric methods to carry out the simultaneous extrapolation in the time and frequency domains using other mathematical principles that need not be specified by certain parameters. For example, Adve and Sarkar [4] developed an alternating projection method that enables a simultaneous extrapolation of the early-time response in the time domain and the low-frequency response in the frequency domain. The late-time domain solution may be extrapolated further using the Matrix Pencil technique or the Hilbert transform. In the present section, it is observed that the alternating iterative method provided by Adve and Sarkar is identical to the Neumann series solution of a second-kind integral equation. Here it is demonstrated that this Neumann series solution is an iterative error-reducing method.

It is anticipated that a more efficient iterative method is the conjugate gradient method. For this procedure, it is shown that the governing operator is self-adjoint and positive. The operator can be implemented by using the FFT and the IFFT to improve the computational efficiency as will be illustrated.

The integral equations to be developed in this nonparametric solution procedure are equivalent to the solution of a set of matrix equations. The extrapolation then can be achieved by directly solving the matrix equation. Although this matrix is ill-conditioned and rank deficient, good performance can be achieved by using the SVD because of the special properties of the extrapolation. Because small matrices are dealt with in this extrapolation procedure, extraction of the early-time and low-frequency data by using the SVD is also computationally efficient.

The problem formulation is provided in section 10.3.1, and then the integral equation of the second kind is derived in section 10.3.2. The three direct methods, Neumann iteration, conjugate gradient, and SVD are discussed separately in sections 10.3.3–10.3.5. Finally, three more examples related to EM scattering are presented to validate the methods, and the performances of the three methods are compared in section 10.3.6.

10.3.1 Formulation of the Problem

Assume that the Fourier transform of a causal function $f(t)$, $t \in \Re^+$, is given by

$$\hat{f}(\omega) = \Im\{f\}(\omega) = \int_{t \in \Re^+} \exp(-i\omega t)\, f(t)\, dt, \qquad \omega \in \Re \tag{10.63}$$

The causal function $f(t)$, is easily found from the inverse Fourier transform,

$$f(t) = \Im^{-1}\{\hat{f}\}(t) = \frac{1}{\pi}\, \mathrm{Re} \int_{\omega \in \mathrm{IR}^+} \exp(i\omega t)\, \hat{f}(\omega)\, d\omega, \qquad t \in \mathrm{IR}^+ \tag{10.64}$$

where $\mathrm{Re}[\bullet]$ denotes real part of. Furthermore, from Parseval's relation for causal functions, the following is obtained:

$$\int_{t \in \Re^+} f(t)\, g^*(t)\, dt \;=\; \frac{1}{\pi}\, \mathrm{Re} \int_{\omega \in \Re^+} \hat{f}(\omega)\, \hat{g}^*(\omega)\, d\omega \tag{10.65}$$

which will be used extensively in this section. The asterisk denotes the complex conjugate. Now consider when $\hat{f}(\omega)$ is known only in a bounded domain Ω, a subdomain of $\Re^+$. Denote the known functional values in Ω as $\hat{f}_\Omega(\omega)$. This results in,

$$\hat{f}(\omega) = \hat{f}_\Omega(\omega), \qquad \omega \in \Omega \tag{10.66}$$

Furthermore, assume that

$$f(t) = f_T(t), \qquad t \in T \tag{10.67}$$

where $f_T(t)$ is a known function in the bounded subdomain T of $\Re^+$. Adve and Sarkar [4] have developed an iterative scheme to determine $f(t)$ for all $t \in \Re^+$, or $\hat{f}(\omega)$ for all $\omega \in \Re^+$. To show that this iterative scheme is nothing but a Neumann expansion of an integral equation of the second kind, the reconstruction problem is formulated through an integral equation of the first kind as,

$$\int_{t \in \Re^+} \exp(-i\omega t)\, f(t)\, dt \;=\; \hat{f}_\Omega(\omega), \qquad \omega \in \Omega \tag{10.68}$$

Because $f(t) = f_T(t)$ is known for $t \in T$, Eq. (10.68) is rewritten as

$$\int_{t \in \Re^+ \backslash T} \exp(-i\omega t) f(t)\, dt \;=\; \hat{h}_\Omega(\omega), \qquad \omega \in \Omega \tag{10.69}$$

with

$$\hat{h}_\Omega(\omega) = \hat{f}_\Omega(\omega) - \Im\{\chi_T f_T(t)\}, \qquad \omega \in \Omega \tag{10.70}$$

The characteristic function of the set T is introduced as

$$\chi_T(t) = \{1, \tfrac{1}{2}, 0\}, \qquad t \in \{T, \partial T, T'\} \tag{10.71}$$

where ∂T denotes the boundary of the domain T, and T' denotes the complement of the union of T and ∂T. Similarly, the characteristic function of the set Ω is defined as

$$\chi_\Omega(\omega) = \{1, \tfrac{1}{2}, 0\}, \qquad \omega \in \{\Omega, \partial\Omega, \Omega'\} \tag{10.72}$$

where $\partial\Omega$ denotes the boundary of the domain Ω and Ω' denotes the complement of the union of Ω and $\partial\Omega$. Eq. (10.69) is an integral equation of the first kind with the known function $\hat{h}_\Omega(\omega)$, the kernel $\exp(-i\omega t)$, and the unknown function $f(t)$. This integral equation is ill-posed. (In section 10.3.5, it will be shown that the SVD provides a good way to solve the discrete version of Eq. (10.69).) Therefore, it is transferred to an integral equation of the second kind for the values of $f(t) \in \Re^+ \setminus T$.

For convenience, the inner products and the norms of two functions are introduced, both in $t-$ space and in $\omega-$ space, through

$$\langle f, g \rangle_{t \in T} = \int_{t \in T} f(t)\, g^*(t)\, dt, \qquad \|f\|^2_{t \in T} = \langle f, f \rangle_{t \in T} \tag{10.73}$$

$$\langle \hat{f}, \hat{g} \rangle_{\omega \in \Omega} = \frac{1}{\pi} \mathrm{Re} \int_{\omega \in \Omega} \hat{f}(\omega)\, \hat{g}^*(\omega)\, d\omega, \qquad \|\hat{f}\|^2_{\omega \in \Omega} = \langle \hat{f}, \hat{f} \rangle_{\omega \in \Omega} \tag{10.74}$$

Note that using Parseval's relation, Eq. (10.65) may be written as

$$\langle f, g \rangle_{t \in \Re^+} = \langle \hat{f}, \hat{g} \rangle_{\omega \in \Re^+} \tag{10.75}$$

10.3.2   Integral Equation of the Second Kind

The integral equation of the first kind given by Eq. (10.69) is written in operator form as

$$Af = \hat{h}_\Omega, \qquad \omega \in \Omega \tag{10.76}$$

where the operator $Af = Af(\omega)$ is given by

$$Af = \Im\{(1-\chi_T)f\} \tag{10.77}$$

The conversion of the first kind integral equation to that of the second kind is similar to the conversion used by Van den Berg and Kleinman [26], Fokkema and Van den Berg ([27], section 2.7) for convolution type operators, and Rhebergen et al. [28]. First note that the operator Eq. (10.76) only holds for $\omega \in \Omega$. Therefore the equation is rewritten to one that holds for all $\omega \in \Re^+$, viz.

$$Af = \chi_\Omega \hat{h}_\Omega + (1-\chi_\Omega)Af, \qquad \omega \in \Re^+ \tag{10.78}$$

Using Eq. (10.77), one can obtain

$$\Im\{(1-\chi_T)f\} = \chi_\Omega \hat{h}_\Omega + (1-\chi_\Omega)\,\Im\{(1-\chi_T)f\}, \qquad \omega \in \Re^+ \tag{10.79}$$

Because, this equation holds for all $\omega \in \Re^+$, the inverse Fourier transformation is taken on both sides to arrive at

$$(1-\chi_T)f = \Im^{-1}\{\chi_\Omega \hat{h}_\Omega\} + \Im^{-1}\{(1-\chi_\Omega)\Im\{(1-\chi_T)f\}\}, \qquad t \in \Re^+ \tag{10.80}$$

By considering $t \in \Re^+ \setminus T$ only, this equation becomes an integral equation of the second kind, and in operator notation, it is written as

$$f = g + Kf, \qquad t \in \Re^+ \setminus T \tag{10.81}$$

where

$$g = \Im^{-1}\{\chi_\Omega \hat{h}_\Omega\} \tag{10.82}$$

is the known function and

$$Kf = \Im^{-1}\{(1-\chi_\Omega)\Im\{(1-\chi_T)f\}\} \tag{10.83}$$

is the kernel operator.

10.3.3 The Neumann Expansion

First, observe that the integral equation of the second kind, Eq. (10.71), and can be solved iteratively using the Neumann iterative solution as

$$f_0 = 0, \qquad f_n = g + Kf_{n-1}, \qquad n = 1, 2, \cdots \tag{10.84}$$

At the nth step, the residual error is defined as

$$r_n = g + Kf_n - f_n, \qquad t \in \Re^+ \backslash T \tag{10.85}$$

and the residual errors in two subsequent iterations are related to each other by

$$r_n = Kr_{n-1}, \qquad t \in \Re^+ \backslash T \tag{10.86}$$

To prove that the iterative Neumann solution leads to an error reduction at each iteration, one observes that

$$\|r_n\|_{t \in \Re^+ \backslash T} = \|Kr_{n-1}\|_{t \in \Re^+ \backslash T} \le \|Kr_{n-1}\|_{t \in \Re^+} \tag{10.87}$$

and using the definition of Kf, Parseval's relation, and Eq. (10.75), one can obtain

$$\begin{aligned}
\|Kr_{n-1}\|^2_{t \in \Re^+} &= \left\| \Im^{-1}\{(1-\chi_\Omega)\Im\{(1-\chi_T)r_{n-1}\}\} \right\|^2_{t \in \Re^+} \\
&= \left\| \Im\{(1-\chi_T)r_{n-1}\} \right\|^2_{\omega \in \Re^+ \backslash \Omega}
\end{aligned} \tag{10.88}$$

Again, using Parseval's relation, it is observed that

$$\|r_{n-1}\|^2_{t \in \Re^+ \backslash T} = \|(1-\chi_T)r_{n-1}\|^2_{t \in \Re^+} = \left\| \Im\{(1-\chi_T)r_{n-1}\} \right\|^2_{\omega \in \Re^+} \tag{10.89}$$

Hence,

$$\frac{\|r_n\|^2_{t \in \Re^+ \backslash T}}{\|r_{n-1}\|^2_{t \in \Re^+ \backslash T}} \le \frac{\left\| \Im\{(1-\chi_T)r_{n-1}\} \right\|^2_{\omega \in \Re^+} - \left\| \Im\{(1-\chi_T)r_{n-1}\} \right\|^2_{\omega \in \Omega}}{\left\| \Im\{(1-\chi_T)r_{n-1}\} \right\|^2_{\omega \in \Re^+}} \tag{10.90}$$

The second quantity in the numerator of Eq. (10.90) is always nonzero (i.e., the energy in any bandwidth in the spectrum of a truncated function is always nonzero), and it can be directly observed that

$$\|r_n\|^2_{t \in \Re^+ \backslash T} < \|r_{n-1}\|^2_{t \in \Re^+ \backslash T} \tag{10.91}$$

This proves that the Neumann iterative scheme reduces the $\mathscr{L}_2$ residual norm in each step. From the operator expression, Eq. (10.83), directly observe that the Neumann scheme converges faster when Ω tends to $\Re^+$ and/or T tends to $\Re^+$. For these cases, $\|Kf\| \to 0$.

A schematic representation of the Neumann iteration is given in Table 10.3 on how to update for f_n, $t \in \Re^+ \backslash T$. The down arrows in Table 10.3 are operations in the same domain. The right and left arrows represent the Fourier and inverse Fourier transforms. The Fourier transform is performed on the late-time data f_n, and add the high-frequency component of the transform with the known low-frequency component. The scheme can be modified to compute updates for f_n, $t \in \Re^+$, by observing that $\tilde{h}_\Omega = \tilde{f}_\Omega - F\{\chi_T f_T\}$ and

$$f_n = (1-\chi_T)f_n + \chi_T f_T, \quad t \in \Re^+ \tag{10.92}$$

This scheme is presented in Table 10.4. Obviously, this latter scheme is identical to the one described in Ref. [2].

Table 10.3 Neumann iterative scheme for f_n, $t \in \Re^+ \backslash T$

t -domain	$\longleftrightarrow$	ω -domain
f_{n-1}		
$\downarrow$		
$\times(1-\chi_T)$	$\longrightarrow$	$\times(1-\chi_\Omega)$
		$\downarrow$
f_n	$\longleftarrow$	$+\chi_\Omega \hat{h}_\Omega$

Table 10.4 Iterative scheme for f_n, $t \in \Re^+$

t -domain	$\longleftrightarrow$	ω -domain
f_{n-1}		
$\downarrow$		
$\times(1-\chi_T)$		
$\downarrow$		
$+\chi_T f_T$	$\longrightarrow$	$\times(1-\chi_\Omega)$
		$\downarrow$
f_n	$\longleftarrow$	$+\chi_\Omega \hat{h}_\Omega$

It is also observed that the Neumann iterative solution can be written as

$$f = \sum_{n=0}^{\infty} K^n g \tag{10.93}$$

which is a representation in the Krylov subspace spanned by $\{K^n g, n = 0, 1, 2, ...\}$. If the error reduction of the Neumann iteration is very slow, then it can be speeded up by using more advanced Krylov-subspace methods (e.g., over-relaxation and successive over-relaxation methods [29]). Experience has taught that a conjugate gradient method can be used advantageously.

10.3.4   The Conjugate Gradient Method

Before the conjugate gradient (CG) method is discussed, the integral equation of Eq. (10.81) can be rewritten as

$$Lf = g, \qquad t \in \mathfrak{R}^+ \backslash T \tag{10.94}$$

with $L = I - K$, where I is the identity operator. Now, observe

$$Lf = \mathfrak{S}^{-1} \{ \chi_\Omega \, \mathfrak{S} \{ (1 - \chi_T) f \} \} \tag{10.95}$$

This operator is equivalent to

$$Lf = \int_{t' \in T} C(t - t') f(t') dt' \tag{10.96}$$

where

$$C(t) = \begin{cases} \mathfrak{S}^{-1} \{ \chi_\Omega \}, & t \geq 0 \\ 0, & t < 0 \end{cases} \tag{10.97}$$

For the CG method to be discussed, the adjoint operator L^H of L is defined through

$$\langle Lf, g \rangle_{t \in \mathfrak{R}^+ \backslash T} = \langle f, L^H g \rangle_{t \in \mathfrak{R}^+ \backslash T} \tag{10.98}$$

Using Eq. (10.95) in Eq. (10.98) and the Parseval's relation results in

$$
\begin{aligned}
\langle Lf, g \rangle_{t \in \mathfrak{R}^+ \backslash T} &= \langle \mathfrak{S}^{-1} \{ \chi_\Omega \, \mathfrak{S} \{ (1 - \chi_T) f \} \}, (1 - \chi_T) g \rangle_{t \in \mathfrak{R}^+} \\
&= \langle \chi_\Omega \, \mathfrak{S} \{ (1 - \chi_T) f \}, \mathfrak{S} (1 - \chi_T) g \} \rangle_{\omega \in \mathfrak{R}^+} \\
&= \langle \mathfrak{S} \{ (1 - \chi_T) f \}, \chi_\Omega \, \mathfrak{S} \{ (1 - \chi_T) g \} \rangle_{\omega \in \mathfrak{R}^+} \\
&= \langle (1 - \chi_T) f, \mathfrak{S}^{-1} \{ \chi_\Omega \, \mathfrak{S} \{ (1 - \chi_T) g \} \} \rangle_{t \in \mathfrak{R}^+} \\
&= \langle f, Lg \rangle_{t \in \mathfrak{R}^+ \backslash T} \tag{10.99}
\end{aligned}
$$

which shows that L is self-adjoint ($L^H = L$). Subsequently, it will be proven that the operator L is positive by showing that the inner product

$$
\begin{aligned}
\langle Lf,f\rangle_{t\in\Re^{+}\backslash T} &= \left\langle \Im^{-1}\{\chi_{\Omega}\Im\{(1-\chi_{T})f\}\}, (1-\chi_{T})f\right\rangle_{t\in\Re^{+}} \\
&= \left\langle \chi_{\Omega}\,\Im\{(1-\chi_{T})f\},\,\Im\{(1-\chi_{T})f\}\right\rangle_{\omega\in\Re^{+}} \\
&= \left\|\Im\{(1-\chi_{T})f\}\right\|^{2}_{\omega\in\Omega} > 0
\end{aligned}
\tag{10.100}
$$

for $f \neq 0$ everywhere in $\Re^{+}\backslash T$. Convergence of the conjugate gradient method (CG) for positive operators has been established in Ref. [30]. The conjugate gradient iterative scheme to solve Eq. (10.94) is given subsequently. It minimizes the inner product $\langle f - f_n, L(f - f_n)\rangle_{t\in\Re^{+}\backslash T}$. The initial estimate is started with

$$
f_0 = 0, \qquad r_0 = h = F^{-1}\{\chi_{\Omega}\hat{f}_{\Omega}\}
$$
$$
\|r_0\|_{t\in\Re^{+}\backslash T} = \|h\|_{t\in\Re^{+}\backslash T}
\tag{10.101}
$$

Next the scheme chooses

$$
\begin{aligned}
v_1 &= r_0 \\
\alpha_1 &= \|r_0\|^{2}_{t\in\Re^{+}\backslash T} / \langle v_1, Lv_1\rangle_{t\in\Re^{+}\backslash T} \\
f_1 &= f_0 + \alpha_1 v_1 \\
r_1 &= r_0 - \alpha_1 Lv_1
\end{aligned}
\tag{10.102}
$$

and computes successively for $n = 2, 3, \ldots$

$$
\begin{aligned}
v_n &= r_{n-1} + \left(\|r_{n-1}\|^{2}_{t\in\Re^{+}\backslash T} / \|r_{n-2}\|^{2}_{t\in\Re^{+}\backslash T}\right) v_{n-1} \\
\alpha_n &= \|r_{n-1}\|^{2}_{t\in\Re^{+}\backslash T} / \langle v_n, Lv_n\rangle_{t\in\Re^{+}\backslash T} \\
f_n &= f_{n-1} + \alpha_n v_n \\
r_n &= r_{n-1} - \alpha_n Lv_n
\end{aligned}
\tag{10.103}
$$

in which

$$
Lv_n = \Im^{-1}\{\chi_{\Omega}\Im\{(1-\chi_{T})v_n\}\}
\tag{10.104}
$$

In this CG scheme, the operator Lv_n only consists of the forward and inverse Fourier transforms of causal functions, defined in Eqs. (10.63)–(10.64). In computer simulations, the FFT and IFFT [31] can be used to implement the Fourier transform. Because it is not possible to deal with FFT or IFFT of the data f, which has infinite length, it is assumed that f has a finite support W_t. For $t > W_t$, it is assumed that $f(t) \approx 0$. It is reasonable to assume that this is true for most EM analysis [13]. Similarly, it also is assumed that f has finite bandwidth so that W_f ($\hat{f}(\omega) \approx 0$ for $|\omega| < W_f$). The latter is always true, as the structures are discretized spatially based on the highest frequency of interest.

10.3.5 Equivalent Matrix Formulations and Solution Using the SVD

It is assumed that discrete samples of the data are available. Apply the discrete Fourier transform (DFT) to the sampled data. The sampling time Δt is assumed to be sufficiently small. Suppose the sampled data sequence of f in the time domain is $\mathbf{x}$, with the length M_t, and the sequence in the frequency domain is $\mathbf{X}$, with the length M_f. The early-time data $\mathbf{x}_1$ is obtained up to M_1 data points. The next step is to extrapolate the late-time data $\mathbf{x}_2$ with the length of $M_t - M_1$. Similarly, extrapolate the high-frequency data $\mathbf{X}_2$ (with the length $M_f - M_1$) using the low-frequency data $\mathbf{X}_1$ (with the length M_1). Ref. [4] discussed how to choose M_1 and M_2 to obtain good performance. Experience shows that a high quality of extrapolation usually can be obtained if we choose the early-time to contain response later in time than the peak in the excitation, and the low-frequency data should be larger in size containing at least the first resonant frequency of the structure. However this restriction is flexible because we can reduce M_1 by increasing M_2. We now define the data sequences as

$$\mathbf{x} = \left\{ f(0), f(\Delta t), \cdots, f((M_1 - 1)\Delta t), f(M_1 \Delta t), \cdots, f((M_t - 1)\Delta t) \right\}^T$$

$$\mathbf{X} = \left\{ \hat{f}(0), \hat{f}(\Delta f), \cdots, \hat{f}((M_2 - 1)\Delta f), \hat{f}(M_2 \Delta f), \cdots, \hat{f}((M_f - 1)\Delta f) \right\}^T \quad (10.105)$$

where Δf is the sampling interval in the frequency domain, and the superscript T denotes the transpose of a vector or matrix. Performing the DFT to a vector $\mathbf{x}$ is equivalent to multiplication of $\mathbf{x}$ with the DFT matrix $\mathbf{K}_{DFT}$. The vectors $\mathbf{X}$ and $\mathbf{x}$ are related by

$$\mathbf{X} = \mathbf{K}_{DFT} \, \mathbf{x} \quad (10.106)$$

where the DFT matrix $\mathbf{K}_{DFT}$ can be defined and partitioned as

$$\mathbf{K}_{DFT} = \begin{bmatrix} 1 & \cdots & 1 & 1 & \cdots & 1 \\ \vdots & \ddots & \vdots & \vdots & \ddots & \vdots \\ 1 & \cdots & e^{-j2\pi t_{M_1-1}f_{M_2-1}} & e^{-j2\pi t_{M_1}f_{M_2-1}} & \cdots & e^{-j2\pi t_{M_t-1}f_{M_2-1}} \\ 1 & \cdots & e^{-j2\pi t_{M_1-1}f_{M_2}} & e^{-j2\pi t_{M_1}f_{M_2}} & \cdots & e^{-j2\pi t_{M_t-1}f_{M_2}} \\ \vdots & \ddots & \vdots & \vdots & \ddots & \vdots \\ 1 & \cdots & e^{-j2\pi t_{M_1-1}f_{M_f-1}} & e^{-j2\pi t_{M_1}f_{M_f-1}} & \cdots & e^{-j2\pi t_{M_t-1}f_{M_f-1}} \end{bmatrix}$$

$$= \begin{bmatrix} \mathbf{B} & \mathbf{A} \\ \mathbf{D} & \mathbf{C} \end{bmatrix} \qquad (10.107)$$

Eq. (10.106) can be rewritten as

$$\begin{bmatrix} \mathbf{B} & \mathbf{A} \\ \mathbf{D} & \mathbf{C} \end{bmatrix} \times \begin{bmatrix} \mathbf{x}_1 \\ \mathbf{x}_2 \end{bmatrix} = \begin{bmatrix} \mathbf{X}_1 \\ \mathbf{X}_2 \end{bmatrix} \qquad (10.108)$$

To reduce the number of unknowns, retain the upper part of Eq. (10.108) and, after some manipulation, obtain

$$\mathbf{A}\mathbf{x}_2 = \mathbf{X}_1 - \mathbf{B}\mathbf{x}_1 \qquad (10.109)$$

It is a matrix equation for the unknown x_2. It is interesting to note that Eq. (10.109) is just the discrete version of the integral equation of the first kind in Eq. (10.76). Similarly, the IDFT matrix can be defined as

$$\mathbf{K}_{IDFT} = \begin{bmatrix} 1 & \cdots & 1 & 1 & \cdots & 1 \\ \vdots & \ddots & \vdots & \vdots & \ddots & \vdots \\ 1 & \cdots & e^{j2\pi t_{M_1-1}f_{M_2-1}} & e^{j2\pi t_{M_1-1}f_{M_2}} & \cdots & e^{j2\pi t_{M_1-1}f_{M_f-1}} \\ 1 & \cdots & e^{j2\pi t_{M_1}f_{M_2-1}} & e^{j2\pi t_{M_1}f_{M_2}} & \cdots & e^{j2\pi t_{M_1}f_{M_f-1}} \\ \vdots & \ddots & \vdots & \vdots & \ddots & \vdots \\ 1 & \cdots & e^{j2\pi t_{M_t-1}f_{M_2-1}} & e^{j2\pi t_{M_t-1}f_{M_2}} & \cdots & e^{j2\pi t_{M_t-1}f_{M_f-1}} \end{bmatrix}$$

$$= \begin{bmatrix} \bar{\mathbf{B}} & \bar{\mathbf{A}} \\ \bar{\mathbf{D}} & \bar{\mathbf{C}} \end{bmatrix} \qquad (10.110)$$

where the overbar marks the matrix and denotes it as a part of the IDFT matrix. Multiplying both sides of Eq. (10.109) with $\bar{\mathbf{D}}$, results in

$$\bar{\mathbf{D}}\mathbf{A}\mathbf{x}_2 = \bar{\mathbf{D}}\mathbf{X}_1 - \bar{\mathbf{D}}\mathbf{B}\mathbf{x}_1 \qquad (10.111)$$

It is the discrete version of Eq. (10.94), and the matrix $\bar{\mathbf{D}}\mathbf{A}$ is exactly the operator L for the discrete sense. Notice that $\mathbf{A}$ has the dimension of $M_2 \times (M_t - M_1)$, $\bar{\mathbf{D}}$ has the dimension of $(M_t - M_1) \times M_2$, so that $\bar{\mathbf{D}}\mathbf{A}$ is a square matrix in $C^{M_2 \times M_2}$. From Eqs. (10.107) and (10.110) we find that $\bar{\mathbf{D}} = \mathbf{A}^*$. It follows that $\bar{\mathbf{D}}\mathbf{A}$ is a Hermitian matrix, which means that the operator $\bar{\mathbf{D}}\mathbf{A}$ is self-adjoint. This already has been shown to be so in section 10.3.4. The eigenvalues of $\bar{\mathbf{D}}\mathbf{A}$ are positive, which follows that the matrix $\bar{\mathbf{D}}\mathbf{A}$ is positive definite. The applicability of the conjugate gradient method thus is established. Furthermore, by looking at any element of the matrix $\overline{\mathbf{D}}\mathbf{A}$, which is given by

$$\overline{d}a_{mn} = \frac{e^{j2\pi\Delta f \Delta t(m-n)M_2} - 1}{e^{j2\pi\Delta f \Delta t(m-n)} - 1} \tag{10.112}$$

it is clear that it is only dependent on $(m-n)$. Hence, $\overline{\mathbf{D}}\mathbf{A}$ is a Toeplitz matrix. Therefore, the CGM-FFT method can be implemented to reduce the computational load. The recipe is similar to that outlined in section 10.3.4, and the result should be exactly the same. Instead of solving Eq. (10.111), in this section, the feasibility of solving Eq. (10.109) is discussed directly.

Matrix $\mathbf{A}$ contains integer powers of the phasor $e^{-j2\pi\Delta t\Delta f}$. It is an ill-conditioned matrix. Furthermore, M_2 necessarily is not equal to $(M_t - M_1)$, so that $\mathbf{A}$ may be rank-deficient. SVD is a stable method for solving ill-conditioned matrix equations in a least-square sense [32]. Among the solutions of Eq. (10.109), SVD will provide the solution of $\overline{\mathbf{x}}_2$ with as small a norm as possible. This property is important for the extrapolation. The feasibility of the extrapolation is based on the assumption that the low-frequency data consists of the late-time information and vice versa. Using SVD, the norm or the energy of $\overline{\mathbf{x}}_2$ will be minimized under the condition of Eq. (10.109). Eq. (10.109) guarantees that $\overline{\mathbf{x}}_2$ contains the information of the low-frequency data, and SVD makes sure that $\overline{\mathbf{x}}_2$ does not contain any information about the high-frequency (SVD minimizes the energy of $\overline{\mathbf{x}}_2$). Hence, the solution of Eq. (10.109) by SVD should be what we want in an extrapolation. Simulation results in section 10.3.6 will validate this explanation.

Another advantage of using SVD is that we can decide when we have sufficient data by analyzing the distribution of the singular values. If the smallest singular value is not of the same order in magnitude as the noise in the data then there is not enough data to carry out the extrapolation. More data must be generated either in the time domain or in the frequency domain, which means either M_1 or M_2 need to be increased.

Although the computational load of a SVD is of the order of O(mn^2) for an $m \times n$ matrix, it will not be much for the extrapolation problem; M_2 and $M_t - M_1$ is of the order of 10^2. The CPU times used for SVD and the CG method are similar. If the data ($\mathbf{x}_1$ and $\mathbf{X}_1$) that is obtained from an EM computation are not accurate (e.g., there are some perturbations in $\mathbf{X}_1$), then stable solutions still can be obtained if some threshold is set for the singular values of $\mathbf{A}$. To avoid computational errors, Eq. (10.109) is separated into real and imaginary parts, and one solves the following real matrix equation:

$$\begin{bmatrix} \text{Re}(\mathbf{A}) \\ \text{Im}(\mathbf{A}) \end{bmatrix} \mathbf{x}_2 = \begin{bmatrix} \text{Re}(\mathbf{X}_1 - \mathbf{B}\,\mathbf{x}_1) \\ \text{Im}(\mathbf{X}_1 - \mathbf{B}\,\mathbf{x}_1) \end{bmatrix} \tag{10.113}$$

10.3.6 Numerical Implementation and Results

In this section, three representative scattering structures are considered to validate the formulations. They are a sphere, an electrically large cube, and a dipole mounted on the top of a car. The goal is to extrapolate the induced surface current or the scattered field by the structure using early-time and low-frequency data and compare it with the results obtained using a MoM [1] solution in the frequency domain and DFT of the MOT solution in the time domain. The extrapolated time-domain response is compared with the results obtained using the MOT [7] and inverse discrete Fourier transform (IDFT) of the complete frequency-domain data. Note that when MOT is used to calculate the early-time data, the mesh size of the structure should be determined at the highest frequency considered in the analysis. The highest frequency is determined by the spatial discretization of the structure and vice versa. In all the figures showing the current or the field data, the vertical line represents the partition to the left of which is the data given and to the right is the extrapolated data.

A Gaussian pulse is used as the incident plane wave in the time domain. It has the following form:

$$s(t) = \frac{1}{\sigma\sqrt{\pi}} U_0 e^{-\left(\frac{t-t_0}{\sigma}\right)^2} \tag{10.114}$$

where σ is proportional to the width of the pulse. The time t_0 is the delay to make $s(t) \approx 0$ for $t < 0$. U_0 is a scalar amplitude. In frequency domain, the pulse has a spectrum

$$S(f) = U_0 e^{-\left|\frac{(2\pi f \sigma)^2}{4} + j2\pi f t_0\right|} \tag{10.115}$$

Note that $S(f)$ is approximately band limited to $1/\sigma$.

To evaluate the performance of the extrapolation, we use the following normalized mean square error (MSE):

$$\text{MSE} = \frac{\|\bar{\mathbf{x}}_2 - \mathbf{x}_2\|_2}{\|\mathbf{x}_2\|_2} \tag{10.116}$$

where $\|\cdot\|_2$ is the $\mathscr{L}_2$ norm of a vector. For the Neumann iteration, up to $n = 1000$ steps was used in Eq. (10.84). For the CG method, the iteration is repeated until the normalized error

$$\text{Er} = \frac{\|g - L(f)\|_2}{\|g\|_2} \tag{10.117}$$

becomes less than 10^{-12}.

Example 3. A conducting sphere: In this example, a conducting sphere with diameter 0.2 m is positioned at the origin as shown in Figure 10.27. An excitation plane wave is incident along the $-x$ direction and is polarized along the y direction. The parameter of the incident Gaussian pulse used in this example is $\sigma = 1.7$ ns and $t_0 = 3.4$ ns. Observe that the induced current in the middle of the shaded plate is considered to be the data of interest. The current is shown in Figure 10.28 in the time domain and in Figure 10.29 for the frequency domain. The sampling time Δt = 0.0293 lm (1 lm = 3.33 ns), and the frequency step $\Delta f = 20$ MHz. The first 35 time domain points are taken (up to 0.996 lm) as the early-time data and the first 35 frequency domain points (up to 0.68 GHz) as the low-frequency data, and we extrapolate them up to 128 points (3.72 lm) in the time domain and 128 points (2.54 GHz) in the frequency domain. It takes 163 iterations for the CG method to converge. Figure 10.28 also illustrates that the extrapolated data agrees well with the original data for all the three methods. The SVD has the minimum MSE and the performance of the Neumann iteration is the worst. The MSEs for the three methods are shown in the legend. The performance by SVD (MSE is of the order of 10^{-3}) is similar to the extrapolation by the orthonormal functions with the optimal scaling factor and optimal number of basis functions [15–17]. The extrapolated data also can be obtained from the FFT of the total data in the time domain.

Example 4. An electrically large cube: In this example, a 1 m × 1 m × 1 m conducting cube is positioned at the origin as shown in Figure 10.30. The incident plane wave is arriving along $-x$ direction and is polarized along the y direction. The parameter of the incident Gaussian pulse used in this example is $\sigma = 0.1335$ ns and $t_0 = 1.068$ ns. Its bandwidth is approximately $1/\sigma = 7.5$ GHz. The wavelength λ in air at 7.5 GHz is 0.04 m. Hence, the cube is approximately 25λ long at the highest frequency. Consider the reflected field along $+x$ direction as the original data. The scattered field (the dominant polarization is in the z direction) is shown in Figure 10.31 in the time domain and in Figure 10.32 for the frequency domain. The sampling time $\Delta t = 0.0073$ lm (1 lm = 3.33 ns), and the frequency step $\Delta f = 20$ MHz. Because the structure is large compared with the wavelength, most of the energy exciting the cube is reflected back from the front plate. The effect is equivalent to taking a time derivative, which dominates the early-time response. However, the resonance effect from the late time data can be observed way out along the time axis. A change from positive to negative "DC" field occurs around 4.43 ns. The delay between this change and t_0 is the round trip time for the surface current from the center to the edge of the front plate (1 m). This is the signature of a conducting plate. A small late pulse appears at 7.76 ns, the delay of which from t_0 is 6.69 ns. During this delay, the wave travels 2 m, which is the round trip time taken to travel from the front to the rear plate. This is the signature of the cube. In the frequency domain, the higher order resonances are small ripples in Figure 10.32. The first 40 time domain samples (up to 0.293 lm) are taken as the early-time data, and the first 120 frequency domain points (up to 2.4 GHz) are taken as the low-frequency data, and we extrapolate them up to 800 points (5.86 lm) in the time domain and to 600 points (12 GHz) in the frequency domain. It takes 277 iterations to make the CG method converge. To validate the extrapolated results, the complete solution has been generated by taking two planes of symmetry, which were not used in the low-frequency and the early-time data. Figure 10.31 shows that the extrapolated data agrees well with the original data for all the three methods. The SVD produces the minimum MSE and the performance of the Neumann iteration is the worst. The frequency domain data is shown in Figure 10.32.

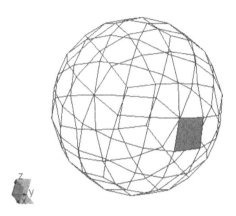

Figure 10.27. The discretization of a sphere.

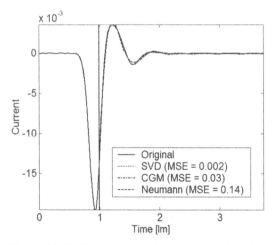

Figure 10.28. The extrapolated results in the time domain.

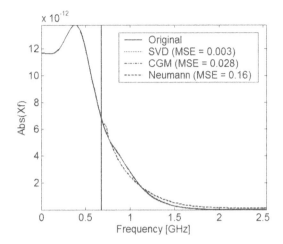

Figure 10.29. The extrapolated results in the frequency domain.

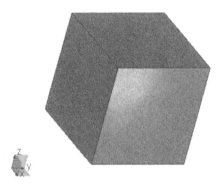

Figure 10.30. The structure of a cube.

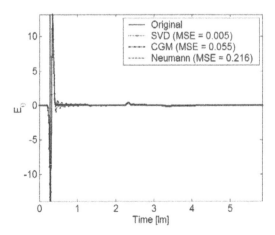

Figure 10.31. The extrapolated results in time domain.

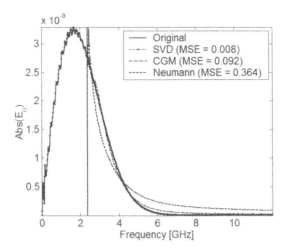

Figure 10.32. The extrapolated data in the frequency domain.

Example 5. A dipole mounted on the top of a car. In this example, an FM wire antenna is mounted on the top of a car. The main dimensions of the car are shown in Figure 10.33. The plane wave is incident along the $-y$ direction and is polarized along the z direction. The parameters of the incident Gaussian pulse used in this example are $\sigma = 10$ ns and $t_0 = 15$ ns. The induced current at the bottom of the antenna is shown in Figure 10.34 in the time domain and in Figure 10.35 in frequency domain. The sampling time $\Delta t = 0.293$ lm, and the frequency step $\Delta f = 2$ MHz. The first 30 time domain points (up to 8.496 lm) are taken as the early-time data, and the first 45 frequency domain points (up to 0.088 GHz) are taken as the low-frequency data, and we extrapolate them up to 256 points (74.7 lm) in the time domain and to 256 points (0.51 GHz) in the frequency domain. It takes 348 iterations to make the CG method converge. Figure 10.34 shows that the SVD has the minimum MSE and the performance of the Neumann iteration is the worst. In this example, the extrapolated range is 5~10 times larger than the range over which the data is known. Moreover, the given low-frequency data, as shown in Figure 10.35, does not cover even up to the first resonant frequency of the structure, but using this extrapolation method, one still can recover the high-frequency data, including the resonant frequency accurately, as the early-time data contain the missing high-frequency information.

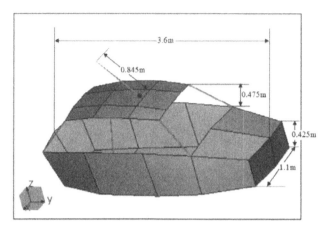

Figure 10.33. The schematic of a dipole on a car.

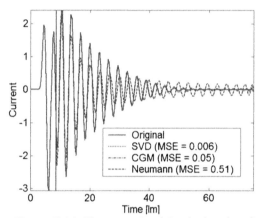

Figure 10.34. The extrapolated data in time domain.

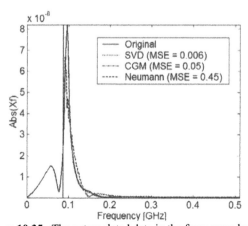

Figure 10.35. The extrapolated data in the frequency domain.

10.4   CONCLUSION

In this chapter, a hybrid method that deals with the problem of simultaneous extrapolation in the time and frequency domains using only early-time and low-frequency data and the ways to obtain the best performance for the extrapolation are presented. It is important to note that the early-time domain data and low-frequency domain data are complementary. This time–frequency domain hybrid methodology extrapolates the missing late-time and high-frequency information. This procedure interestingly does not create any new information. The early-time data provides the missing high-frequency information, and the low-frequency data provides the late-time information. Both parametric and nonparametric methods have been presented to carry out this extrapolation.

Factors influencing the performance of the parametric method are studied, and a practical search method using only the available data to find the optimal parameters is outlined. Finally, a comparison of the performances using Hermite, Laguerre, and Bessel–Chebyshev polynomials is discussed. The performance of the extrapolation is sensitive to two parameters — the scaling factor and the order of polynomials. Hermite functions show the advantage of having the same functional form both in the time domain (TD) and the frequency domain (FD). The upper bound and the optimal choice for the scaling factor is easiest to locate for Bessel–Chebyshev basis functions. The Laguerre basis functions have the widest range of convergence. The computational load of this method is not large because we only deal with the solution of small matrix equations and the computation of the DFT/IDFT of a data set. This procedure is easy to program in practice. The range of convergence of this method mainly is dependent on the time support and bandwidth of the given data. Three kinds of basis functions have been analyzed. Use of this hybrid technique may reduce the computational cost of solving electrically large problems that may be difficult to solve by exclusive use of either the TD or the FD codes, and it also solves the problem of late-time instability in the TD. Typically, for good reconstruction it seems that one needs a time-bandwidth product of the order of 1.5–3.0. However, this factor is dependent on the quality of the time and frequency domain data. Because two smaller problems are solved, the computer resources required are modest. Even though the starting information is obtained from techniques that have the potential to become unstable in late times, the extrapolation scheme does not exhibit such behavior.

In addition, three direct methods to solve the extrapolation problem using early-time and low-frequency data also are discussed and presented in the form of a nonparametric scheme. The process of extrapolation can be derived by the solution of integral equations of the second kind. The Neumann iteration and the conjugate gradient method can be used to solve the equations. However, the integral equations used to set up the problem are equivalent to a set of matrix equations in a discrete sense. The feasibility and efficiency of the SVD are discussed for solving the integral equation of the first kind directly. Electromagnetic scattering examples are presented to validate these procedures, and the performances of the three methods are compared. Direct methods are more straight forward, and we do not need to worry about the effect of the scaling factor. They perform very well even for complex electrically large problems that cannot be solved exclusively in either the time or in the frequency domain, even when using a supercomputer.

REFERENCES

[1]   Y. Zhang and T. K. Sarkar, *Parallel Solution of Integral Equation-Based EM Problems in the Frequency Domain*, John Wiley and Sons, New York, 2009.

[2]   R. S. Adve, T. K. Sarkar, S. M. Rao, E. K. Miller, and D. Pflug, "Application of the Cauchy method for extrapolating/interpolating narrowband system responses," *IEEE Trans. on Microwave Theory Tech.* Vol. 45, pp. 837–845, 1997.

[3]   E. K. Miller and T. K. Sarkar, "an introduction to the use of model-based parameter estimation in electromagnetics," in W. R. Stone (ed.) *Review of Radio Science 1996-1999*, Oxford University Press, N.Y., pp. 139-174. 1999.

[4]   R. S. Adve and T. K. Sarkar, "Simultaneous time and frequency domain extrapolation," *IEEE Trans. on Antennas Prop.*, Vol. 46, pp. 484–493, 1998.

[5]   S. Narayana, S. M. Rao, R. S. Adve, T. K. Sarkar, V. Vannicola, M. Wicks, and S. Scott, "Interpolation/Extrapolation of frequency responses using the Hilbert transform," *IEEE Trans. on Microwave Theory Tech.*, Vol. 44, pp. 1621–1627, 1996.

[6]   S. Narayana, T. K. Sarkar and R. S. Adve, "A comparison of two techniques for the interpolation/extrapolation of frequency domain responses," *Digit. Signal Proces.*, Vol. 6, pp. 51–67, 1996.

[7]   T. K. Sarkar, W. Lee, and S. M. Rao, "Analysis of transient scattering from composite arbitrarily shaped complex structures," *IEEE Trans. on Antennas and Prop.*, Vol. 48, pp. 1625–1634, 2000.

[8]   S. M. Rao and T. K. Sarkar, "Implicit solution of time domain integral equations for arbitrarily shaped dielectric bodies," *Microwave Opt. Techno. Let.*, Vol. 21, pp. 201–205, 1999.

[9]   R. S. Adve, T. K. Sarkar, O. M. Pereira-Filho, and S. M. Rao, "Extrapolation of time domain responses from three dimensional conducting objects utilizing the matrix pencil technique," *IEEE Trans. on Antennas Prop.*, Vol. 45, pp. 147–156, 1997.

[10]   T. K. Sarkar and O. M. Pereira-Filho, "Using the matrix pencil method to estimate the parameters of a sum of complex exponentials," *IEEE Trans. on Antennas Prop.*, Vol. 37, pp. 48–55, 1995.

[11]   M. M. Rao, T. K. Sarkar, T. Anjali, and R. S. Adve, "Simultaneous extrapolation in time and frequency domains using Hermite expansions," *IEEE Trans. on Antennas Prop.*, Vol. 47, pp. 1108–1115, 1999.

[12]   M. M. Rao, T. K. Sarkar, R. S. Adve, T. Anjali, and J. F. Callejon, "Extrapolation of Electromagnetic responses from conducting objects in time and frequency domains," *IEEE Trans. on Microwave Theory Tech.*, Vol. 47, pp. 1964–1974, 1999.

[13]   T. K. Sarkar, J. Koh, W. Lee, and M. Salazar-Palma, "Analysis of electromagnetic systems irradiated by ultra-short ultra-wideband pulse," *Meas. Sci. & Tech.*, Vol. 12, pp. 1757–1768, 2001.

[14]   T. K. Sarkar and J. Koh, "Generation of a wide-band electromagnetic response through a Laguerre expansion using early-time and low-frequency data," *IEEE Trans. Microwave Theory Tech.*, Vol. 50, pp. 1408–1416, 2002.

[15]   M. Yuan, A. De, T. K. Sarkar, J. Koh, and B. H. Jung, "Conditions for generation of stable and accurate hybrid TD-FD MoM solutions," *IEEE Trans. Microwave Theory Tech.*, Vol. 54, pp. 2552–2563, 2006.

[16]   M. Yuan, J. Koh, T. K. Sarkar, W. Lee, and M. Salazar-Palma, "A comparison of performance of three orthogonal polynomials in extraction of wide-band response using early time and low frequency data," *IEEE Trans. Antennas and Prop.*, Vol. 53, pp. 785–792, 2005.

[17]   M. Yuan, T. K. Sarkar, B. H. Jung, Z. Ji, and M. Salazar-Palma, "Use of discrete Laguerre sequences to extrapolate wide-band response from early time and low frequency data," *IEEE Trans. on Microwave Theory Tech.*, Vol. 52, pp. 1740–1750, 2004.

[18]   M. Yuan, P. M. van den Berg, and T. K. Sarkar, "Direct extrapolation of a causal signal using low-frequency and early-time data," *IEEE Trans. on Antennas Prop.*, Vol. 53, pp. 2290–2298, 2005.

[19]   G. H. Golub and C. F. Van Loan, *Matrix Computations*, Johns Hopkins University Press, Baltimore, M.D., 1991.

[20]   D. Slepian, "Prolate spheroidal wave functions, Fourier analysis, and uncertainty-V: The discrete case," *Bell Sys. Tech. J.*, Vol. 57, pp. 1371–1430, 1978.

[21]   D. Slepian, "On bandwidth," *Proc. IEEE*, Vol. 64, pp. 292–300, 1976.

[22]   M. A. Masnadi-Shirazi and N. Ahmed, "Optimum Laguerre networks for a class of discrete-time systems," *IEEE Trans. Signal Processing*, Vol. 39, pp. 2104–2108, 1991

[23]   A. D. Poularikas, *The Transforms and Applications Handbook*, IEEE Press, Washing, D.C., 1996.

[24]   B. H. Jung and T. K. Sarkar, "Transient scattering from three-dimensional conducting bodies by using magnetic field integral equation," *J. Electromagn. Waves Appl.*, Vol. 16, pp. 111–128, 2002.

[25]   B. H. Jung, Y. S. Chung, and T. K. Sarkar, "Time-domain EFIE, MFIE, and CFIE formulations using Laguerre Polynomials as temporal basis functions for the analysis of transient scattering from arbitrarily shaped conducting structures," *Progress in Electromagnetics Research*, Vol. 39, pp. 1-45, 2003.

[26]   P. M. van den Berg and R. E. Kleinman, "The Conjugate gradient spectral iterative technique for planar structures," *IEEE Trans. Antennas and Prop.*, Vol. 36, pp. 1418–1423, 1988.

[27]   J. T. Fokkema and P. M van den Berg, *Seismic Applications of Acoustic Reciprocity*, Elsevier, Amsterdam, The Netherland, 1973.

[28]   J. B. Rhebergen, P. M. van den Berg and T. M. Habashy, "Iterative reconstruction of images from incomplete spectral data," *Inverse Probl.*, Vol. 13, pp. 829–842, 1997.

[29]   R. E. Kleinman and P. M. van den Berg, "Iterative methods for solving integral equations," *Radio Sci.*, Vol. 26, pp. 175–181, 1991.

[30]   T. K. Sarkar (ed.), *Application of Conjugate Gradient Method to Electromagnetics and Signal Analysis*, Elsevier, New York, 1991, pp. 67–102.

[31]   E. O. Brigham, *The Fast Fourier Transform*, Prentice-Hall, Englewood Cliffs N.J., 1974.

[32]   L. N. Trefethen and D. Bau III, *Numerical Linear Algebra*, SIAM, Philadelphia, P.A., 1997.

APPENDIX

USER GUIDE FOR THE TIME AND FREQUENCY DOMAIN EM SOLVER USING INTEGRAL EQUATIONS (TFDSIE)

A.0   SUMMARY

In this appendix, the user manual for the *Time and Frequency Domain EM Solver using Integral Equations* (TFDSIE) is presented. How to download the code to one's personal computer and then execute it, is described in this section. The user-oriented computer code has a Matlab-based graphical user interface for both the input and the output parameters.

A.1   INTRODUCTION

The beginning of the twenty-first century has witnessed a rapid expansion both in computer capabilities and in the availability of efficient and highly stable numerical techniques for solving electromagnetic (EM) problems. Naturally, this growth has resulted in an increased interest in developing general purpose software codes for computing electromagnetic responses from arbitrarily shaped structures. One of the most notable achievements in this direction is the implementation based on integral equations in both the time and frequency domains.

Since the beginning of the development of computational electromagnetics from the early 1960s, most researchers have focused their attention in the frequency domain because it was analytically more manageable than the time domain techniques. Recently, the transient analysis of electromagnetic scattering has received a great deal of attentions. There are various reasons as to why one should be interested in calculating and predicting transient electromagnetic behavior from arbitrarily shaped structures. First, many systems including short-pulse radar systems, which are being used for target identification, high-range resolution and wide-band digital communications, are becoming popular. Also, there is a great interest in determining and correcting the vulnerability and susceptibility of electronic systems to lightning and electromagnetic pulses caused by nuclear explosions. Basically there are two approaches to obtain the transient response from an arbitrarily shaped structure. They are: (1) obtaining the frequency response of the structure excited by time-harmonic sources and using the inverse Fourier transform techniques to calculate the required transient response and (2) formulating the problem directly in the time domain and then numerically solving it. In this book, both formulations in the frequency and time domains are included, which are based on the integral equations so that the validity of the solutions can be checked by using these two different approaches.

To help the reader to better understand the contents of this book, a self-contained EM solver, called TFDSIE, is presented and aimed to provide the reader with a tool to analyze some examples in this book or their own scattering and radiation problems. The primary objective is to make the reader familiar with time domain integral equation solvers. However, a frequency domain code using the same discretizations for the structure also is provided so that one can check the time domain results with the inverse Fourier transform of a frequency domain integral equation solution or vice versa. Accuracy of the solution is of course the primary objective followed by computational efficiency. A free version of TFDSIE can be downloaded from http://lcs.syr.edu/faculty/sarkar /softw.asp, which is for the analysis of arbitrarily shaped conducting bodies. Any questions or comments about TFDSIE are welcome and can be sent to the authors. For a more sophisticated version of the EM tools, please contact the authors. In this appendix, a simple introduction to this program and its applications are described.

A.1.1   What is TFDSIE?

TFDSIE, which stands for time and frequency domain EM solver using integral equations, is a self-contained program to analyze electromagnetic problems in both time and frequency domain for arbitrarily shaped conducting bodies. The core of the program is implemented from the formulations introduced in Chapters 2 and 8 for the frequency and the time domain techniques. These formulations are developed in FORTRAN, which is a common programming language in the area of computational electromagnetics, and embedded into a Matlab graphical user interface. Users can build their structure of interest using the built-in components and then select the required types of analyses and outputs. TFDSIE can provide the simulation results related to the surface currents or far-field distributions either in a text format or through animated graphs.

A.1.2   Features of the Code

The features of TFDSIE can be summarized as follows:

- Analyze conducting structures in both the time and frequency domains
- Flexible geometrical model for arbitrarily shaped structures
- Analyze a closed or open structure
- The density of patches can be controlled by the user
- The incident wave can be a Gaussian-shaped pulse or its derivative
- The output can be represented by a graph or by a text-based data file for post-processing

A.1.3   Scope of the Analysis

TFDSIE can be used to compute the following:

- Scattering and radiating from arbitrarily shaped conducting structures in the frequency domain
- Scattering and radiating from arbitrarily shaped conducting structures in the time domain for an incident Gaussian or the derivative of a Gaussian pulse
- Input impedance of arbitrarily shaped conducting structures
- Far field response from arbitrarily shaped conducting structures

A.1.4   Hardware and Software Requirements

TFDSIE is developed and tested on Intel 32-bit x86 machines with Windows 9X or XP operation systems and Matlab V5.3 or later versions. Computers equipped with at least 100 MB free space and 64 MB memory, operating a suitable version of Windows operation system and Matlab can execute TFDSIE. For analysis of larger structures, computers with more memories are required.

A.1.5   Limitations of the Downloaded Version of TFDSIE

TFDSIE is an EM solver based on the techniques introduced in the previous chapters, and this small version of TFDSIE is limited the following:

- *Patch size*: The patch dimension should not be larger than 1/10 of the wavelength; otherwise it may lead to an inaccurate or unstable result. Particularly in time domain, users should consider the bandwidth of the incident Gaussian pulse to discretize the structure at the highest frequency of interest.
- *Number of variables*: The upper limit of the number of nodes, edges, and patches is 3500. The upper limit for the number of sampling points in the far field is 3000. These limits are set to avoid computer memory overflow.
- *Junctions*: This code cannot handle any junction of more than two metal plates or three triangular patches belonging to the same edge.

A.2   GETTING STARTED

Before starting to use TFDSIE, users are recommended to check the hardware and software requirements mentioned in section A.1.4 and to download TFDSIE from the website. The file is compressed, and it can be unzipped by any common available file compression/decompression tools.

A.2.1   Program Installation

No complicated installation procedure is needed for TFDSIE. Users only need to extract the files into a destination folder where you would like to execute TFDSIE, for example, `c:\tfdsie\`. Please note that no blank is allowed in the path because FORTRAN will not recognize the blank in many platforms. For illustration purposes in this chapter, it is assumed that the code TFDSIE is located in the folder `c:\tfdsie\`.

A.2.2   Start the Program

The interface of the TFDSIE program is developed in Matlab, so it must be run under a Matlab environment to use that interface to create the model of the structure and input all related parameters for the analysis.

1. Run Matlab
2. Type `cd c:\tfdsie\` in Matlab command window to go to the folder where the program TFDSIE is installed
3. Type `tfdsie` in the Matlab command window to execute TFDSIE, then a welcome window will pop up as shown in Figure A.1
4. Click the ` OK ` button, and the main window as shown in Figure A.2 will appear

Figure A.1. Welcome window of TFDSIE.

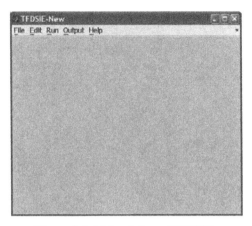

Figure A.2. Main window of TFDSIE.

Any EM problem to be analyzed by TFDSIE is organized into a project. For the general case, a project contains both the input and output data. The input data defines the problem to be analyzed. For example, it may contain the structure of the objects, whereas the output data defines the types of output desired, which can be the current or the far field distribution. All available functions of TFDSIE can be accessed by selecting the commands in the pull-down menu of the main window. The menu consists of five categories: **File**, **Edit**, **Run**, **Output**, and **Help** and is used for project file management, generation of the structure related to the model, setup and execution of the analysis, resultant output, and help, respectively.

A.3   A QUICK TOUR

In this part, we will go through some main functions of TFDSIE using an example. Here, the TFDSIE is used to create a project for analyzing the scattering from a conducting sphere in the time domain.

A.3.1   Create a Sphere Object

Follow the steps as outlined to create a model for a conducting sphere using the triangular patches:

1.   In TFDSIE Main Window, click "**Edit**" menu in the main menu bar and choose the "**Sphere**" option under the "**Add Face**" function, as shown in Figure A.3. **It is important to point out that we are using the word** *face* **to represent an** *object*.

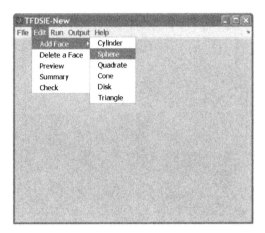

Figure A.3. The built-in sphere model.

2. A dialog box as illustrated by Figure A.4(*a*) will pop up for defining the dimensions of the conducting sphere. In this example, a sphere with a radius of 1 m, located at the origin (0, 0, 0) is to be analyzed. The dimensions of the sphere can be input according to the data shown in Figure A.4(*a*). At the top of the dialog box, the number of the subsections along theta (θ) or phi (ϕ) direction can be chosen freely. In this example, eight subdivisions are used in both the directions.

3. Click the ⬚Create⬚ button, and the sphere is generated accordingly. Note that the dialog box then is updated. It also activates the other buttons and changes the number in blue displayed at the upper right corner to 1. The newly activated button is used to preview or delete this sphere component, and the number at the upper right corner is used to indicate the total number of complete faces existing in the structure.

4. Click the ⬚Disp_Data⬚ button to view the graphical patch model of the sphere, as shown in Figure A.5(*a*).

5. Click the ⬚Disp_face⬚ button to show the data for all the patches, as shown in Figure A.5(*b*).

6. Up to this step, a triangular patch model of a sphere has been generated. Return to the main window and close the "Display Patches of Face" window and the "Sphere" window.

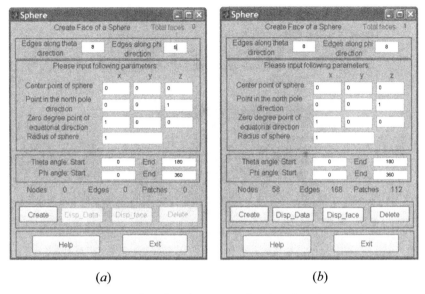

Figure A.4 Setting of a sphere: (*a*) Before the sphere is created.
(*b*) After the sphere is created.

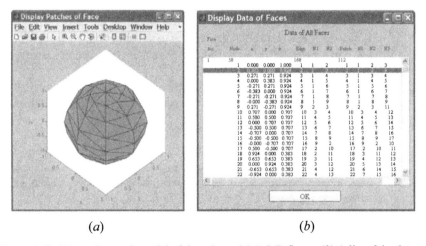

Figure A.5. Triangular patch model of the sphere: (*a*) A 3-D figure. (*b*) A list of the data.

A.3.2 Save the Sphere Object as a Project

You may notice that the title of the main window is still "TFDSIE-New," which is the same as when the program started. The default project name is "New," and it can be changed to any other name by saving the project. A project can be saved using the following steps:

- Go to "**File**" menu and choose "**Save**" or "**Save as**" option, a "Save as" dialog box will pop up (Figure A.6)
- Choose the directory and type a project name. Here, this sphere project is named as "test.prj," and you may observe that the title of TFDSIE changes into "TFDSIE-test"

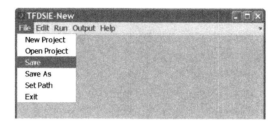

Figure A.6. Save the structure into a project.

A.3.3   Simulation Setup

The next step is to set up the type of analysis one needs. In this example, we intend to calculate the transient electromagnetic scattering from the sphere.

1. As shown in Figure A.7, go to the "**Run**" menu in the main menu bar and choose the "**Run Setup**" function. A dialog box as shown in Fig. A.8 will pop up, (Note: All projects should be set up first before initiating the computations)
2. In "**Running setup**" dialog, select the options "Scattering" and "Time-domain." (To simplify the presentation, the far field calculation is not included in this example)
3. Select the incident field as a "Gaussian pulse," and set all pulse parameters as shown in Figure A.8. (Note: The time domain solver makes use of the MOD algorithm, "Sample time" and "Stop time" only are used to generate output results in a predefined time scale, but they are not used in this computation)

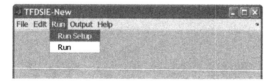

Figure A.7 The "**Run**" menu.

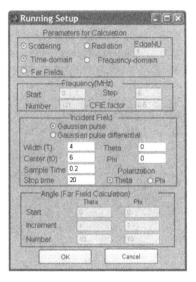

Figure A.8 The "**Running setup**"
dialog box.

A.3.4   Start Simulation

1. Go to the "**Run**" menu again and choose the "**Run**" option; a dialog box,
   as shown in Figure A.9, will pop up. Click the Check button, and the
   input data will be checked for consistency, and the results then will be
   displayed. This checking procedure will determine whether there is any
   common node or edge, the structure is open or closed, and so on. For a
   structure to be analyzed in the frequency domain if it is an open structure
   then the EFIE solution method will be applied; otherwise, the CFIE will
   be used. Also, it is shown in Figure A.9 that the sphere model has 58
   nodes, 168 edges, and 112 patches.

Figure A.9 An active window in
the execution mode.

2. After checking, the [Run] button will be activated. Click the [Run] button, and TFDSIE will execute the corresponding algorithm. The intermediate computational steps are displayed in Figure A.10. After it is finished, close the active window and return to the main window.

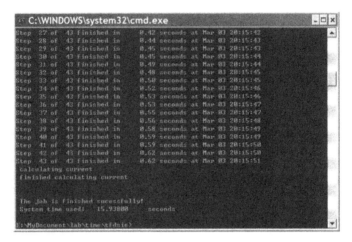

Figure A.10. An active window in the execution mode.

A.3.5   View Simulation Results

After executing the computation, the simulation results are stored in the appropriate project folder. The currents flowing across each edge can be viewed either in a static or dynamic display. To view the current statically, follow the steps as outlined.

1. Go to the "**Output**" menu in the main menu bar and choose the "**Graph**" function and then the "**Current**" option, as shown in Figure A.11.
2. After loading the data, a window with a plot of the current will pop up. As shown in Figure A.12, the current on the eighteenth edge is presented.

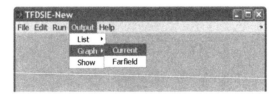

Figure A.11. The "**Output**"menu.

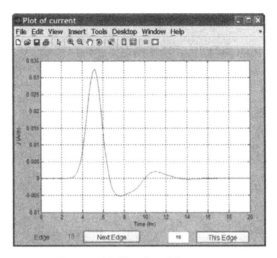

Figure A.12. The plot of the current.

To view the current in a dynamic way, apply the following steps:

1. Go to the "**Output**" menu and choose the "**Show**" option. A window called "Output Viewer," as shown in Figure A.13 will pop up.
2. In the "Output Viewer" windows, go to the "**File**" menu and choose "**Open structure …**" to load the results of the project.
3. Go to the "**File**" menu again and choose "**Open currents …**," and select the file named "testN1.cur" to be opened.
4. Then click the Play button, and the current distribution on the sphere is displayed animatedly as time progresses.
5. The current on the surface of the sphere has been computed, and the analysis of this project is completed. Users can return to the main window for other simulations.

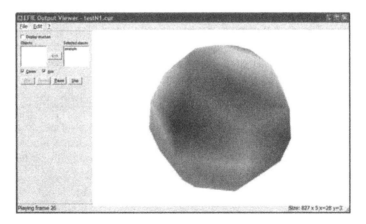

Figure A.13. Displaying the currents on the surface by an animated figure.

A.4   A TUTORIAL GUIDE OF THE CODE

The basic steps, from creating a structure to the viewing of the computational results have been illustrated by an example in the previous section. This section will complement the functions of the code that not yet have been described in the previous example, as well as provide more details about the basic principles and commands for using this software.

A.4.1   Terminologies

- *Node*: A node is defined in the Cartesian coordinates using three variables x, y, and z.
- *Edge*: An edge is a link connecting two adjacent nodes, node(1) and node(2), represented by a vector in the program. Each edge will be assigned an edge number.
- *Patch*: A patch is a triangle that is described by three nodes: node(1), node(2), and node(3). Each patch will be assigned a patch number.
- *Face:* A face is the surface comprising the patches. It can be the surface of a cylinder, sphere, or a combination of planar triangular patches approximating the actual surface. TFDSIE can deal with any arbitrary shaped conducting surfaces, and they can be constructed by a combination of the six types of built-in surfaces: cylinder, sphere, cone, quadrilateral, disk, and triangle. Users can use them to define their own structure of interest.
- *Common nodes and edges*: When a face is connected to another face, there may exist some common nodes and edges. They should be removed before running the code by using the "**Check**" function under the "**Edit**" menu.
- *Structure*: A structure consists of faces. It is the object to be analyzed. A structure may be open or closed. If in the specification of all faces for this structure, every edge has a pair of triangles attached to it, then the structure is said to be closed, and the corresponding edge is called a closed edge. Otherwise, the structure is open if there is an edge only belonging to one triangular patch, and the edge is called an open edge. A current flowing across a closed edge whereas an open edge is a boundary edge and has no currents associated with it.
- *Current*: Current is the induced electric current density on the surface of the structure.
- *Farfield*: Farfield is the electric field in the far zone of the structure. It is described by the angular directions. When a particular value of θ and ϕ are specified, TFDSIE provides the far field result along that angular direction.

A.4.2   Units

The units used in the TFDSIE are described in Table 4.1.

Table A.1 Units in TFDSIE

Name	Unit
Coordinate (x, y, z) or distance	meter (m)
Angle	degree
Current density	A/m
Electric field	V/m
Frequency	MHz
Time	lm (light meter $= 1/3 \times 10^8$ sec)

A.4.3   File Types

The file types in the TFDSIE software are listed in Table 4.2. Most of these files are text files so that it is convenient to export the computation results to other software for post-processing. Assuming the project is saved with a name "demo," all related files associated with this project are as follows:

Table A.2 Files related to the project "demo"

File Name	Type	Use
demo.prj	Text	Main project file
demo.fac	Text	Structure data file
demo.par	Text	Parameters for running
demo.cem	Text	Structure data for showing results
demo.cut	Text	Time-domain current
demo.fat	Text	Time-domain far field
demo.ndd	Text	Data of nodes and edges
demo.cuf	Text	Frequency domain current
demo.faf	Text	Frequency domain far field
demo1E.cur	Text	Current across edge for showing results
demo1N.cur	Text	Number of node for showing results
demo1N.avi	Video	Recorded movie file

A.4.4   Functions

Some of the functions in TFDSIE have been introduced in the previous example. In Table 4.3, all functions in the menu bar are listed.

Table A.3. Menu Structure

Main menu	Submenu	Sub–options	Function
File	New project	No	Create new project
	Open project	No	Open an existing project
	Save	No	Save current project
	Save as	No	Save current project using a name to be specified
	Set path	No	Set the file path for running the code
	Exit	No	Exit the program
Edit	Add a face	Cylinder, Sphere, Quadrate, Cone, Disk, and Triangle	Add an additional component of the specified surface to the structure
	Delete a face	No	Delete a face from the structure
	Preview	No	Visualize the structure
	Summary	No	Summarize information of the structure
	Check	No	Check whether the structure is open or closed and eliminate redundant (common) nodes and edges
Run	Run setup	No	Set up the analysis parameters
	Run	No	Run the simulation
Output	List	Current, farfield	Show the result as a list of data
	Graph	Current, farfield	Show the result graphically
	Show	No	Show the transient current by animation
Help	Help tips	No	The items of help
	About…	No	Information about the software

A.4.5   Load a Project

The functions of making a new project and saving a project have been introduced in section A.3. To open a project that already exists, the project can be loaded into the TFDSIE environment using the following steps:

- As shown in Figure A.14, go to "**File**" menu in the main menu bar and choose "**Open Project**" function, and a dialog box for selecting the project to be opened will appear. Once the project is opened, the project name will be displayed in the title of the main window.

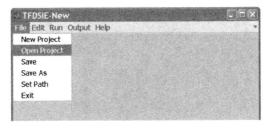

Figure A.14. Open an existing project.

A.4.6   Build a Model in TFDSIE for a Structure

A structure may consist of several faces. In TFDSIE, a structure can be built by adding or deleting the faces given by the six types of built-in components, e.g., cylinder, sphere, quadrilateral, cone, disk, and triangle.

A.4.6.1 Add a Face
- In section A.3, a simple structure with a sphere object only is created; and now we consider adding a cylinder with this sphere object. First, we load the project "test.prj" according to the steps in section A.4.5. To avoid the simulation results of the sphere from being rewritten, users can save the project with another name.
- Then go to the "**Edit**" Menu and choose the "**Cylinder**" option under the function of "**Add a Face**," a dialog box for inputting the parameter of a cylinder, which is shown in Figure A.15 will appear.
- Set the dimension of the cylinder according to the parameters of Fig. A.15, and click the Create button to create the cylinder and add to the original structure. Note that the right upper corner shows the total number of faces in the structure and is now equal to 2. Obviously, one is for the original sphere and the other is for the cylinder. The structure now can be viewed by clicking the Disp_face button, and both the sphere and the cylinder are displayed as in Figure A.16.

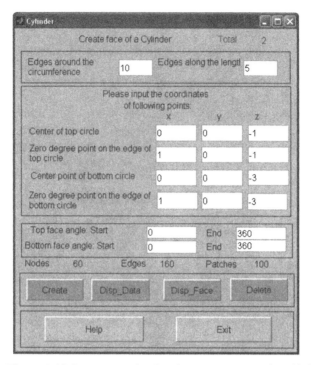

Figure A.15. Parameters related to the new structure to be added.

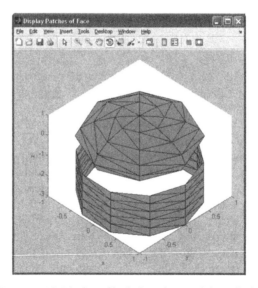

Figure A.16. Display of both the sphere and the cylinder.

A.4.6.2 Delete a Face. In contrast to adding a face, a face can also be deleted. Go to the "**Edit**" menu and choose the "**Delete a Face**" function, and a dialog box lists the information of all the faces, as shown in Fig. A.17. Choose a face and click the ⸢Delete⸣ button, the corresponding face will be deleted. As shown in Figure A.17; there are two faces, the first face contains 58 nodes, 168 edges, and 112 patches and is corresponding to the sphere that has been constructed in section A.3.1, whereas the second one has 60 nodes, 160 edges, and 100 patches and is the cylinder that is just created. Here, when selecting the first item in the list and clicking the ⸢Delete⸣ button, the sphere will be deleted. **It is important to point out that we are using the word** *face* **to represent an** *object.*

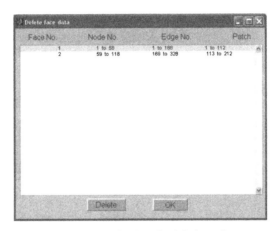

Figure A.17. Dialog box for deleting a face.

A.4.6.3 Preview the Structure. The updated structure can be viewed by the preview function at any time. Go to the "**Edit**" menu in the main menu bar and choose the "**Preview**" function. A three-dimensional figure of the updated structure pops up as shown in Figure A.18. It is shown that there is only one cylinder, as the sphere has been deleted.

The structure displayed in the "Display Patches of Face" window can zoom in, zoom out, or rotate. In addition,

- Click the ⸢See Edge Number⸣ button, and all the edge numbers are displayed.
- Input an edge number in the "edge number" box and click the ⸢See a Edge⸣ button, and the corresponding edge and connected patches will be highlighted.

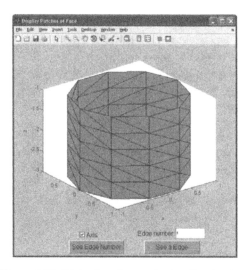

Figure A.18. Dialog box for previewing the structure.

A.4.6.4 Get Parametric Data for the Structure. The structure information, such as the number of nodes, edges, and patches of each face, can be obtained by going to the "**Edit**" menu and choosing the "**Summary**" function. Then a dialog box pops up, and a summary of the whole structure is presented as shown in Figure A.19. In this example, the summary shows that there is only one face containing the current structure with 60 nodes, 160 edges, and 100 patches.

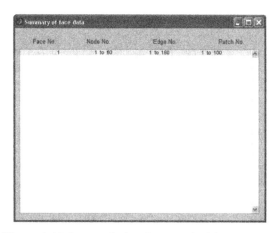

Figure A.19. Parametric data (summary) of the structure.

A.4.6.5 Checking the Data of the Structure. When the model of the structure is generated, a *checking* function can be performed. Go to the "Edit" menu in the main menu bar and choose the "**Check**" function, and a dialog box pops up. Click the [Check] button, and the checking process starts, and the information is presented as shown in Figure A.20.

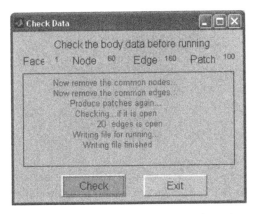

Figure A.20. Performing checking of the input data.

The main functions in the check procedures contain the following:

- Removal of redundant/common nodes. When one face is connected to another face, there may be some nodes that may have the same coordinates. The duplicate nodes are deleted.
- Removal of an edge that may be specified twice.
- Regeneration of the nonredundant patches. Because the total number of nodes and edges has been changed as a result of the removal of duplicated nodes and edges, a recalculation for the patches is required.
- Checking whether the structure is open or closed and the total number of open edges. For an open structure, only the EFIE can be applied because the MFIE and the CFIE can only be applied to closed structures.

A.4.7 Simulation Options

After the structure has been created, the analyses can be performed in both the time and the frequency domains. Before running, some parameters should be set up. Go to the "**Run**" menu in the main menu bar and choose the "**Run Setup**" function, and a dialog box about the analysis parameters pops up. In this dialog, the following setting can be manipulated:

- Select the computation for either in the time or in the frequency domain
- Include or exclude the far field calculations in the computation process
- For frequency domain problems, the incident wave is always a plane wave. You can set the direction of arrival in terms of theta (θ) and phi (ϕ)

- For time domain computations, the incident wave can be a pulse with a Gaussian shape or its derivative, which does not have a DC component. The incident Gaussian pulse can be represented mathematically by

$$\mathbf{E}(\mathbf{r},t) = \mathbf{E}_0 \frac{4}{T\sqrt{\pi}} e^{-\gamma^2} \tag{A.1}$$

where

$$\gamma = \frac{4}{T}(ct - ct_0 - \mathbf{r}\cdot\hat{\mathbf{k}}) \tag{A.2}$$

The parameter T is the width of the pulse, and t_0 is the time delay at which the pulse reaches its peak. Both these quantities are defined in light meters (lm). The parameter c is the velocity of the wave. $\hat{\mathbf{k}}$ is the unit vector along the direction of wave propagation. t is the time variable. The amplitude of the incident electric field $\mathbf{E}_0$ is assumed to be unity. The derivative of the Gaussian pulse is given by

$$\mathbf{E}(\mathbf{r},t) = \mathbf{E}_0(-\frac{32\gamma}{T^2\sqrt{\pi}})e^{-\gamma^2} \tag{A.3}$$

The polarization of the incident field is defined relative to the incident direction. All appropriate directions are defined in Figure A.21. For far field calculation, you can choose the angle of observation by inputting the angles of observation theta (θ) and phi (ϕ).

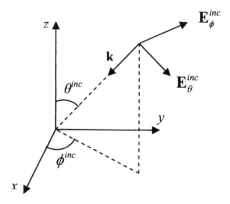

Figure A.21 The direction of the incident wave and its polarizations.

A.4.8 View Simulation Results

TFDSIE provides three ways to view the results: list of text data, static figure, and animated figure. After the computation is completed, you can choose one of the following ways to view the simulation results.

A.4.8.1 List

- Go to the "**Output**" menu and choose the "**List**" function, then you can choose "**Current Data**" or "**Farfield Data**". Some examples for the data of the current and the far field are shown in Figures A.22(*a*) and A.22(*b*). Note that the far field contains two components of the field, namely the theta component (E_θ) and the phi component (E_ϕ), and the radar cross section (RCS) is defined by

$$RCS = |\mathbf{E}_\theta|^2 + |\mathbf{E}_\phi|^2 \qquad (A.4)$$

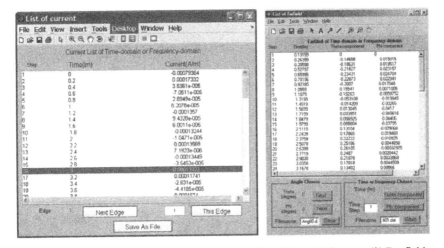

Figure A.22. Computation of the result shown as a list of data: (*a*) Current. (*b*) Far field.

A.4.8.2 Graph

- Go to the "**Output**" menu and choose the "**Graph**" function, and you can select either "**Current**" or "**Farfield**" to be displayed by a figure. Examples for the plots of far field are shown in Figures A.23(*a*) and A.23(*b*). The plots for different angles can be selected in the "**Run Setup**" menu. Note that you also can observe the far field at a specific time or frequency in three dimensions as in Figure A.23(*b*) by providing the number of the time or frequency step.

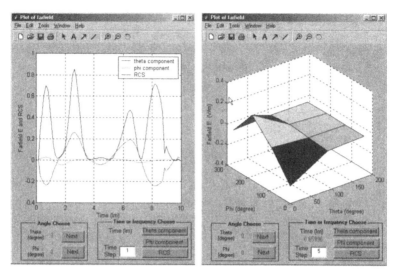

Figure A.23. Far Field plots: (*a*) Time-domain. (*b*) Theta component at a specific time instant.

A.4.8.3   Show

An animated figure of the current only can be displayed for time domain computations, as mentioned before. Follow the steps as presented in section A.3.5, to display the current on the surface as a sequence of time steps.

A.5   EXAMPLES

Two examples are included in this version of TFDSIE, One of the examples, called demo1.prj, is a simple cylinder, and the other, called, demo2.prj, is an aircraft as shown in Figures A.24 and A.25. These two examples are presented to make a user familiar with the code.

A.6   GETTING HELP

Finally, there are some helping tips under the "**Help**" menu to assist the users. Go to the "**Help**" menu, as shown in Figure A.26. Select "**Help Tips**", and a help screen will appear, as shown in Figure A.27, which provides the explanation of the terms used in TFDSIE.

    Also, in most windows or dialog boxes there is a [Help] button. Some information about the corresponding function is explained by clicking this button. Moreover, an illustrative message will be displayed by pointing the mouse over an item in the TFDSIE windows.

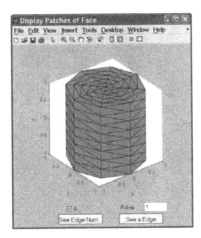

Figure A.24 Demo related to the analysis of a cylinder.

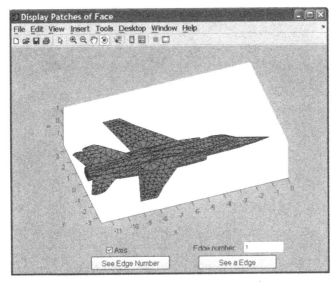

Figure A.25. Demo related to the analysis of an aircraft.

Figure A.26. Help menu.

Figure A.27. Screen shot dealing with Help.

INDEX

W

ABOUT THE AUTHORS

BAEK HO JUNG received B.S., M.S., and Ph.D. degrees in electronic and electrical engineering from the Kyungpook National University, Taegu, Korea in 1986, 1989, and 1997, respectively. From 1989 to 1994, he was a researcher with the Agency for Defense Development in Korea. Since 1997, he has been a lecturer and is currently a professor with the Department of Information and Communication Engineering, Hoseo University, Asan, Chungnam, Korea, He was a visiting scholar with Syracuse University, Syracuse, NY from 2001 to 2003. His current research interests are computational electromagnetics and wave propagation.

TAPAN K. SARKAR is a professor in the Department of Electrical and Computer Engineering at Syracuse University, NY. USA. His current research interests deal with numerical solutions of operator equations arising in electromagnetics and signal processing with applications in system design. He has authored or coauthored more than 300 journal articles, numerous conference papers, and 32 book chapters. He is the author of 15 books, including *Smart Antennas*, *History of Wireless*, and *Physics of Multi-Antenna Systems and Broadband Processing* (all published by Wiley).

SIO-WENG TING received B.Sc., M.Sc., and Ph.D. degrees in electrical and electronics engineering from the University of Macau, Macau, China in 1993, 1997, and 2008, respectively. From 2009 to 2010, he was a postdoctorate research associate at Syracuse University, Syracuse, NY. He is now an assistant professor in the Department of Electrical and Electronics Engineering at University of Macau. His current research interests are computational electromagnetics and passive microwave circuits.

YU ZHANG is with Xidian University and currently is a visiting scholar at Syracuse University. He authored two books, namely *Parallel Compuation in Electromagnetics* and *Parallel Solution of Integral Equation-Based EM Problems in the Frequency Domain*, as well as more than 100 journal articles and 40 conference papers.

ZICONG MEI received a B.S. degree from the Department of
Electronic Engineering and Information Science at the University
of Science and Technology of China in 2008. He currently is
working toward a Ph.D. degree in the Department of Electrical
Engineering and Computer Science at Syracuse University.

ZHONG JI received both B.S. and M.S. degrees in electronic
engineering from Shandong University, China, in 1988 and 1991,
respectively. He received a Ph.D degree from Shanghai Jiao Tong
University, China in 2000. He is currently an antenna engineering
designer with Research In Motion, Irving, TX. His research
interests are in the areas of antenna design for smart phone
and time domain analysis of electromagnetic fields.

MENGTAO YUAN received his B.S. degree in information and
electronic engineering and his M.S. degree in information and
communication systems from the Zhejiang University, China in
1999 and 2002, respectively, and the Ph.D. degree in electrical
engineering from Syracuse University, NY in 2006. From 2002 to
2006, he was a research assistant with the Electromagnetic and
Signal Processing Research Lab at Syracuse University. From
2006 to 2008, he was with Cadence Design Systems, Inc. Tempe,
AZ. Since 2008, he has been a principal engineer at Apache
Design Solutions. His current research interests include the
electromagnetic simulation on power/signal integrity and
RF/mixed-signal circuits.

ARIJIT DE received a B.Tech. degree (with honors) in
electronics and electrical communication engineering and a minor
in computer science and engineering, from the Indian Institute of
Technology, Kharagpur, India in 2004. He currently is working
towards a Ph.D. degree in the Department of Electrical
Engineering at Syracuse University, Syracuse, NY. He is a
National Talent Scholar of India. His research interests are in the
areas of electromagnetic field theory and computational methods
arising in electromagnetics and signal processing.

MAGDALENA SALAZAR-PALMA is a professor in the Department of Signal Theory and Communications at Universidad Carlos III de Madrid, Spain. Her research interests, among others, deal with numerical methods in electromagnetics and advanced network and filter design for communications systems. She has authored or coauthored more than 60 journal articles, more than 220 conference papers, more than 20 book chapters, and 5 books, among them are *Smart Antennas*, *History of Wireless*, and *Physics of Multi-Antenna Systems and Broadband Processing* (all published by Wiley).

SADASIVA M. RAO is, at present, working as a research scientist at Naval Research Laboratory, Washington DC. Previously, he was a professor in the Department of Electrical and Computer Engineering at Auburn University. His research interests include acoustic and electromagnetic scattering, antenna analysis, numerical methods and other related areas. Dr. Rao is a fellow of the IEEE. He has published/presented around 150 papers in international journals/conferences.

WILEY SERIES IN MICROWAVE AND OPTICAL ENGINEERING

KAI CHANG, Editor
Texas A&M University

ACTIVE AND QUASI-OPTICAL ARRAYS FOR SOLID-STATE POWER COMBINING • *Robert A. York and Zoya B. Popović (eds.)*

OPTICAL SIGNAL PROCESSING, COMPUTING AND NEURAL NETWORKS • *Francis T. S. Yu and Suganda Jutamulia*

ELECTROMAGNETIC SIMULATION TECHNIQUES BASED ON THE FDTD METHOD • *Wenhua Yu, Xiaoling Yang, Yongjun Liu, and Raj Mittra*

SiGe, GaAs, AND InP HETEROJUNCTION BIPOLAR TRANSISTORS • *Jiann Yuan*

PARALLEL SOLUTION OF INTEGRAL EQUATION-BASED EM PROBLEMS • *Yu Zhang and Tapan K. Sarkar*

ELECTRODYNAMICS OF SOLIDS AND MICROWAVE SUPERCONDUCTIVITY • *Shu-Ang Zhou*

Printed and bound by CPI Group (UK) Ltd, Croydon, CR0 4YY

16/04/2025